WEEKEND MECHANIC'S HANDBOOK

Complete Auto Repairs You Can Make

Paul Weissler

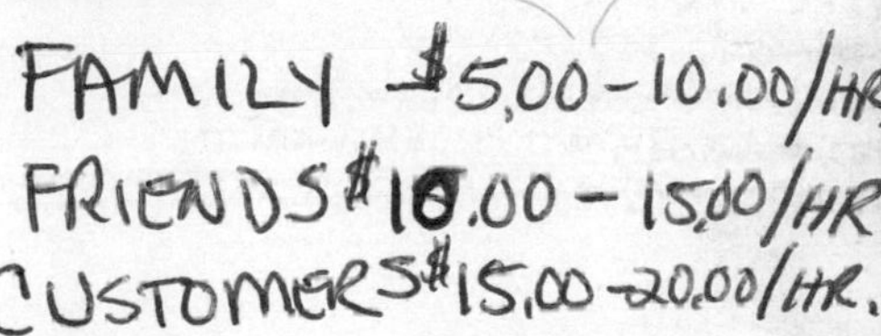

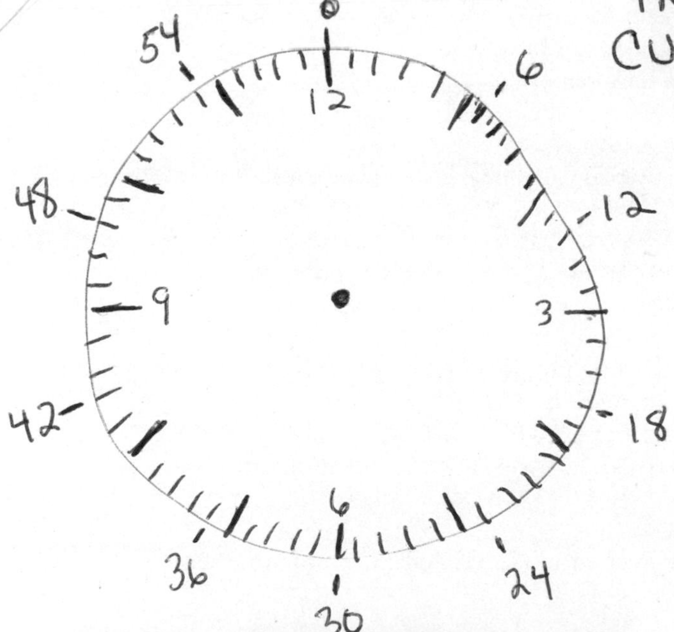

ARCO PUBLISHING, INC.
NEW YORK

The author has made every effort to provide accurate information. However, changes in cars, specifications, and service procedures, as well as typographical errors and omissions, may occur. The author and publisher do not assume responsibility for such changes, errors, and/or omissions.

Published by Arco Publishing, Inc.
215 Park Avenue South, New York, N.Y. 10003

Library of Congress Cataloging in Publication Data

Weissler, Paul.
Weekend mechanic's handbook.

Includes index.
1. Automobiles—Maintenance and repair—Amateurs' manuals. I. Title.
TL152.W428 1982 629.28'722 82-3992
ISBN 0-668-05379-8 (Cloth Edition)
ISBN 0-668-05384-4 (Paper Edition) AACR2

Printed in the United States of America

CONTENTS

ACKNOWLEDGMENTS

The author wishes to thank the following companies for their contributions of illustrations for this book: American Motors, Champion Spark Plug Co., Chrysler Corp., Ford Motor Co., General Motors, Prestolite Division, Eltra Corp., and the 3M Co.

INTRODUCTION

As THE COST of professional auto repair service rises, more and more people are saving big money by doing most of the work themselves. Although some jobs require major investments in equipment, the overwhelming majority can be done with inexpensive tools. Further, although some jobs require considerable experience to diagnose, in most cases a simple test or two and visual inspection will isolate the problem.

Weekend Mechanic's Handbook has been written to enable you to diagnose and do the jobs that are within your ability to handle. In each section you will be shown what you can do and how to do it, or when you would be better off leaving the work to a professional. You probably will find that you can take care of virtually all the work your car needs.

If you have done some work on your car before, you can proceed directly into engine tuneup, the subject of Chapters 1 through 4. Engine tuneup is becoming so popular as a weekend mechanic's service that you can readily obtain the inexpensive, light-duty testers and tools you need for a tiny fraction of the cost of professional equipment. You'll recoup the investment with the very first tuneup you do.

If you are a rank beginner, start with the easiest jobs, such as oil and filter change (see *Chapter 5*), air filter replacement (see *Chapter 3*), and cooling system service (see *Chapter 7*).

Most important, approach each job with self-confidence but not cockiness. Auto repair requires attention to detail, and if you have the desire and the willingness to work carefully you'll consistently get results of which you can be proud.

METRIC CONVERSIONS

As the conversion to the metric system continues, you may find that you must convert American/English dimensions to metric. It's particularly easy with a pocket calculator, and here are the formulas you will need.

INCHES TO MILLIMETERS: Multiply by 25.4. Ex.: .001-inch equals .0254–mm.
MILLIMETERS TO INCHES: Multiply by .04. Ex.: .05-mm equals .002-inch.
INCHES TO CENTIMETERS: Multiply by 2.54. Ex.: 3.41 inches equals 8.66 cm.
CENTIMETERS TO INCHES: Multiply by .4. Ex.: 8.4 cm equals 3.36 inches.
CUBIC INCHES TO LITERS: Divide by 61. Ex.: 122 cubic inches equals 2 liters.
LITERS TO CUBIC INCHES: Multiply by 61. Ex.: 3 liters equals 183 cubic inches.
LITERS TO CUBIC CENTIMETERS: Multiply by 1000. Ex.: 1.4 liters equals 1400 cc.
PRESSURE—POUNDS PER SQUARE INCH (P.S.I.) TO KILOPASCALS (kPa): Multiply by 6.9. Ex.: 10 p.s.i. equals 69 kPa.
PRESSURE—KILOPASCALS TO POUNDS PER SQUARE INCH: Divide by 6.9. Ex.: 138 kPa equals 20 p.s.i.
TORQUE—POUNDS-FEET (LBS.–FT.) TO NEWTON–METERS: Multiply by 1.356. Ex.: 100 lbs.–ft. equals 136 n–m.
TORQUE—NEWTON–METERS TO POUNDS–FEET: Divide by 1.356. Ex.: 225 lbs.–ft. equals 166 n–m.
QUARTS TO LITERS: Multiply by .946. Ex.: 16 quarts equals 15.1 liters.
LITERS TO QUARTS: Divide by .946. Ex.: 10 liters equals 10.6 quarts.
DEGREES FAHRENHEIT TO DEGREES CELSIUS: Subtract 32, then multiply by .5555. Ex.: 212 degrees F. equals 100 degrees C.
DEGREES CELSIUS TO DEGREES FAHRENHEIT: Multiply by 1.8, then add 32. Ex.: 80 degrees C. equals 176 degrees F.
OUNCES TO MILLILITERS: Multiply by 30. Ex.: 2 oz. equals 60 ml.
MILLILITERS TO OUNCES: Divide by 30. Ex.: 45 ml equals 1.5 oz.

These conversions are based on the use of new metric standards. If you have an old imported-car shop manual, it may refer to pressure in terms of atmosphere (atm). One atmosphere equals 15 p.s.i. Or it may refer to torque in kilograms–meter (kgm). One kgm equals 7.2 lbs.-ft.

1 A GUIDE TO BASIC TUNEUPS

VIRTUALLY EVERY MOTORIST knows that an automobile needs a periodic engine *tuneup*. In general, a tuneup is a group of services—replacement of certain parts, some testing of the engine's general mechanical condition, and a few mechanical adjustments. There is no uniform list throughout the industry, however. Mechanics don't agree because they are at the pricing end, and what they put into a tuneup varies according to the package price they feel is most marketable. If the situation is competitive, they make the package (and the price) smaller; if there is a market that is not price-conscious, the package is bigger.

What this all comes down to is that some services yield more obvious and immediate results than others, and the typical tuneup package always includes the services that the customer is more likely to notice when he or she drives the car out of the shop.

If you do your own tuneup, however, you can make it as extensive as you wish. Although some services do not produce immediately noticeable results, they do provide long-term benefits or assurances of reliable performance. By doing these jobs at little or no cost, you get a tuneup that most professional shops could not sell except to those few customers to whom price is no object.

Doing your own tuneup work provides another important benefit: you don't have to replace parts that are still serviceable. If you bring your car into a garage and ask for a tuneup, whatever is on the list will usually be replaced, even if it has a few thousand miles of potential service remaining.

Engine Systems

The tuneup of a gasoline engine customarily involves five areas:

• *The ignition system.* It has the job of producing high-voltage electricity and delivering it to each cylinder of your engine. Threaded into each cylinder is a spark plug, a device that turns the electricity into a spark that ignites the fuel mixture. The electricity must be delivered to each cylinder in precisely-timed fashion, coordinated with other actions taking place, so that the burning of the fuel mixture produces maximum power.

• *The fuel system.* It must combine fuel and air in the right proportions to produce a mixture that will burn readily when ignited by a spark. It then must deliver the gasoline mixture to each cylinder of the engine at the proper time.

• *Pollution controls.* So that the engine releases a minimum amount of gasses and fumes that could pollute the air we breathe, every car is equipped with a variety of control devices. Most of these devices are designed to regulate the way the engine operates, so checking their operation and performing certain services has two results:

(1) The engine produces lower levels of polluting emissions.

(2) Engine performance is maintained according to car manufacturer's specifications.

• *The engine itself.* Actually, very few engine tuneups involve the engine per se, but there are a few. In no case, however, is any real disassembly of the engine required. The parts you must reach all are external, or just under a sheet metal cover held by a few screws (see Figs. 1–1, 1–2).

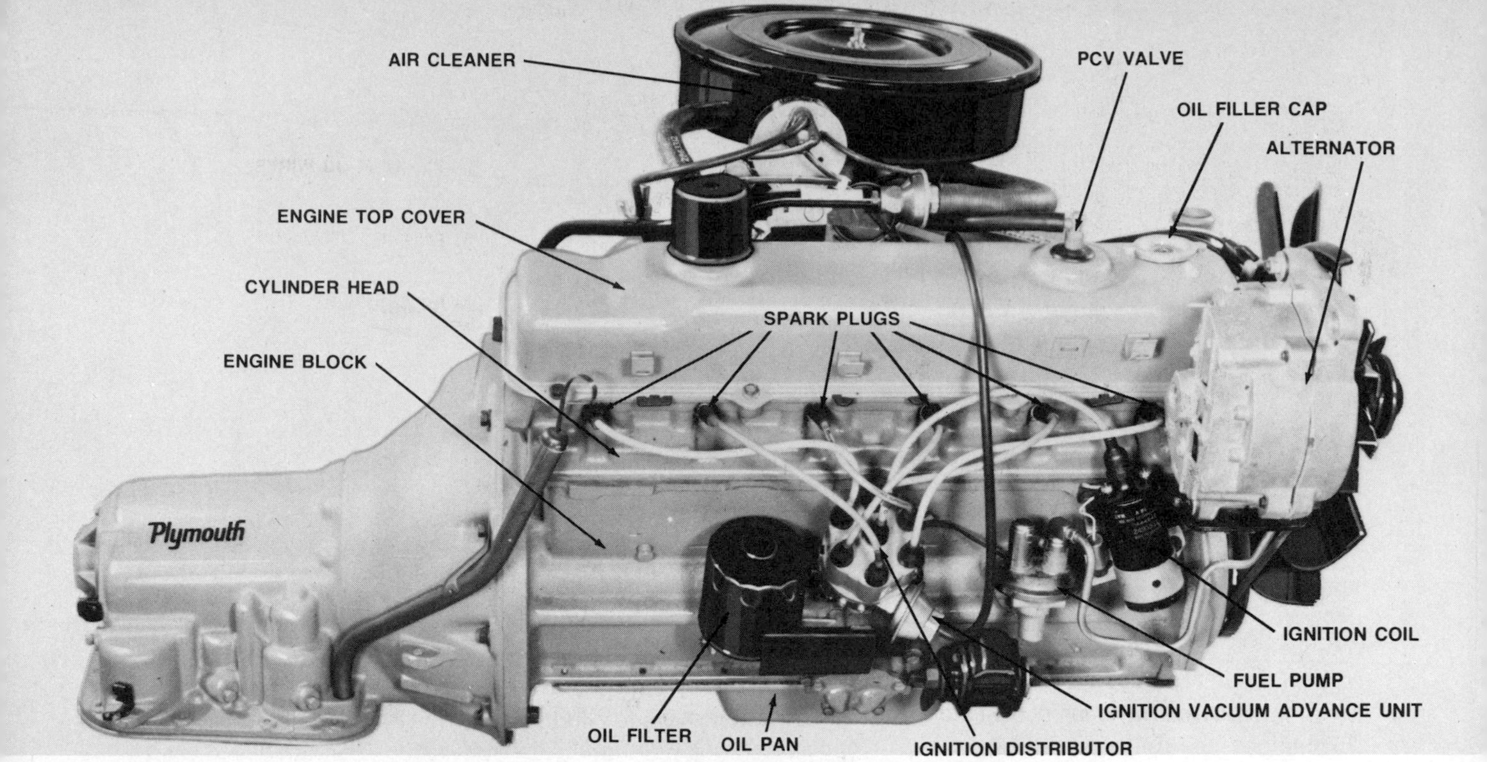

AIR CLEANER
PCV VALVE
OIL FILLER CAP
ALTERNATOR
ENGINE TOP COVER
CYLINDER HEAD
SPARK PLUGS
ENGINE BLOCK
Plymouth
IGNITION COIL
FUEL PUMP
IGNITION VACUUM ADVANCE UNIT
OIL FILTER
OIL PAN
IGNITION DISTRIBUTOR

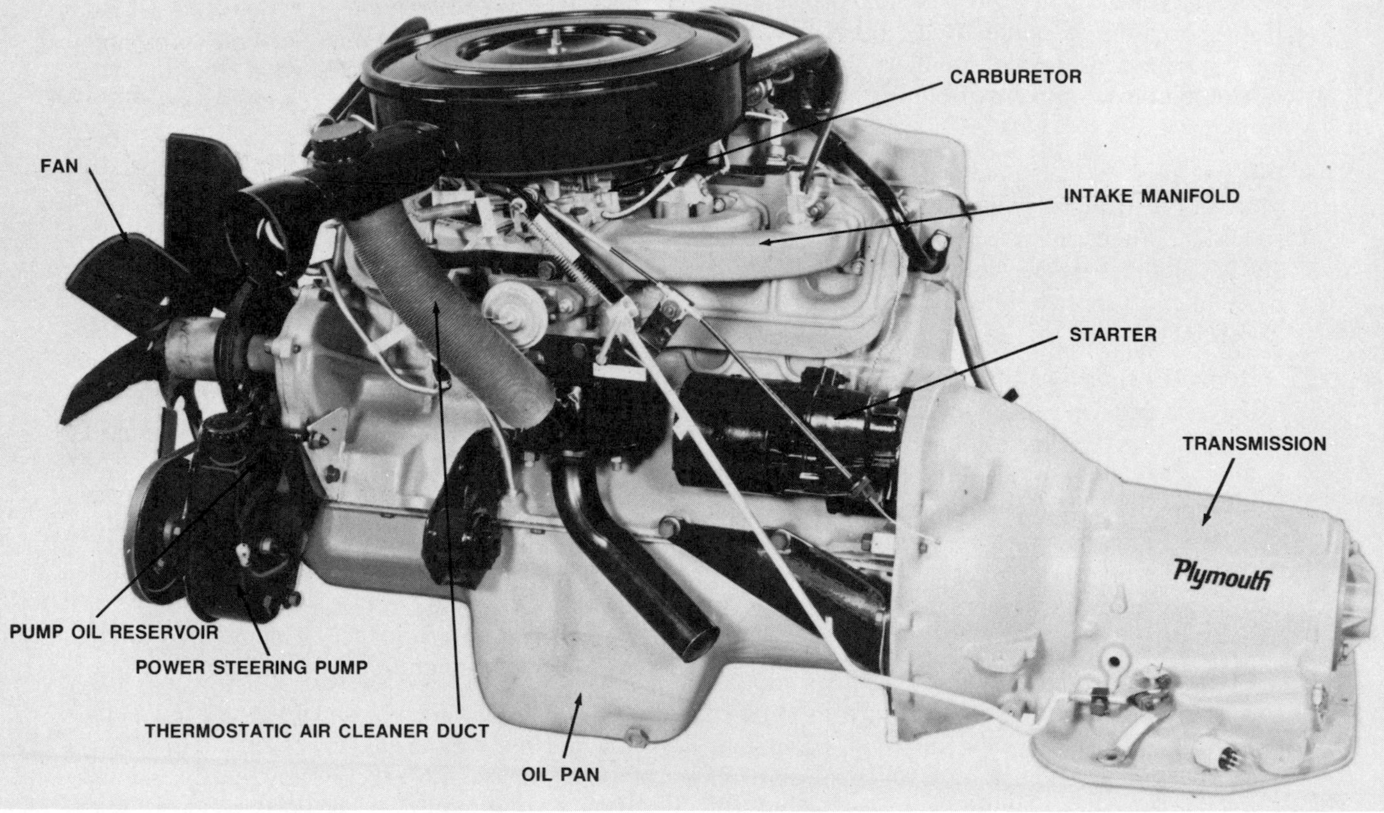

CARBURETOR
FAN
INTAKE MANIFOLD
STARTER
TRANSMISSION
Plymouth
PUMP OIL RESERVOIR
POWER STEERING PUMP
THERMOSTATIC AIR CLEANER DUCT
OIL PAN

Fig. 1–1. Side views of an in-line six-cylinder engine. Although many of the parts are unfamiliar to you, you will understand their function as you perform the jobs covered in this book.

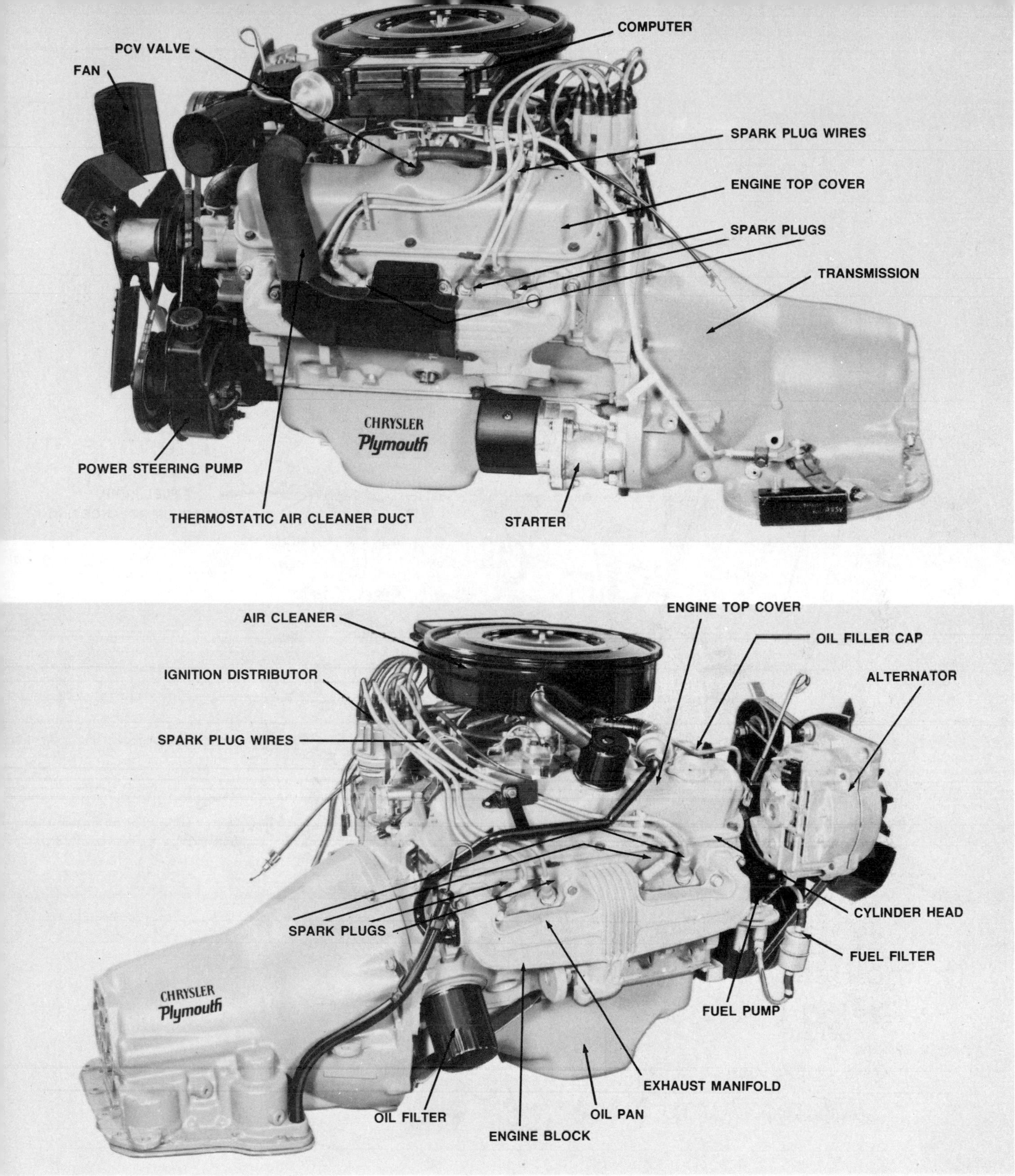

COMPUTER
PCV VALVE
FAN
SPARK PLUG WIRES
ENGINE TOP COVER
SPARK PLUGS
TRANSMISSION
CHRYSLER
Plymouth
POWER STEERING PUMP
THERMOSTATIC AIR CLEANER DUCT
STARTER

AIR CLEANER
ENGINE TOP COVER
OIL FILLER CAP
IGNITION DISTRIBUTOR
ALTERNATOR
SPARK PLUG WIRES
CYLINDER HEAD
SPARK PLUGS
FUEL FILTER
CHRYSLER
Plymouth
FUEL PUMP
EXHAUST MANIFOLD
OIL FILTER
OIL PAN
ENGINE BLOCK

Fig. 1–2. Side views of a V-8 engine.

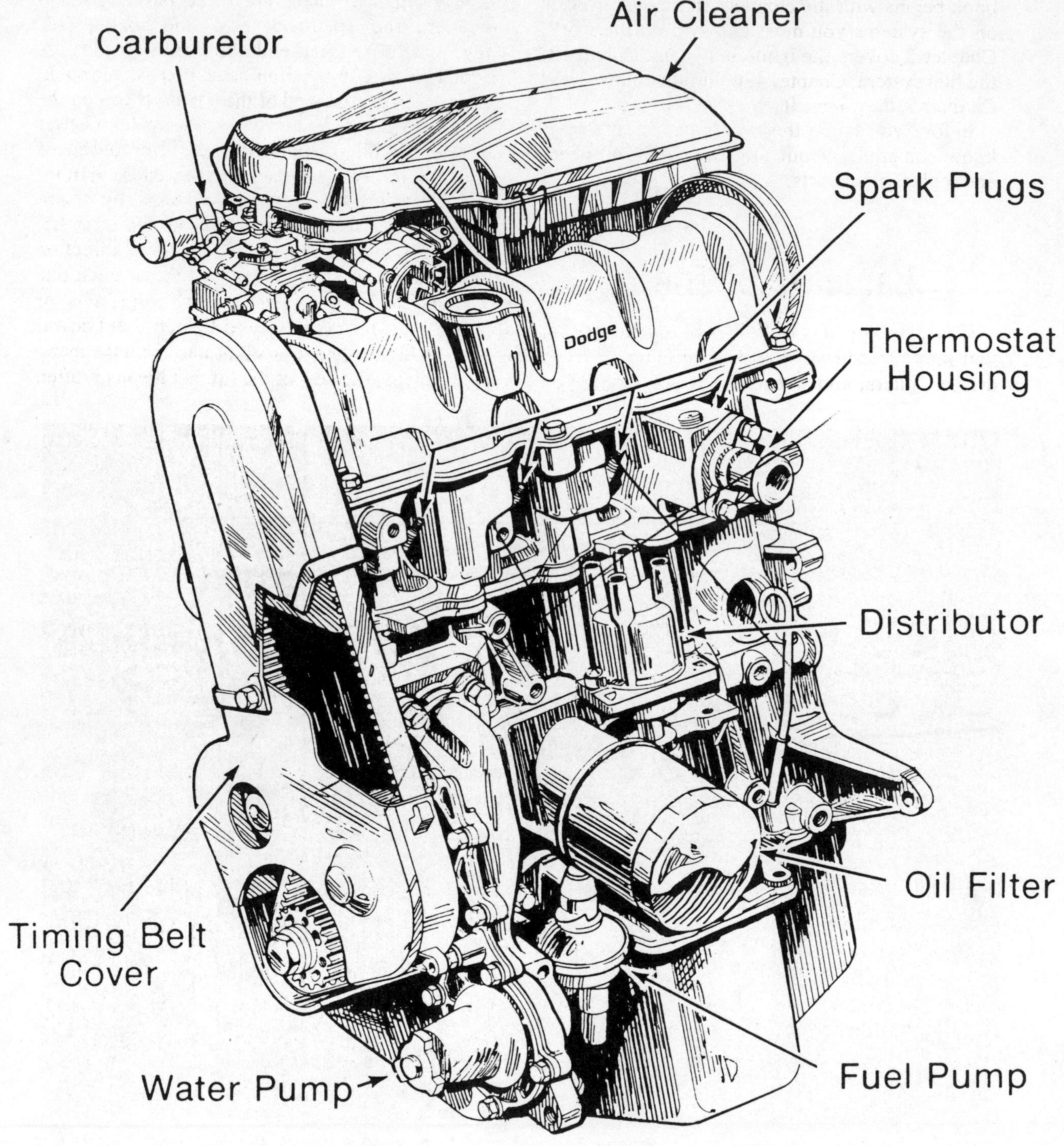

Fig. 1–3. Side view of a four-cylinder engine.

Because the tuneup is one of the most popular weekend mechanic's areas of operation, this book begins with the chapters that cover service on the systems you must know to do the work. Chapter 2 covers the ignition system, Chapter 3, the fuel system, Chapter 4, pollution controls and Chapter 5, the engine itself.

Before you begin these chapters, you should know something about the tools you will need, and replacement parts.

Basic Hand Tools

If you do any sort of work around the house, you may already have an assortment of screwdrivers, pliers, and wrenches. If you don't, you will need them to do a tuneup and other work on your car. Here are some buying guidelines:

• *Wrenches.* There are three basic types of wrench, the open-end, box, and socket (see Fig. 1–4). The open-end and box wrenches are available in combination sets; that is, an open-end design on one end of the shank, a box on the other. Although the box wrench provides a better grip on the nut or bolt, there are often problems of accessibility when only an open-end will fit in.

Socket wrenches cannot be used by themselves, but instead fit into a tool called a ratchet wrench, which can be set to lock in one direction and free-wheel in the other. Set the wrench one way and it is used to remove spark plugs, nuts, or bolts; set it the other way to tighten them down. Although the socket is small and fits into many tight places, the size of the ratchet means it often

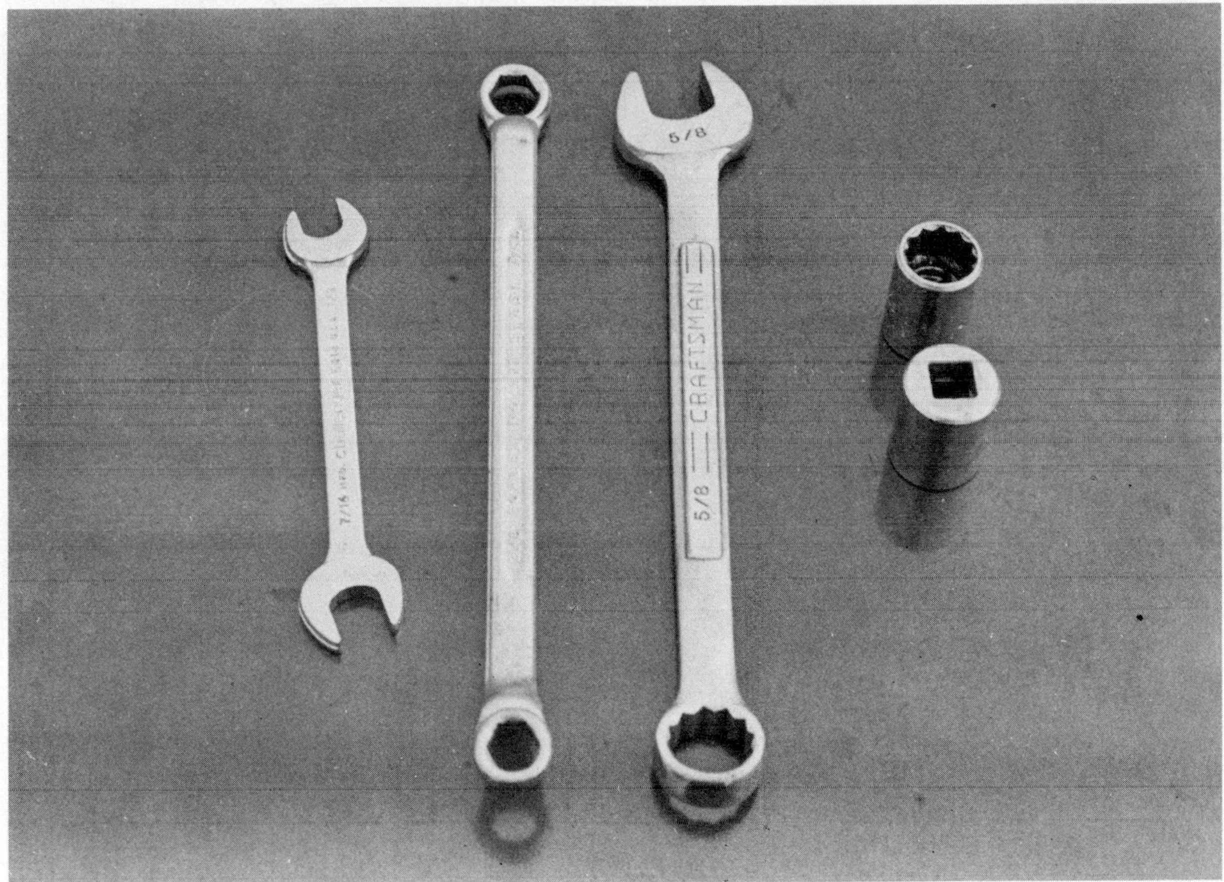

Fig. 1–4. Shown are the basic types of wrenches. From left to right, they are: double open-end, double box, combination box-open end, and sockets. Note that a wrench may be a box-type on one end and an open-end on the other. Also note that two sockets are shown. The square end shown up fits into a ratchet wrench; the other end shown up fits onto the nut or bolt.

cannot fit. To permit using the socket under such conditions, you can attach extension rods and flexible universal joints between socket and ratchet (see Fig. 1–5). The size of the male and female squares in the tools is called the drive, and socket sets for automotive use are commonly available in ½- and ⅜-inch drive. The ⅜-inch drive type is the one most suitable for weekend mechanics.

If you have an American car with American-size nuts and bolts, these are the size wrenches you will find most useful for automotive service: 5/16, 3/8, 7/16, 1/2, 9/16, 5/8, 11/16, 3/4 and 7/8-inch, in box, open-end, and socket styles. In addition, you will need a 5/8- or 13/16-inch deep socket (depending on your car) to fit the spark plugs. A deep socket is one that is taller than a standard socket, and one made for spark plug use will usually have a sponge rubber insert to grasp the body of the plug (see Fig. 1–6). In addition, you will find a few short extension rods (two to six inches) useful, which if necessary can be locked together to form one long rod. A universal joint also is helpful.

If your car has metric-size nuts and bolts, the key sizes are: 6, 7, 8, 9, 10, 11, 12, 13, 14, 15, 17, and 19 millimeter, plus a spark plug socket.

• *Screwdrivers*. Most screwdriver slots are either the simple slot or one of two Phillips types, the straight Phillips cross or a variation called Posi-driv. You can tell Posi-driv by what appears to be hairline cracks at each right angle of the cross. Screwdrivers are not interchangeable between the two, and in fact, you must have an assortment of sizes to properly fit the cross of each. Use of the wrong size screwdriver may

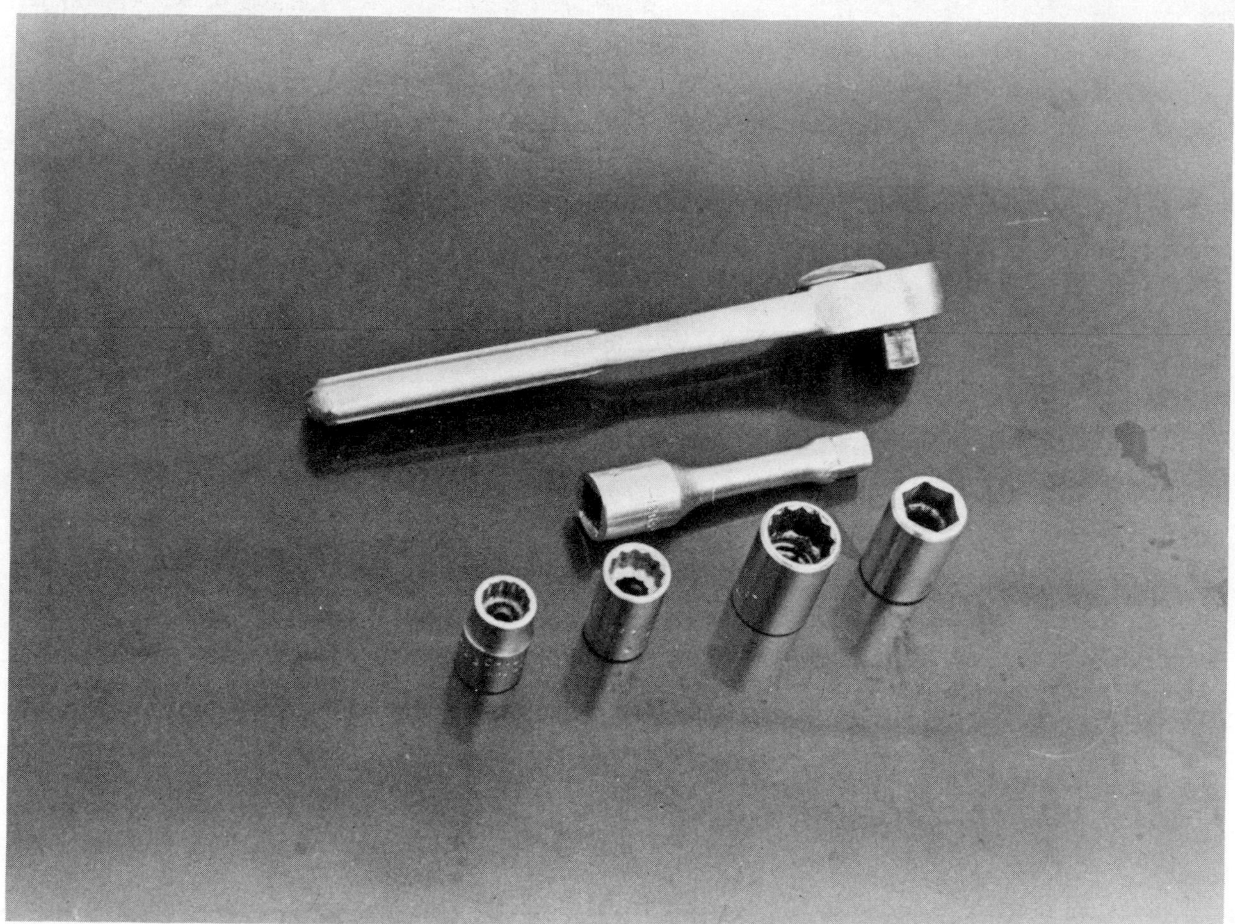

Fig. 1–5. Ratchet, extension rod, and socket wrenches. The square hole in the end of the socket or extension rod fits into the male square on the ratchet or on an extension rod. A universal joint with one male and one female end also may be used for work in tight quarters.

Fig. 1–6. Spark plug socket has sponge rubber insert (pulled halfway out for illustration) to grip ceramic insulator of spark plug during removal or installation.

gouge the cross; use of the wrong type screwdriver (particularly a Phillips in a Posi-driv slot) will create removal problems and may result in damage to the cross. Look carefully at the assortments in the stores, and try to find a small, medium, and large in both slot and Phillips and Posi-driv types. Select a quality screwdriver made of chrome vanadium steel. This designation will be found on the screwdriver shank. On slot screwdrivers, look for fine vertical lines at the tip, an indication of good machining. Make sure the screwdriver has a comfortable handle, whether it's wood or plastic.

Note: many late-model cars have screws with hex or six-pointed-star slots cut into the heads. The hex-type is called an Allen head, the star a Torx head. In addition, some Torx-head screws have a head with an external star-shape. Special socket bits are available for these screws, although in an emergency you can use an Allen wrench in a female Torx screw. With the male Torx head, you must use the proper wrench.

• *Pliers*. The vise-type locking pliers that most people have for work around the house are the most useful. A small and a large pair will cover your needs. In addition, you can consider: needle-nose pliers (for tight quarters), cutting pliers for cutting wires and other jobs, and hose clamp pliers for easy release of the spring-type clamps used on most cooling system hoses (see Fig. 1–7).

Specific Tuneup Tools

In addition to standard wrenches, pliers, and screwdrivers, you'll need the following tools specifically for tuneup work:

• *Tachometer*. This is a meter that indicates the speed at which the engine is running. It's

Fig. 1–7. Special hose clamp pliers permit squeezing together the ears of a spring-type hose clamp.

usually combined with a dwellmeter, which is necessary for ignition system work on cars without electronic ignition. Look for a tachometer with at least two scales: one low and one high, such as 0–1000 or 0–2000 revolutions per minute (r.p.m.) or somewhere in between, plus 0–5000 or 0–8000 r.p.m. or somewhere in between. The low scales should be easily readable to within 25 r.p.m. for precise work. The dwell scale is typically 0–90°, and so long as you can read it to within two degrees, it will do.

Note: Some tachometers will not work properly on cars with electronic ignition. Be sure the one you buy is guaranteed by the manufacturer to record accurately on electronic systems, for even if your present car has breaker points ignition, you surely will eventually have one with electronic ignition.

Reasonably-priced, digital-display electronic meters, some of which are available in combination with other instruments described here, are becoming a major factor in the market. Digital meters, of course, are easier to read than dial-type, but make sure you have one that works properly. There are some low-quality imports also available.

- *Timing light*. This is a strobe-type light used to check ignition timing, which means the arrival of the high-voltage electricity at the spark plug. Timing is adjustable on all but some Ford computer-equipped cars, to match factory specifications, by loosening a lockbolt and turning the distributor body. First, however, it is checked, which is done by connecting the light and aiming it at a pair of reference marks on the engine (see *Chapter 2*). There are many types of timing lights, including an inexpensive neon-tube design. However, the only type that will really provide a bright light to permit an accurate check is the "Power Timing Light." For ease of use,

Fig. 1–8. These are two types of combination tachometer, dwellmeter and timing light, but they differ widely in capability and price. The dial-meter type is inexpensive, but it is not easy to connect for a timing check, and it may not work on some electronic ignition systems. In addition, it can only flash when the spark plug fires. On many cars with computer-controlled spark timing, the firing may be well off the readable scale. The type shaped like a large gun is a computerized test instrument. It has a trigger wire that simply clamps to the plug wire. It has a digital display and push-button controls in the back, so you can read numbers for engine r.p.m. and dwell (on cars without electronic ignition) with complete accuracy. Further, you can

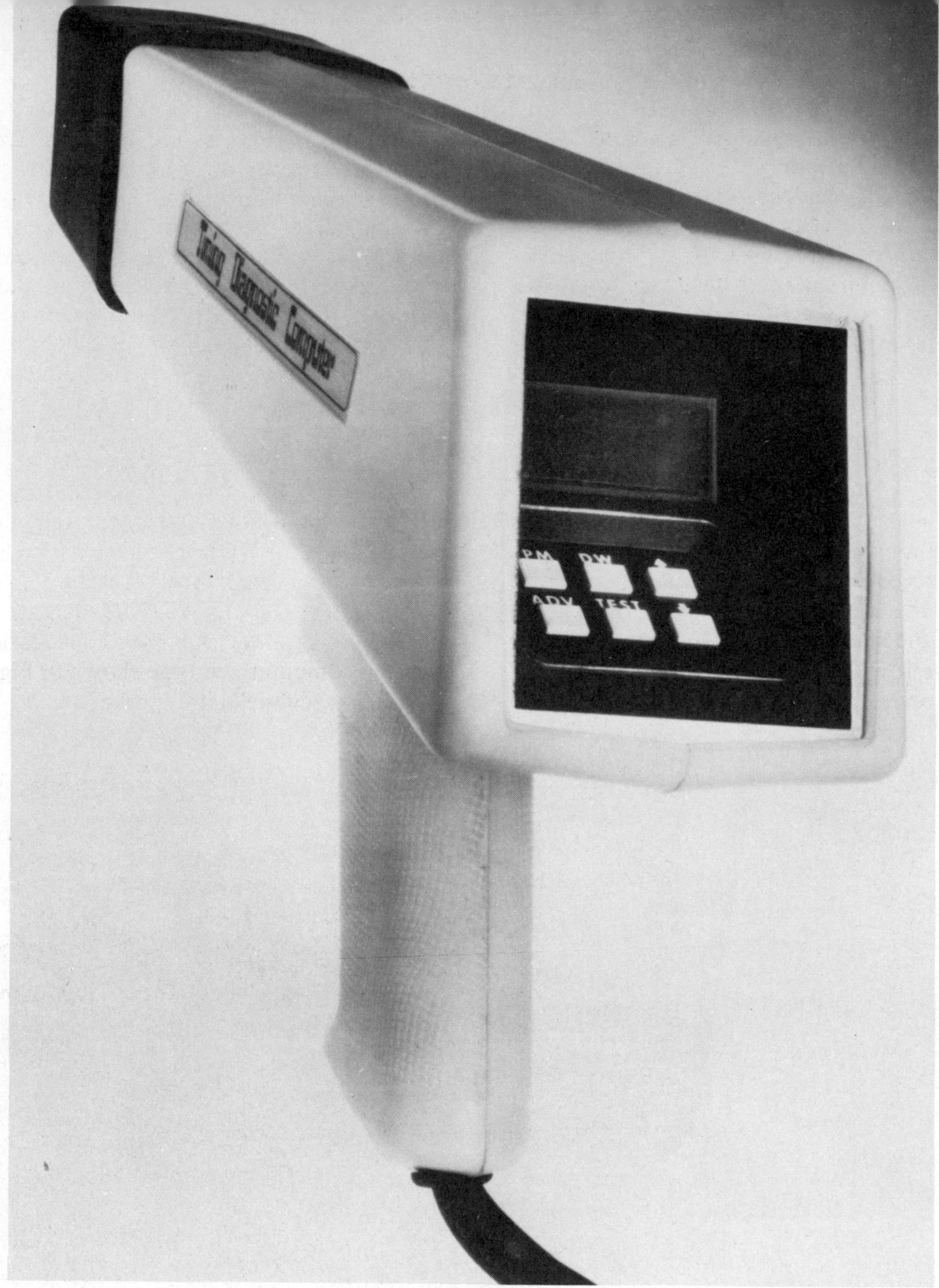

push buttons to preset any number of degrees of spark advance, and the timing light will delay that number before flashing. Therefore, if the number is very high, well off the readable scale on the engine, just dial in that number, and the timing light should flash at zero. Example: timing should be 14 degrees advance; you dial 14 on the digital display and the timing light will flash 14 degrees later. If the timing is correct, it will flash at zero. The computer tester shown, marketed by Chrysler through parts stores, also provides greater precision in r.p.m. and dwell readings than the dial meter.

buy a 12-volt power timing light, which connects directly to the car's battery.

Note: For convenience you may consider a combination tach-dwellmeter and timing light, a single instrument that covers all the tuneup electrical meter requirements a beginner will need (see Fig. 1–8).

Two additional buying factors you should consider are: the type of trigger wire connector on the timing light and where the timing marks are on your car.

The trigger wire is connected to a spark plug and/or plug wire so that the light is triggered when the spark plug is firing. There are many types of trigger mechanisms, but the only type that is suitable for all types of ignition systems is the "inductive pickup," which clamps onto the exterior of the spark plug wire (see Fig. 1–9). Do not confuse it with the capacitive pickup, which physically resembles it. Ask before you buy.

If the timing marks on your car are really buried, as they are on some late-model cars, the brightest timing light won't help. To cover this situation, all late-model cars have special holders on the front of the engine for an electromagnetic probe, part of a timing indicator that reads at the meter (you don't have to look at marks). This type is often combined with a special universal tachometer suitable for diesel engines (see *Chapter 6*). The price of this type of timing indicator is not low, but it can save you the inconvenience of making your own timing marks on the engine.

Or you may find the computerized instrument shown in Fig. 1–8 suitable for this problem. Because it can delay the flashing of the light any number of degrees, it may permit you to set it to fire at a point you can see. The magnetic probe, however, is definitely the wave of the future, and even the computerized type shown in Fig. 1–8 is available in a model with a probe.

Fig. 1–9. Clamp from timing light easily attaches to spark plug wire. This is unquestionably the easiest type to use.

• *Feeler gauges.* The term feeler gauge is what the name implies. It's a set of strips of metal or pieces of wire of different thicknesses that permit you to measure gaps by feel. Why would you want to measure a gap? The answer is that in automotive work there are certain gaps—clearances—between parts that must be precisely correct (to within a thousandth of an inch in some cases) for the engine to perform correctly. If a gap is supposed to be .016 inch (sixteen thousandths of an inch), for example, you select the gauge marked .016 inch, hold it in your hand, and insert it in the gap between the parts. If the gauge can be inserted and withdrawn with just light to moderate drag, which you can easily distinguish, the gap is correct. If the gap is too large or too small, you make an adjustment on the parts to bring it to the specified number. A set of feeler gauges is very inexpensive, and you can buy either a combination set (of strip and wire types) or individual sets (just wire or just strips). If you have a car with electronic ignition, you need only a gauge for spark plugs, and that is commonly made with strips of wire (see Fig. 1–10). If your car does not have electronic ignition, you will also need the set with metal strips (see Fig. 1–11).

• *Compression gauge.* To understand the function of a compression gauge, you need a basic understanding of the piston engine and its four-stroke cycle (see Fig. 1–12). When the piston rises on the compression stroke, it squeezes the air-fuel mixture. The more tightly the mixture is squeezed, the greater is the power developed when it is ignited. The compression pressure is determined by these factors: (1) the design of the engine, including the length of the cylinder, the shape of the top of the piston, and the size of the combustion chamber above the piston; (2) the condition of the seals—called the piston rings—

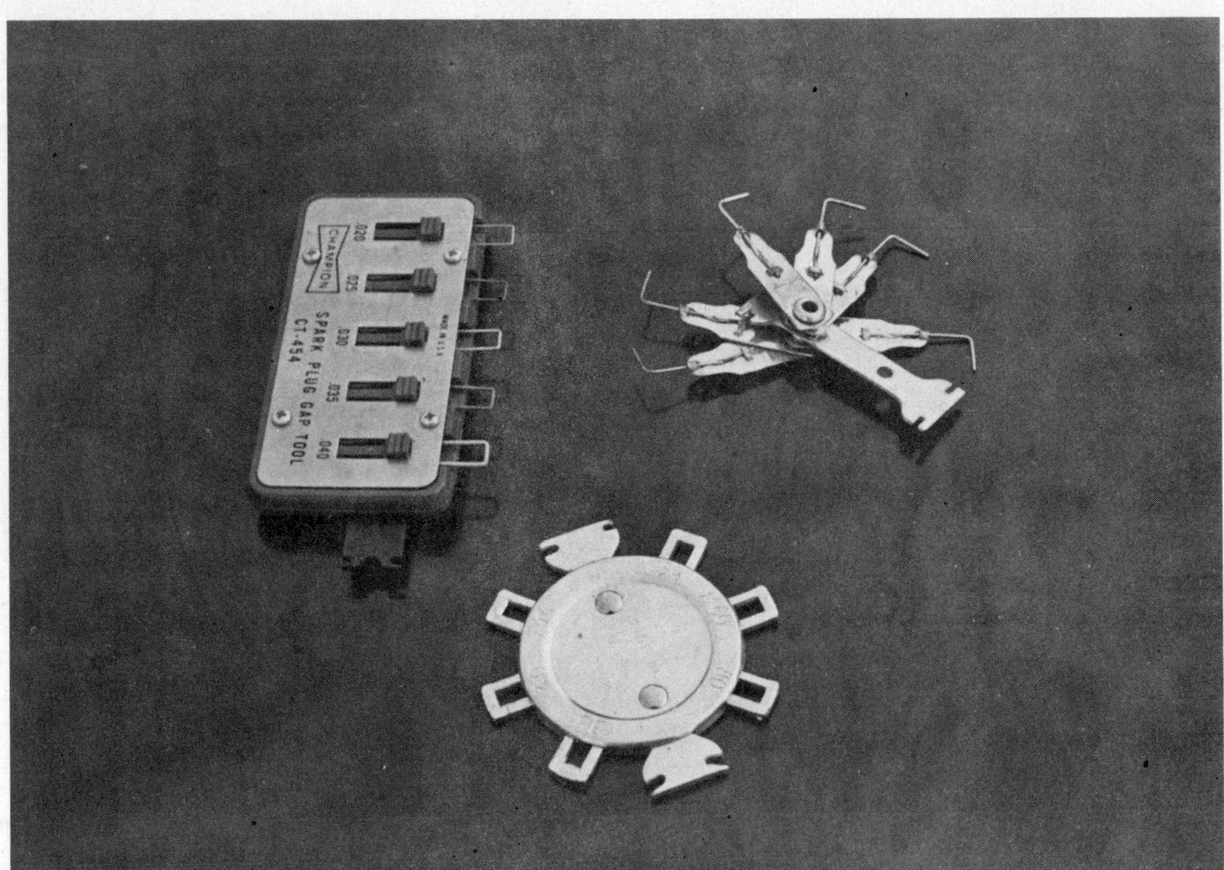

Fig. 1–10. These are three of the most common types of spark plug gauges. The rectangular one with U-shaped wires, the pocket-knife type with L-shaped wires, and the circular one with U-shaped bars are all readily available.

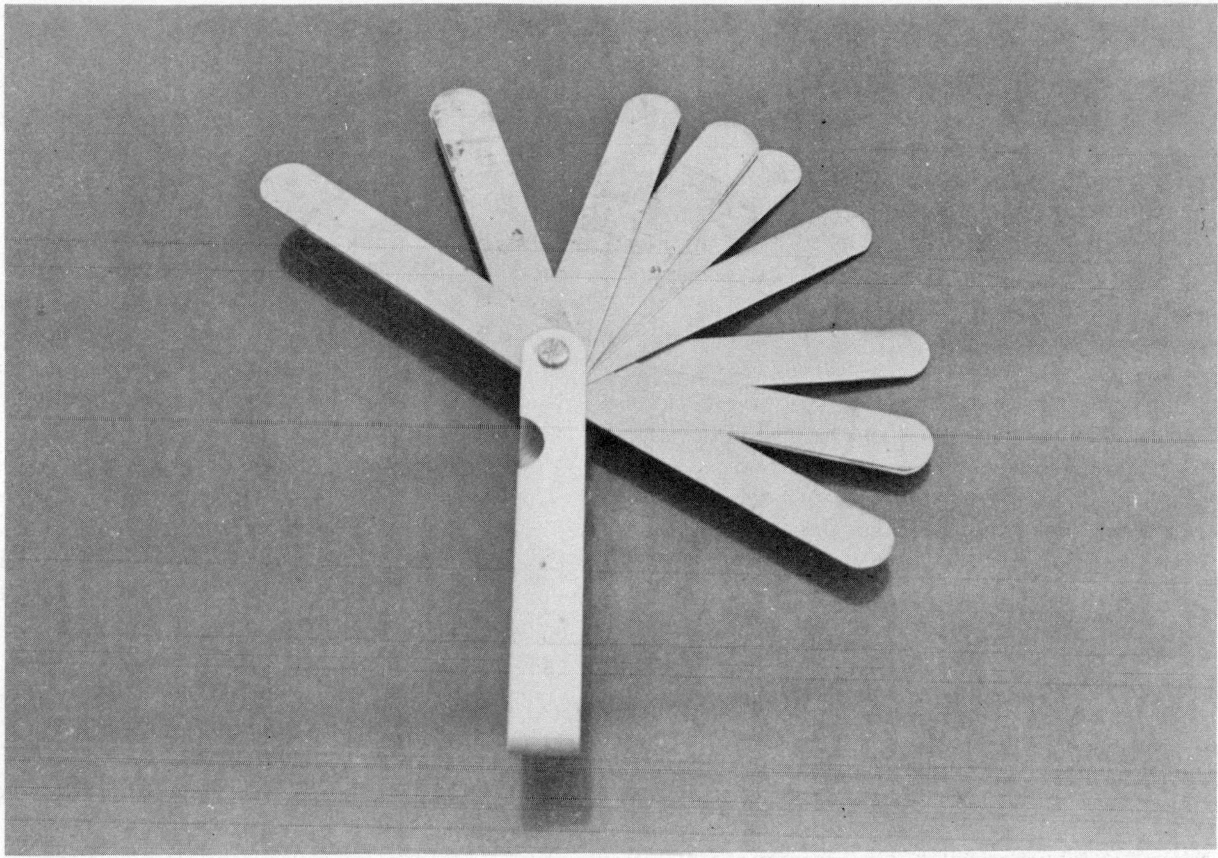

Fig. 1–11. This feeler gauge with metal strips of varying thicknesses is used for adjusting ignition breaker points.

that fit around each piston. If they do not seal tightly between piston and cylinder, some of the compressed fuel mixture can leak down from the cylinder into the bottom of the engine; (3) the condition of the intake and exhaust valves. If the valves do not seal tightly, some of the compressed fuel mixture can leak past them; and (4) the condition of the gasket (called the head gasket) that seals the joint between the combustion chamber (in the cylinder head) and the cylinder itself (in the engine block).

By testing, the car manufacturer determines what pressure the engine should develop when it is in good condition, with only a minimum amount of leakage past the piston rings and valves, and provides this information for test purposes. The compression gauge is the tool used to make the test. With a spark plug removed, the gauge is inserted into the plug hole and the engine is cranked (see Fig. 1–13). The compression pressure in the cylinder builds up a reading on the compression gauge, and this reading is compared with the manufacturer's specification. If the readings in all cylinders are acceptable, the engine can be given a tuneup. If the compression readings are too low, performance will be very poor and a tuneup will not correct the problem. The compression test is normally performed immediately before new spark plugs are installed; for further details, see Chapter 2, *The Ignition System*.

You can buy either of two basic types of compression gauges for gasoline engines: one that is hand-held in the spark plug hole, or one that is connected to a hose that can be threaded into the spark plug hole (see Fig. 1–13). The second type is the only choice if access to any of the spark plug holes is not good. Although it costs more and is not sold in most discount houses, it may be the only type you can use.

• *Ohmmeter*. Resistance to the flow of electricity is an important factor in the operation of

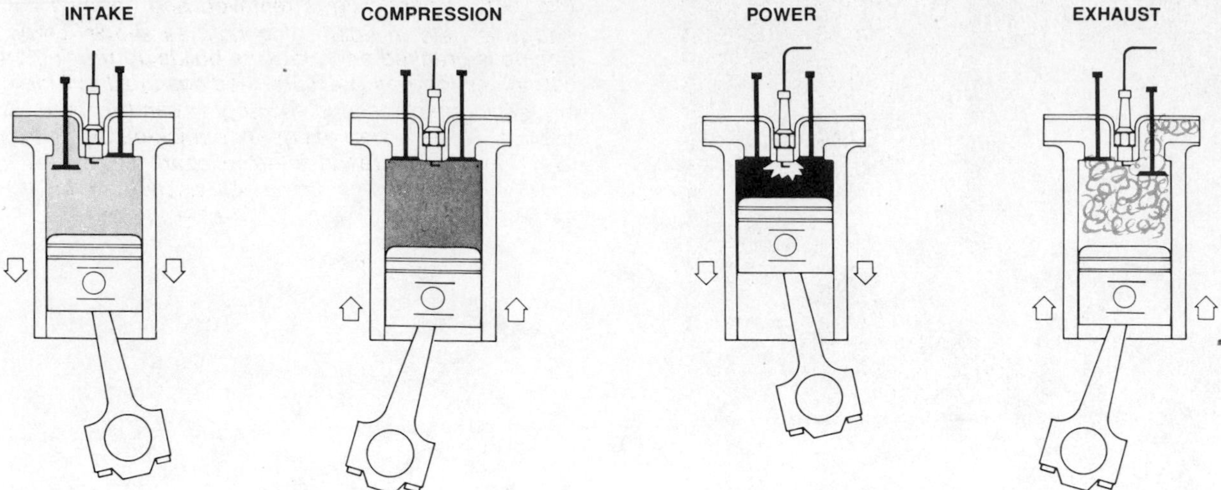

Fig. 1–12. The four-stroke cycle of the gasoline engine. As the engine crankshaft turns, a piston drops in the cylinder and the intake valve opens, allowing fuel mixture to flow into the cylinder (intake stroke); the piston rises and squeezes the mixture as the intake valve closes (compression stroke); the spark plug ignites the fuel mixture, which explodes and pushes the piston down (power stroke); the piston rises and the exhaust valve opens, permitting the burned gases to flow out of the cylinder into the exhaust system (exhaust stroke). The cycle repeats over and over. Each spark plug fires in a prescribed sequence, so there is always power from burning gases pushing down on a piston somewhere in the engine.

automobile ignition systems, particularly electronic ignition systems. An ohmmeter, which reads out in units called "ohms," is the instrument that measures resistance, and if you wish to be able to do detailed troubleshooting you may want to add this to your toolbox. An ohmmeter that covers all needs will have the following scales: 0–50, 0–0500, 0–5000, 0–50,000, and 0–500,000. The ohmmeter is not a beginner's instrument, but as you advance to the more sophisticated work described in Chapter 2, you may wish to acquire it (see Fig. 1–14).

• *Multi-meter*. A multi-meter is a single instrument that combines the function of several. For example, it will often have scales that read engine speed (tachometer), dwell, resistance, volts, and amps. The price will be little more than that for just a tach-dwellmeter; however, it is not always the bargain it seems. The multi-meter may not have all the useful scales, such as both low and high engine speed scales, all the resistance scales, and all the volt and amp scales you need for general electrical service. Some multimeters will meet many of your requirements, and if you see one with the right scales at the right price, it might not be a bad purchase.

Note: Electronic meters with digital displays are available that offer many meter functions and a wide range of scales. These are perhaps a best buy for a really enthusiastic weekend mechanic.

Special Wrenches

The general purpose wrenches described earlier in this chapter may be all you need for most jobs on most cars, but on some cars you may need the following:

• *C-shaped distributor wrench*. The C-shaped distributor wrench permits easy access to the lockbolt that secures the distributor, the key component in the ignition system. The lockbolt must be loosened for a key adjustment—called ignition timing—so if you can't reach this bolt with a conventional wrench, the C-shape is required (see Fig. 1–15).

• *Allen wrench*. If you own a General Motors V-8 car (Chevrolet, Buick, Oldsmobile, Pontiac, or Cadillac) without electronic ignition, you need a ⅛-inch Allen wrench (see Fig. 1–16). The Allen wrench is a male hex that fits into a special screw with a female hex cut into its head, and is used

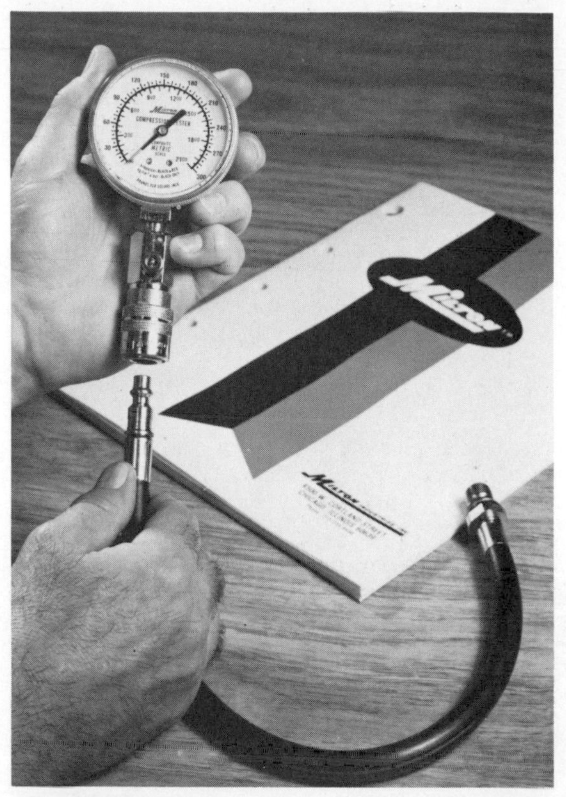

Fig. 1–13. Spark plug is removed and compression gauge is held in spark plug hole as shown below. Engine is cranked and pressure builds up reading on gauge. As long as pressure readings in all cylinders are within specifications, the engine can be tuned. At left is a compression gauge that quick-connects to a long hose that threads into the spark plug hole. If spark plug holes are not easily accessible for the gauge shown below, this type is what you need.

Fig. 1–14. Shown is a dial-type ohmmeter (digital types also are available). The knob at right permits you to multiply the dial reading as required for measuring high resistance.

for a key ignition adjustment. Although Allen wrenches are sold in sets, you can get a ⅛-inch Allen type made just for General Motors V-8s.

• *Spark plug ratchet.* If the spark plugs in your engine are not particularly easy to reach, a special ratchet is usually helpful. It's called a spark plug ratchet and it features a curved handle and a univeral joint in its head. This permits it to work in very tight areas and around obstructions, as shown in Chapter 2 (see Fig. 1–17).

• *Torque wrench.* Certain nuts and bolts must be tightened very precisely and if a part is held by many of them, all must be tightened evenly. The torque wrench is the tool that is required. It's a

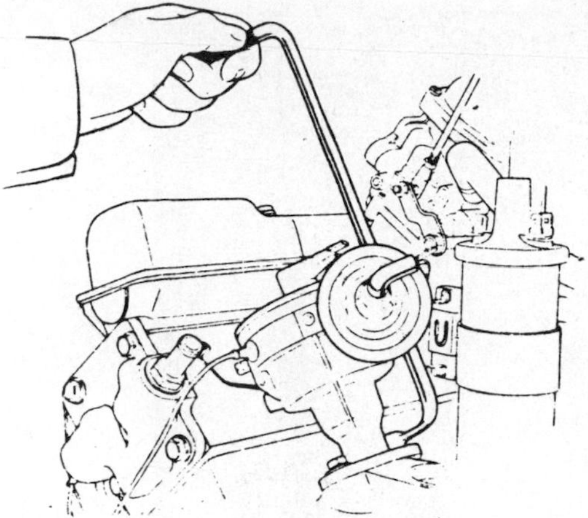

Fig. 1–15. If the lockbolt that holds the distributor in the engine is not accessible with a conventional wrench, the C-shaped wrench is used. It is readily available in auto parts stores.

wrench with a dial gauge (see Fig. 1–18) that accepts a socket. As you torque (apply tightening force) the nut or bolt, the dial needle moves to indicate exactly how tight the nut or bolt is. Torque is measured in pounds-feet (abbreviated lbs.-ft.), or pounds-inch (lbs.-in.). On a metric scale, torque is measured in pascals or kilograms-meter (kg-m). The modern metric measurement is pascals, actually kilopascals, abbreviated kPa. However, on older imports the specifications in the shop manuals may be given in kg-m.

Specifications

The object of a tuneup is to install certain replacement parts, and in many cases to make adjustments to factory numerical specifications. The most common specifications are:

gaps, measured with feeler gauges in thousandths of an inch or (metric) tenths of a millimeter.

torque, in lbs. ft., kPa or kg-m, measured with a torque wrench as you tighten.

engine speed (r.p.m.) measured with a tachometer.

dwell, on non-electronic ignition systems, measured with a dwellmeter in degrees.

ignition timing, read with a timing light on an indicator marked in degrees.

The specifications are supplied for each make and year car, engine size, and also according to some special characteristics of that engine, such as the type of carburetor and whether or not it has a manual or automatic transmission. Most public libraries have reference manuals that contain specifications sections, or the counter salesmen in many parts stores will provide a tuneup specifications chart you can consult.

Note: Some key tuneup specs are printed on an underhood decal (see Fig. 1–19); if your repertoire is limited, they may be all you need.

Buying Tuneup Parts

When you buy the tuneup parts for your car, you must be sure to get exactly the ones specified for your car. Even if the wrong parts happen to fit, the performance of the engine will be adversely affected.

You can buy your parts from a wide variety of outlets, including discount houses, department stores, home-and-auto stores, car dealers, service stations, and the parts-only stores that serve primarily professional mechanics. This final choice, called a parts jobber, also is a good one for the beginning weekend mechanic, for these reasons: (1) The counter salesmen are more knowledgeable than at most other outlets, will look up the parts number for you and hand you the parts. Although they may make a mistake, this is less likely than at outlets where you have to read a catalog, or where there are less knowledgeable personnel; (2) They are more likely to be able to supply advice on installation and look up service specifications; (3) They carry name brands, the top-quality lines used by service stations and garages.

Caution: Look closely at the package when buying a name brand. Some spurious brands are put in packages designed to closely resemble the name brand.

The only disadvantage of the parts jobber is that prices usually are higher than at discount houses and department stores such as Sears. Most parts jobbers will give you a discount from list automatically (and if they don't, ask for it),

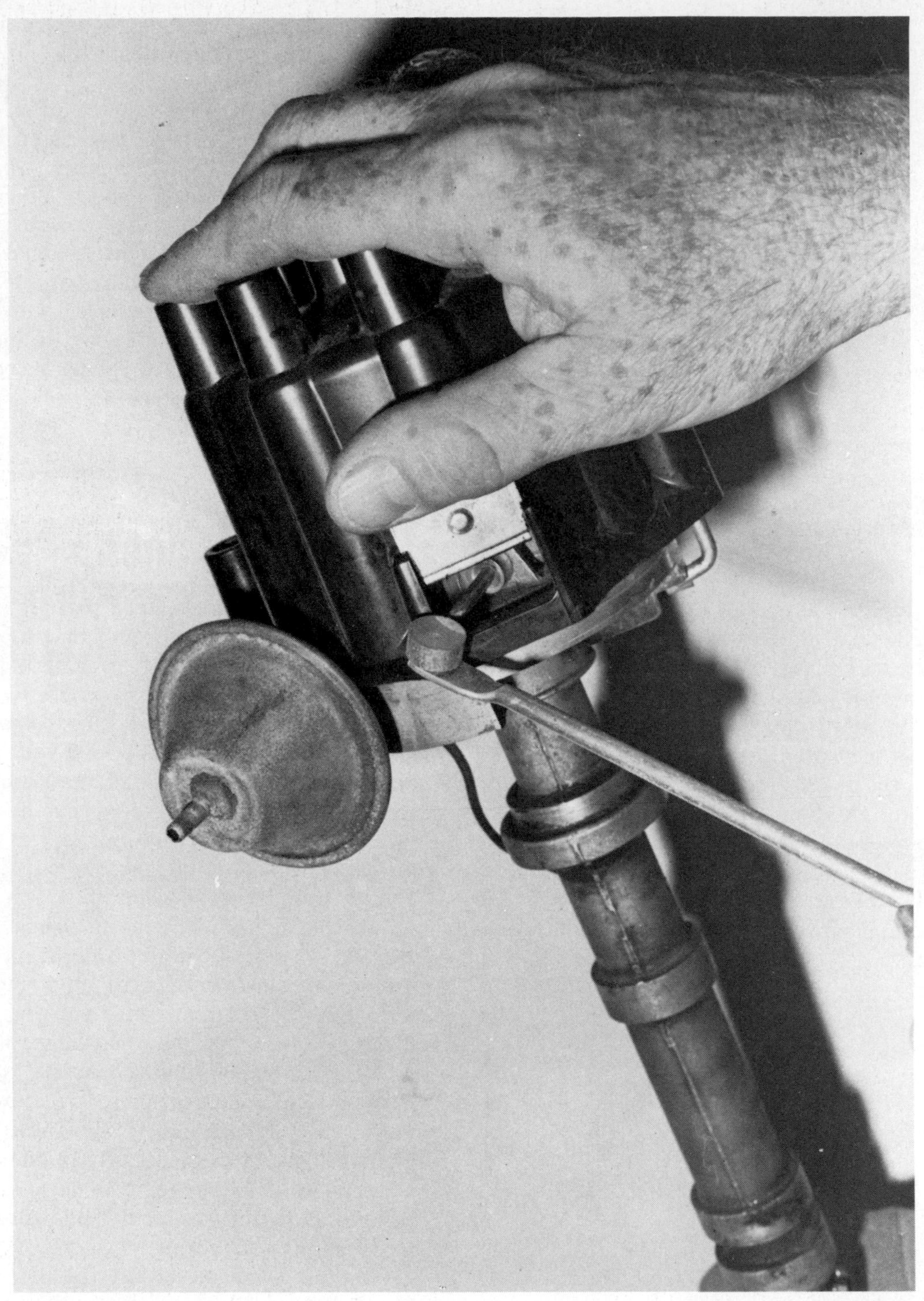

Fig. 1–16. If you have a General Motors car with a V-8 engine and breaker points, you need an Allen-type wrench, such as the one shown, to reach in through the access hole shown (metal window lifted up) to turn on adjuster.

Fig. 1–17. Ratchet with curved handle and universal joint in the head permits you to work around obstructions.

Fig. 1–18. Two popular types of torque wrenches are shown. One has a dial gauge and a needle that deflects as a nut or bolt is tightened. The other has a torque-setting adjuster in the end of the handle. When you reach the pre-set torque, you hear a click.

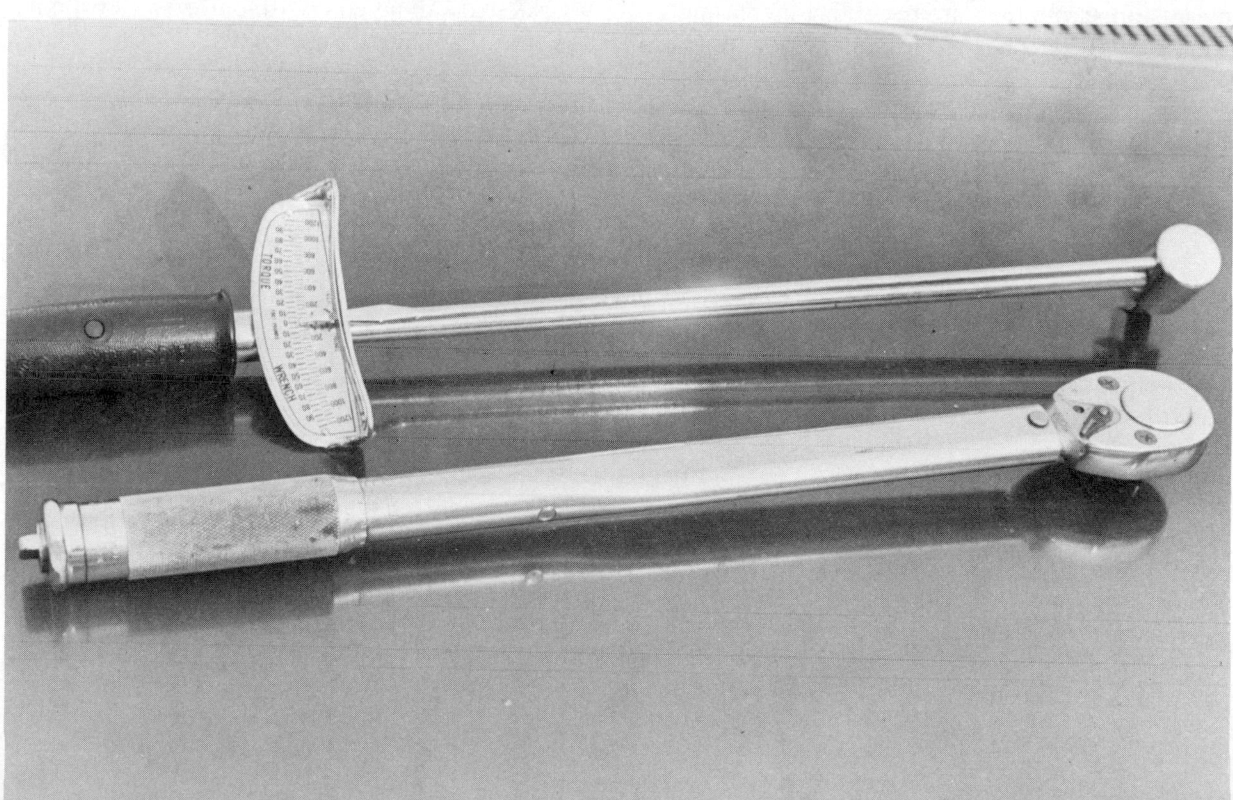

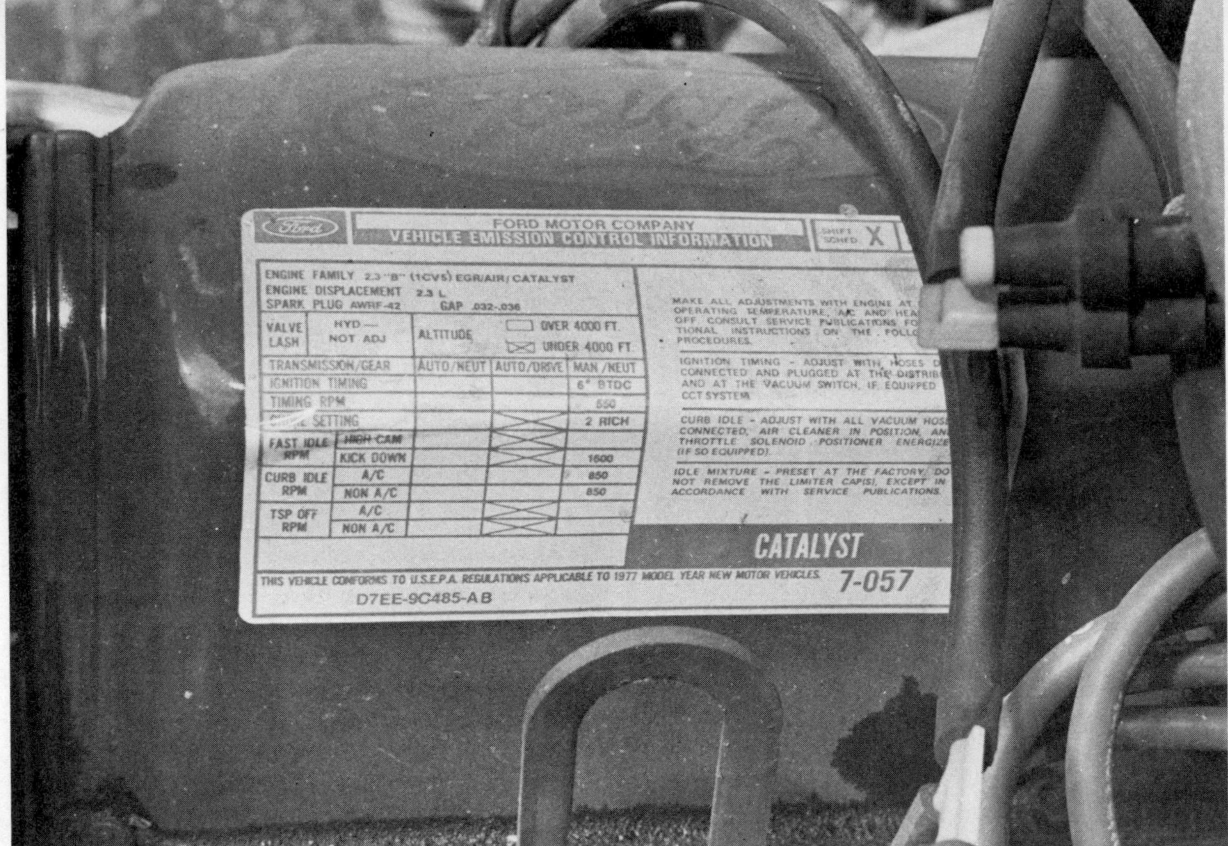

Fig. 1–19. Closeup look at the tuneup decal on all late-model cars. The decal, which lists important tuneup specifications, is located somewhere under the hood, in this case on the engine top cover.

but it won't be as good as you'll get elsewhere. However, at least at the beginning, the extra service is worth the difference. The weekend mechanic's trade is of increasing importance to these stores, and they are becoming increasingly competitive on price, particularly for tuneup parts.

For details on general parts buying, see Chapter 16.

The Way to Begin

Before you buy anything, tools or parts, read *Chapters 2* through *5*, with periodic references to your actual engine compartment, so you become basically familiar with what you will be doing. The general understanding will help you pick the most suitable tools and limit your parts purchases to what you will actually need.

2 THE IGNITION SYSTEM

MUCH OF THE TUNEUP is concentrated around the ignition system, and for good reason: Unless there is a consistent, healthy spark to the right cylinder at the right time, the engine will suffer. It may not start, and even if it starts, it will run poorly, suffer from stalling, and have many other problems.

With the exception of a few makes (primarily Chrysler, Plymouth, and Dodge since 1973), automobiles until 1975 were equipped with what is called a breaker points ignition system. In 1975, the rest of the American car industry switched to electronic ignition; most imports did not begin to follow this lead until years later. If you are among the millions whose car is equipped with the breaker points system, this opening section of Chapter 2 is for you. If your car has electronic ignition, it does not have breaker points. Read the following descriptions of the basic electrical circuit and how breaker points ignition works, so you can also understand electronic ignition; then proceed to the next section in this chapter.

The Electrical Circuit—A Simplified Description

The ignition system is a battery-powered electrical circuit and, as such, it bears certain resemblances to any battery-powered electrical circuit, including the lights on your car. The basic principles of the circuit are that it must have a current source (the battery) and that the circuit be completed back to the battery. Each battery has two posts or terminals, one from which the current can flow, and a second to which it returns, depleted of its energy (see *Ch. 9,* Fig. 9–1).

It is impractical and expensive to have two wires for each electrical device in a car, so the electrical ground principle is employed (see *Ch. 9,* Fig. 9–2). The car body and the engine are metal, and therefore can conduct electricity. One post of the battery is connected by a wire, called a ground cable, to the car body and/or engine. The second post is connected by wiring to all the electrical accessories, and the circuits are completed by connecting all the accessories to the body or engine with metal mounting screws, or wires connected to the body by screws. The use of the engine or car body to complete a circuit is called "grounding," and it is employed throughout the car, including the ignition system.

A final principle of general electrical circuits you must understand is the function of the switch (see *Ch. 9,* Fig. 9–3). You know that when you flip a light switch from off to on, a bulb lights. The switch is actually a mechanical device that, when turned "off," breaks the completed electrical circuit by separating two contact discs (or something similar). When the switch is turned on, the discs touch, and current can flow through them. The switch, therefore, permits you to make or break the electrical circuit.

Now let's see how all this applies to the ignition system.

Breaker Points Basics

The source of current for the ignition system is the car's 12-volt battery. The following—like the description of the general electrical circuit—is a simplified explanation, but it will give you the basic understanding you need to service the system.

When you turn the ignition key on, current flows from the 12-volt battery into a device called a resistor (on most cars it is just a special piece of wire), which reduces the voltage to somewhere between 6 to 9 volts. The lower-voltage current then flows into the ignition coil, a high-voltage transformer that when triggered will build it up to as much as 35,000 volts or even more.

The current circuit is completed (see Fig. 2–1) by a wire from the coil to the breaker points (also called contact points or just points), which are located on a plate in the distributor, the heart of the ignition system. Opening the points triggers the coil, so let's look at them.

The points assembly consists of two parts, one a spring-loaded movable arm on a pivot, the second a stationary part held by a screw. When the contact discs of the points touch, the circuit from the battery through the coil is complete, just as if a switch were closed (see Figs. 2–2, 2–3).

The movable arm has a fiber block that bears against the distributor cam, which is the upper part of a two-section shaft that runs through the center of the distributor into the engine (which

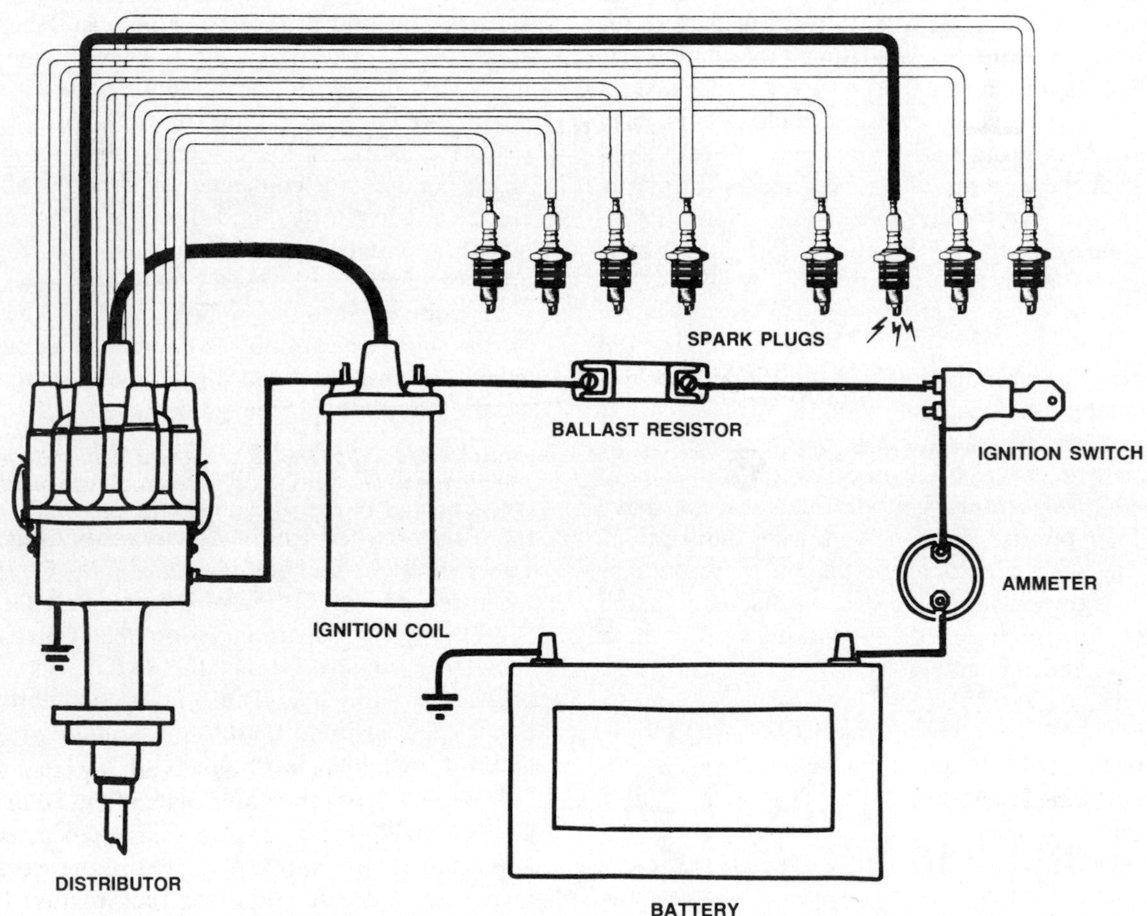

Fig. 2–1. This is the basic ignition system using breaker points. When you turn the ignition key, you close the ignition switch and complete a circuit from the 12-volt battery through a resistor to the ignition coil, in which a magnetic field builds up. When the ignition breaker points ("contact") in the distributor open, the coil builds up high-voltage electricity that exits through the center wire in the top of the coil, into the cap on the distributor, into a rotor in the distributor, then from the rotor out the distributor cap, through a spark plug wire to the plug, which fires the air-fuel mixture in the cylinder. On many cars a ballast resistor as shown is not used; instead there is a strand of resistance wire between the ignition switch and the coil.

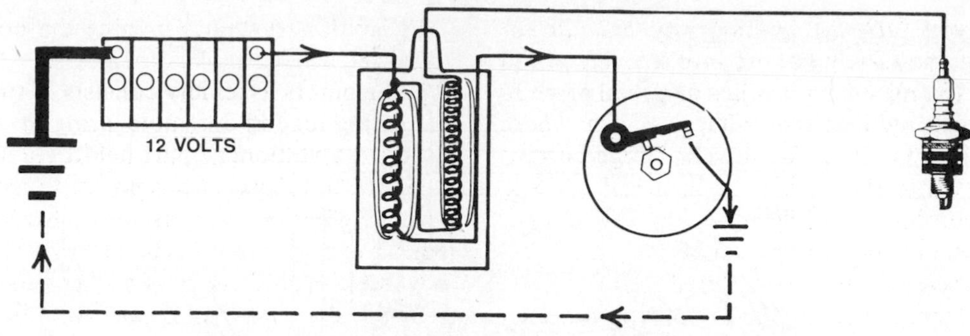

WHEN THE DISTRIBUTOR CONTACTS ARE CLOSED, THE MAGNETIC FIELD BUILDS UP IN THE IGNITION COILS.

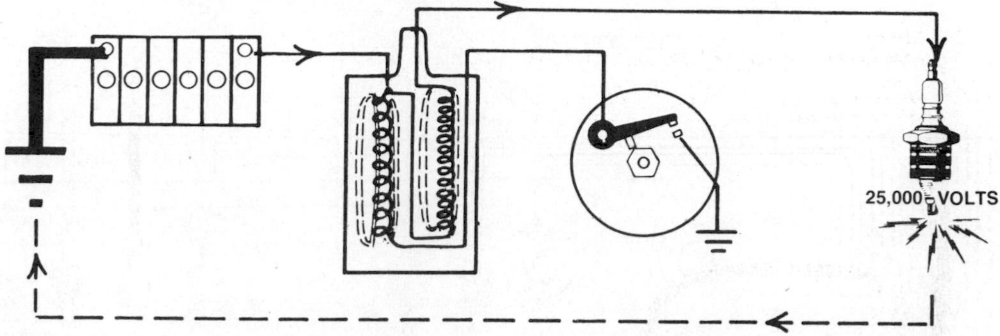

WHEN THE DISTRIBUTOR CONTACTS OPEN, THE MAGNETIC FIELD COLLAPSES AND A SPARK IS PRODUCED AT THE SPARK PLUG.

Fig. 2–2. This simplified illustration shows how the opening of the contact points creates a high-voltage spark. Notice that the coil has two windings, one in which the magnetic field builds up, a second (with much finer wire windings) onto which the electromagnetic field collapses when the points open. Because of the greater number of windings in the secondary windings, the voltage (electrical pressure) increases. In this illustration, the high-voltage electricity is shown going directly to the spark plug. Actually, it goes to the distributor cap in the multi-cylinder engine.

supplies power to spin it whenever the engine is running or being cranked).

The cam is a section with evenly-spaced bumps (called lobes) around it, one lobe for each cylinder of the engine. As the cam spins, a lobe comes up against the fiber block and pushes on it, moving the arm away from the fixed point to trigger the coil. The opening of the points is the same as opening a switch, except it's done automatically.

The high-voltage electricity flows from the center of the coil into the center of a plastic cap on the distributor, and through that onto a part called the rotor. The rotor is attached to the top of the distributor cam, so it spins with it, and as it does, its tip aligns with electrical contacts that pass through from the outer circumference of the cap. Each of these contacts contains a wire connected to the spark plug, the part threaded into the cylinder to ignite the fuel mixture (see Figs. 2–3, 2–4).

The distributor is so designed that whenever the points are opened, triggering the coil, the rotor tip is aligned with the distributor cap terminal for the spark plug of the cylinder that contains a fuel mixture ready for ignition.

The spinning of the distributor cam carries the lobe past the fiber block of the movable point, and the spring on that movable point pushes it closed—into contact with the fixed point. The

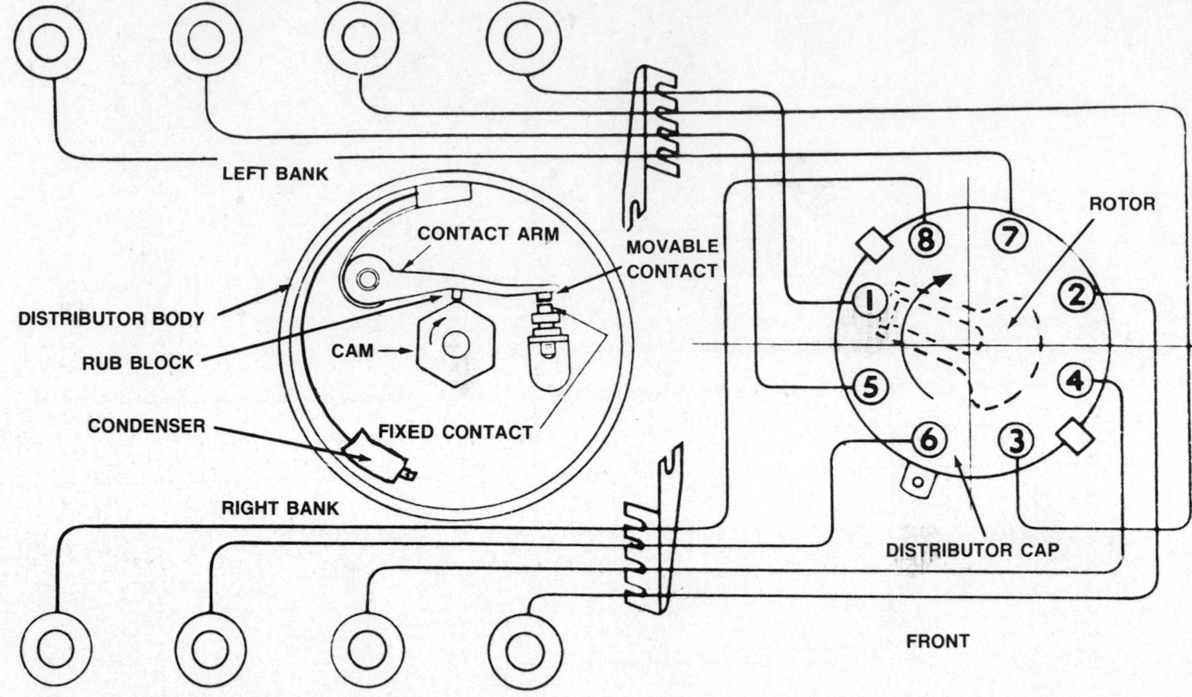

Fig. 2–3. This drawing shows the function of the rotor and distributor in the transmission of the high-voltage electricity to the correct spark plug (the one whose cylinder has the compressed air-fuel mixture ready for ignition). The cap and rotor are shown removed. The rotor is on the same shaft as the cam so both distributor cam and rotor are turned simultaneously by the engine. When the rotor aligns with a spark plug wire terminal, this coincides with the opening of the contact points by the distributor cam, and therefore, with the discharge of high-voltage electricity from the coil, into the cap, to the rotor, and from the rotor to a spark plug wire and spark plug.

circuit is complete once more, the coil gets ready, and in another instant the next lobe will push open the movable point and trigger the coil. This action occurs thousands of times per minute (see Figs. 2–2, 2–3).

Condenser

There's so much high-voltage electricity in the coil that some of it could flow through to the breaker points and arc across them while they are open. This would have two undesirable effects: (1) The contact discs would be burned very quickly, and soon fail to complete the circuit; (2) Since electricity takes the path of least resistance most of the high voltage would leak across the points rather than go to the spark plugs. In either case, the engine might not start.

To prevent this, the condenser is used (see Fig. 2–3). It's a small cylindrical can that is wired to the movable point, and it absorbs any stray high voltage before it can arc across the points. When the points close, it discharges the electricity safely.

Ignition Advance

It is desirable to slightly change the instant the spark fires the fuel mixture—called ignition timing—in accordance with changes in engine speed and engine load (how hard the engine is working).

The typical distributor has two automatic controls:

• *The mechanical advance*. See Fig. 2–5. The distributor cam is actually not a part of a solid shaft that goes into the engine, but really just fits over the shaft and is connected to it by a pair of

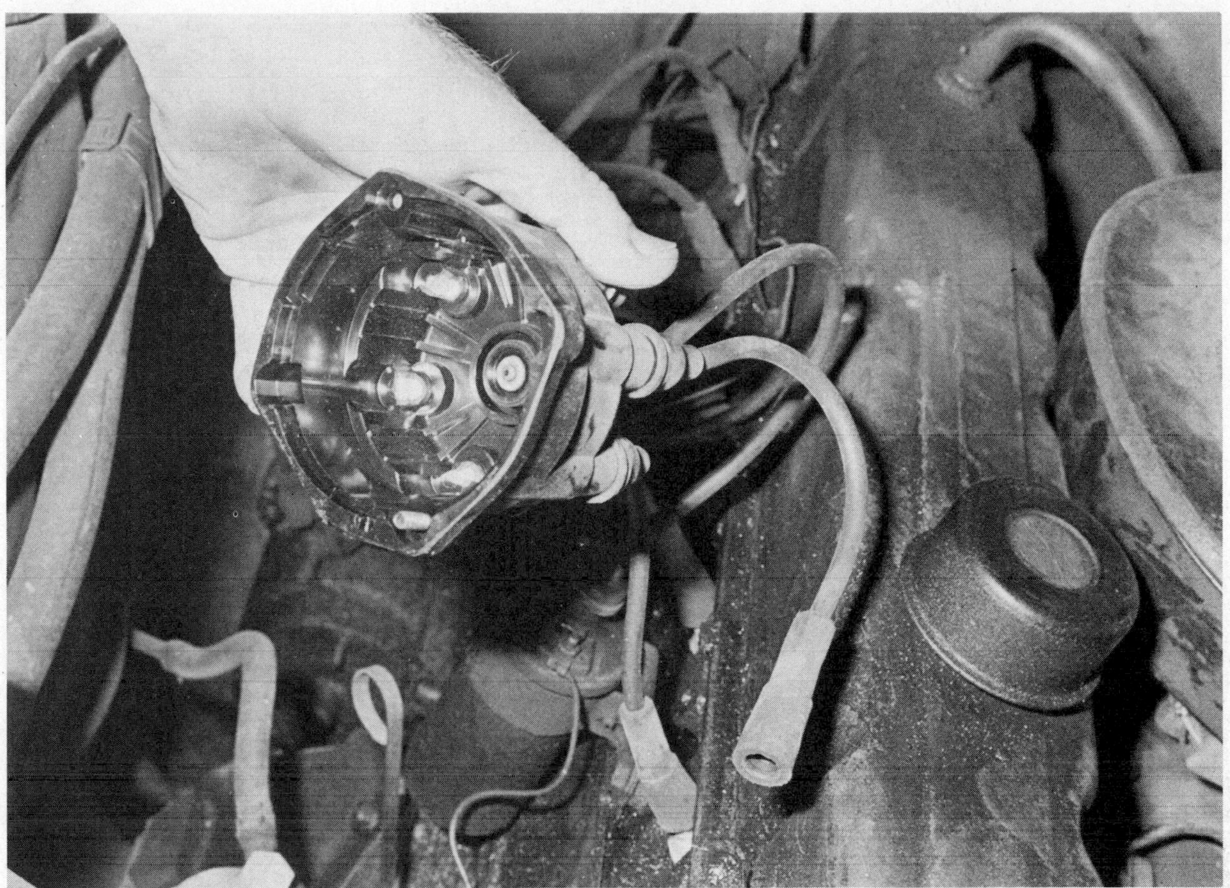

Fig. 2–4. A look inside a distributor cap shows the button in the center and the buttons around the inside circumference. These buttons are electrical terminals that project through the cap to the spark plug wires and the center wire from the ignition coil. The center button inside the cap transfers the high-voltage electricity to the rotor, and the circumferential buttons take that electricity from the rotor to a spark plug wire.

weights and springs. As engine speed increases, centrifugal force moves these weights outward against the spring pressure and the position of the distributor cam is shifted slightly. Moving the cam (clockwise or counterclockwise, depending on car) changes the time the lobes come up against the movable point's fiber block, so that the ignition coil is triggered at a different time.

• *The vacuum advance*. See Figs. 2–6, 2–7. The engine develops a substantial amount of vacuum, the exact amount depending on the engine load. Therefore vacuum is used as the control for changing ignition timing as the load changes. The distributor has a diaphragm device with an arm connected to the breaker plate, the part on which the points are mounted. The breaker plate looks as if it's just bolted down, but actually it (or a section of it) can be pivoted back and forth. When vacuum is applied to the diaphragm unit, it sucks on the diaphragm, moving the arm and the breaker plate. This shifts the position of the points relative to the distributor cam, which also changes the instant the coil is triggered. Vacuum usually is supplied to the diaphragm unit by a hose or piece of tubing from a part on the engine called the intake manifold (a set of chambers that carry air and fuel from the carburetor to the cylinders). Or the hose may be from the base of the carburetor, which is bolted to the intake manifold.

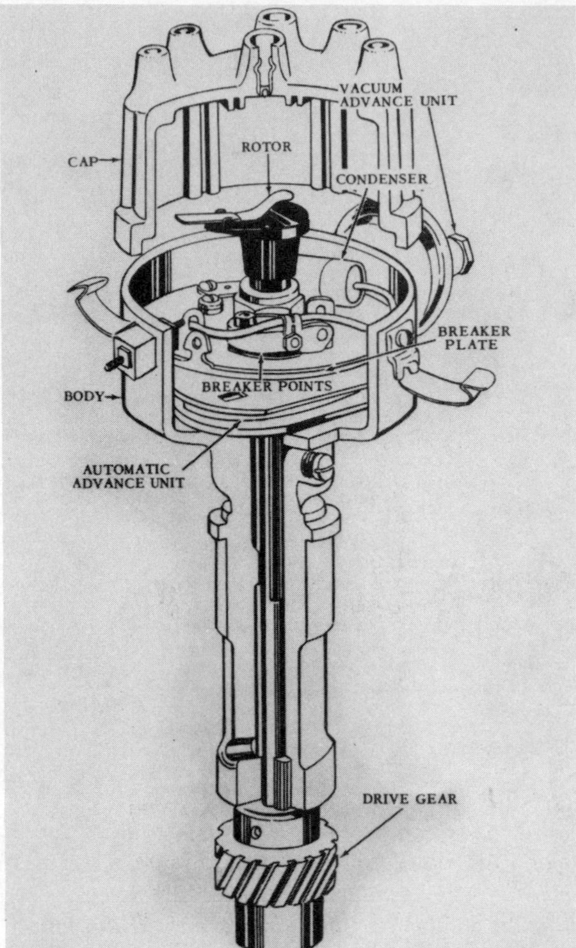

Fig. 2–5. Here's a cutaway view of a breaker points distributor. You can see the metal strip that transfers the high-voltage electricity from the coil wire to the appropriate spark plug wire on top of the rotor. You also can see the "automatic advance unit," a set of centrifugal weights that at high engine speed shift the position of the breaker points plate to change the time at which the points open and therefore the time the high-voltage electricity is transferred to the plug wires.

Servicing the Distributor

Now that you have a basic understanding of the ignition system, you can service the breaker points and condenser in the distributor. This is a key job in any tuneup.

Begin by locating the distributor (see Fig. 1–1, 1–2). This should be easy—it's the part with the cap into which all the thick spark plug wires are connected. The job should be done with the distributor in the car, so if your car's distributor is not conveniently located, and you must drape your body across the engine. Use a foam cushion for chest protection.

Remove the distributor cap. In most cases it isn't necessary to disconnect any wires from it, but if it is, label each wire and the terminal from which you remove it with pieces of masking tape (marked with letters or numbers for easy identification).

The cap is held in one of three ways:

- *C-shaped spring clips.* Just pry them off the squared sections of the cap on which they grip, using a screwdriver or your fingers (see Fig. 2–8).
- *Screws.* Just undo the screws and lift up on the cap. The screws are installed so they won't come out of the cap (to prevent you from losing them).
- *L-shaped locking rods.* If you look at the cap, these appear to be screws, but if you check further, you'll see that the rods are L-shaped and lock under the main section of the distributor body (see Fig. 2–9). To release them, insert a screwdriver into the slot at the top, push down (against spring pressure), and turn counterclockwise.

Most distributors have just two clips, screws or locking rods, but a few (primarily ones with electronic ignition systems) have four.

With the distributor cap up and out of the way, you can see the points and condenser. On General Motors V-8 distributors, you also can see the mechanical advance weights and springs, for they are just under the rotor (on other distributors, the mechanical advance is hidden under the breaker plate). See Figs. 2–10, 2–11.

Inspect the points before you disconnect anything. Just push the movable arm away and look at the contact faces, using a magnifying glass if necessary. If the points are burned black, they must be replaced. If there is a metal transfer between the two resulting in a hill on one and a valley in the other or pockmarks in both, a replacement also is called for. If, however, the contact faces are clean, or have no more than an even gray coating, leave them in place. They may need an adjustment, but that's all.

Note: It is possible that the fiber block of the movable point has worn down so much that you

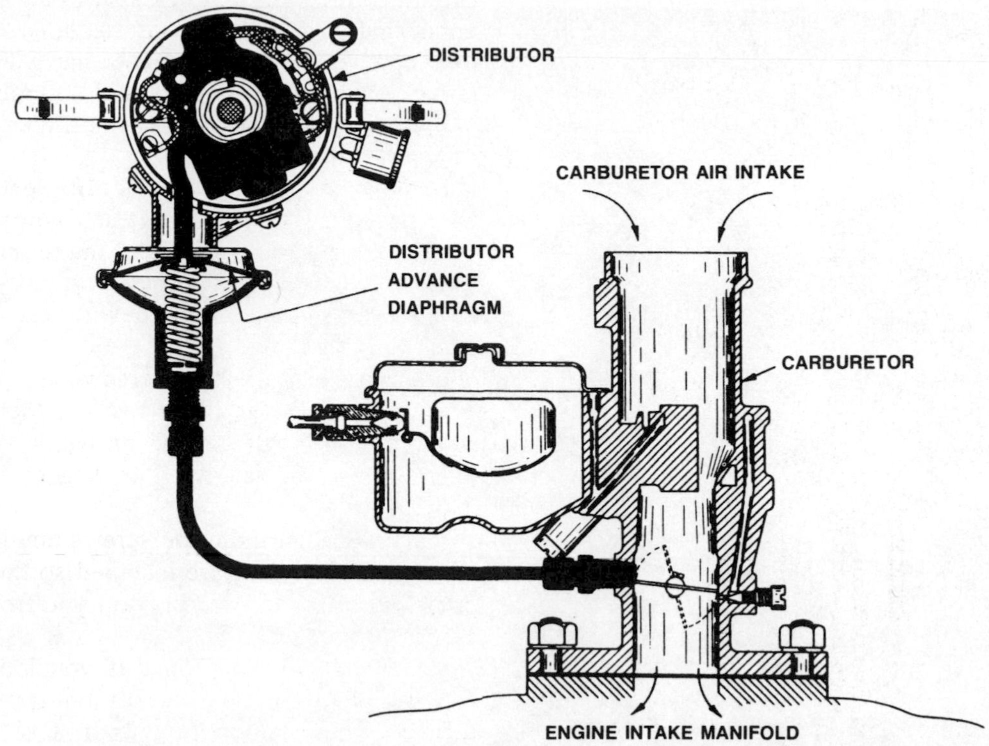

Fig. 2–6. This is the vacuum advance mechanism. At low engine speed it changes the time the points open. The unit is a spring-loaded diaphragm, connected by a hose to a source of engine vacuum (usually at the base of the carburetor). Vacuum sucks on the diaphragm, which is connected to the breaker points' plate by a link. The diaphragm moves against spring pressure and turns the breaker plate so the distributor cam lobe opens the points earlier than otherwise. When vacuum drops as the throttle opens wide, the spring returns the diaphragm, link and breaker plate. However, at this time a mechanical advance system sensitive to engine speed takes over.

will not be able to readjust the points satisfactorily. In this case, the points will have to be changed even if the contact faces look good.

If you see an oily film in the distributor, this can cause premature failure of the points or other ignition problems. It should be cleaned off all parts, particularly the points' contact faces. The problem usually is caused by a clogged Positive Crankcase Ventilation system, discussed in Chapter 4.

Adjusting the Points (Except on General Motors V-8s)

If the points are in good condition, the points' adjustment should be checked and corrected if necessary. Adjustment refers to the gap between the two contact faces when the movable point is pushed away by a cam lobe. The check is made when the fiber block is against the very peak of the cam lobe, and is done with a feeler gauge. You merely insert a feeler gauge (or combinations of a few feeler gauge strips) until you find the size that can be inserted and withdrawn with light-to-moderate drag. If the gauge just slides in and out easily, try a thicker strip. If you have to push against the movable point to get it in at all, try a thinner strip. No set of feeler gauges will have all sizes, but to measure precisely, combine two thin strips to get a thicker size. Example: The set has a .016-inch strip and a .020-inch strip. If you need something in between, perhaps you can combine .008- and .010-inch strips for a total of .018 inch.

The major problem is getting the fiber block against the peak of the cam lobe. There are many methods, but this is the one that will give the beginner the least difficulty:

Fig. 2–7. Here's an in-the-metal look at the distributor and the vacuum advance unit. Fingers are holding the vacuum hose connected to it.

First, locate the lockbolt at the end of the distributor body, where it mates with the engine (see Fig. 2–12). If you need a C-shaped wrench to reach it, obtain one.

Second, with chalk, tire marking crayon, nail polish, etc., make a thin mark on the distributor body and another just opposite on the engine. These are alignment marks.

Third, slacken the lockbolt with the wrench and turn the distributor body clockwise or counterclockwise until the fiber block is against the peak of the nearest cam lobe.

Now, you're ready for breaker points adjustment. The principle of the adjustment is this: the swing of the movable arm is determined by the distributor cam lobe; therefore the gap is determined by the position of the fixed point. Although the fixed point is held to the breaker plate by a screw, it can be moved toward or away from the movable point when the screw is loosened.

Therefore, loosen the screw and move the fixed point toward or away from the movable point, to decrease or increase the gap. On most fixed points there is a little notch, plus another in the breaker plate, into which you can insert a thin screwdriver. Just turn the screwdriver one way or the other and the fixed point will move. Insert the feeler gauge and when you feel the correct gap (light to moderate drag as you insert and withdraw it) tighten the screw (see Fig. 2–13). Recheck the gap after tightening the screw.

The maximum gap between the two points is extremely critical. The factory specification will give you a few thousandths of an inch of leeway (it might be .014 to .019 inch), but never exceed those limits. The reason is that the gap adjustment also determines how quickly the points will close when the lobe passes the fiber block, and how quickly they will open when the next lobe

Fig. 2–8. Using a slot-screwdriver to pry the spring clips off the distributor cap.

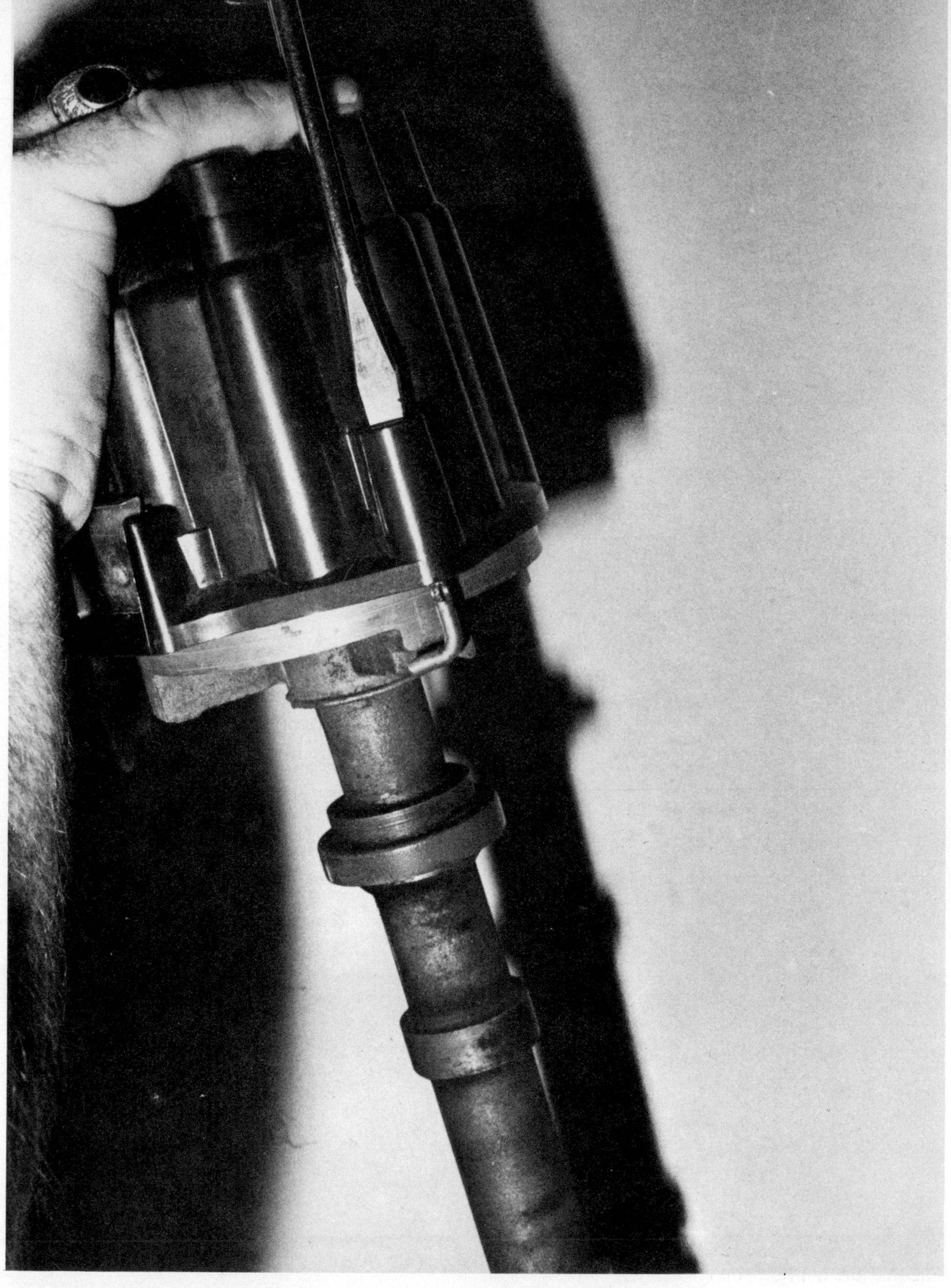

Fig. 2–9. This distributor cap has locking rods. Insert screwdriver as shown, push down against spring pressure, then turn and L-shaped bottom of rod will turn out of the way, permitting you to lift the cap.

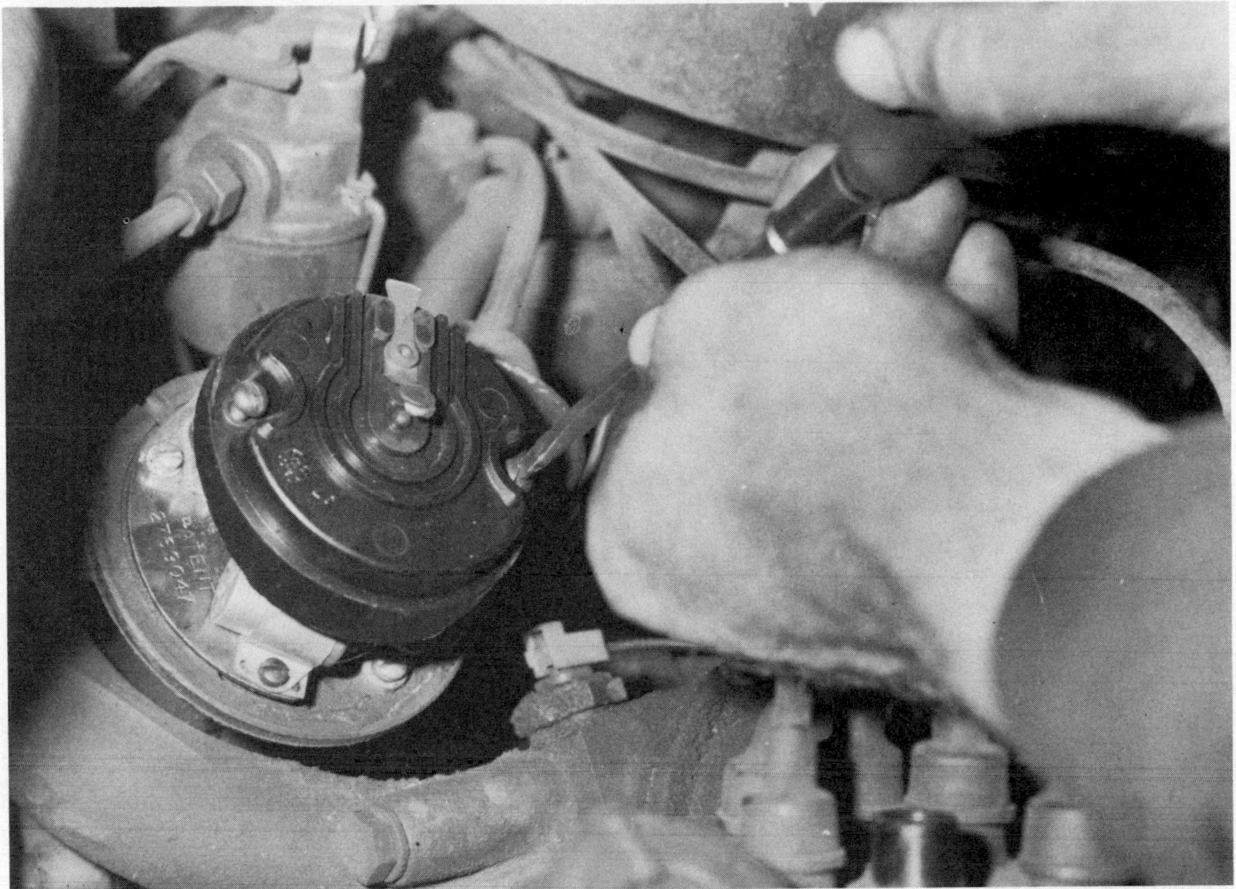

Fig. 2–10. This is a General Motors V-8 distributor. Centrifugal advance weights are on top of the distributor, just under the rotor. The rotor is held by a pair of easy-to-remove screws.

comes around. The closing of the points completes the circuit, so the charging up of the coil is affected. The opening of the points causes the coil to trigger, so the ignition timing is affected.

Next, turn the distributor body back until the alignment marks you made earlier are aligned once more. Tighten the lockbolt securely.

Checking Points Dwell

The opposite of the points being open is when they are closed, which is called dwell. Whenever the lobes on the distributor cam do not hold the points open, they spring closed. The time they are closed (measured in degrees of rotation of the distributor cam) is important, for it is during that time that the ignition coil charges up. Just as the maximum gap between the points must be within specifications, so must the dwell.

Checking dwell is done with the engine running at idle, so begin by reinstalling the rotor (push it down firmly onto the top of the distributor cam, until it seats), then refit the distributor cap and engage the clips, screws, or locking rods.

You're ready to use the dwellmeter. If it is a combination tach-dwellmeter, turn the knob to dwell, and the cylinders' knob to the correct number (four, six, or eight). Note: On some dwellmeters there is no four-cylinder knob position; just set it for eight and double the reading you get.

Before you make the connections, look at the top of the ignition coil. You'll see three wires, one wire with the thick jacket that goes to the top of the distributor cap, and one on each side of it. Of the two additional wires, one is to a terminal marked "SW" (for switch) or plus, the other "CB" (for contact breakers) or minus. Connect the test meter wire with the red-covered clip (or from the plus terminal of the meter) to the ignition coil terminal marked "CB" or minus.

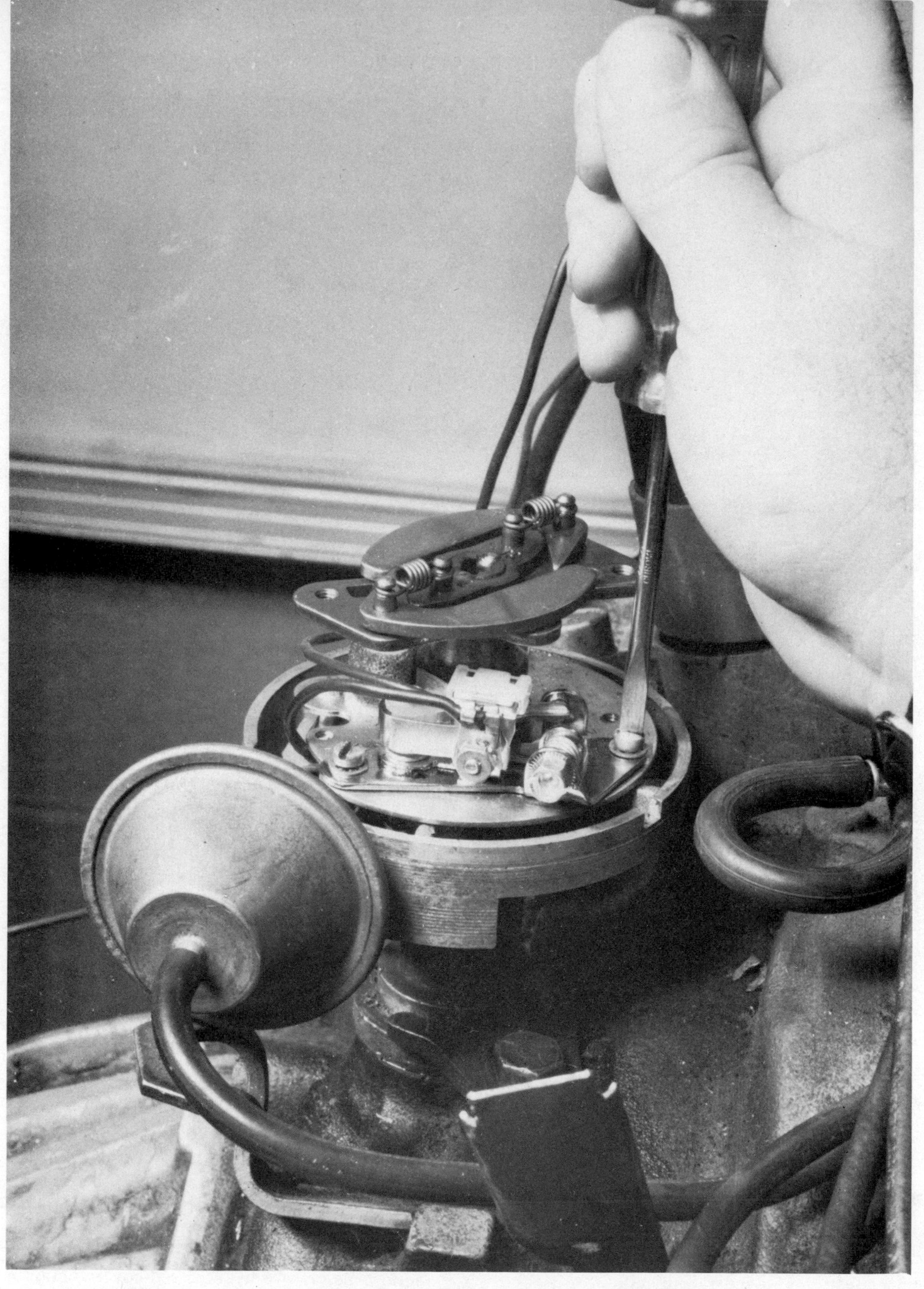

Fig. 2–11. With the rotor off, you can clearly see the centrifugal weights held by little springs.

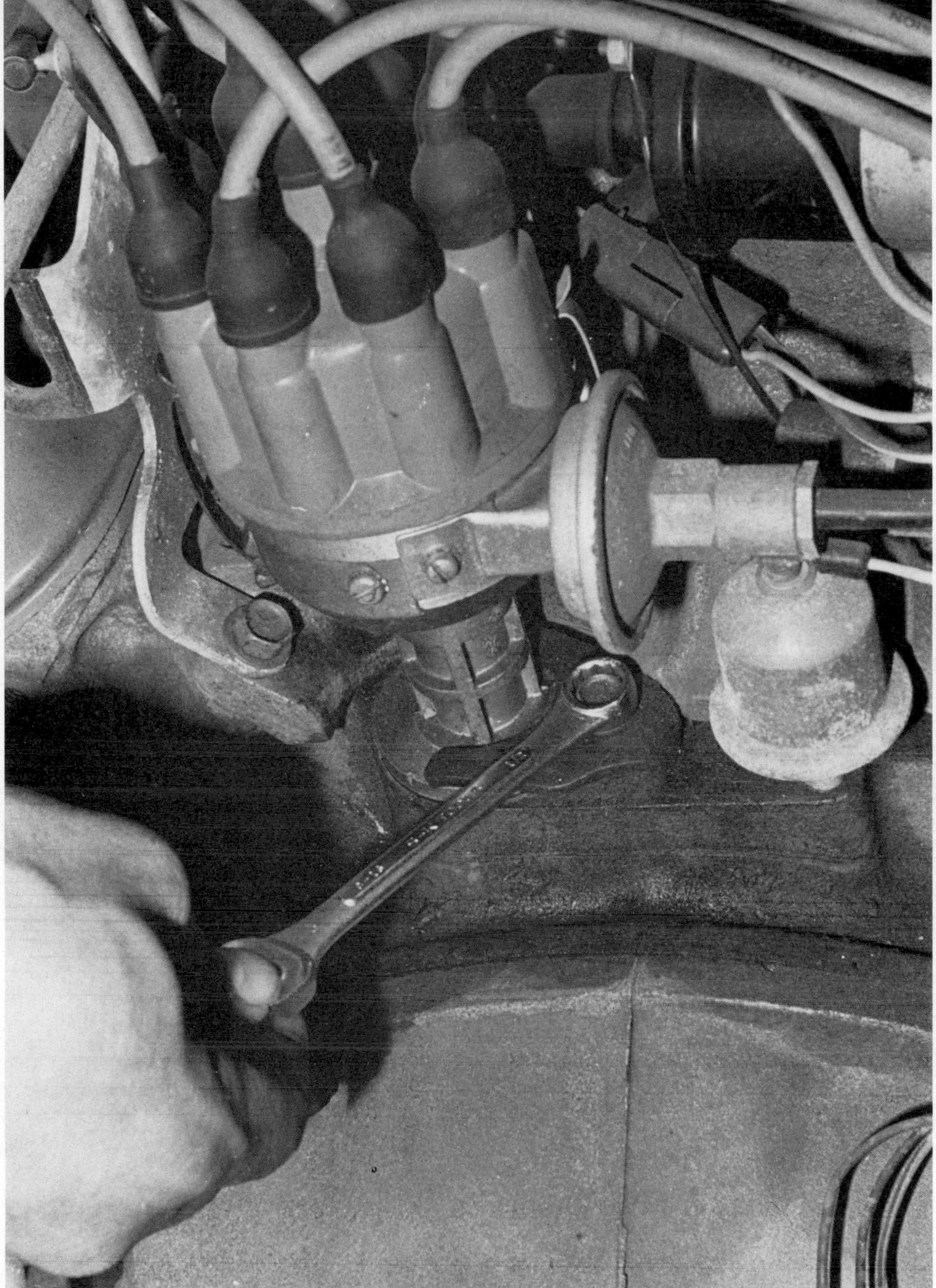

Fig. 2–12. Shown is what the distributor lock looks like. It's just a clamp part held by a bolt. Once you slacken the bolt, you can turn the distributor body. When preparing for a points adjustment, remove the distributor cap so you can see the points, then turn the distributor body so the points' fiber block is up against the peak of a cam lobe. Always make alignment marks on the distributor body and the engine so you can return the distributor to the same position after the adjustment.

Fig. 2–13. *Checking gap between the breaker points with a feeler gauge of the specified thickness. Once the gap is set so the feeler moves in and out with light-to-moderate drag, tighten the screw as shown.*

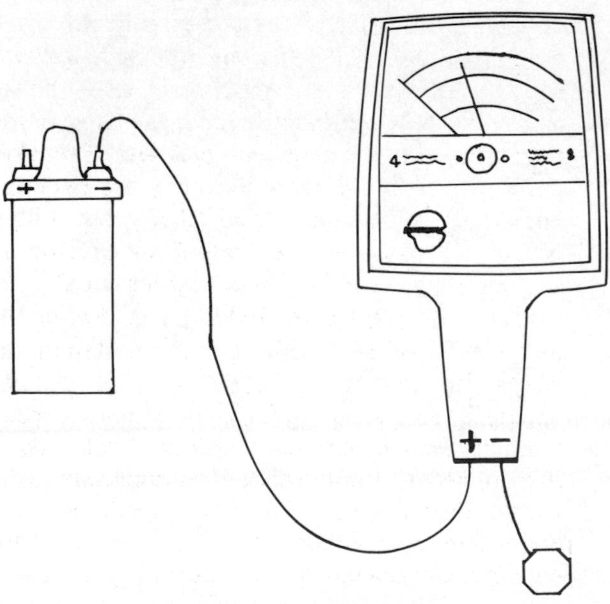

Fig. 2–14. How to connect a dwellmeter. The wire from the dwellmeter positive terminal is connected to the negative terminal of the ignition coil (also may be marked SW or CB). The wire from the dwellmeter negative terminal (usually black) is connected to an electrical ground, such as an engine or body bolt.

Fig. 2–15. To inspect points, remove distributor cap (on GM V-8 distributors, also remove the rotor). Then push movable point away with a screwdriver so you can look closely at the contact faces.

Connect the other test meter wire (with a black clip or from the minus terminal of the meter) to an electrical ground, such as an engine bolt, or the metal body of the distributor (see Fig. 2–14).

Start the engine and let it idle. Read the meter and compare the number of degrees shown with factory specifications. If you have set the breaker points accurately, the dwell reading should be within specifications. If the dwell reading is too high, increase the gap between the breaker points; if it's too low, reduce the gap. In either case, keep the gap within specifications; aim toward the high or low end of the car manufacturer's range.

• *Dual points*. Some older cars are equipped with two sets of points, which overlap each other and permit only one distributor cam lobe for each two cylinders. On this design, which improved spark at high engine speed, you must measure dwell both separately and with both points operating. To measure it separately, merely insert an insulator (such as a piece of cardboard) between the contacts of one set, and the dwell reading you get will be for the other set. Reverse the procedure for the second set. Note: Although this book does not include Japanese cars, if you have a Japanese Datsun with two sets of breaker points, you should know that they are operated and adjusted differently from other types. Either obtain a Datsun service manual for specific instructions or leave this job to a professional.

Next, perform a dwell-holding test. Switch the meter to "TACH" and have a helper step on the gas pedal—and hold it steady—so as to raise engine speed 1000 r.p.m. over the idle speed reading. Switch the meter back to dwell, and at this higher speed the dwell reading should be within three degrees of the idle speed reading. If the difference is greater, it can be responsible for poor performance. The only cure, if you have a performance problem, is to replace the distributor, a job for a professional.

Notes: On some engines, the dwell variation may exceed three degrees without affecting performance noticeably. The criterion for replacement, therefore, is engine performance.

When you tighten down the fixed point screw(s), be sure to use the correct size screwdriver and tighten securely. If it is not tight, engine vibration allows it to loosen, and the point gap adjustment will be lost. In some cases, the initial movement may cause the engine to fail the dwell-holding test.

Adjusting Points—General Motors V-8s

On its V-8 engines, General Motors (producer of Pontiac, Chevrolet, Buick, Oldsmobile, and Cadillac) used what was called a Peek-A-Boo distributor with breaker points until 1975, when it switched to electronic ignition.

This distributor differs from others in several important ways. There is a metal window in the side of the distributor cap. Instead of removing the cap for points' adjustment, you just lift the window, insert a ⅛-inch Allen wrench into an adjuster and turn it to change the point gap (see Fig. 1–15). The point gap is adjusted with the engine running, using a dwellmeter.

Although you can, you need not remove the distributor cap and adjust the points with a feeler gauge. You should, however, take off the cap to inspect them (see Fig. 2–15). On some distributors there is a two-piece metal shield (for radio interference suppression) on top of the breaker plate covering the points and condenser. Remove it and you can see the points. If they need only an adjustment, you can do either of the following:

Reinstall the metal shield (if used) and the distributor cap. Connect a dwellmeter as with other types of distributors and turn the adjuster whichever way is required to bring the dwell reading to within specifications.

As with other distributors, make alignment marks on the distributor body and engine, slacken the lockbolt at the base of the distributor, and turn the distributor body until the fiber block is against a peak of a cam lobe. Insert a ⅛-inch Allen wrench into the adjuster and turn to set the point gap to specifications (Fig. 2–16).

Replacing Breaker Points and Condenser

The condenser may last tens of thousands of miles. If you don't see metal transfers on the points (hill on one contact face, valley in the other), the condenser is apparently doing its job and need not be replaced. However, the easiest points and condenser to install is a combination assembly, in which the condenser is already a permanent part of the movable point rather than a separate item wired to it. With the combination assembly, you simply discard the old condenser and the job of installation is the same as if the condenser did not exist.The combination costs a little more than separate sets of points and condenser, and you may be replacing a condenser that's still serviceable. However, the assurance of a good condenser-to-points connection and the security of a brand-new condenser is well worth it (see Fig. 2–17).

Begin the replacement job by removing the distributor cap and, on GM V-8 engines, the radio suppression shield if used. Observe how the wiring is connected to the terminal for the movable point, and check if there are any plastic or fiber sleeves or washers used (they are insulators) that must be reinstalled. Typically, the terminal will have a lockscrew to hold the wires from the condenser and the coil; there may be a nut threaded on a stud (see Fig. 2–18); or finally, the wires could have push-in spades that are held by a spring lock in the terminal. With this design, you push on a tang on the terminal as you pull on the wire and the wire will come out. Whatever arrangement is used (see also Fig. 2–19), disconnect the two wires from the movable point terminal. If you have an access problem, perhaps you will find it easier if, as the next step, you unscrew the points and lift them up.

The points themselves are usually held in place by a single screw in the slot hole of the fixed point (see Fig. 2–20). Or there are two screws, such as on some Ford and all GM V-8 distributors (see Fig. 2–11). Note: On the Ford distributors, the second screw will have a thin wire strap attached to it. This is a ground strap and must be reconnected.

If you have a magnetized screwdriver, remove the points' screws with it, so you don't drop the little screw and the washers.

To install the new set of points (see Fig. 2–21) just put it in place (see Fig. 2–22), install the

Fig. 2–16. If you wish, you can adjust point gap on GM V-8 distributor with engine off and distributor cap removed. Just slacken lockbolt and turn distributor body until fiber block is against peak of cam lobe. Insert specified feeler gauge between the points and turn the Allen head adjuster with Allen wrench, as shown, until feeler moves in and out with light-to-moderate drag.

retaining screw(s), and reconnect the wire from the ignition coil. If you are installing a separate set of points and condenser, make the second connection from the condenser and tighten the condenser mounting screw securely.

Apply a thin film of white grease using a soft fiber brush or your finger (see Fig. 2–23) to the circumference of the distributor cam where it comes in contact with the fiber block. This grease is sold in tubes, or a small amount may be included with the set of points. If the point set comes with a lubricating wick built in, applying the grease is unnecessary.

Finally, adjust the points and check the dwell as explained earlier.

Other Distributor Checks

While you're replacing or adjusting the breaker points, there are several other checks you should make:

(1) Look at the rotor. If the metal tip is burned, try to clean it with a wire brush or fine sandpaper. If the burn is severe, and a lot of sanding or brushing would be necessary, the metal removed from the tip might cause ignition problems. In this case, replace the rotor. Caution: Don't sand the rotor tip on a car with electronic ignition. What seems to be a brown mark on the rotor tip may actually be a special coating that should not be removed. However, if there's a clear-cut burn mark replace the rotor. In fact, you should apply a fresh coat of that grease (called silicone dielectric grease and sold in tubes) to the rotor firing tip(s) and on Ford products with electronic engine control (Fig. 2–62), to the rotor contact plate tips in the distributor cap too.

(2) Disconnect one spark plug wire at a time from the cap and look at the terminal into which it fits. If you see whitish corrosion deposits, clean them out with a wire brush (see Fig. 2–24). Also check the coil wire terminal in the cap.

(3) Wipe the inside of the cap with a dry, clean rag and look for cracks (including fine hairlines). If you see any, replace the cap (see Fig. 2–25).

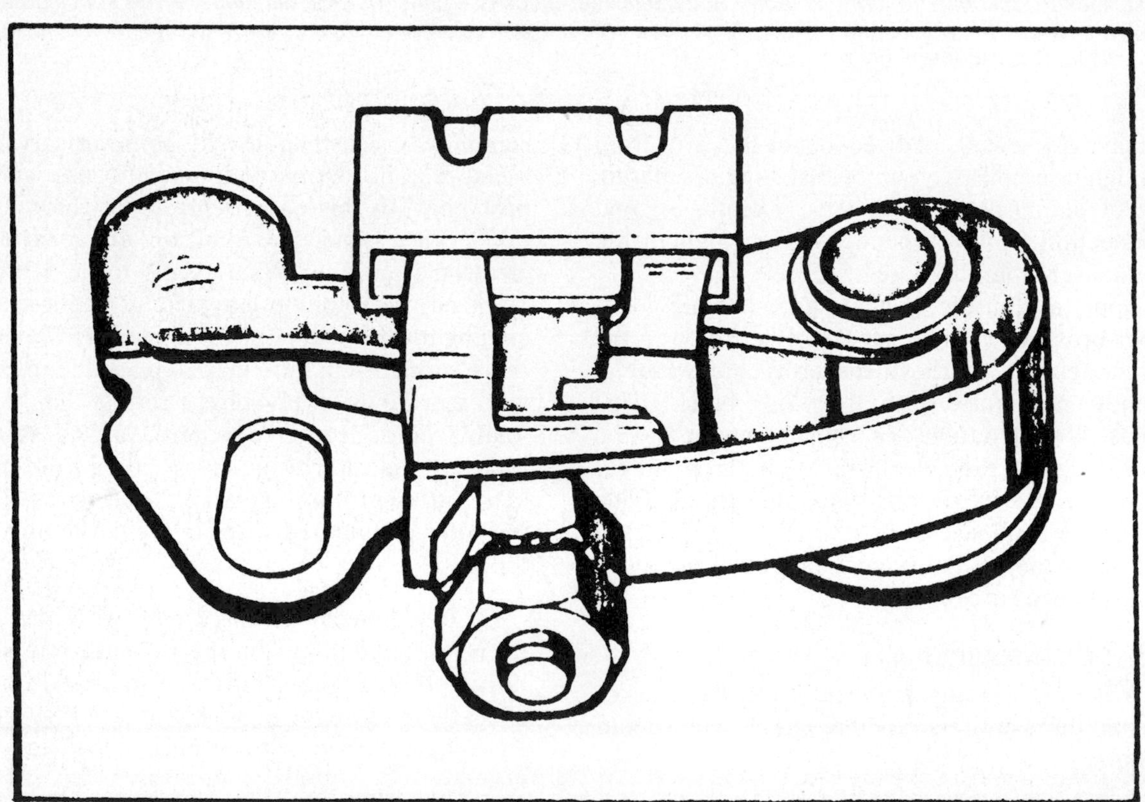

Fig. 2–17. Installing a set of points with a built-in condenser. The rectangular segment on top of the points holds the condenser, eliminating the need for an extra connection. Bottom illustration is a closeup drawing of the unified point set.

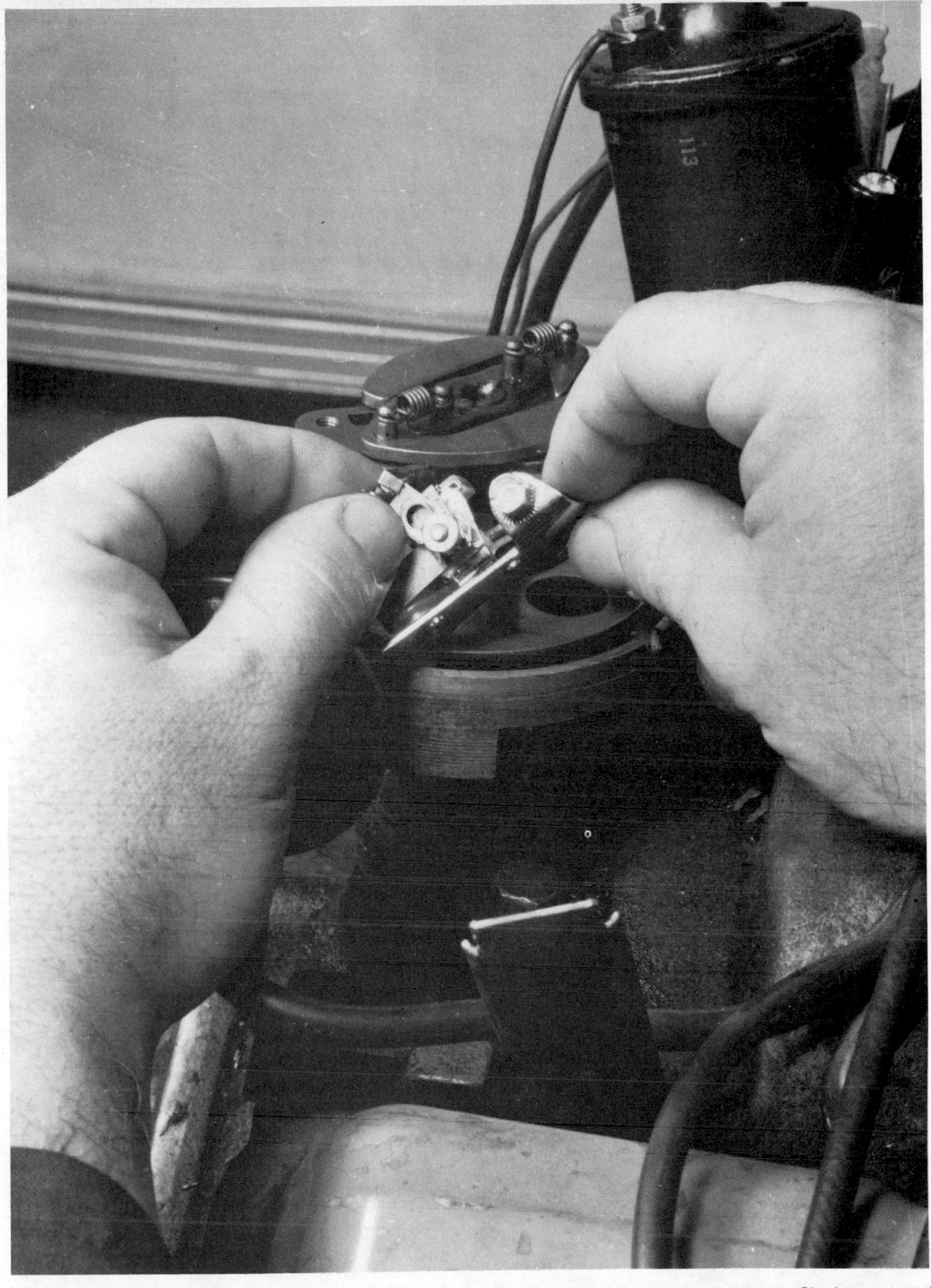

Fig. 2–18. Nut threaded on stud is the terminal for the wires from the coil and the condenser. Slacken nut and disconnect the wires as shown.

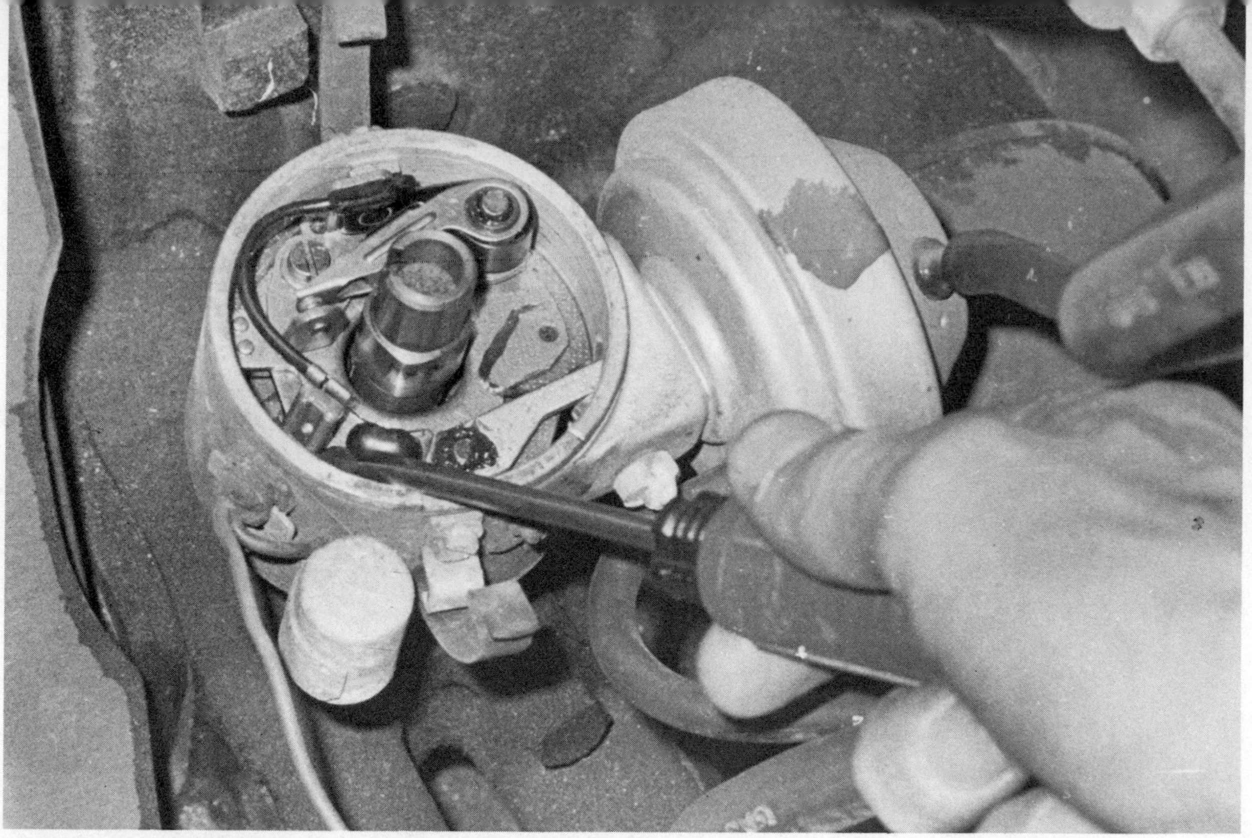

Fig. 2–19. This distributor has a points' terminal with a male spade. Push up on the wire with the female spade, using a screwdriver as shown, to disconnect it.

Fig. 2–20. A single screw in a slot hole holds this set of ignition points.

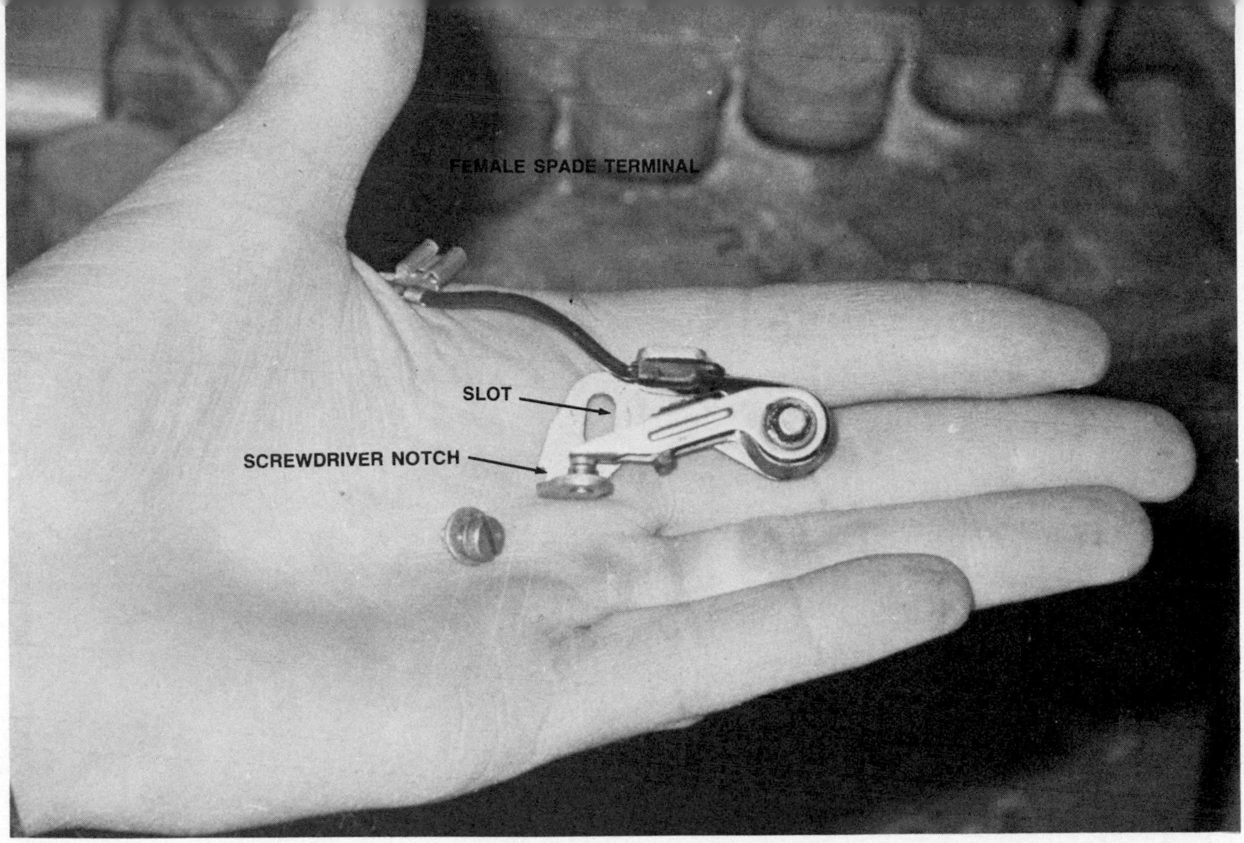

Fig. 2–21. Here's another set of points with a lead wire that has a female spade terminal. You can see the slot hole and the notch for the screwdriver.

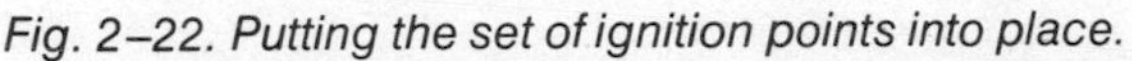

Fig. 2–22. Putting the set of ignition points into place.

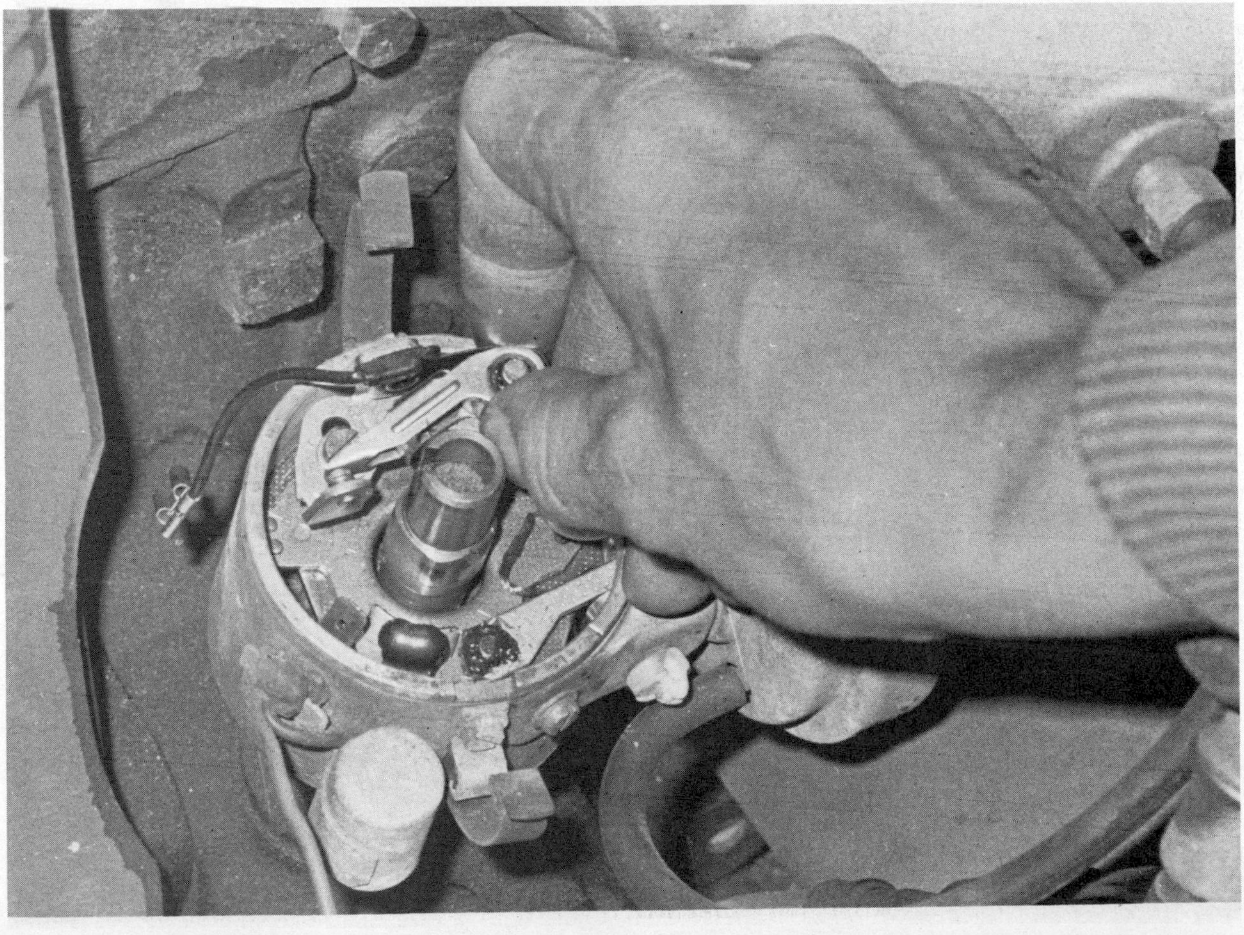

Fig. 2–23. A dab of white grease, about the size of a match head, is all you need to lubricate the cam. Smear the grease evenly around the cam, using a fiber brush or your finger.

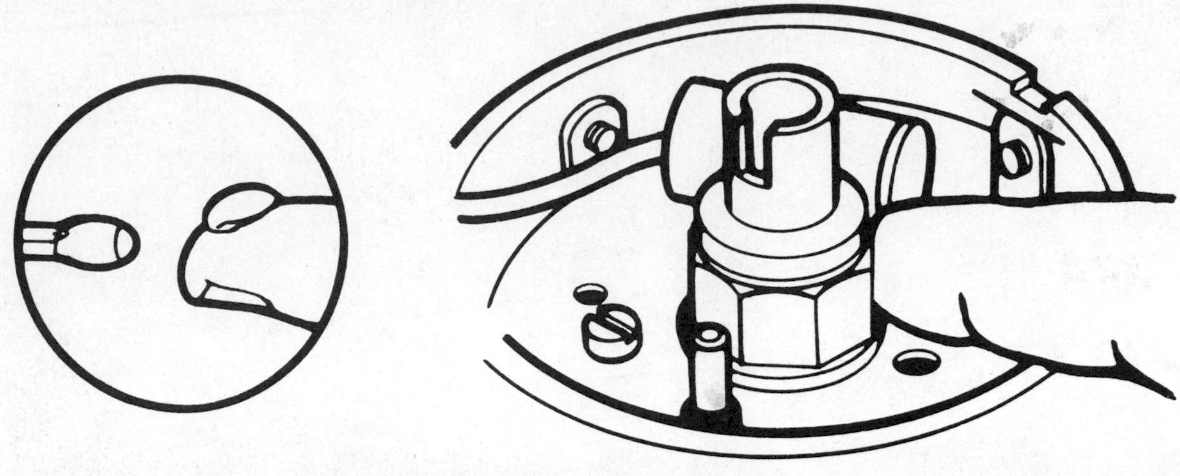

Fig. 2–24. Cleaning out distributor cap terminals with a wire brush. If cap has male terminals, wire brush their exterior. Also clean the spark plug wire terminals, male or female, with a wire brush, if required.

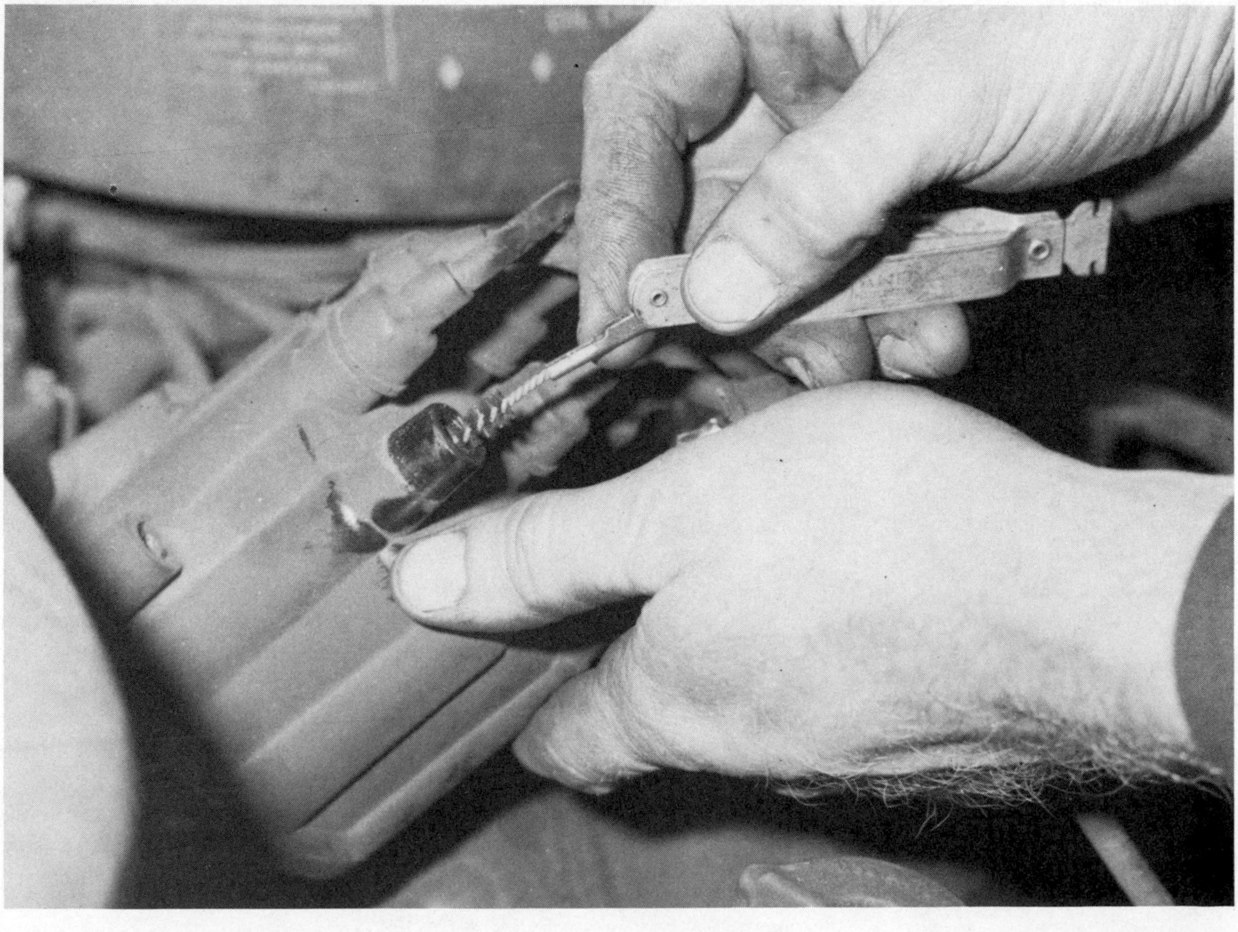

Fig. 2–25. After cleaning, inspect the inside of the distributor cap, using a flashlight if necessary. If you see any cracks, including fine hair lines, replace the cap, as explained later in this chapter under the subject of spark plug wires.

(4) On all but General Motors V-8 Peek-a-Boo distributors, which have it on top (and most computer-controlled electronic spark advance systems, which do not have it at all), the mechanical advance system is under the breaker plate, out of sight. You can quick-check it with the rotor in place. Just turn the rotor—it will go one way, about a quarter-inch. Release it and it should spring back. If it doesn't, the mechanical advance is defective. Servicing it is a job for a professional.

(5) On GM distributors with the centrifugal advance on top—that's the Peek-a-Boo—and on electronic ignition systems (all without a computer), make a careful inspection of the centrifugal advance. Look for enlargement of the weights' holes and eating away of the posts on which the weights swing. If there is none, just spray on silicone lubricant. If there is erosion of either or both parts (Fig. 2–26), you can obtain repair kits at auto parts stores to correct the problem. If the erosion is severe, a professional can repair the distributor by replacing the shaft-and-weights as an assembly. Detail parts are not available.

Erosion of weights and posts is rare on breaker points systems, but all too common on GM electronic ignition. The reason is the very high voltage the system develops. High voltage arcing in the distributor cap forms corrosive gases, which accelerate the erosion.

You also may see erosion of the distributor cap rotor button and the cap in the area around the button itself. In severe cases, high-voltage arcing from the coil to the rotor button can completely destroy the cap. Change spark plugs at reasonable intervals (maximum 24,000 miles) and apply

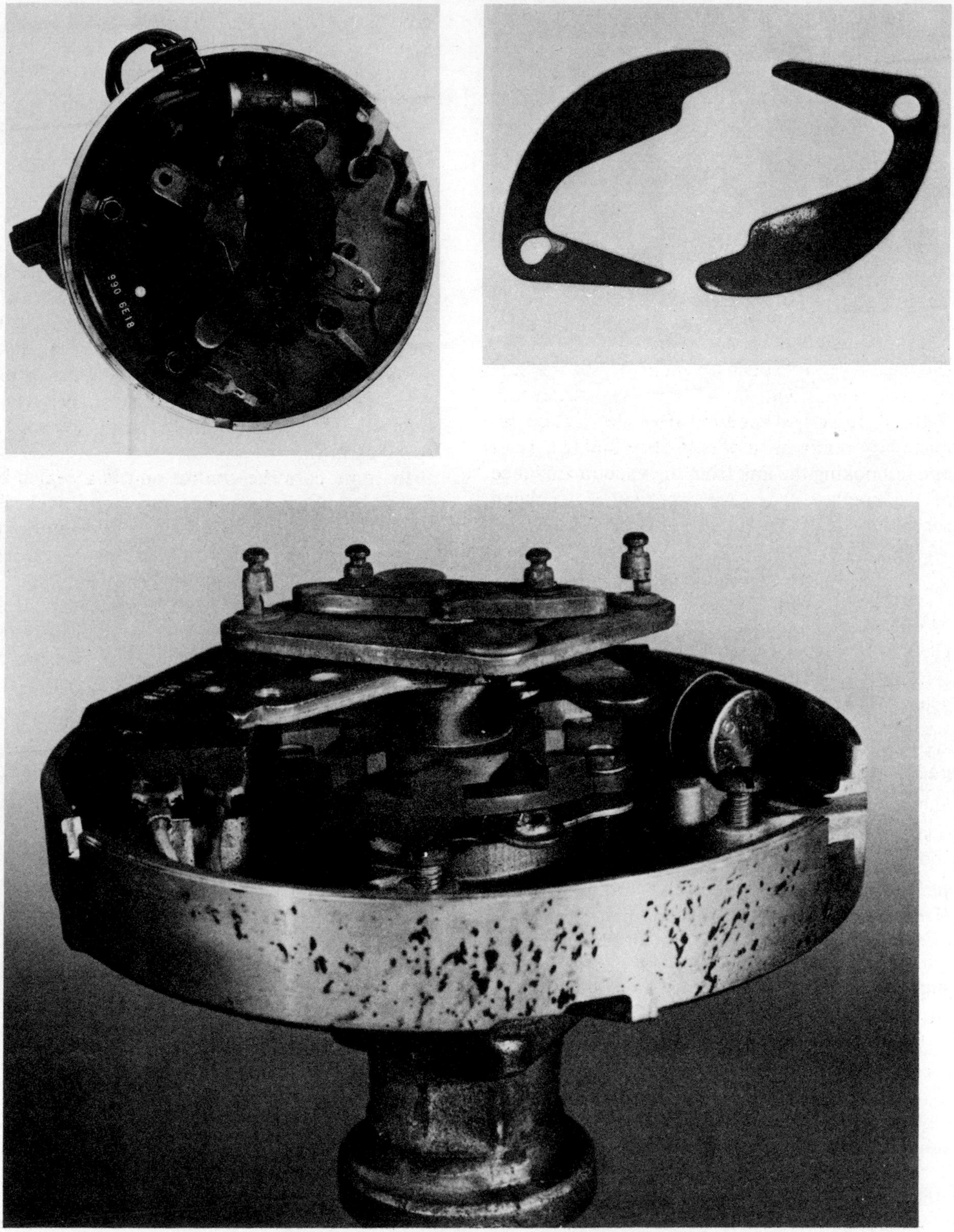

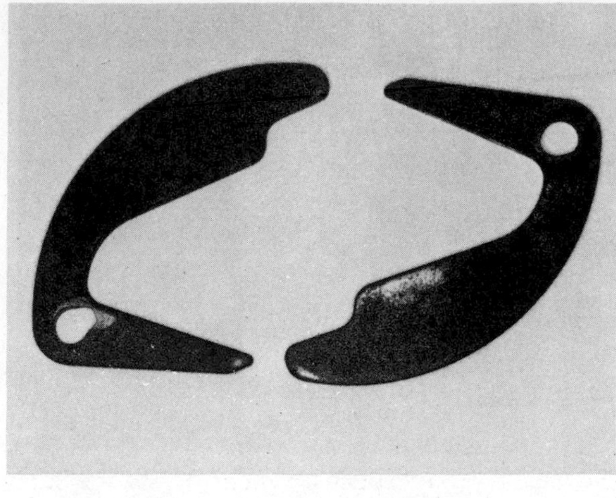

Fig. 2–26. Centrifugal advance mechanism of GM electronic ignition (HEI) distributors may wear away at the points shown at top left. Top right shows how holes in centrifugal weights wear into slots. Bottom photo shows how posts wear away near base. Repair kits are available that permit installation of oversize plastic bushings to correct the problem if it is caught before excessive wear has occurred.

silicone dielectric grease to the rotor tip. Both these steps will reduce firing voltages and the corrosive gases that result. Also, spray silicone lubricant on the centrifugal advance mechanism at least once a year.

(6) Check the vacuum spark advance, a feature on all engines except those with computer-controlled electronic spark advance mechanisms. With the engine running at 2000 r.p.m. (fast idle), disconnect the vacuum hose from the advance unit (see Fig. 2–7) and put your finger over the hose end. Let engine speed stabilize, then reconnect the hose. When you reconnect, the engine should speed up slightly if the vacuum advance is working. On all but General Motors V-8s, a defective vacuum advance can be replaced by removing a couple of retaining screws and unhooking the link from the vacuum advance at the breaker plate. An experienced weekend mechanic may attempt this. On the GM V-8 type, it may not be so easy on many models to unhook the link, so if you are not experienced, leave the job to a professional.

Note: If there are two vacuum hoses connected to the vacuum unit, it is a dual-action device that serves to both advance and retard the spark. With the engine warm and running at 2000 r.p.m. (fast idle), disconnect both hoses and plug the hose ends. Reconnect the inner hose and the engine should respond by slowing down slightly. Have a helper hold down the gas pedal slightly, then reconnect the outer hose, and the engine should respond by speeding up. If there is no response, particularly from the outer hose reconnection, and you have an engine performance problem that other tuneup services do not correct, have the distributor checked by a professional.

Electronic Ignition

Electronic ignition is standard equipment on all American cars today. If the thought of electronics frightens you, keep this in mind—the electronic circuitry does nothing more than replace the breaker points and condenser.

Here are the basics you need to know:

• *Pickup coil.* Inside the distributor, where the points used to be, is an electromagnet called a pickup coil. Do not confuse it with the ignition coil. The pickup may also be called a sensor or stator.

• *Reluctor, trigger wheel, or armature.* In the center of the distributor, replacing the cam and its lobes is a wheel with teeth (one for each cylinder). The wheel may be called a reluctor, trigger wheel, or armature (see Fig. 2–27).

As the reluctor is spun by the distributor shaft, its teeth pass the electromagnet, breaking into the magnetic field around it. This disruption in the magnetic field creates an electronic signal to a transistorized circuit board called the control unit. When it gets this signal, the control unit electronically breaks the ignition coil circuit (just as the breaker points did it electromechanically). The coil discharges and the high-voltage electricity goes to the rotor, the cap and the appropriate spark plug wire.

In some cars the control unit is a sealed box bolted somewhere under the hood (see Fig. 2–28). On others, primarily General Motors cars, it is miniaturized and placed inside the distributor, on the same plate as the pickup coil.

General Motors cars have another design wrinkle all their own. On most of their electronic ignition systems, the ignition coil is built into the top of the distributor cap (see Fig. 2–29).

Servicing the Electronic Ignition

The electronic ignition components themselves require no periodic service, with the exception of some Chrysler Corp. designs. The pickup coil position on those is adjustable (the part is held by a single screw, which passes through a slot hole in the pickup coil, as shown in Fig. 2–30). The pickup coil must be a specified distance from the teeth of the reluctor, a gap which is measured with a feeler gauge. Because the pickup is magnetic, a non-magnetic (brass) feeler must be used. A steel feeler (the conventional kind) would be drawn to the pickup and a false feeling of drag would be provided, even if the gap were too great.

• *Dwell.* Dwell is not measured on an electronic ignition system. It is a designed-in function of the electronic control unit and, unlike breaker points systems, it normally will vary with engine speed.

• *Engine r.p.m.* The "hot" tachometer connection (the one that isn't to the electrical ground) on some cars with electronic ignition is not quite the same as on breaker points ignition

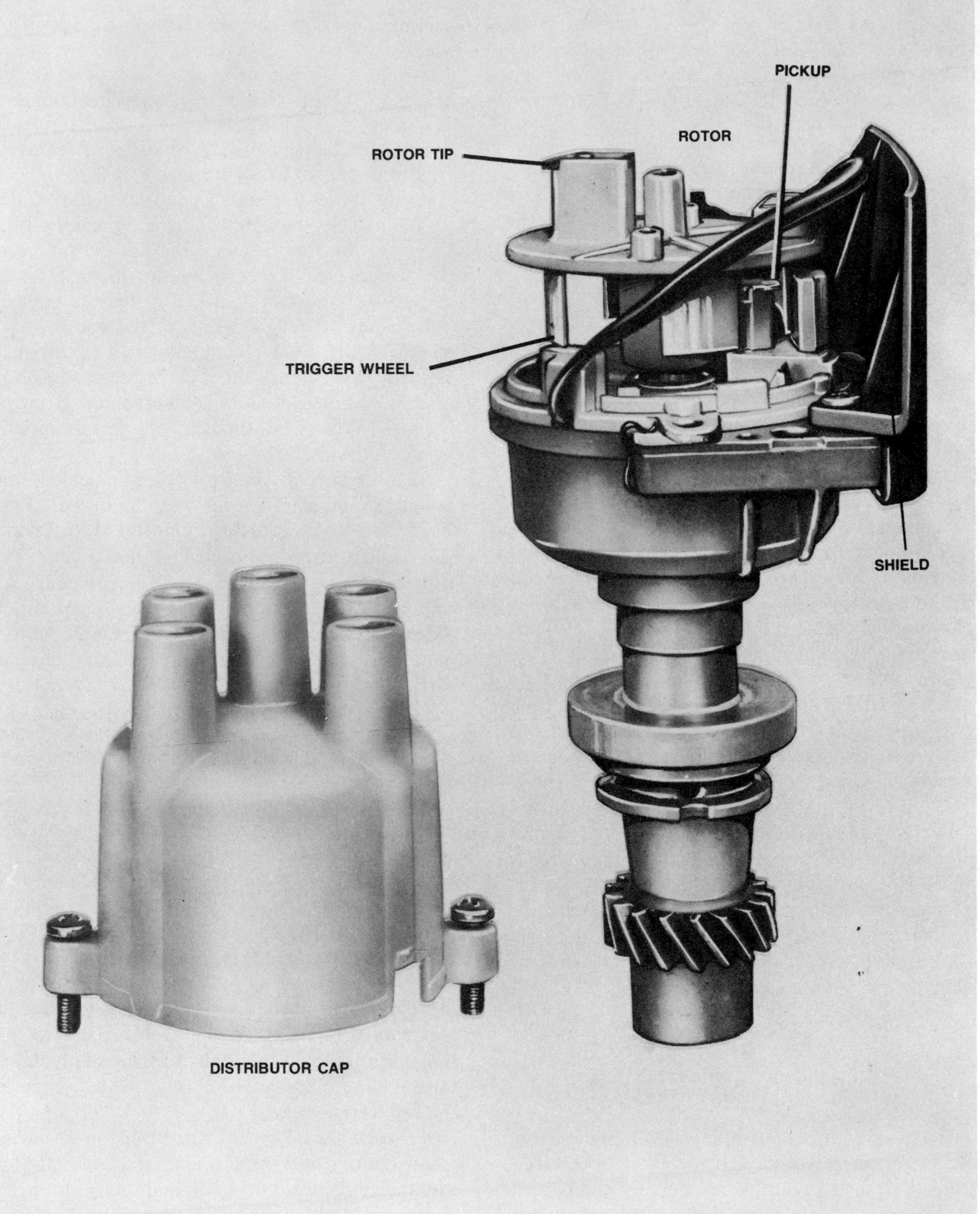

Fig. 2–27. Here is one type of electronic ignition distributor. As the teeth of the trigger wheel pass the electromagnetic pickup, they create a signal in the pickup that is transmitted to a transistorized control unit that triggers the ignition coil. Notice that the rotor is combined with the trigger wheel. If you have this design (Chrysler front-drive cars), you will notice a contact tab on the underside that grounds the trigger wheel. The tab must make good contact with the distributor shaft (or the car will not start).

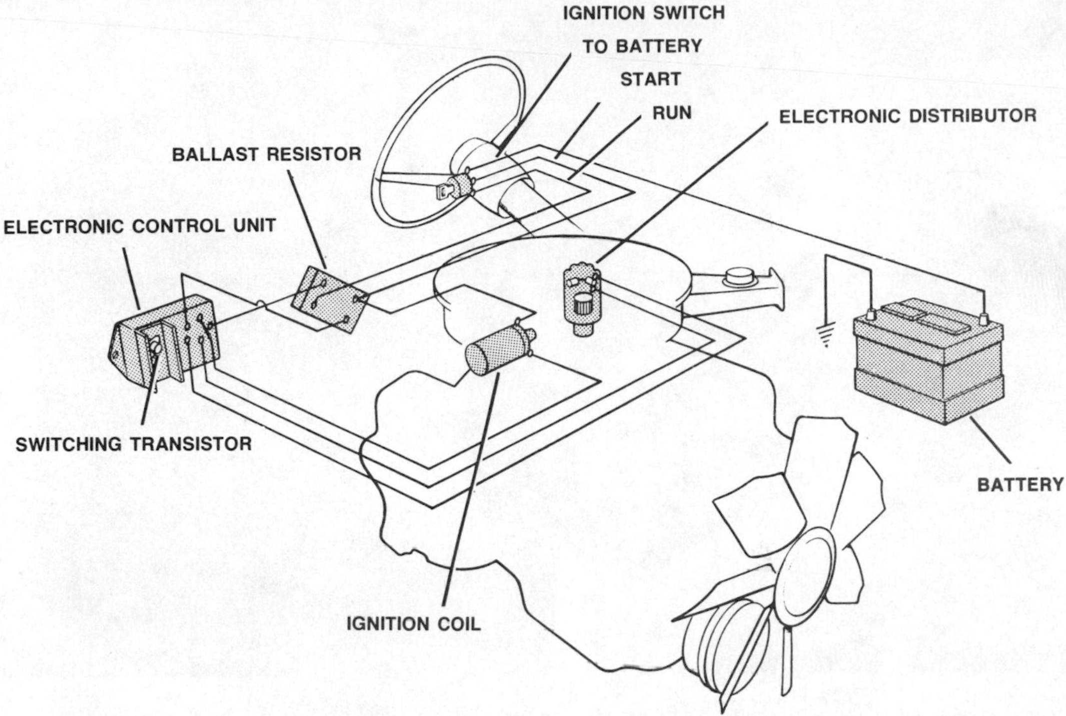

Fig. 2–28. Schematic of an electronic ignition system showing the electronic control unit with its switching transistor, a part that opens the coil electromagnetic circuit just as if the breaker points were opening.

Fig. 2–29. Here's a typical General Motors High Energy Ignition distributor. Ignition coil is built into top of cap. Transistorized control unit is miniaturized and located next to pickup on what used to be called the breaker plate. Notice that plug wires are held in place by the plastic cage on top and that the cap itself has male plug wire terminals.

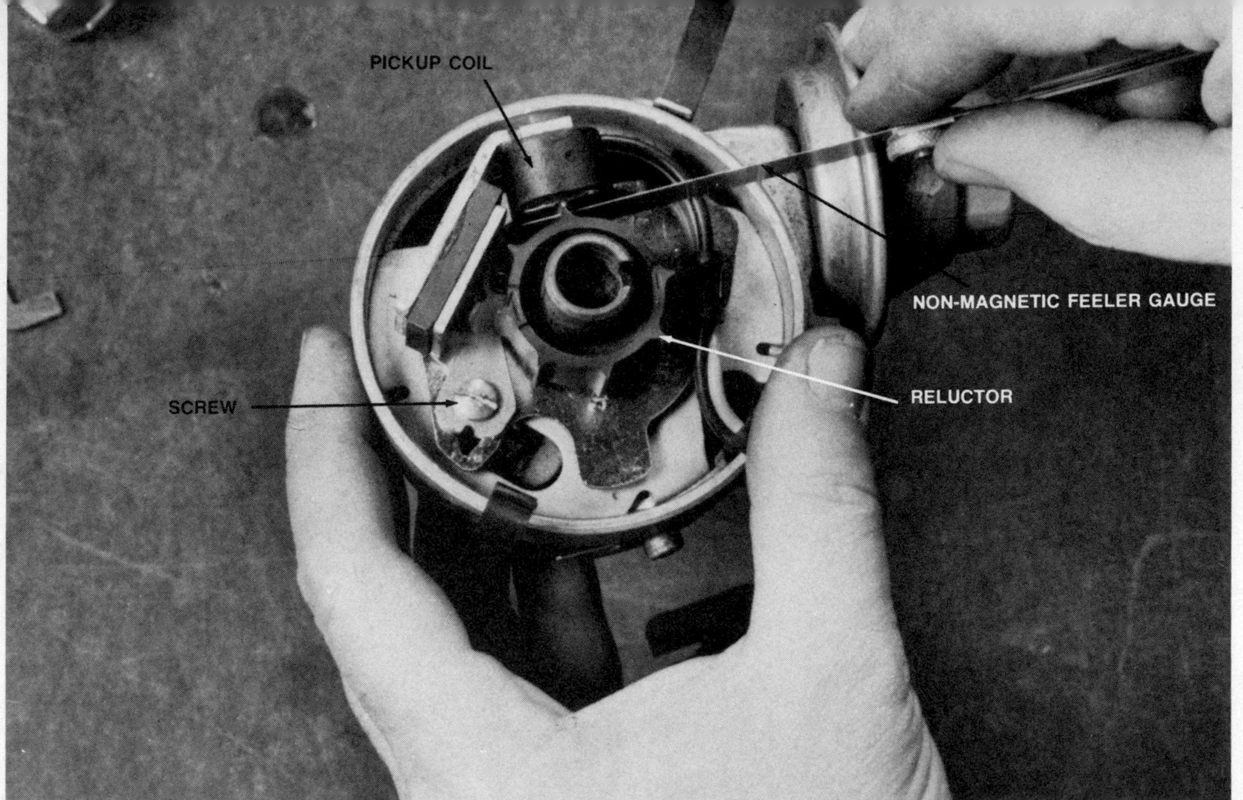

Fig. 2–30. Some Chrysler distributors have adjustable pickup coil (screw goes through slotted hole). Line up reluctor tooth with pickup coil as shown, insert specified non-magnetic feeler gauge and, if necessary to change gap, slacken screw and move pickup as required.

Fig. 2–31. Here's a look at the underside of a GM High Energy Ignition distributor cap. Wire connects to battery terminal from the car's harness. The male spade above is for a tachometer connection; however, you will probably need a special adapter, available in auto parts stores, for alligator clip on tachometer is too big to attach to spade. Adapter has female spade that plugs in, and adapter itself hangs down for easy tachometer connection.

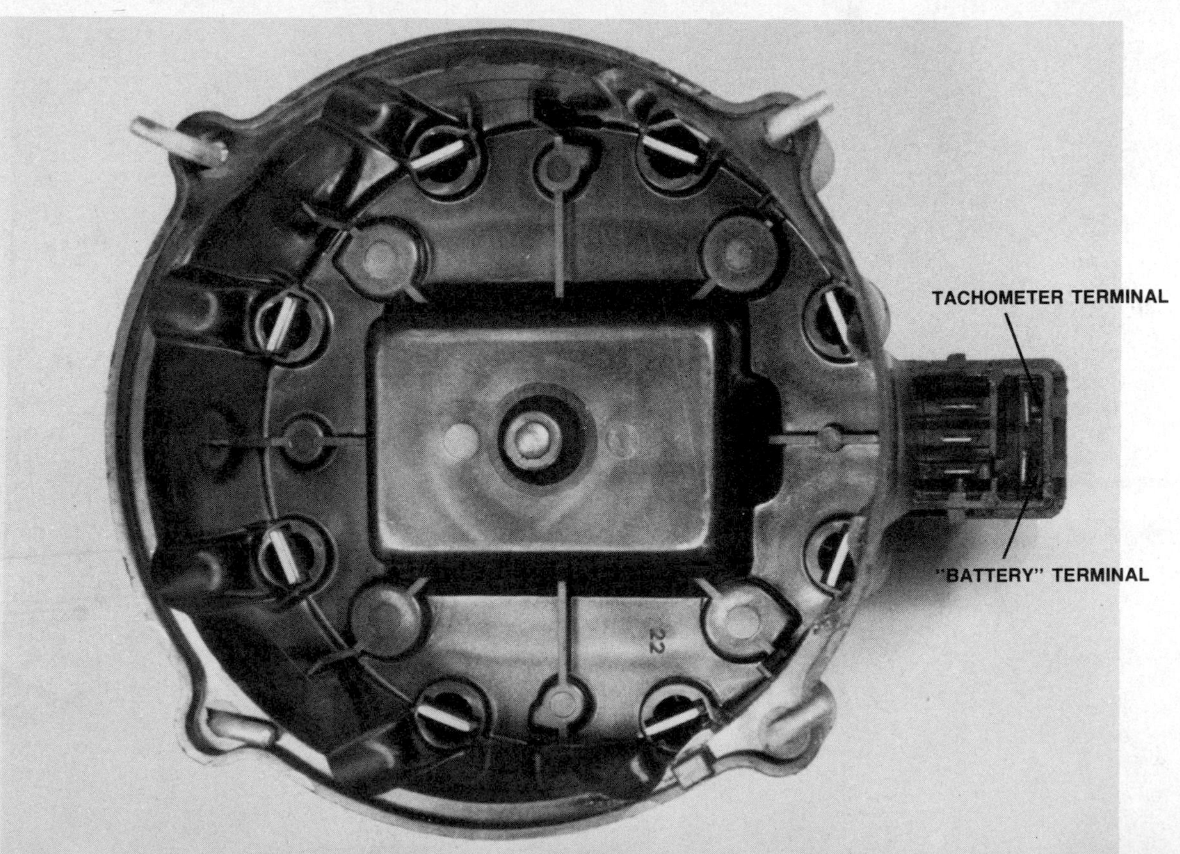

Fig. 2–32. Here's a Ford ignition coil with its special plastic connector. Insert male spade adapter or thin nail into location called "Tach Test" as indicated by arrowhead, then connect tachometer to adapter or thin nail.

Fig. 2–33. This is one type of electronic ignition analyzer. It has light-emitting diodes (electronic bulbs) at left, and chart tells you what's wrong, according to which light-emitting diode(s) light up during test.

cars, particularly General Motors products with ignition coils built into the distributor cap. The hot connection must be made to a male spade terminal in the underside of a flange on the cap of GM cars (see Fig. 2–31). It must be made to the "TACH" terminal of a plastic connector on the ignition coil of most Ford Motor Co. cars (see Fig. 2–32). Note: On some tachometers made for electronic ignition systems, special connections are required. For example, you may have to connect to the positive battery terminal, not to an electrical ground. Check the instructions furnished with your tachometer very carefully.

• *Electronic testers*. The best way to test an electronic ignition system is with an electronic tester made specifically for the purpose. Today there are many electronic testers that will check out virtually all cars, such as the inexpensive ones shown. The pocket-size tester (see Figs. 2–33, 2–34) has light-emitting diodes, little solid-state electronic "bulbs." When the tester is properly connected and some simple procedures are followed, certain light-emitting diodes will light up to tell you what is wrong. Then you replace the defective part.

The tester shown (see Fig. 2–35) has light-emitting diodes, but basically works on a different principle. When you connect it per the instructions, it substitutes for the pickup in the distributor. The tester includes a special high-voltage coil wire that substitutes for the conventional one. If the system is functioning when you activate the tester, you actually see a spark jumping through a clear plastic section of the special high-voltage coil wire.

With this arrangement, you quickly know if the electronic control unit and ignition coil are functioning, for if they are not, there's no spark. If there's a spark with the tester, but no spark to the plugs without it, either the pickup is defective, or you have a bad distributor cap, rotor, or plug wire (the latter items are all things you can easily check).

This type of tester, which is inexpensive and available in auto parts stores, permits running the electronic ignition system for as long as you wish, with the engine off. Therefore, if you have a problem that develops only after the engine has been running a while, you can simulate it. Heat is often a cause of an electronic failure, so you can simulate this problem by holding a hair dryer close to the electronic control unit, pickup, or ignition coil. If the spark in the clear plastic section of the coil wire suddenly stops when you apply heat to the electronic control unit, that's the problem part. The tester does not distinguish between an ignition coil or electronic control unit failure if there is no spark, but light-emitting-diode tests you make afterward will pinpoint this.

• *Testing with ohmmeter*. If you have a multi-scale ohmmeter, you can make some important electronic ignition tests, and they may be conclusive.

The key test is of the pickup coil. All the pickups have two-wire connectors except Ford, which has three (disregard the third, a black wire that serves as an electrical ground). On all but General Motors cars (or American Motors cars with GM systems), you can undo the connector near the distributor quite easily (see Fig. 2–36). On GM systems, remove the two wires (from the pickup) at the electronic control unit (see Fig. 2–36).

Check manufacturer's specifications and set the ohmmeter for the appropriate scale. Connect the two ohmmeter wire leads to the two terminals for the wires that go to the distributor (do this with engine and ignition turned off). Get your ohmmeter reading (see Fig. 2–38) and compare with specifications. If the reading is off and you have an ignition system problem, it apparently is in the pickup coil.

Tip: On GM cars, disconnect and hold the other end of the hose from the vacuum advance unit. Suck as hard as you can on the hose end (you'll draw about 10 inches of vacuum), and if the ohmmeter reading changes, the pickup is defective, even if it passed the previous ohmmeter test. Note: A minor change in resistance may be noted in this test (10–20 ohms or less) if you are using a digital ohmmeter, even if the pickup coil is good. Unless you have a performance problem, therefore, do not condemn the pickup. If you do have a performance problem, it might be advisable to repeat the test with an analog (needle and dial) ohmmeter to double check.

Leave one of the ohmmeter wire leads in contact with one of the wire terminals and touch the other ohmmeter lead to a metal part of the distributor. You should get a reading of infinity, which indicates that the pickup is electrically insulated from the distributor. If you get any reading short of that, the pickup is electrically grounded to the distributor, which it should not

Fig. 2–34. Tester also comes with this spark-plug-like part that clips to an electrical ground as shown. A plug wire is disconnected, attached to this part, and the engine is cranked. If you see spark jumping inside the "spark plug," the ignition system is okay. This is ideal for electronic ignition systems, as it permits a spark check without the need to hold the plug wire, which could be dangerous on the ultra-high-voltage electronic ignition system.

be. The pickup probably will have to be replaced.

On Chrysler products (Dodge, Plymouth, Chrysler), most of the electronic ignition systems have a dual resistor (see Fig. 2–39). One section of this resistor is to reduce voltage to the ignition coil, as in a conventional ignition system; the other is wired to the electronic control unit. The resistance of each must be within specifications, as measured with a multi-scale ohmmeter. If either resistance section is outside specifications, the complete resistor must be replaced.

• *Other checks*. As with a breaker points ignition system, you must check the distributor cap for cracks, the rotor for burns and, on non-computer systems, the advance mechanisms. Details on these checks are at the conclusion of the previous section.

Replacing Components

The control unit on virtually all systems is a simple unbolt-and-disconnect proposition. The only special procedure is on the modules in the General Motors distributors. A small capsule of special heat-sink grease should come with a replacement module, and you should smear this grease on the metal base of the module before installation. If you don't, the module will overheat and fail prematurely. The pickup coil on all but General Motors cars (and American Motors cars with GM electronic ignition) is easy to remove from the distributor. On General Motors systems, the distributor must be removed and partly disassembled, a job for a professional. On pre-1978 American Motors cars (Prestolite system), pickup coil service also should be left to a professional. Most AM cars from 1978 to the present have the Ford system.

Ignition System Quick-Check

When a car won't start, but cranks normally, the ignition system is the first place to check, to

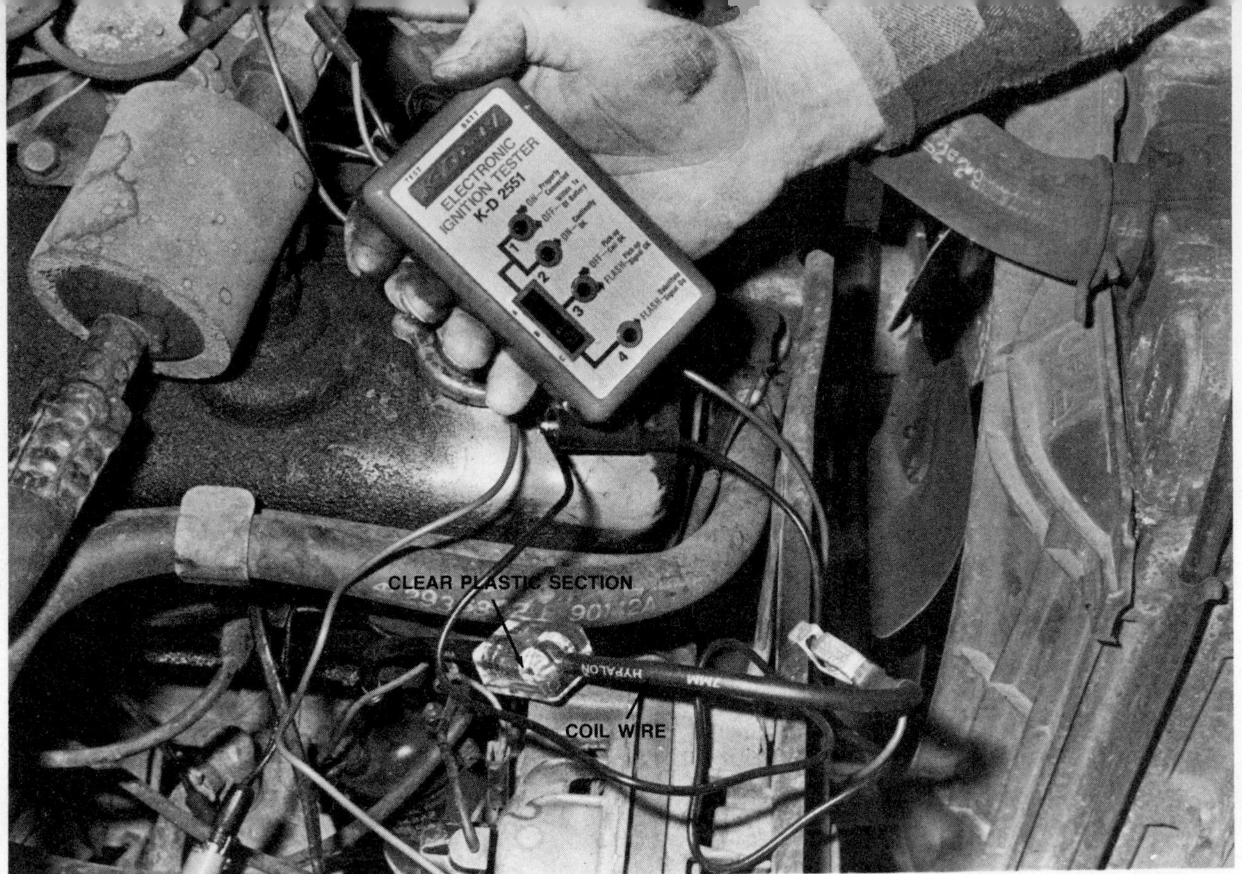

Fig. 2–35. Here's another electronic ignition tester. This one includes a replacement for the normal ignition coil wire. The replacement has a clear plastic section. The tester permits you to activate the ignition system with the engine off. You can look in the clear plastic section and see if the system is producing a spark. Light-emitting diodes also permit testing of individual components of the system.

determine if there is a spark to the plugs.

All Systems

(1) Disconnect the most accessible spark plug wire, insert a key or paper clip into the wire terminal, and hold the end of the key or clip within ¼ to ⅛ inch of a metal part of the engine (see Fig. 2–40).

(2) Have a helper crank the engine and you should see a steady spark from the key or clip to the engine. If you do, the ignition system is working, although worn spark plugs may be responsible for hard starting.

(3) If there's no spark, disconnect the thick-jacketed wire from the center of the coil to the center of the distributor cap, at the cap. Hold the end of this wire within ¼ inch of a metal part of the engine (see Fig. 2–41). Have a helper crank the engine and you should see a steady spark. If you now do, the problem is in the distributor cap, rotor, or spark plug wires.

On General Motors High Energy electronic ignition, with the ignition coil built into the cap, the test in No. 3 is impractical. Instead, inspect the cap for cracks and make sure the pink wire in the cap flange is securely connected (see Fig. 2–42).

(4) If you still have no spark on an electronic ignition car, remove the distributor cap and look for physical damage to the rotor, the pickup, or the reluctor (timing armature). If there is nothing obvious, proceed to a test instrument checkout as described elsewhere in this section. If you have a car with breaker points ignition, proceed to No. 5.

(5) With the ignition turned on, flick the points open (see Fig. 2–43) and allow them to close, using an insulated-handle screwdriver. You should see a small electric arc. Or connect a test lamp to the movable point and electrical ground. Push open the points with a wooden stick and the test lamp should light (see Fig. 2–44). Let the points close and the test lamp should go out. If the test lamp does not light in the first case, there is no current to the points (go to step 6). If it doesn't go out completely in the second case, the

points are defective (go to step 7). If the test lamp results are normal, go to step 8. Note: Points should be closed and touching for this test. Crank engine to allow this, if necessary.

(6) Reconnect the test lamp to the negative terminal of the ignition coil, with the ignition on and the points pushed open. If the test lamp now lights, the wire from coil to points is defective. If the test lamp doesn't light, switch it to the positive terminal, and if it now lights, the ignition coil is defective. If it still doesn't light, the resistor, ignition key switch, or wiring is defective, a repair job for a professional.

(7) Replace the breaker points and adjust to specifications.

(8) The condenser apparently is defective. If the problem is a grounded condenser, you can check it with the test lamp (see Fig. 2–45). Disconnect the condenser wire, touch one lead of the test lamp to the battery starter terminal and the other to the end of the disconnected condenser wire. If the lamp lights, the condenser is grounded and must be replaced. Even if the condenser does not fail this test, test it by substitution with a new one.

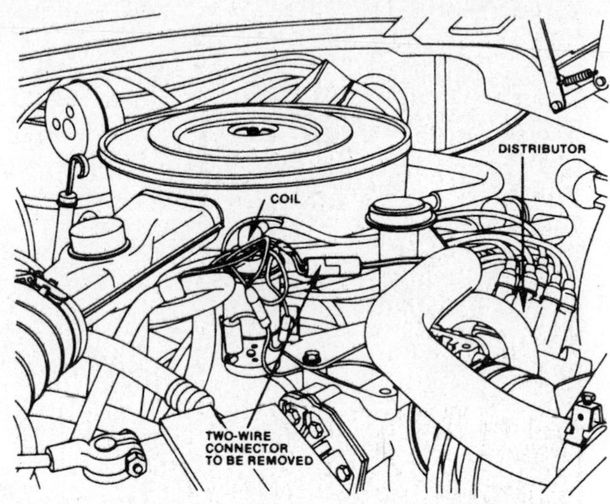

Typical V-8

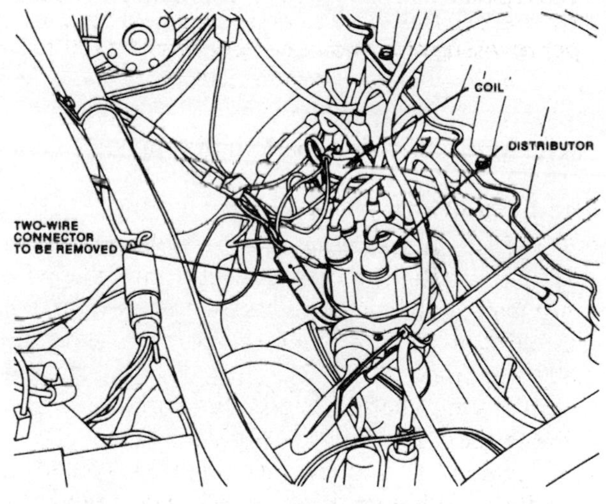

Typical 6 Cylinder

Fig. 2–36. Illustrations show typical two-wire connectors from pickup coil. Undo connector and check resistance across the dual-terminal arrangement of the wire section from the distributor.

Ignition Timing

Ignition timing—the arrival of the high-voltage electricity at the plugs—is one of the most important tuneup adjustments you make. On cars with breaker points, it follows the points adjustment, for points gap changes can affect it.

All ignition systems have provided for automatically changing the timing according to engine operating conditions—speed and load for example—but it is important that the engine begin with a basic setting that is correct. Note: If your car has a computer that regulates spark timing, special procedures in addition to (or replacing) those given in this section are required. Computer-controlled spark advance is covered in the next section of this chapter.

Here's how to check the basic timing and make adjustments if necessary.

Timing Marks

Ignition timing is checked by aiming the timing light (see Fig. 2–46) at the timing marks. There are two marks, one of which is fixed to the

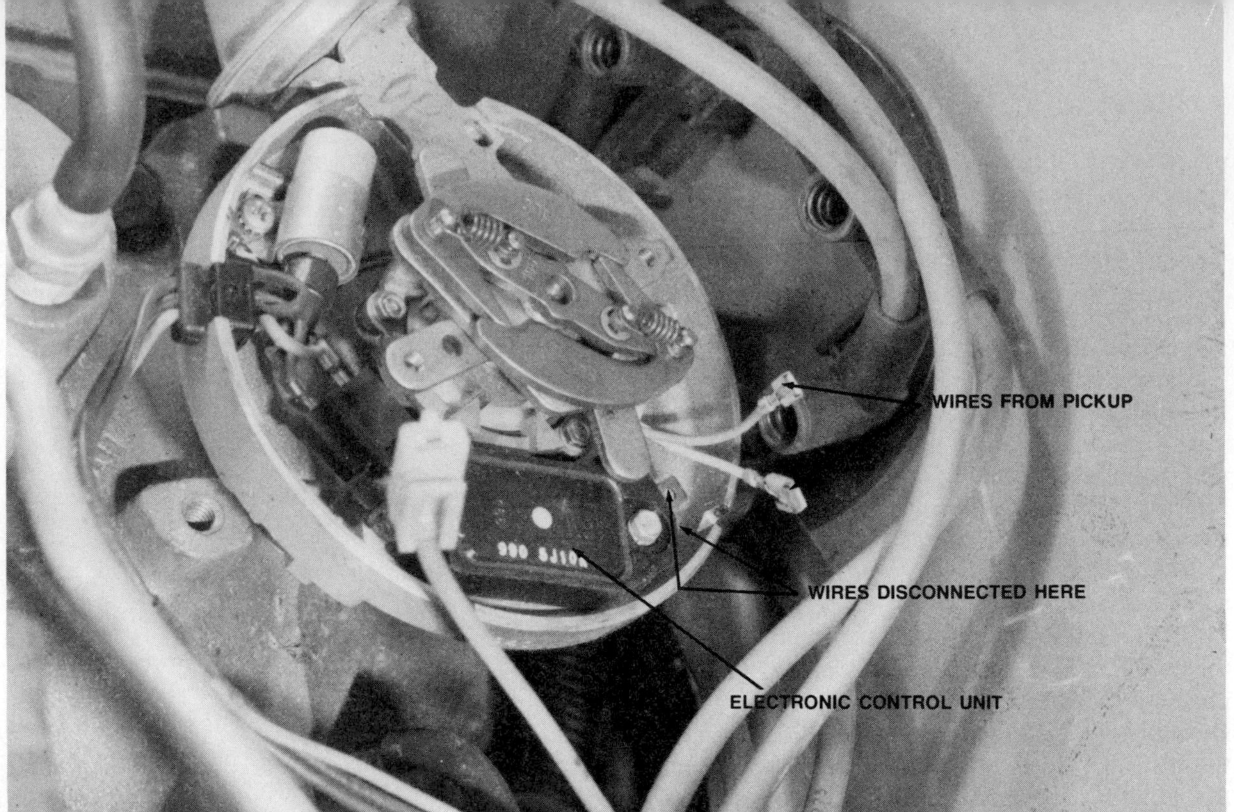

Fig. 2–37. A look inside a GM High Energy Ignition system distributor. Wires from pickup coil, which is hidden underneath, have been disconnected from electronic control unit terminals. Connect ohmmeter across the two wires that are from the pickup.

Fig. 2–38. This is a Ford distributor on a car with electronic ignition system. Terminal from distributor has three wires. Disregard black wire and connect ohmmeter as shown, across the terminals for the other two wires.

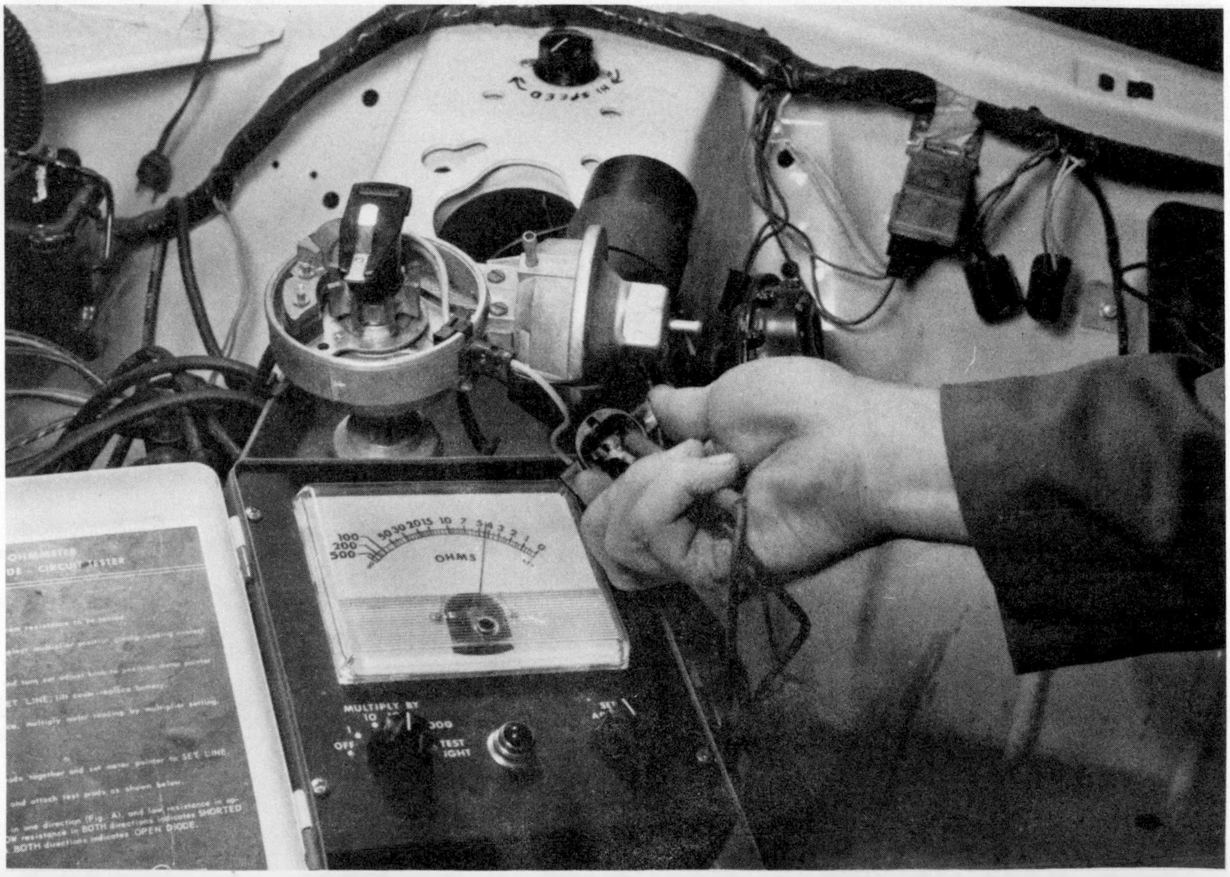

Fig. 2–39. Checking the resistance across one of the two resistance sections of the Chrysler ballast resistor. The resistance being checked is for the wiring to the electronic control unit. Upper section is for ignition system resistance to the ignition coil. Use notch in end of resistor assembly as key, for resistor may be mounted upside down or vertically on some models. The resistor was discontinued in 1981.

engine, the other of which is on a rotating part, such as the belt pulley assembly of the crankshaft, at the front of the engine, or the flywheel at the rear, visible through a hole in the flywheel housing. On transversely-mounted engines, the marks are at one side (see Figs. 2–47 to 2–50).

These marks are so located that when they line up with each other, the spark plug on either one of two cylinders will be firing. All cylinders are numbered by the car manufacturer, but you won't find the numbers on the engine. Instead you must consult a reference chart (see Fig. 2–51), which shows how the numbers are assigned.

It is customary for the car maker to position the timing marks so that they align when the No. 1 cylinder's plug is firing. If the timing is correct for the No. 1 cylinder, it will be correct for all.

With the timing light hooked up and the engine idling, the light will flash on when the spark plug to which it is connected is firing. If you aim the light at the marks, they should be aligned whenever the light goes on.

• *Finding the marks.* By reference to Figs. 2–47 to 2–50, you should be able to find the fixed timing indicator. It may be just a pointer, or a dial with numbers. The rotating mark will usually be a notch, perhaps a dial, or occasionally a number. If you can't conveniently see it, crank the engine in very short bursts until it comes up into view.

• *Hard to see marks.* The fixed timing indicator may be badly obscured, such as by an alternator or air conditioning compressor. This is a common problem on cars with the marks at the front of the engine, the usual location on American cars.

If you can't see the marks, you can't aim the timing light at them. There are two practical answers to this problem:

Most late-model engines with timing marks at the front have a fixed indicator with a holder for a pencil-like probe (see Fig. 2–52). You obtain a

Fig. 2–40. Key is inserted into plug wire terminal to provide a projection from rubber nipple. If you hold edge of key to electrical ground while a helper cranks the engine, you can check for spark. Note: If the plug wire is even slightly deteriorated, do not hold it or you could get a shock. Use insulated handle pliers, electrical pliers, a couple of sticks, anything that will hold and insulate.

Fig. 2–41. If there's no spark at the end of the plug wire, disconnect wire from center of distributor cap (not possible on most General Motors High Energy Ignition systems). Push back nipple and hold terminal close to electrical ground while helper cranks engine. As with plug wire, use special insulated pliers, sticks, etc. to hold wire if it's deteriorated.

special timing indicator with the probe, push the probe into the holder, and read the timing on a meter (see Fig. 2–53). This type of timing indicator is much more expensive than a conventional power timing light. Note: If you can see part of the timing indicator, you can use a special timing light as shown in Fig. 1–8.

You can make a set of timing marks elsewhere, in a better viewing location. In some extreme cases, the only location may be from under the car, with the front jacked up (see *Chapter 5* for jacking instructions).

• *Making new marks*. To set up the engine for a set of new marks, you must properly align the existing marks once. This may pose some difficulties, but even if you have to use a flashlight and mirror to do it, just realize that after you make the new marks you won't have to struggle again.

Crank the engine in very short bursts until the rotating indicator (usually a notch on the crankshaft pulley) is as close as possible to the fixed indicator. Then grasp a drive belt that wraps around the crankshaft pulley at two locations midway between the crankshaft pulley and another pulley. Push the two sections together and simultaneously pull up on one section, down on the other. The crankshaft pulley will turn and you'll be able to get perfect alignment of the marks. Or if you can get a wrench on the crankshaft pulley bolt, you can use a wrench to turn it.

If the specification reads, for example, 8 BTDC (before top dead center), that means that the notch on the pulley must align with the eight on the degree dial of the fixed indicator. If there are two eights, align with the first as you turn the engine clockwise (viewed from the front).

Now look for an open area around the circumference of the pulley, somewhere close to an engine bolt, to which you can attach a fixed pointer (see Fig. 2–54). Make the pointer from a piece of metal, something substantial, at least ⅛ inch thick. Make the pointer L-shaped (or as

Fig. 2–42. Shown is pink "battery" wire in distributor cap flange of GM High Energy Ignition system.

Fig. 2–43. You should see an arc across points when you flick them open with screwdriver with ignition on.

Fig. 2–44. Connect test lamp to ground and hold its probe on movable point as shown. With ignition on, push away movable point with stick as shown, and test lamp should light. Let points close and test lamp should go out.

Fig. 2–45. Testing condenser with test lamp. Wire from condenser is disconnected and test lamp alligator clip is hooked to it as shown. Test lamp probe is touched to movable point, which is insulated from fixed point with piece of cardboard between the points. Turn on ignition, and if test lamp lights, it is because condenser is providing an electrical ground, which it should not. *Replace condenser.*

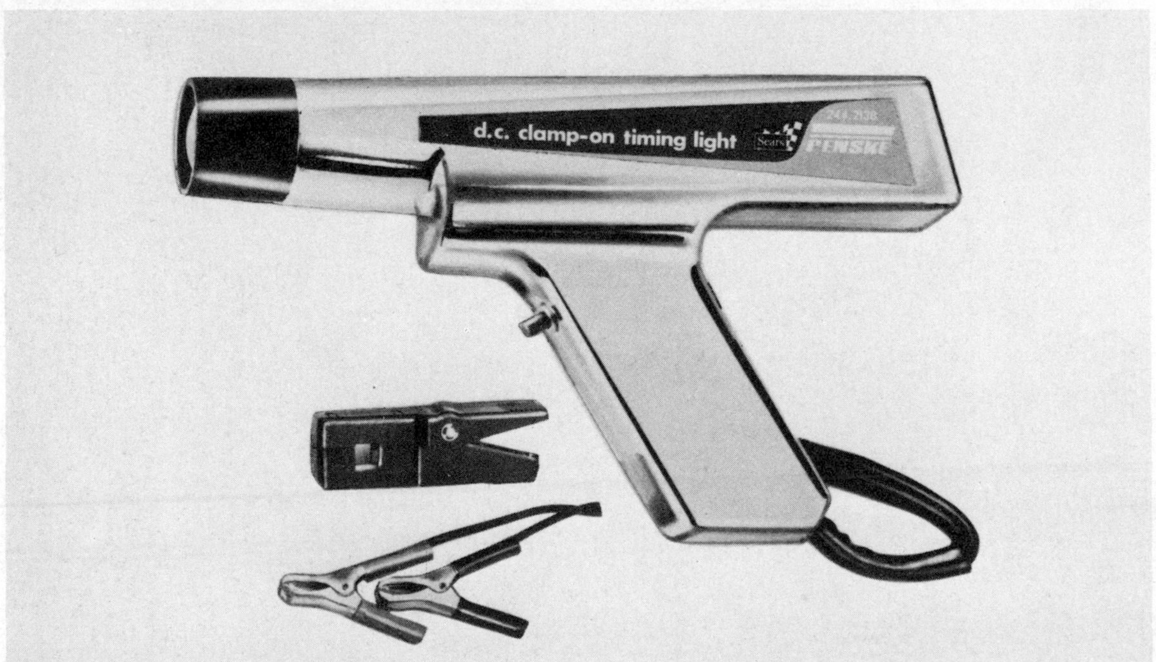

Fig. 2–46. Here is a typical timing light. Two alligator clips at bottom are connected to car battery and clamp above them is clipped onto No. 1 spark plug wire.

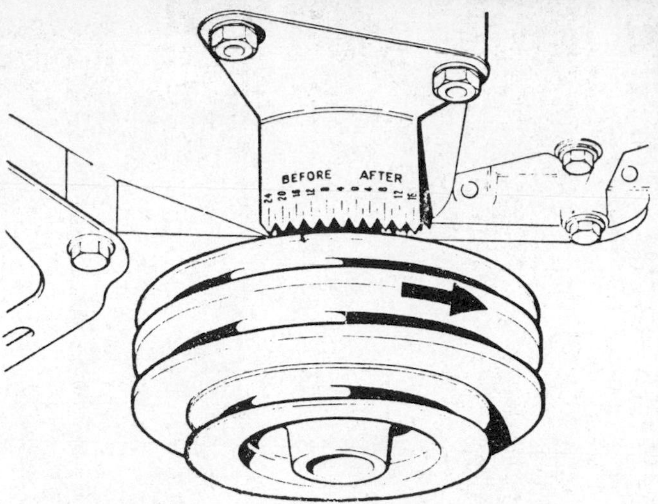

Fig. 2–47. This is a typical timing mark arrangement at front of engine. Degree indicator is fixed to front of engine and there is a notch on the crankshaft pulley that rotates adjacent to it. Although this degree indicator has many marks and numbers, plus the words before and after, the typical indicator may have no more than a few lines and perhaps just a couple of numbers.

Fig. 2–48. This is the VW Beetle air-cooled engine. There is a notch on the pulley (if more than one notch, check specifications to see which one to use) and the fixed indicator is simply the joint between the two halves of the engine.

JOINT

Fig. 2–49. Here's a fixed indicator with a degree dial and a probe holder, for ignition timing with an electromagnetic probe instead of a timing light.

Fig. 2–50. Many cars have timing marks at the flywheel. On this particular one, a VW Rabbit, you remove the computer-timing sensor and either aim a timing light into the hole (the mark, a number, should show up in the middle of the hole when the light goes on) or use a probe-type timing indicator. On later VWs there is no sensor, but there may be a plastic plug in the hole. Remove it for a timing check.

close as is practical) and drill a hole in the pointer for the bolt. The part of the pointer that has the bolt hole should be long enough so it can be bent around some part of the engine to hold it in position. Then tighten the bolt to secure it.

Next, file a mark in the pulley, aligned with the pointer, and you have a suitable set of marks. Because you have lined up the manufacturer's timing marks, these additional marks also are in alignment. You need never use the hard-to-see original marks.

Note: In selecting the alternate location for the timing marks, check to determine if the wire leads of your timing light will reach. If they will not, you must obtain extensions, which may be difficult for the trigger wire to the No. 1 spark plug if it is the clamp-on type that attaches to the plug wire. Or you may choose a location for the new marks that may not be quite as convenient, but which will not require modification to the wire leads of your timing light.

Specifications

Timing is set to specifications provided by the car maker. In most cases the information is on an underhood decal. On a car with breaker points ignition you must set the breaker points first (or be assured by a check that the gap is correct), for changes in point gap change ignition timing.

You must understand what the specifications mean and how to interpret them. If a specification reads 8 BTDC, that means eight degrees before top dead center, or before the piston reaches the top of its stroke in the cylinder. You can't look inside the engine, but there is a degree dial at the timing marks, as shown in Fig. 2–47. As the engine rotates clockwise (as you look at it from the front), the first set of numbers before the zero or TDC (top dead center) mark is BEFORE. Those that follow are ATDC (after top dead center). On some cars, timing may be ATDC, but that is rare. In some cases the specifications may read BTC and ATC (before top center and after top center). If neither ap-

FIRING ORDER AND CYLINDER NUMBERING

MAKE OF CAR	FIRING ORDER	CYLINDER NUMBERING*	
American Motors			
V-8	1-8-4-3-6-5-7-2	D: 1-3-5-7	P: 2-4-6-8
four	1-3-4-2	1-2-3-4	
six	1-5-3-6-4-2	1-2-3-4-5-6	
Chevrolet, Buick, Olds, and Pontiac			
V-8	1-8-4-3-6-5-7-2	D: 1-3-5-7	P: 2-4-6-8
V-6**	1-6-5-4-3-2	D: 1-3-5	P: 2-4-6
four**	1-3-4-2	1-2-3-4	
six (in line)	1-5-3-6-2-4	1-2-3-4-5-6	
Cadillac			
V-8 (except 350)	1-5-6-3-4-2-7-8	D: 2-4-6-8	P: 1-3-5-7
350 V-8	1-8-4-3-6-5-7-2	D: 1-3-5-7	P: 2-4-6-8
Chrysler Corp.			
V-8	1-8-4-3-6-5-7-2	D: 1-3-5-7	P: 2-4-6-8
six	1-5-3-6-2-4	1-2-3-4-5-6	
four	1-3-4-2	1-2-3-4	
Ford			
V-8 (351, 400)	1-3-7-2-6-5-4-8	D: 5-6-7-8	P: 1-2-3-4
V-8 (other)	1-5-4-2-6-3-7-8	D: 5-6-7-8	P: 1-2-3-4
four (1600)	1-2-4-3	1-2-3-4	
four (2000)	1-3-4-2	1-2-3-4	
four (2300)	1-3-4-2	1-2-3-4	
V-6	1-4-2-5-3-6	D: 4-5-6	P: 1-2-3
six (in-line)	1-5-3-6-2-4	1-2-3-4-5-6	
Volkswagen			
Beetle	1-4-3-2	D: 3-4	P: 1-2***
Rabbit, Dasher, or Scirocco**	1-3-4-2	1-2-3-4	
Datsun			
four	1-3-4-2	1-2-3-4	
Toyota			
four	1-3-4-2	1-2-3-4	

*D indicates driver's side and P indicates passenger's side.
**On transversely-mounted engines, numbers start at crankshaft pulley (drive belt end) of engine.
***On rear engine, reverse the numbers if you are counting from the rear bumper: i.e. D: 4-3 becomes P: 2-1 counted from the rear bumper.

Fig. 2–51. Chart gives the firing order and method of cylinder numbering of popular car engines.

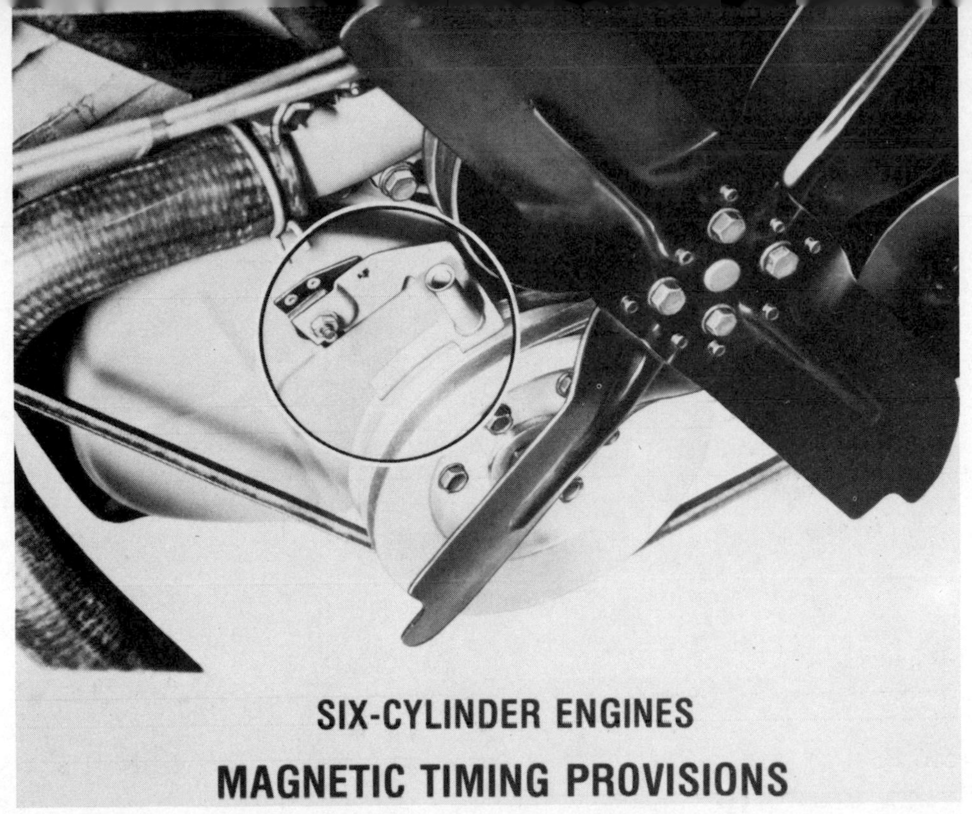

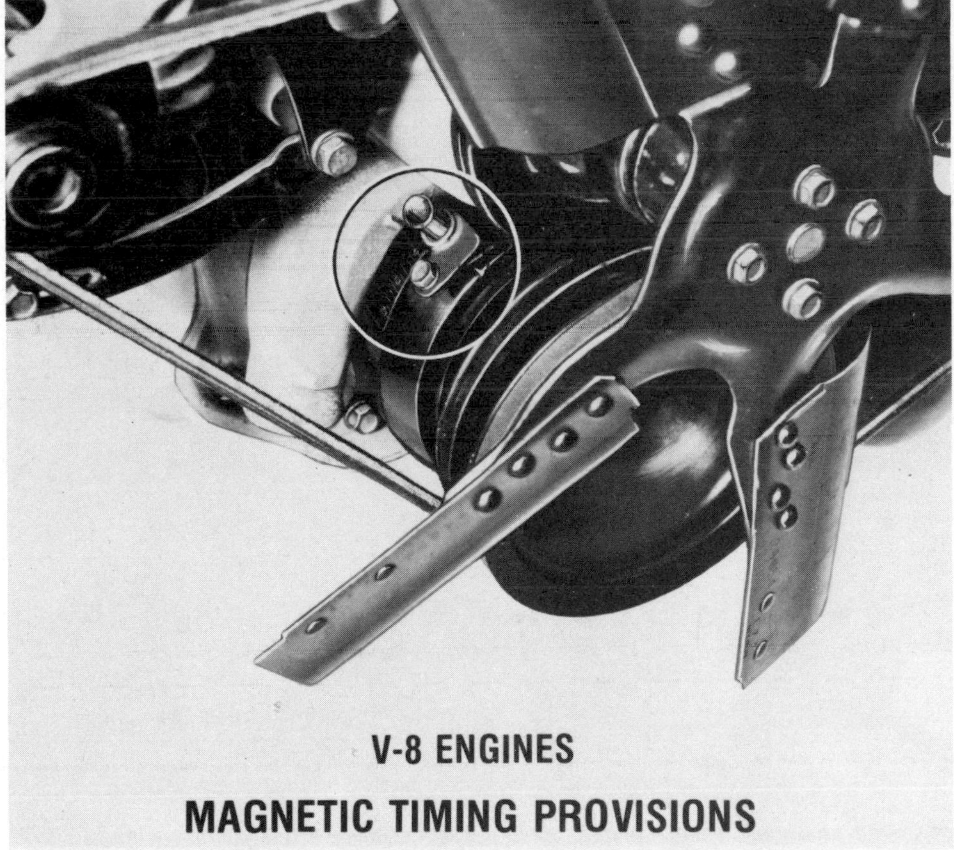

Fig. 2–52. This illustration shows the magnetic timing probe holds built into the fixed timing indicators on Chrysler engines.

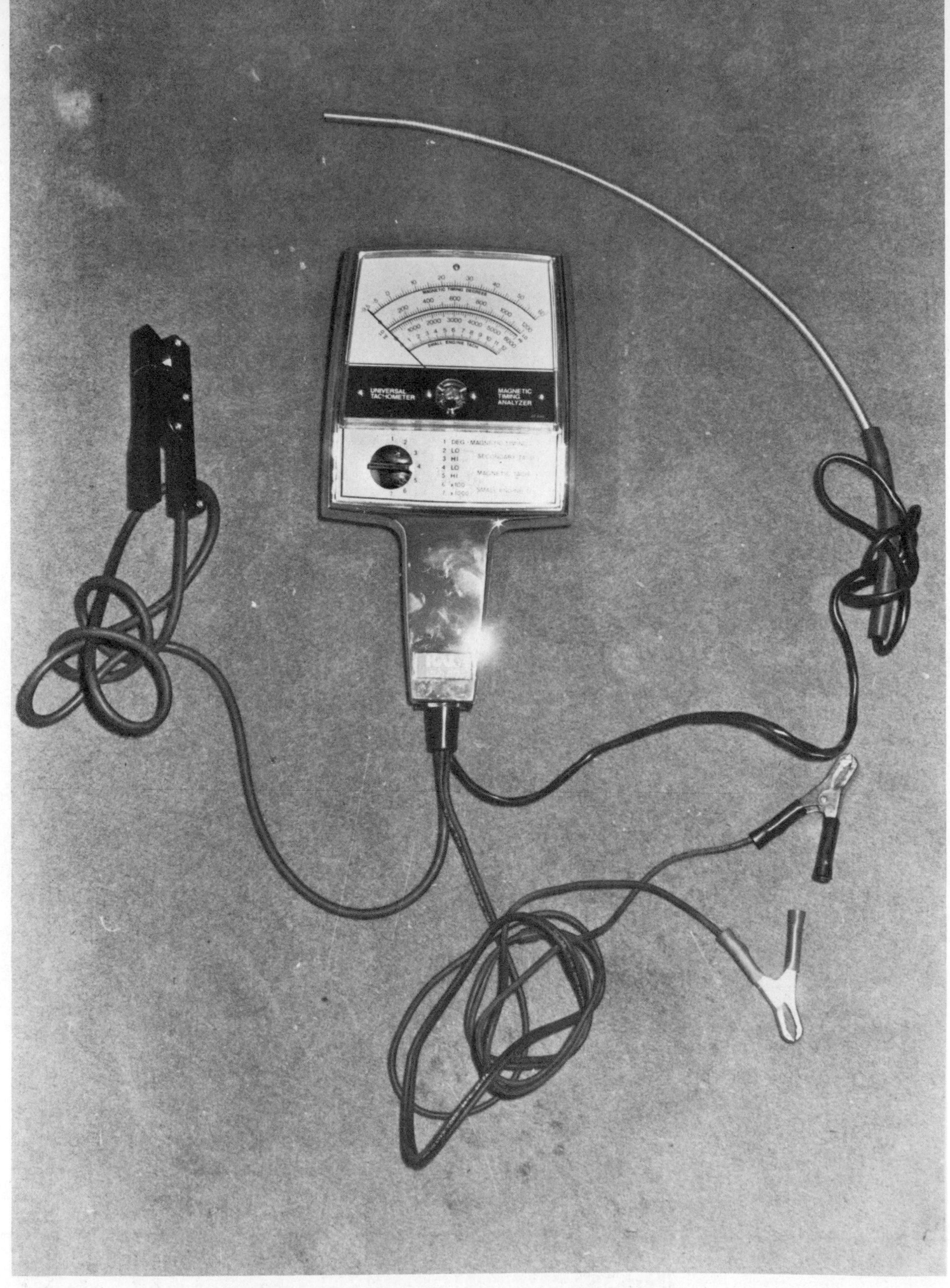

Fig. 2–53. Here's what a magnetic timing indicator looks like. Long probe is flexible and goes into hole. Two clips at lower right are connected to car battery and clamp at upper left clips onto No. 1 spark plug wire. This device also can be used for tachometer readings on diesel engines with probe holders, and on small gas engines such as on lawnmowers.

Fig. 2–54. If timing marks are inaccessible, make your own elsewhere on engine. This pointer is attached to an oil pan bolt and the crankshaft pulley has had a notch filed into it, on the underside of engine.

pears, the timing should be set for before top dead center.

Checking Ignition Timing

To check ignition timing, follow these steps:

Warm up the engine.

Disconnect the vacuum hose or hoses at the ignition distributor's vacuum control unit (see Fig. 2–55), and plug the hose(s) end(s) with a pencil(s).

Connect the power timing light. Clip the red wire to the battery's positive terminal, the black wire to the negative terminal and the remaining wire, called the trigger, to the No. 1 spark plug, or its wire. Exactly how you connect the trigger wire depends on the make of timing light you have purchased, so you must follow the instrument maker's instructions (see also *Chapter 1* and Figs. 2–56, 2–57).

As explained earlier, the timing marks line up when No. 1 cylinder spark plug is firing, so connect the trigger wire to No. 1 plug or its wire, in accordance with the instructions.

Start the engine and let it idle. Note: On some cars timing is specified at a speed other than idle. Check the underhood decal and if necessary, operate the throttle to obtain the specified speed. You will have to connect a tachometer to be sure. If you find it difficult to hold the specified speed, you can have a helper at the gas pedal, or obtain an inexpensive special tool that is installed to bear against the throttle linkage in any position and hold a desired engine speed.

Aim the timing light at the marks. If they are perfectly aligned, timing is correct. If they are not, loosen the lockbolt or nut at the base of the distributor (see Fig. 2–58) and turn the distributor body in whichever direction is required to bring the marks into alignment. Tighten the lockbolt or nut and the adjustment is complete.

Computer Spark Advance

Computer-controlled spark advance first appeared in 1976, when Chrysler Corp. introduced its version, originally called Lean Burn (see Fig. 2–59). General Motors followed with one called MISAR (used on 1977–78 Olds Toronado only) and Ford came in with Electronic Engine Control (see Fig. 2–60), which it has been installing on many of its V-8s. In 1979 General Motors began installing a computer in some of its cars (primarily for California), but these controlled only fuel mixture and some emission controls, not spark timing. In 1981 this computer was installed on all GM cars and was changed to also control spark timing. Chrysler also changed its computer systems, not only going to a different type of computer (from analog to digital on virtually all models) but adding fuel mixture control to the computer functions.

All these systems have certain things in common:

A miniature computer that signals the electronic ignition control unit, to tell it when to trigger the ignition coil.

A group of electronic and electro-mechanical devices called sensors wired to the computer and feeding it the information needed to make a decision as to the optimum ignition timing for each situation.

These computer systems were introduced because it was found that if the ignition timing could be changed radically according to a range of engine operating conditions, two things could be accomplished:

Exhaust emissions could be reduced.

Engine performance and gasoline mileage would be improved.

Prior to these computer systems, ignition timing was automatically controlled by a set of centrifugal weights on springs in the distributor (responding to engine speed) and a vacuum diaphragm unit bolted to the distributor (responding to changes in engine load). These mechanical devices are not nearly as precise as a computer, not nearly as fast, and they cannot make very radical changes in timing (see Figs. 2–61, 2–62).

Sensors

The sensors commonly used include the following:

engine speed

coolant temperature

throttle position (see Figs. 2–63, 2–64)

engine vacuum (also measured as a form of pressure in the General Motors and the Ford systems, as in Fig. 2–60)

crankshaft position, so the computer knows

Fig. 2–55. Disconnect and plug the vacuum hose from distributor before checking ignition timing. If there are two hoses, disconnect and plug both.

Fig. 2–56. Trigger wire on this timing light has alligator clip which attaches to spring. Disconnect No. 1 plug wire, push one end of spring onto end of spark plug, and reconnect plug wire to other end of spring.

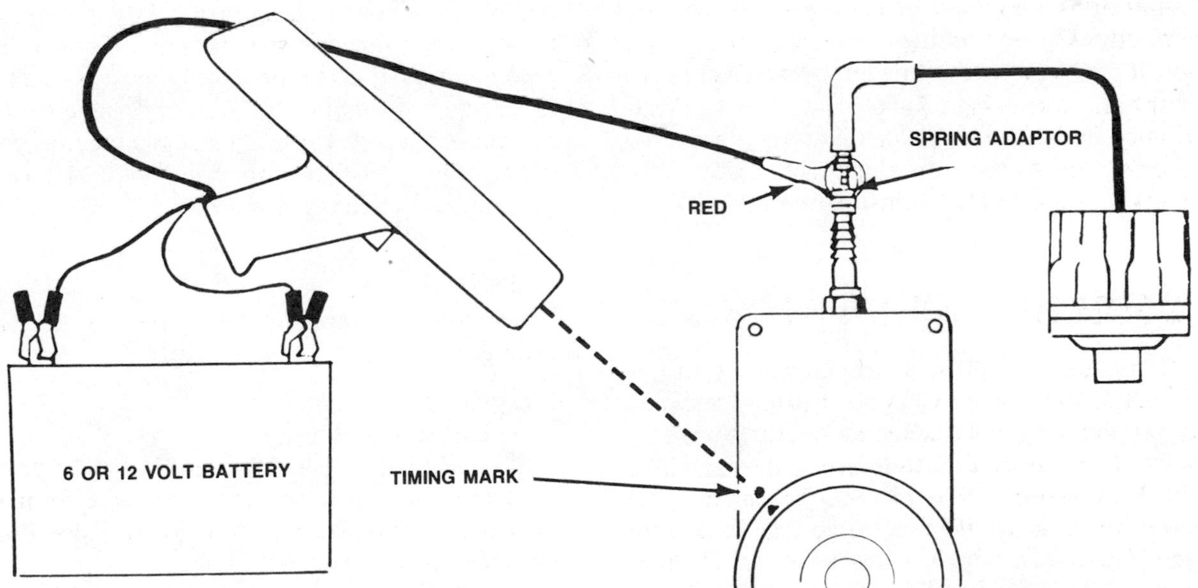

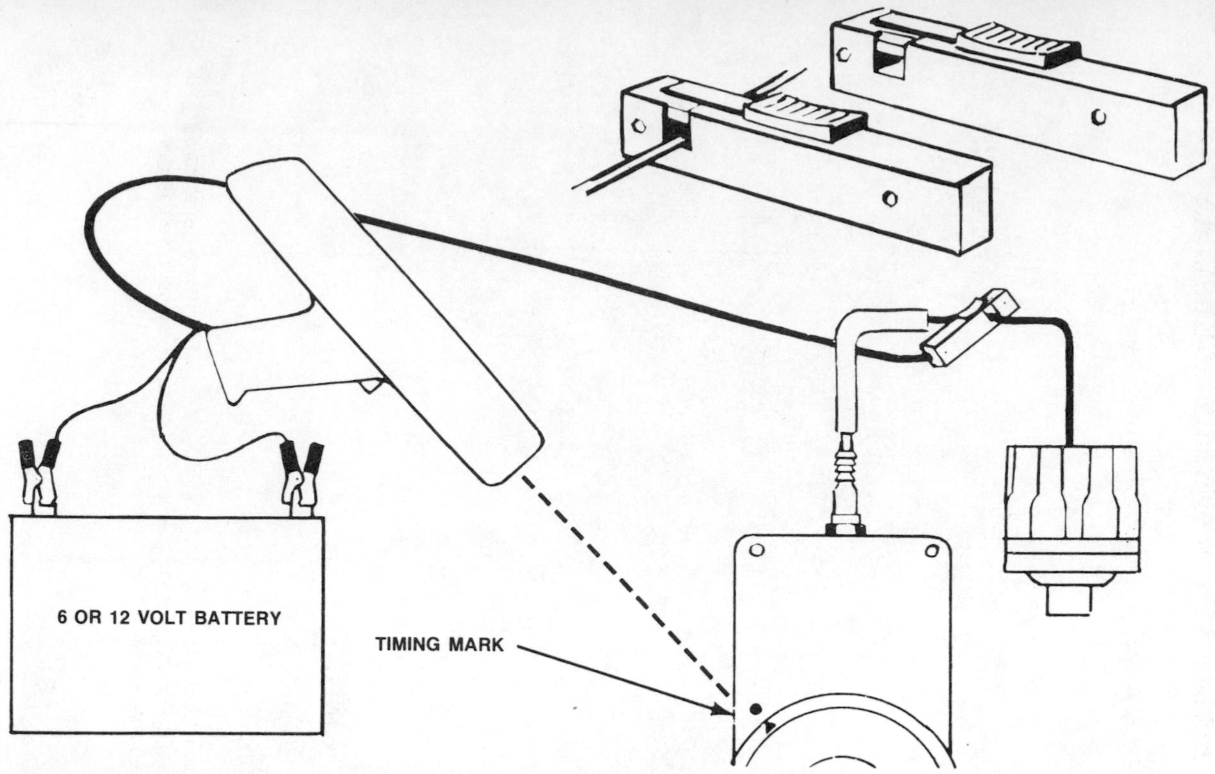

Fig. 2–57. This timing light has clamp-on pickup on the end of the trigger wire.

Fig. 2–58. If timing is off, slacken lockbolt or nut at base of distributor and turn distributor body as necessary to bring timing marks into alignment; then retighten bolt or nut to secure the timing adjustment.

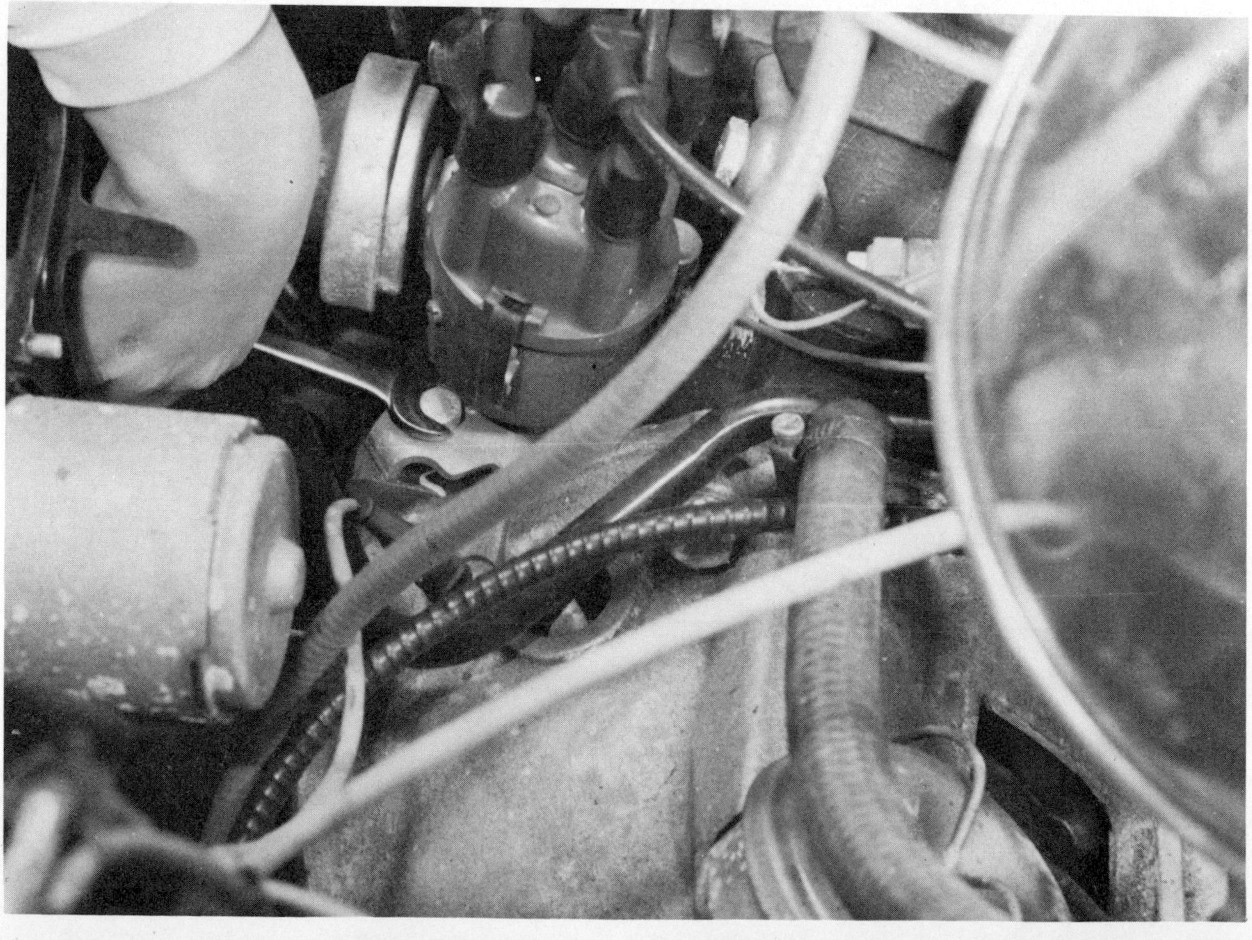

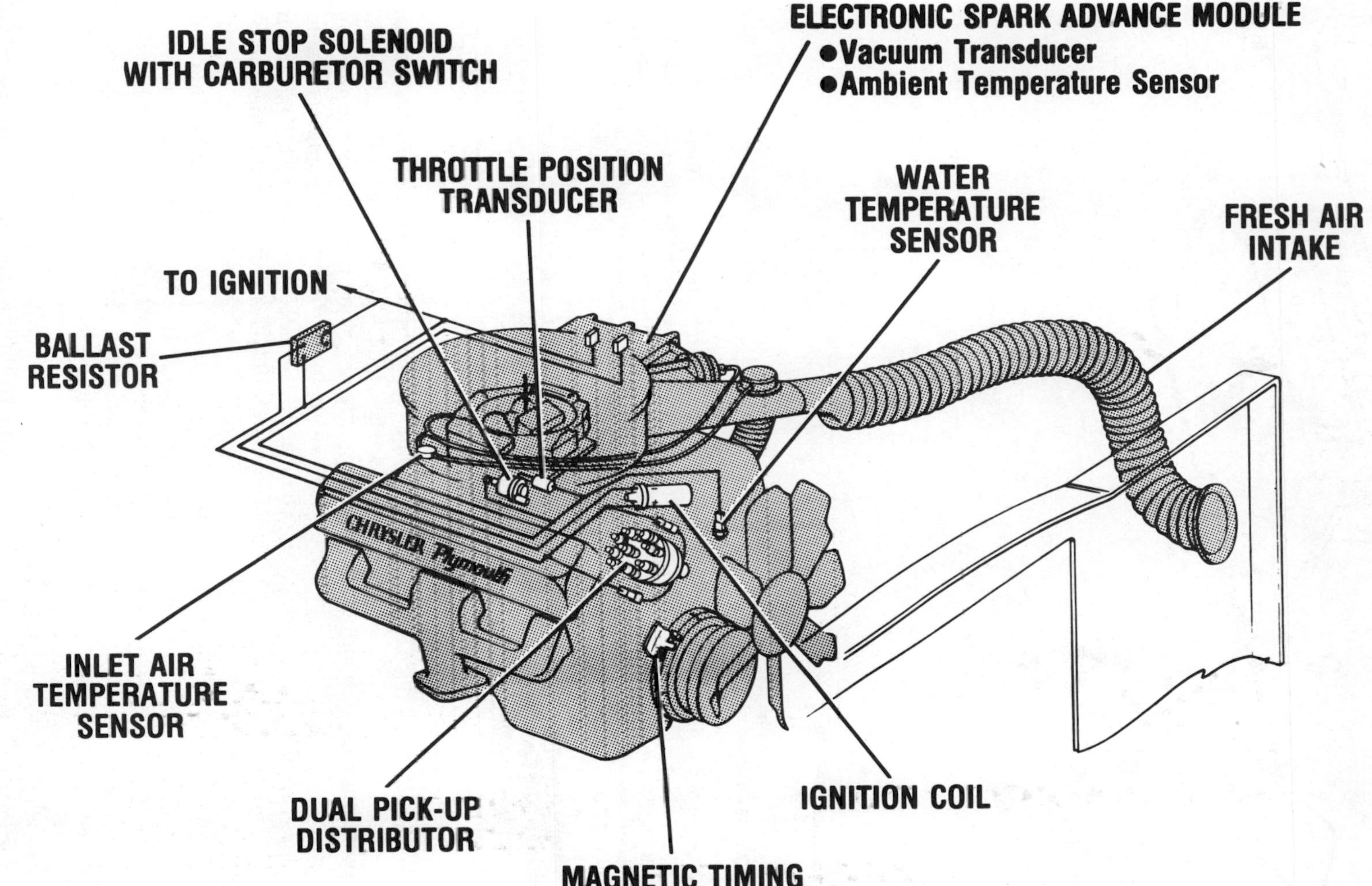

Fig. 2–59. This is a Chrysler electronic spark advance system. Vacuum transducer converts engine vacuum reading into an electronic engine load signal for computer; carburetor switch on idle stop solenoid tells computer if engine is idling or not; other sensors measure water temperature, ambient air temperature, throttle position, and inlet air temperature in air cleaner.

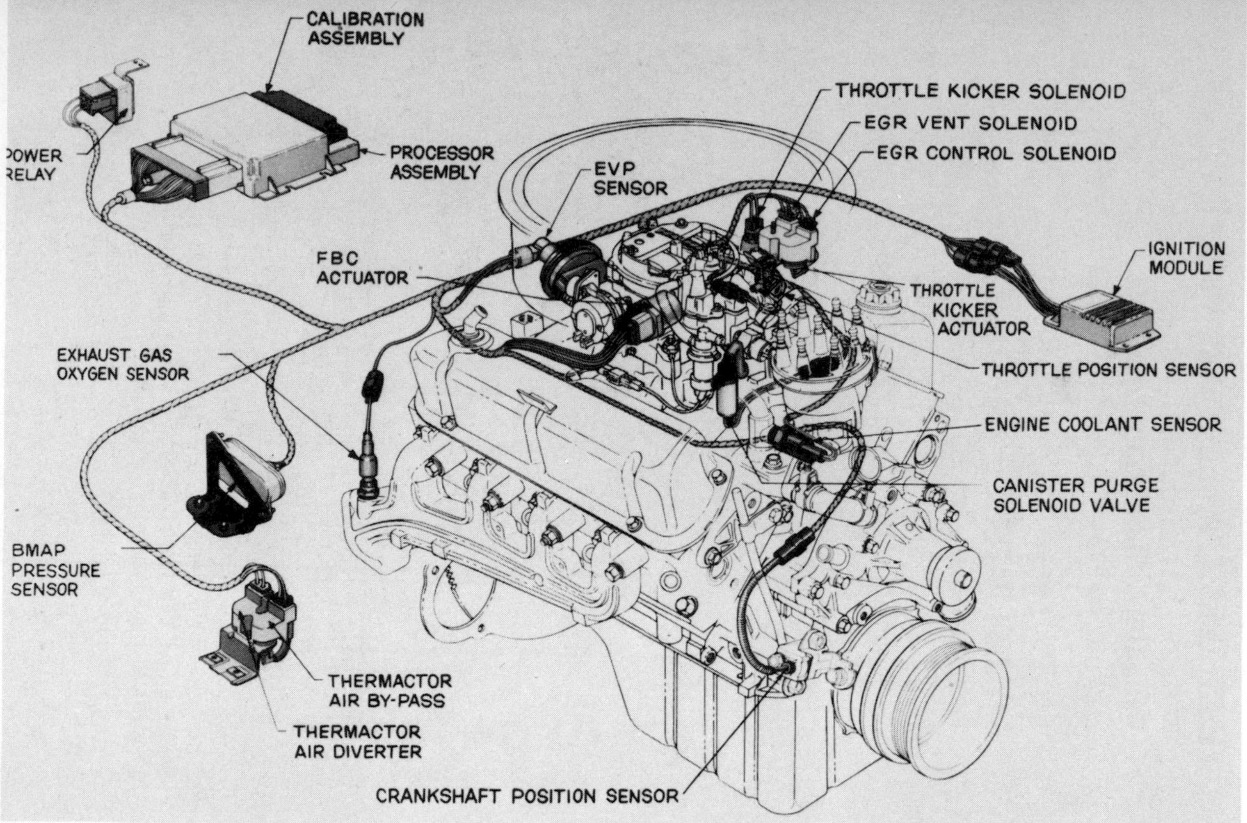

Fig. 2–60. This is the Ford computer system that controls spark advance, fuel mixture by the carburetor, the exhaust gas recirculation valve, the air pump system, and the fuel vapor control cannister. The composite pressure sensor measures both atmospheric and intake manifold pressure (actually a vacuum in the manifold). The solenoids at the lower left are triggered by the computer as necessary to control the output of the air pump, a pollution control. The EGR solenoids regulate vacuum flow to the exhaust gas recirculation valve when they are triggered by the computer. The exhaust gas oxygen sensor tells the computer how rich the fuel mixture is, and if necessary the computer moves a device in the carburetor called a stepper motor to richen or lean out the mixture. The computer also uses information from the various sensors to control ignition timing and, if necessary, to operate a throttle kicker solenoid to increase idle speed.

which cylinder will be the next to require a spark, and when.

air temperature inside the engine compartment (see Fig. 2–59)

Computer Failure

If the computer fails, the systems all have a limp-in feature that permits you to continue to start and operate the car, but with performance and gas mileage often so poor you'll bring the car in for service.

If you believe that a sudden substantial drop in performance may be caused by a computer failure, you can make this simple check: Connect your timing light, aim it at the marks, and start the engine. Have a helper operate the gas pedal, light on the gas, then heavy, then moderate. As the pedal position changes, you should see substantial changes in ignition timing, but if the computer has suffered a catastrophic failure, you'll see none—the timing will remain unchanged. Note: Many cars have a warning lamp that will light in case of a computer failure, eliminating the need for this test. On General Motors cars, the limp-home feature does include provision for some changes in spark timing, built into the distributor electronic module. Therefore, you may see some changes (up to as much as nine degrees) on GM cars even after the computer has failed and the check engine light in the dashboard has gone on.

Checking Computer-Controlled Timing

All three systems have provisions for checking ignition timing, but on Ford products it cannot be adjusted. If it is incorrect, some part is defective and must be replaced, a job for a professional.

Fig. 2–61. This is the distributor of a Ford with a computer-controlled spark advance system. With the rotor removed and at left, you can see there's nothing inside except the distributor shaft.

The basic timing check is done in a manner similar to that on other ignition systems, with several important differences. Read the underhood decal and don't disconnect anything unless the underhood tuneup decal specifically tells you to. If timing checks at other than idle speed are prescribed, connect a tachometer and use the prescribed speed(s).

On the General Motors MISAR system, look up under the dashboard to find the computer and locate a purple wire hanging down unconnected (see Fig. 2–65). You must connect this wire to electrical ground, using a jumper wire. The jumper is a wire that is normally about 14 or 16 gauge, with an alligator clip at each end. Force one clip onto the metal terminal of the purple wire, and connect the other to a grounding point, such as a metal bracket under the dashboard, a body bolt, etc. If you have it grounded, a "check ignition" dashboard warning light will go on when you start the car. Remove the jumper wire after you're done.

Check a MISAR car's ignition timing with the timing light. If timing requires adjustment, it is done by turning the distributor body in the conventional way, except on 1977 Olds Toronado, the model on which MISAR was introduced.

On that car, you must get underneath the car, locate the crankshaft position sensor, slacken its mounting bolts, and then turn an adjusting bolt on top to change the timing (see Figs. 2–66, 2–67). Once the timing is correct, tighten the mounting bolts. Note: This job should be done with a helper so that one person can make the adjustment while the other reads the timing marks. Be very careful, if you are working alone, to keep your hand away from spinning pulleys as you move underneath to reach the sensor's adjusting bolt. It may be prudent to shut down the engine while making an adjustment.

On Chrysler products with Lean Burn or a subsequent computer system, begin by grounding the carburetor switch with a jumper wire, which is a wire that has an alligator clip at each end (sold in auto parts stores). Connect one clip to the metal terminal of the wire, the other clip to an electrical ground (such as the engine top cover or any other metal part of the engine). (See Fig. 2-68.) Then hook up the timing light and have a helper start the engine. Blip the throttle, apply the brakes and put the transmission in Drive. Watch the timing marks and on most engines you

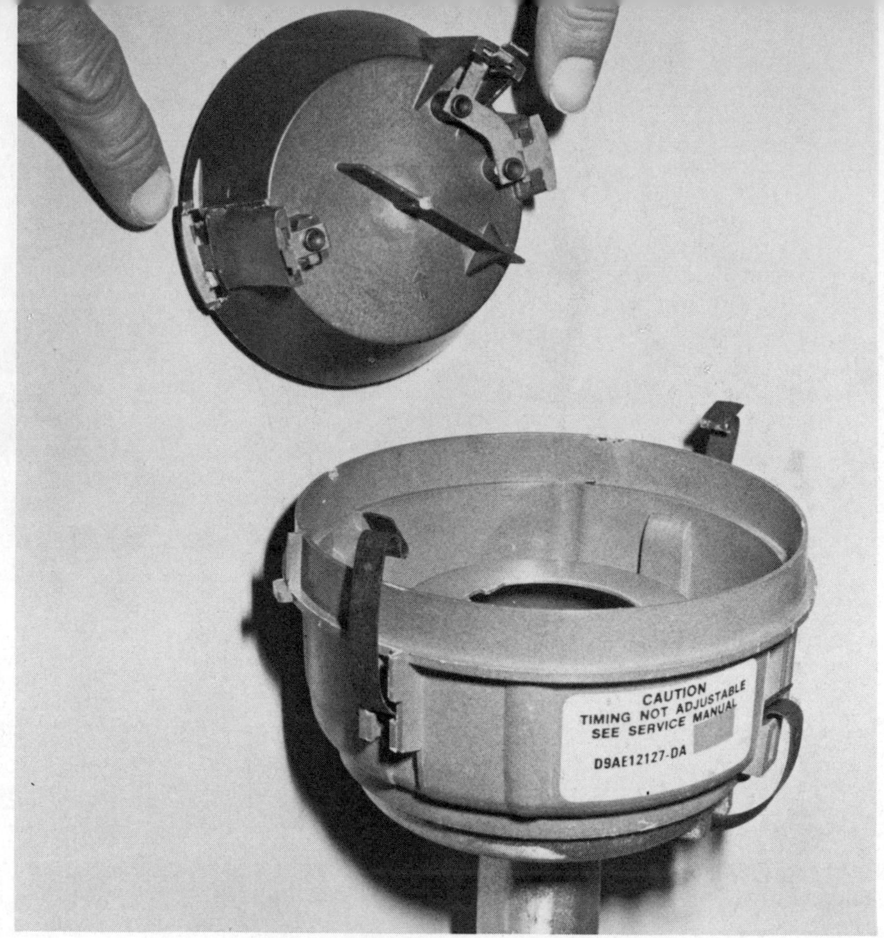

Fig. 2–62. Notice that the rotor of the Ford distributor on a computer system has two firing tips, one high and one low. The inserts in the distributor cap are also at two levels. This provides increased separation, so there is less danger of a high-voltage spark going astray. Only one rotor tip fires at a time.

Fig. 2–63. Throttle position is an important consideration in ignition timing and when a computer knows where the throttle is, it can make precise adjustments. This Chrysler device is a miniature electromagnet containing a movable iron core that is connected by a wire to the throttle linkage. When the throttle is opened the wire is pulled, moving the iron core. This alters the magnetic field and sends a signal to the computer.

will see some amount of timing advance greater than specifications. Watch for ninety seconds and you should see the timing drop (if it was over-advanced) to specifications. Watch for an additional thirty seconds to be sure. If a timing adjustment is necessary, loosen the distributor lockbolt and turn the distributor as required to bring the reading to within specifications.

On Ford products with Electronic Engine Control (some V-8s since 1978), there is no timing adjustment. You check timing and if it is not correct, there is a malfunction in the computer system (possibly a defective sensor). The system is warranteed for five years or 50,000 miles, and it's something you should leave to the dealer for service. To make a basic timing check on early-1978 to early-1980 systems (silver underhood tuneup label in 1980), warm up the engine and run it at idle. With a timing light that is adjustable or a timing indicator, you should get a reading of 28–32 degrees before top dead center. If you get this reading, have a helper step on the gas pedal and you should see the timing advance. Satisfactory results from both these tests indicate the system is probably operating correctly.

The adjustable timing light or timing indicator is needed because 28–32 degrees is well off the timing marks scale on the engine. If you don't have such a timing light, locate the computer under the dashboard on the driver's side, near the support for the brake pedal. There is a rectangular box, called the calibration unit, plugged into the computer and held by two screws. Remove the screws and lift out the calibration unit. Timing should read 10 degrees before top dead center at idle and it should not change (the system is in the limp-home mode). Reinstall the calibration unit and the timing should advance, going off the scale. This isn't as good a test as with an adjustable timing light or timing indicator, but at least it tells you something.

If you have an EEC III system (mid-1980 introduction, had a gold underhood tuneup decal that year), you have a system with built-in diagnosis; that is, the computer can help with troubleshooting. Timing on this car can only be checked when the computer is in the self-diagnosis mode, using an adjustable timing light

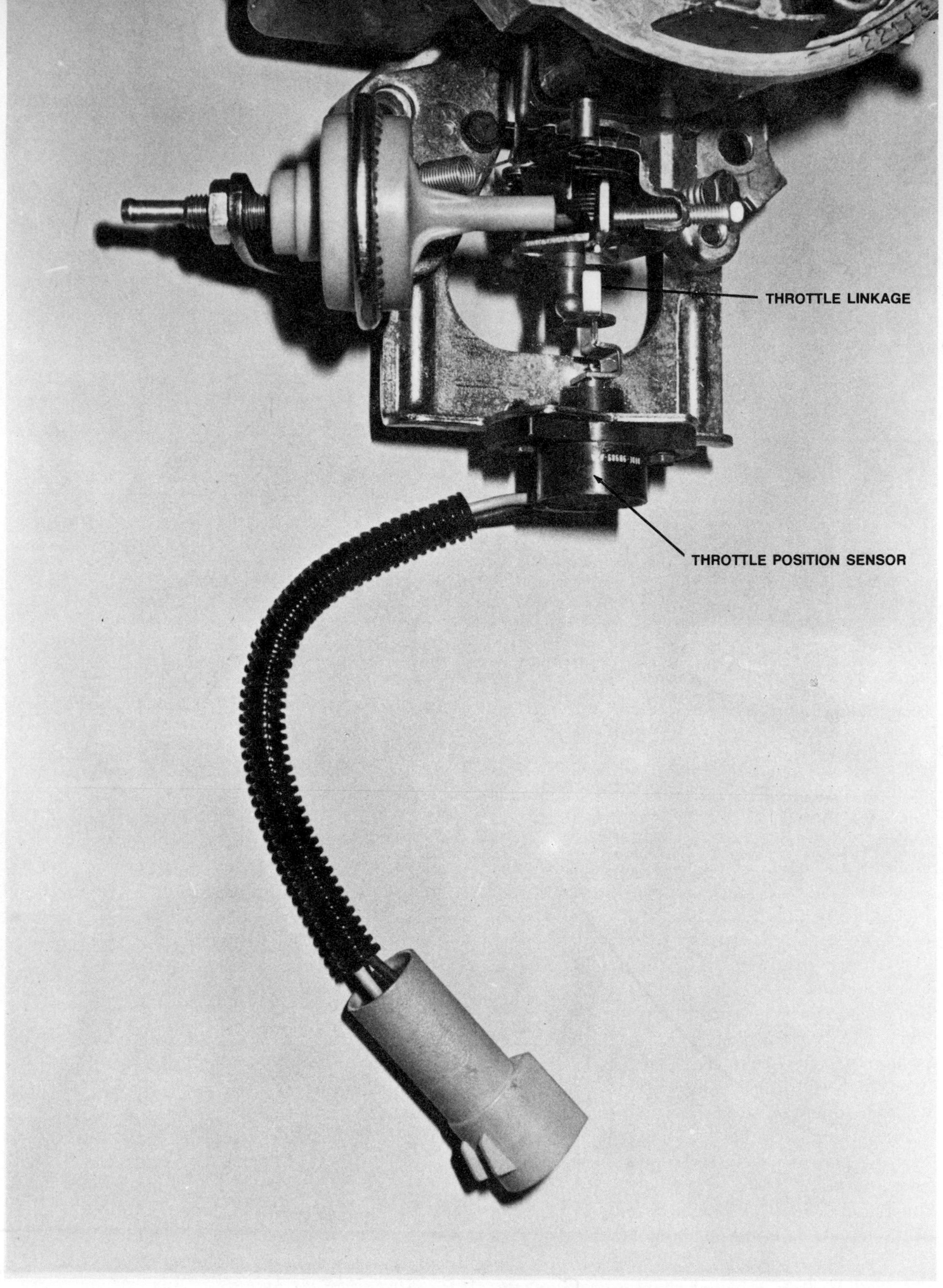

Fig. 2–64. This Ford throttle position sensor is a variable resistor. As the throttle linkage is operated, it also moves a contact inside the sensor, changing the resistance of a current flow transmitted by the computer. When the computer senses a change in this current flow, it translates this information into a throttle position indication.

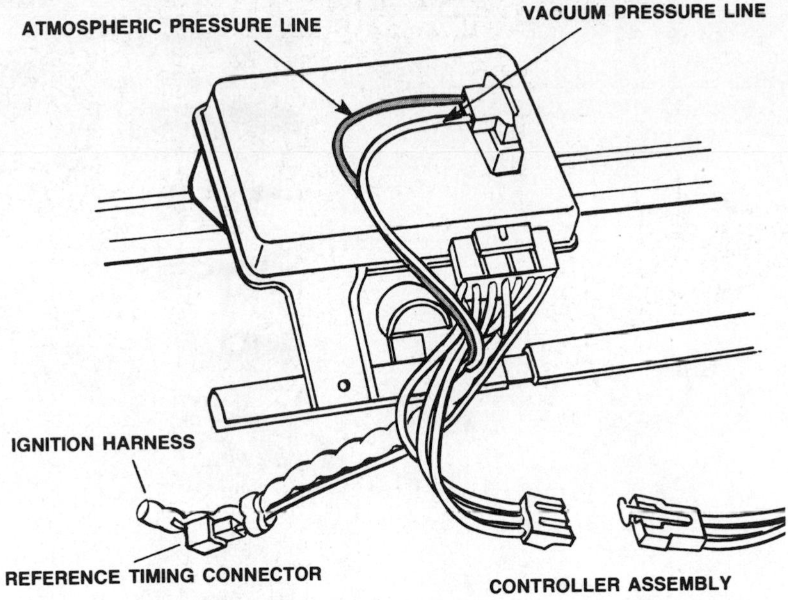

Fig. 2–65. Here is the heart of the GM MISAR system, used on 1977–78 Olds Toronados. Below is the crankshaft (position) sensor, aligned with a toothed pulse generator disc on the front of the engine. Each tooth triggers the crankshaft sensor that transmits a signal to an under-dash computer, shown above. The computer then triggers the High Energy Ignition distributor. Note the timing adjuster bolt and the clamp bolts on the sensor. Also notice the reference timing connector wire, an unconnected wire hanging down from the computer. This connector must be grounded before a timing check can be made. If timing must be changed on this design, slacken clamping bolts and turn adjuster. In 1978, this system was changed so timing changes could be made in the conventional way, by turning the distributor body.

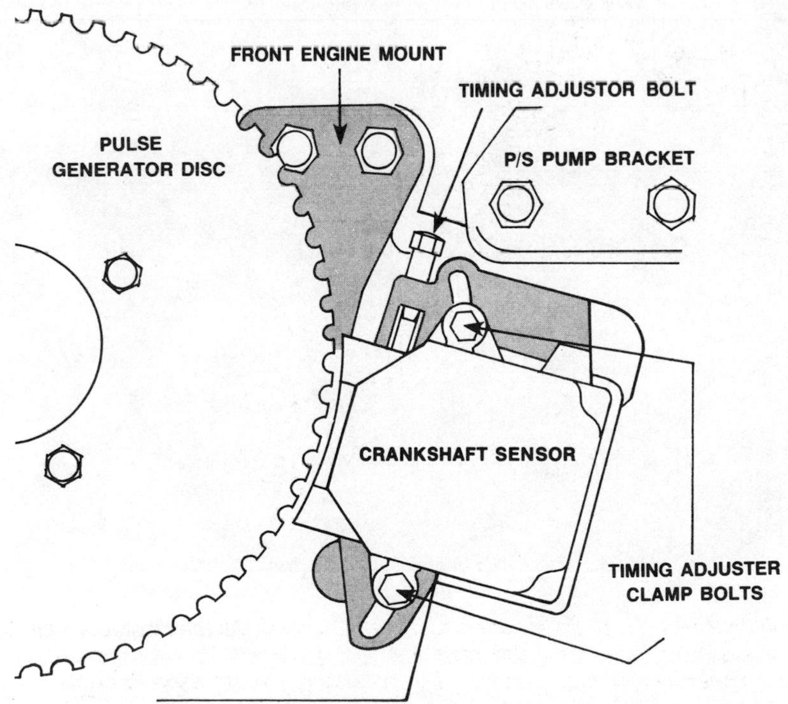

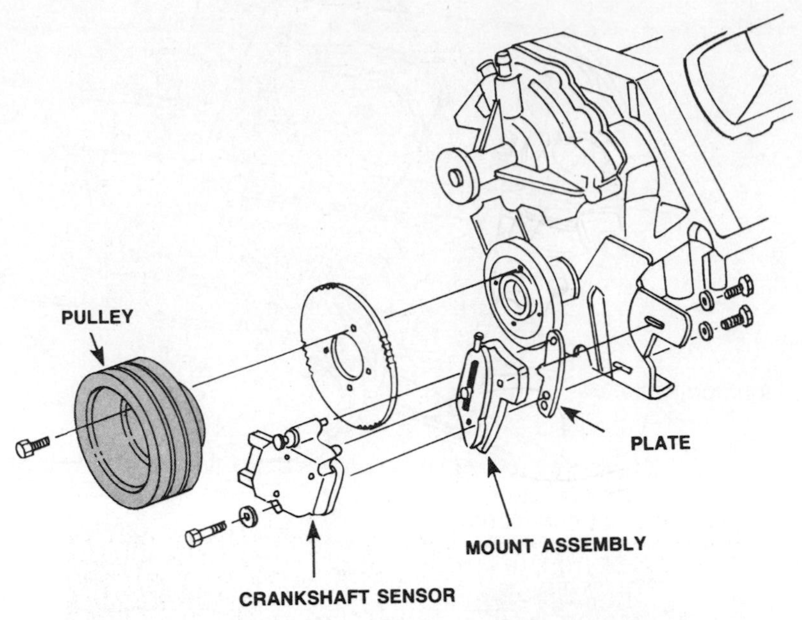

Fig. 2–66. View of front of 1977 Olds Toronado engine shows how toothed pulse disc was mounted on crankshaft, just back of crankshaft pulley.

Fig. 2–67. A look at the actual parts of the 1977 MISAR system.

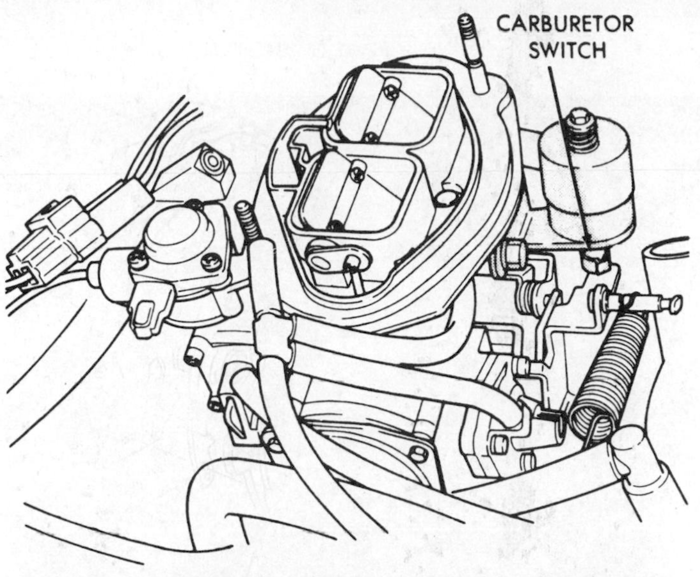

Fig. 2–68. The Chrysler carburetor switch, which signals the computer when the throttle linkage is closed or open, should be grounded with a jumper wire as shown before attempting to check ignition timing. With the carburetor switch grounded, the computer will not change the basic ignition timing. The carburetor switch has a rod that bears against the throttle linkage and a wire from it to the computer. It is typically opposite the idle speed solenoid on front-drive Chrysler products.

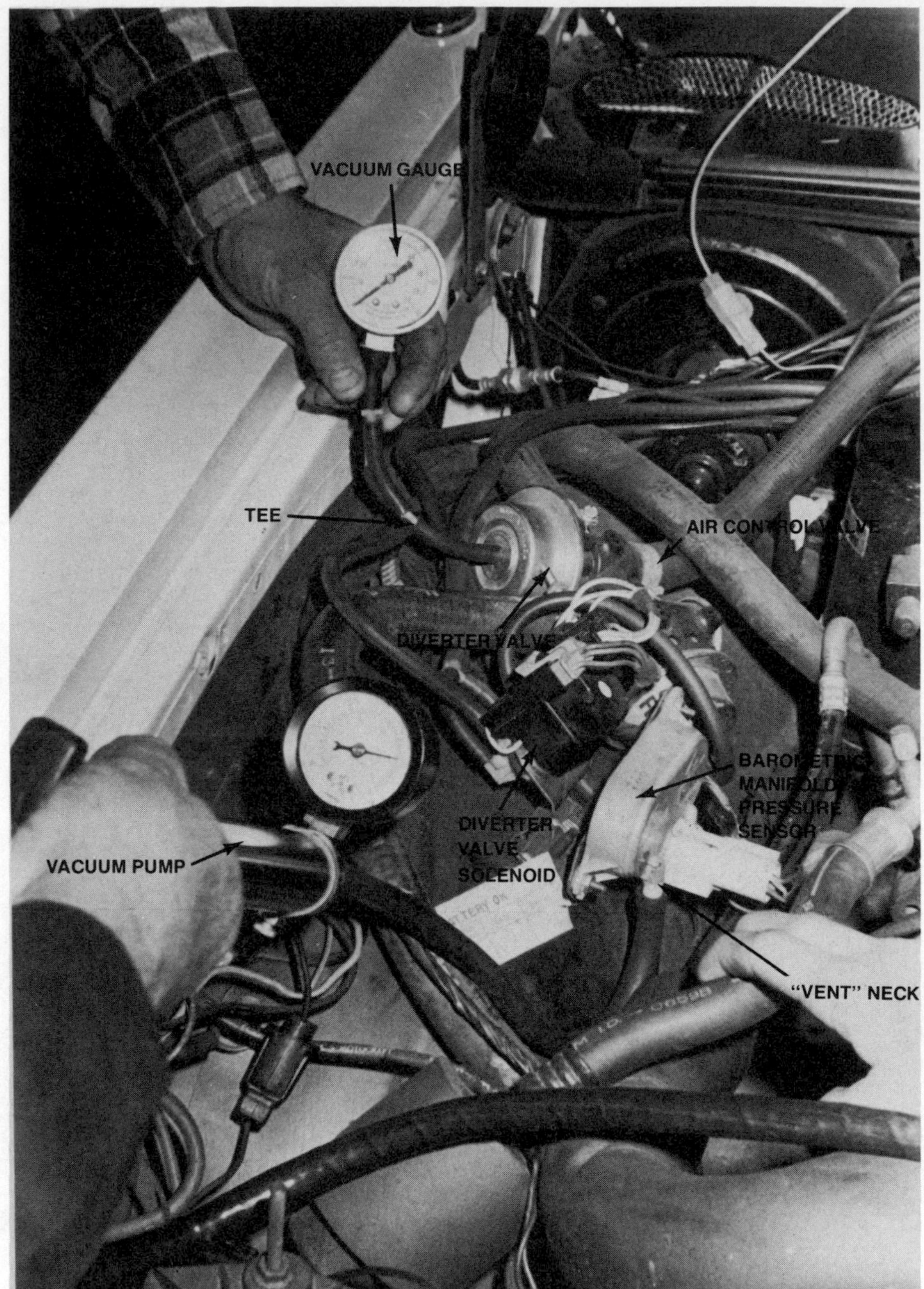

Fig. 2–69. To trigger the EEC III self-test, connect a vacuum gauge with a tee and spare piece of hose into the vacuum hose from the solenoid to the diverter valve section of the air pump control valve as shown. Then attach the hose from the manual vacuum pump securely into or onto the "Vent" neck of the barometric/manifold pressure sensor, also as shown.

or timing indicator. To activate the self-test, you need a manual vacuum pump and a simple vacuum gauge.

Install the gauge with a plastic tee (inexpensive part sold in auto parts stores) as shown in Fig. 2–69. The tee is installed in the thin (vacuum) hose from the air pump bypass valve (see *Chapter 4*). Then force the hose end of the manual vacuum pump into or onto the neck marked "VENT" on the pressure sensor, also shown in Fig. 2–69. Both parts are on the passenger's side of the engine compartment, on the sheet metal of

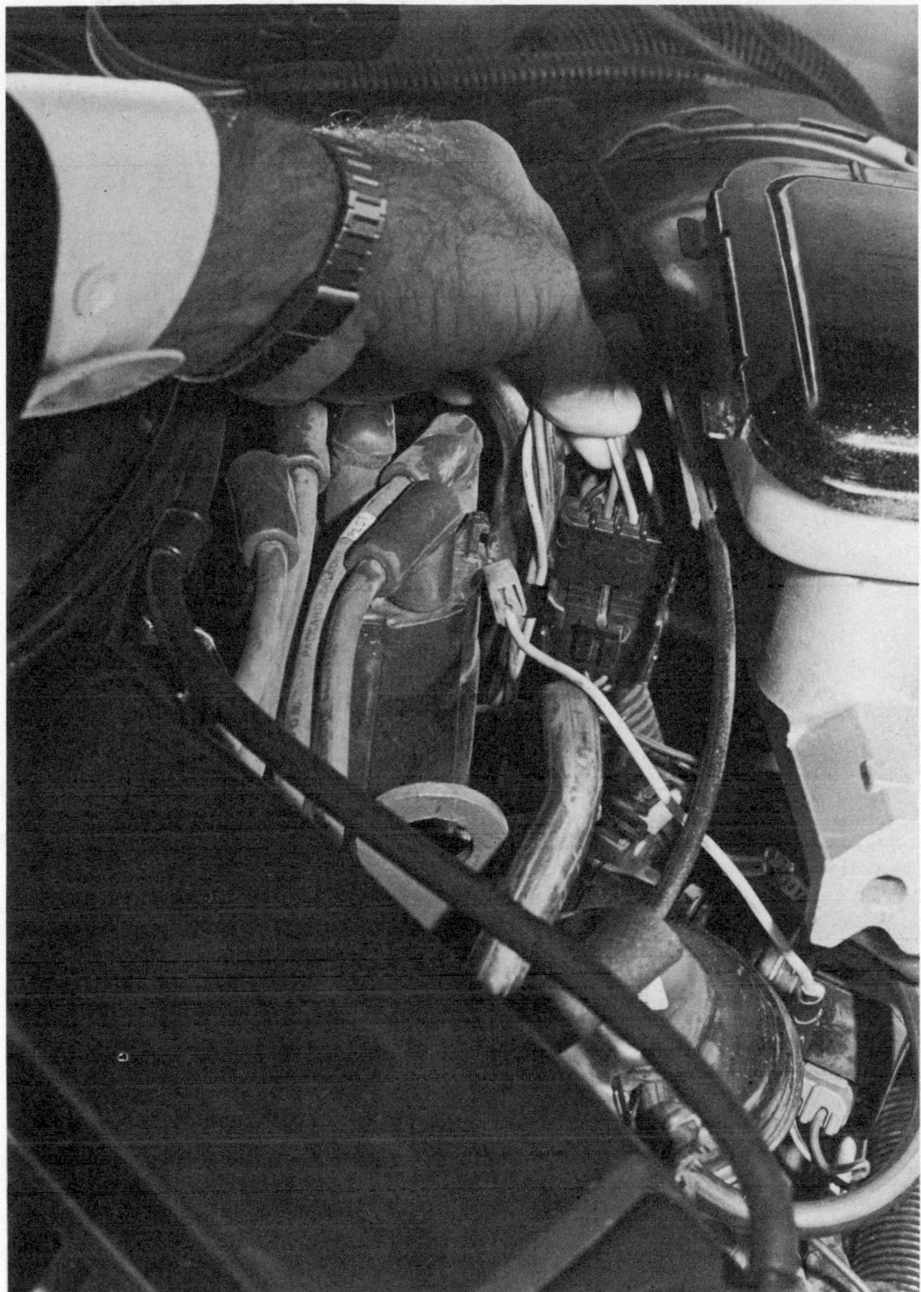

Fig. 2–70. Before checking basic timing on a General Motors computer-equipped car, unplug the four-wire connector from the electronic ignition distributor. Do not unplug any other connector. Unplugging this one will keep the computer from adding any advance to the ignition timing.

the fender well.

Warm up the engine, run it at idle and operate the vacuum pump to draw over 20 inches of vacuum. Hold this reading for a minute, then release. The vacuum gauge needle should pulse two or four times (depending on the engine). This tells you the self-diagnosis mode has been activated.

As you draw vacuum, the throttle kicker should go on automatically. This is a device on

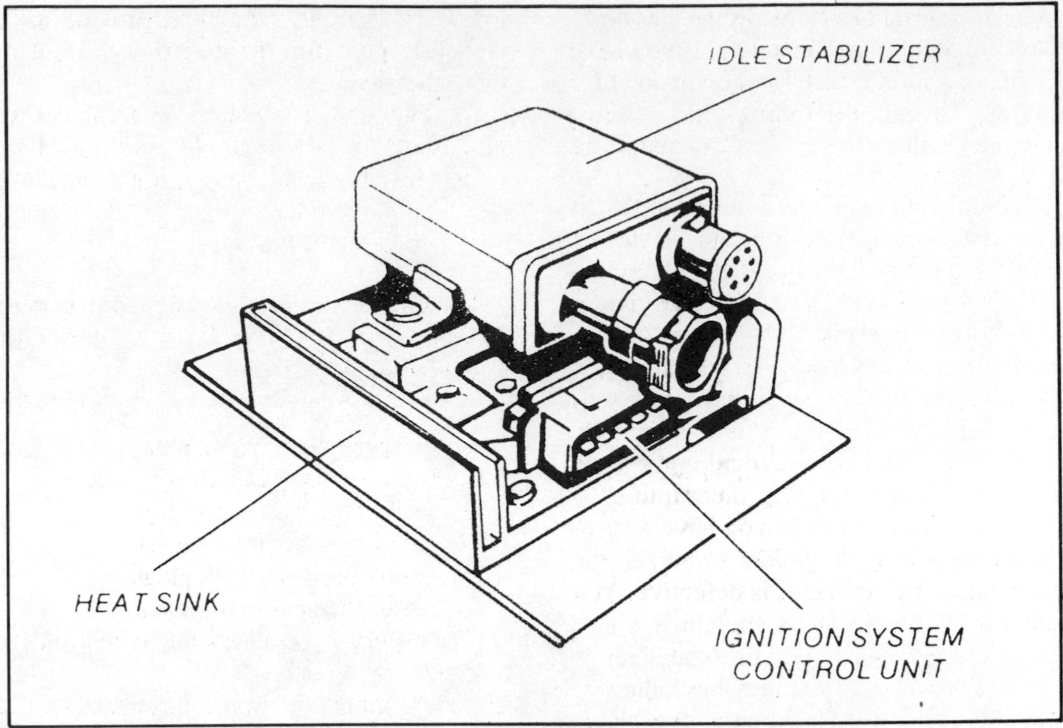

Fig. 2–71. If there is an idle stabilizer atop the electronic control unit on a VW Rabbit or other VW water-cooled model, remove both wiring connectors from it and plug them into each other before checking basic ignition timing on cars with electronic ignition. The electronic control unit is under a plastic shield on the driver's side, just back of the engine compartment in the air intake area. It is accessible with the hood up.

the throttle linkage that normally pushes open the linkage to boost idle speed when accessories are turned on. When you release vacuum, the kicker should go off, then come on again and stay on for the duration of the self-diagnosis, which is about two minutes. You can tell when the kicker goes on (engine speed suddenly rises audibly) and off (engine speed drops). Only while the throttle kicker is on should you check ignition timing, and with an adjustable timing light, you should get a reading of 27–33 degrees. If you do, the computer apparently is controlling spark timing properly. But wait.

After a minute-plus, the vacuum gauge reading should go up and hold steady. Then the gauge needle should pulse two or three times (depending on model). Soon after that, the throttle kicker should disengage (you'll hear idle speed drop), and now you should watch the vacuum gauge. If you see a needle pulse, pause, then another needle pulse, the computer apparently is performing properly in all respects. If you get another code, there is apparently a malfunction in the computer system.

Note: This test should be used in case there is any performance problem with the car, to determine if the problem is in the computer system or not.

On General Motors cars with computers starting in 1981 (except Cadillac), make a basic timing check by unplugging the four-wire connector (no other connector) from the distributor (Fig. 2–70). This eliminates the computer and the timing should be the basic setting, usually 10 degrees before top dead center. Reconnect the wiring and operate the throttle. The timing should advance substantially, off the scale. If it does, you may assume the computer is doing its job.

On 1980 Cadillacs with throttle body fuel injection, there's an unconnected pink/black wire coming out of the harness near the engine top cover on the passenger's side rear of the engine. Ground this wire with a jumper wire, then check timing as for other GM cars.

In 1981, timing is checked from the driver's

seat by pressing certain buttons on the air conditioning panel in a complex sequence. If you have such a Cadillac, obtain a factory shop manual for the procedure. The ignition timing will be shown on the digital display for the A/C temperature setting.

On most 1980 and some subsequent models of Volkswagen with electronic ignition, a device called a digital idle stabilizer is used. This gadget (Fig. 2–71), mounted atop the electronic ignition unit, makes continuous small adjustments in ignition timing to maintain a consistent idle speed. To check timing, therefore, you must take this device out of the picture, which is done simply this way: Remove the two electrical connectors from the idle stabilizer and plug them into each other (they're made for this). If you have a starting problem on a VW with the idle stabilizer, the cause may be that the stabilizer is defective. You can isolate it from the circuit as explained, and if the car now starts, you know the stabilizer is defective and the cause of the starting failure.

On '82 AMC six-cylinder, unplug the black/yellow wire connector and hook up a jumper across the male terminals.

Spark Plugs

The spark plugs are the terminals of the ignition system (see Fig. 2–72). When the rotor aligns with a terminal in the distributor cap, the high-voltage electricity goes into the terminal, along a heavily-insulated wire, and into a metal rod (called the center electrode) embedded in the center of the plug. The electricity travels down that rod and when it comes to the end (inside the cylinder) it looks for a place to go. There is only one: An L-shaped metal prong that is anywhere from .025 to .080 inch away. That prong, called the side electrode, is part of the lower end of the plug that is threaded into the engine and is in metal-to-metal contact with it. The side electrode, therefore, is an electrical ground just like the engine itself. The electricity is sufficiently high-voltage to jump the air gap between center and side electrodes, and so it does, completing a circuit and forming an electric arc. That electric arc is what we call the spark.

Two things can happen to the plug to prevent that electric arc:

(1) The inside of the plug can become fouled with deposits from gasoline or, if the engine is burning oil, from the oil. If the deposits coat the inside of the plug, they can provide an easier, alternate path for the electricity, so it doesn't jump the gap.

(2) The constant electrical arcing across the electrodes causes them to erode and the gap between them gets larger. If it gets too large, the high-voltage electricity may not be high enough in voltage to jump across.

Either of these conditions has the same effect: The mixture isn't ignited and we notice that the engine is misfiring.

Servicing the spark plugs is a multi-step job:

(1) Disconnect the spark plug wires.
(2) Remove the old spark plugs.
(3) Inspect the old spark plugs.
(4) Perform a compression test.
(5) Adjust the new spark plugs.
(6) Install the new spark plugs.
(7) Inspect the spark plug wires and replace any that are defective.
(8) Reconnect the spark plug wires.

Let's take a look at each step. Note: Before you start you should know that a compression test is best performed with the engine warm, so it's best to begin spark plug removal with a warmed-up engine. Wear work gloves to prevent burning your hands.

Servicing Spark Plugs

Disconnecting the spark plug wires. The spark plug wires on all modern cars are the resistance type, for radio suppression. The conductor is not metal wire, but a fiber impregnated with carbon. Therefore, you must handle the wires with care to prevent them from breaking internally, which is damage you can't see, but which can cause the engine to miss.

Grasp the wire at the spark plug terminal's rubber nipple, twist back and forth to break the heat seal, then pull. Never pull on the wire itself (see Fig. 2–73).

If the nipple is not easy to grasp, obtain a special tool (sold at auto parts stores) that is made to do the job. There are several types and all are inexpensive (see Fig. 2–74).

To assure that you will reconnect each wire to the correct plug, label each wire. A simple way is to wrap a piece of masking tape around the wire and put a number on the tape, starting from the front of the engine. First plug is No. 1, second is

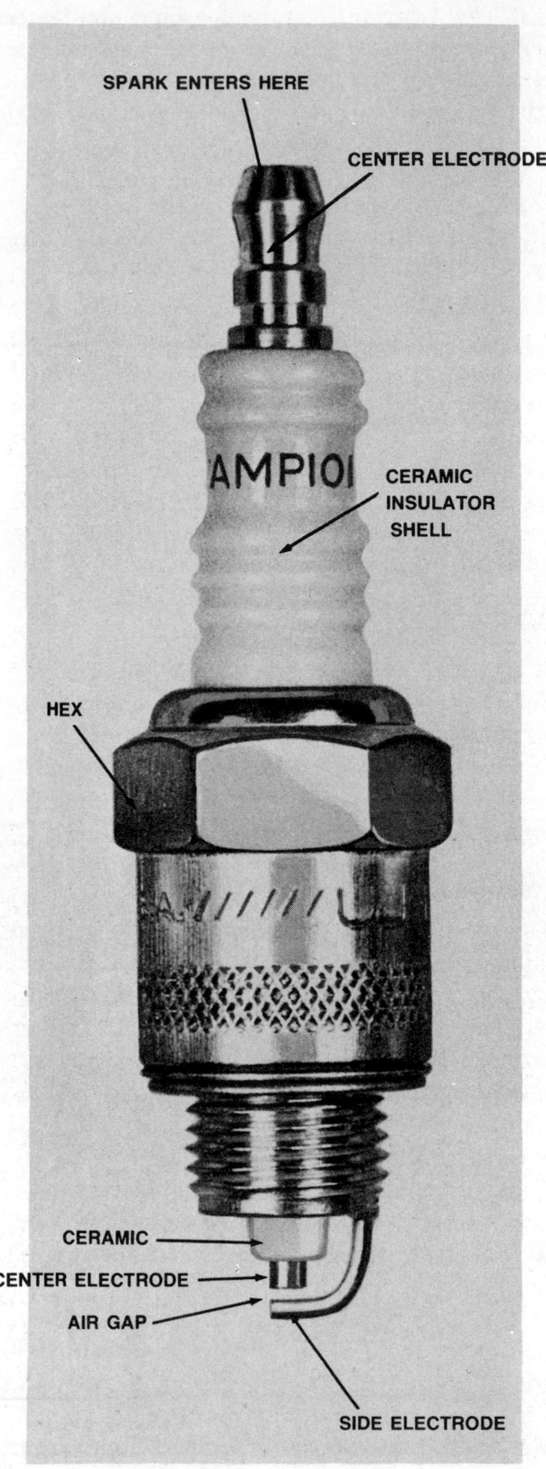

Fig. 2–72. This is a typical spark plug. When threaded into the engine its side electrode is an electrical ground. Spark enters center electrode at top, travels down electrode to inside the cylinder, and jumps air gap to side electrode. As the electricity jumps the air gap, it creates what we call the spark. Circuit is complete when electricity reaches the electrical ground of the side electrode. The ceramic insulator prevents high-voltage electricity from leaking away to ground via another path, which would eliminate the spark that fires the air-fuel mixture in the cylinder.

No. 2, etc.

• *Removing the old plugs.* To remove the old plugs, you'll need a spark plug socket, which is a deep (tall) socket made to fit the hexagonal rim of the plug (see Fig. 2–75). Most spark plug sockets are 13/16 inch, although there are some that are ⅝ inch.

If the spark plug is wide open, all you need to remove it is the socket and a ratchet wrench with the knob set to loosen. This means it will lock when counterclockwise force is applied, and freewheel when the handle is turned clockwise. Just put the socket over the plug and move it down until it fits on the hex section securely. If you're not sure how this should feel, just take a new plug from the box and try the socket on it.

Move the ratchet handle into an open area, then apply steady counterclockwise pressure with one hand while you push down (to keep the socket in position) on the head of the ratchet. If the spark plug is very tight, whack the end of the ratchet handle with your palm to shock it loose.

Less-accessible plugs. It is almost unheard of for all spark plugs to be so accessible that the ratchet and socket is all you need. At the very least, placing the ratchet on the socket will bring the ratchet so close to the engine that you won't be able to get a good grip on the handle. If this is the case, just place an extension rod between the ratchet and socket (see Fig. 2–76).

If there is some underhood accessory in the way of a straight-on approach with the extension rod, there are other possible methods. Use a universal joint, inserted between socket and extension, or between extension and ratchet. That will permit you to work at an angle, around the obstruction.

Or purchase a spark plug ratchet, the type with a universal joint built into the head and a curved handle; it should be even better at clearing obstructions (see Fig. 2–77).

If the problem is lack of clearance between the spark plug and the side sheet metal of the engine compartment, perhaps you can use the socket without the ratchet. Just fit it onto the spark plug and put a box wrench over the hex end of the socket. Put a finger on the end of the socket to hold it in place and use the box wrench to turn the socket.

To some extent you have to look at the spark plug and see what tool arrangement is necessary. In some cases, a plug that looks buried may be reached with a combination of extension rods and a universal joint. Just remember, the spark plug must be removable, even if it isn't easy.

Don't fret if you crack or break the ceramic outer shell of the plug. Sometimes the wrench will slip and that's exactly what will happen during removal. Just reposition the socket on the hex section and get out the broken remainder. You're throwing it out and installing a replacement, so you don't need it for anything.

Inspecting the old spark plugs. As you remove spark plugs, keep them in order. Inspect them, looking for the conditions shown in Figs. 2–78 to 2–81. If a mechanical problem is indicated in only one cylinder, you'll know which cylinder it is. If only one spark plug is oil-fouled, for example, the problem might be a broken valve stem seal, something you can replace yourself, as explained in Chapter 5.

Compression test. Once the spark plugs are out, you can make a compression test, which will indicate if the engine can be tuned successfully. The gauge measures the pressure developed by the piston as it rises in the cylinder to compress the fuel mixture. If compression is good it means that the sealing rings around the piston (the piston rings) and the valves that admit the fuel mixture to the cylinder and allow the burned exhaust gases to escape are not leaking, and the cylinder head gasket, the seal between the engine block and head, is holding. If they leak the pressure is reduced, and so is the force of the burning fuel mixture. Less force means less power. A tuneup will not restore compression; that's a mechanical job for a professional shop.

The compression test requires a helper—someone to sit behind the wheel, hold the gas

Fig. 2–73. To remove a plug wire, twist the nipple to break heat seal, then pull off. Do not grasp wire itself or you could damage it internally.

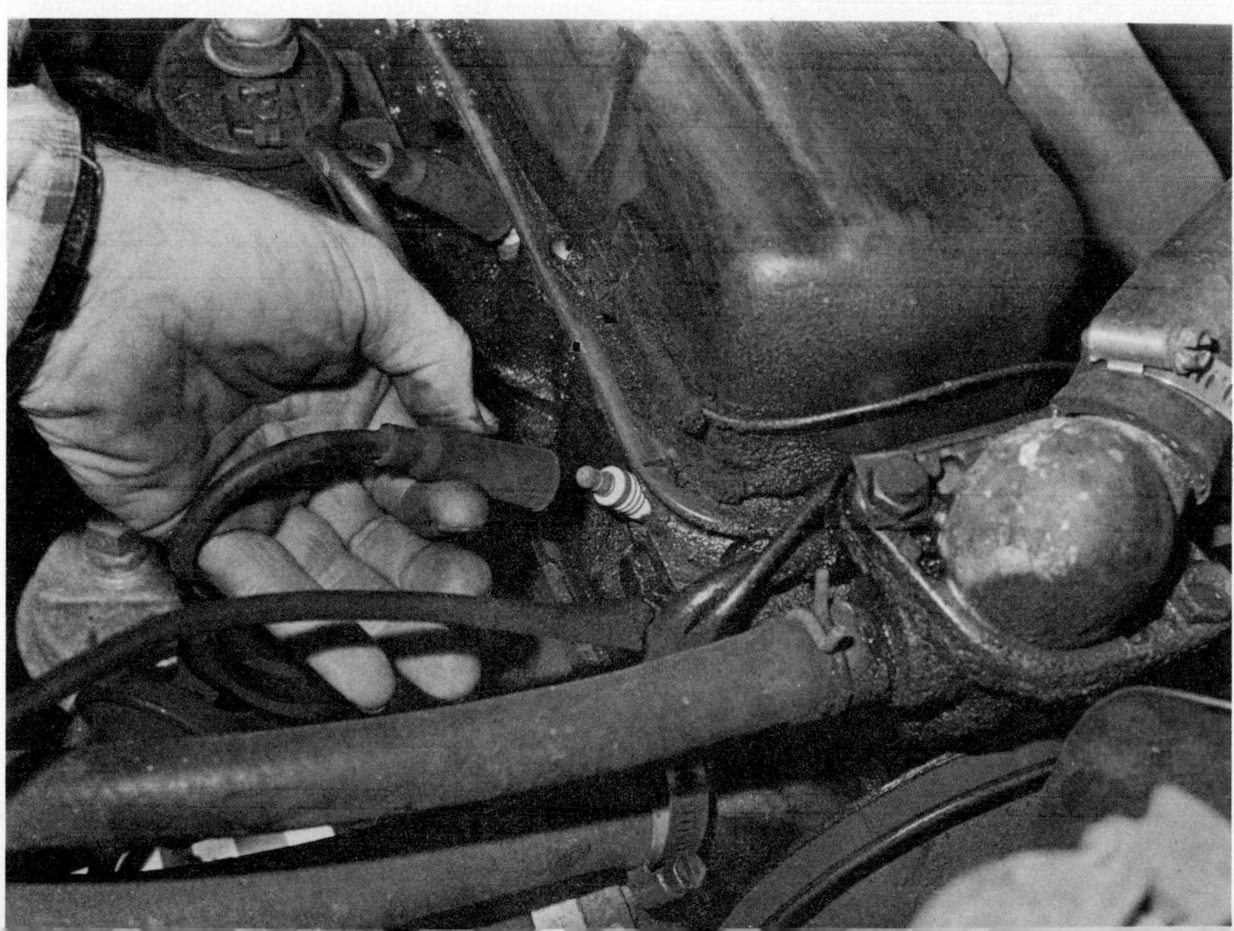

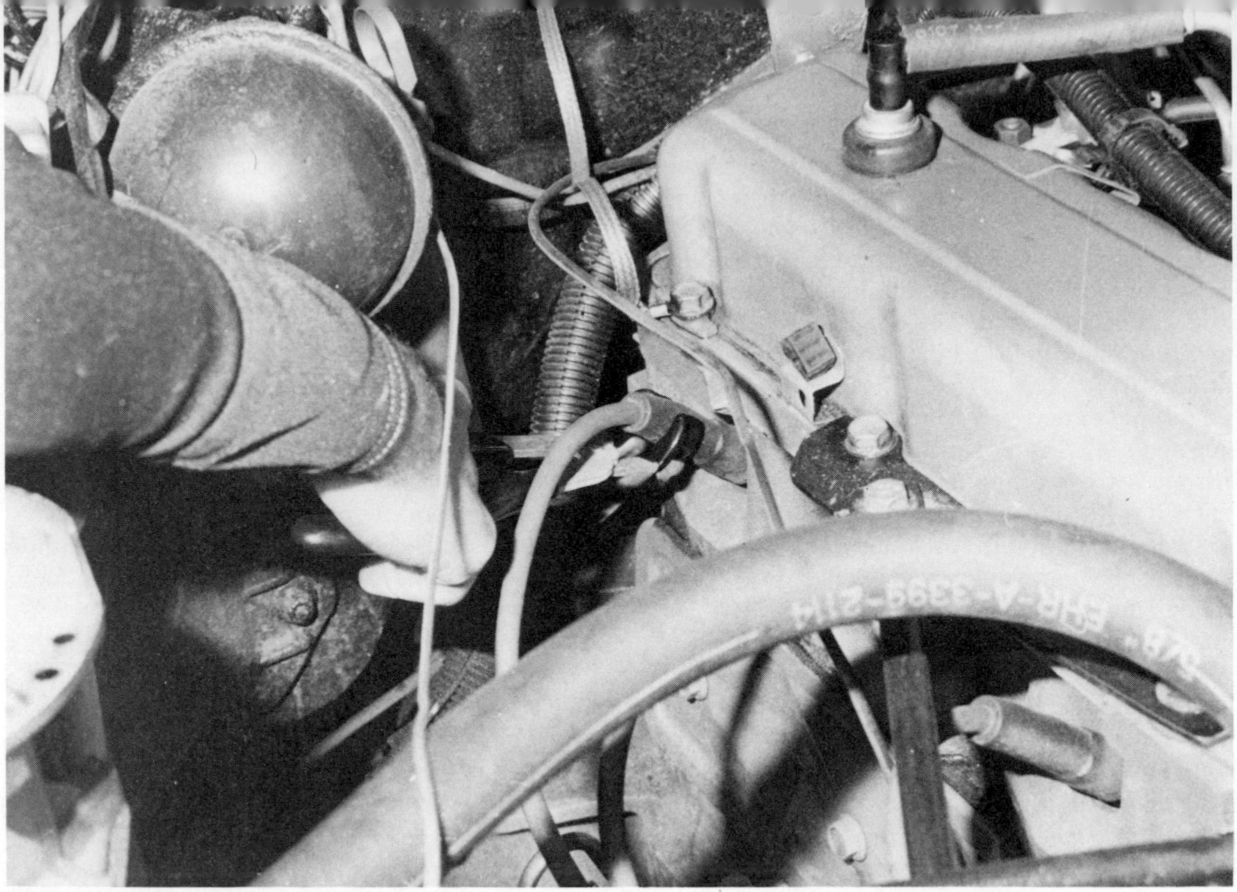

Fig. 2–74. If plug wire is in a tight area use a plug wire removal tool, such as these insulated pliers.

Fig. 2–75. With the plug wire off you can see the exposed hex in the middle of the plug, flush with the exterior surface of the engine's cylinder head.

pedal to the floor, and crank the engine while you hold the gauge in place. You must make simple disconnections to prevent the system from building up high voltage while it is being cranked. If there is a separate ignition coil, remove the two thin wires attached to the top of the coil. On General Motors High Energy Ignition, disconnect the pink wire in the distributer cap flange (see Fig. 2–42). If the pink wire is difficult to reach and there is a separate ignition coil (not built into the distributor cap as on most GM HEI systems), you may be able to reach and unplug all the wiring at the ignition coil.

As shown in Fig. 1–13, hold the gauge with the nipple in the spark plug hole. Push firmly, because the compression pressure will try to force the gauge nipple out. If you have a gauge with a long hose, just thread it into the spark plug hole and simply hold the gauge itself. This type is ideal for those spark plug holes in positions that make the other type impractical to hold.

Have the helper crank the engine for about three or four seconds, then remove the gauge and record the reading. Repeat this procedure for each cylinder. Reset the needle to zero after each reading.

Look at the readings. The lowest should be within 25 percent of the highest, and all readings should be at least 95–100 p.s.i. If the highest is 160 p.s.i. for example, the lowest should be at least 120 p.s.i. If any one reading is unusually low, repeat the test on that cylinder. A low reading in adjacent cylinders only is normally caused by a defective cylinder head gasket. In any case of low compression, have the engine repaired before you tune it.

If the compression readings are within the acceptable range, proceed to adjust and install new spark plugs.

Adjusting spark plugs. The gap between the center and side electrode of a spark plug (see Fig. 2–82) must be adjusted to factory specifications before installation, using a feeler gauge with a bending bracket.

First, pick the feeler gauge with the gap specified for your engine and insert the feeler

Fig. 2–76. If the plug is reasonably accessible, just the plug socket, an extension rod, and a ratchet will permit you to reach it.

Fig. 2–77. If access is a bit difficult, perhaps this spark plug ratchet will help. It has a long curved handle and universal joint in the head, so it can work in tight places.

Fig. 2–78. This spark plug was damaged by pre-ignition, which usually is indicated by a pinging sound from the engine. Some pinging is normal in a modern engine, but an excessive amount can damage the plug. Often this type of damage is caused by a plug with too high a heat range, so check the spark plug application chart carefully.

Fig. 2–79. This spark plug has been fouled by sooty carbon deposits. Because the spark plug is too low a heat range for the engine, the fuel system is maladjusted, or spark plug or plug wire is defective, the plug has been misfiring.

Fig. 2–80. This spark plug is covered with oil deposits. Check for worn piston rings or defective valve stem seals or guides, as explained in Chapter 6.

between the two electrodes. It should go between with light-to-moderate drag. If it does, the plug is ready for installation. If the gap is too large, tap down on the top of the side electrode with a wrench, light hammer, etc. If the gap is too small, position the bending bracket on the side electrode (see Fig. 2–83) and gently apply bending force to increase it. You may have to bend and tap a few times to get the gap just right.

Note: If your spark plugs come with threaded caps for the outer ends, thread them on and tighten with pliers. The plugs now are ready for installation.

Installing spark plugs. If the cylinder head (the part of the engine with the spark plug holes) is aluminum, apply a thin film of "anti-seize compound" to the threads of the new plug. This compound, sold in squeeze tubes, is available in auto supply stores. If you aren't sure if the cylinder head is aluminum, put a magnet on it: the magnet should not adhere if it's aluminum.

Note: If the old spark plugs were very slow to unthread (constant wrench effort was necessary, as contrasted with a loose plug just spinning out), obtain and use a spark plug hole *thread chaser*. This is an inexpensive tool that screws into the hole and cleans deposits from the threads.

Start each spark plug in by hand, so you can feel if you have engaged the threads of the plug and the hole properly. If you feel resistance, remove the plug and try again. On many engines, you will not be able to see the plug hole and, working purely by feel, you may have the plug at a slightly incorrect angle. Just keep experimenting and eventually the plug will thread in smoothly. Once you have it in as tight as you can with your fingers, you are ready for final tightening with a wrench.

If the plug hole is so inaccessible that you can't hold the plug with your fingers, make a plug-holding tool in either of two ways:

Obtain a short (about four inches long) piece of vacuum hose, a narrow diameter hose used to connect the engine (the source of vacuum) to various accessories. Push one end of the hose onto the shank of a Phillips screwdriver, the other end onto the exterior of the center electrode of the plug (see Fig. 2–84). The screwdriver serves as a handle and the hose as a flexible connector. Line up the plug with the hole and thread it in by turning the screwdriver.

If you have an old spark plug wire, push the terminal onto the plug and use the wire as a flexible tool, in basically the same manner as the screwdriver-and-vacuum-hose previously explained.

Tightening the spark plugs. Some spark plugs come with a metal gasket, others do not (see Fig. 2–85). Those that don't have a tapered section above the top line of the threads that seats the plug in the hole when the plug is tightened. The tightening requirement for gasketed and tapered-seat plugs is different.

For a gasketed plug, tighten it with a wrench one-quarter to one-half turn past the finger-tight point. If a fractional turn is hard for you to visualize, just think of the wrench handle as the hand of a clock. A quarter turn is three hours on a clock face, a half turn six hours.

With the tapered seat plug, just a very slight tightening will secure it, normally no more than a couple of hours on the imaginary clock face.

Note: These references to tightening assume that the plug has been threaded all the way down by hand (or with one of the tools you have). It is not uncommon for a plug to bind on a slight burr on the threads of the plug hole, particularly if you didn't clean them with a thread chaser. Therefore, if you encounter slight binding and must put the wrench to the plug before it is threaded all the way, you must "feel" the finger-tight position through the wrench. It should be quite obvious.

Inspecting and testing spark plug wires. Before you reconnect the plug wires, you should inspect them carefully. If their exterior jackets are cut, oil-soaked, or heat-hardened and somewhat brittle, you should replace them.

You should also check the resistance of the plug wires to make sure it is within specification. If it is, the plug wires will make a valuable contribution to radio suppression without affecting engine performance. If the resistance is excessive, the plugs may not fire consistently.

The plug wires can be tested in either of two ways:

• *Ohmmeter*. With an ohmmeter set on the high scale (or with a multi-purpose meter that includes an ohms scale for plug wires), just disconnect a plug wire from the distributor cap and connect one ohmmeter lead wire to the metal terminal on each end of the plug wire (see Fig. 2–86). The reading should not exceed 15,000

Fig. 2–81. This plug is just worn out from excessively long use. Notice that center electrode is worn down from erosion.

Fig. 2–82. Checking gap between center and side electrode, using a wire-type feeler gauge. Gauge of specified thickness should move between electrodes with light-to-moderate drag.

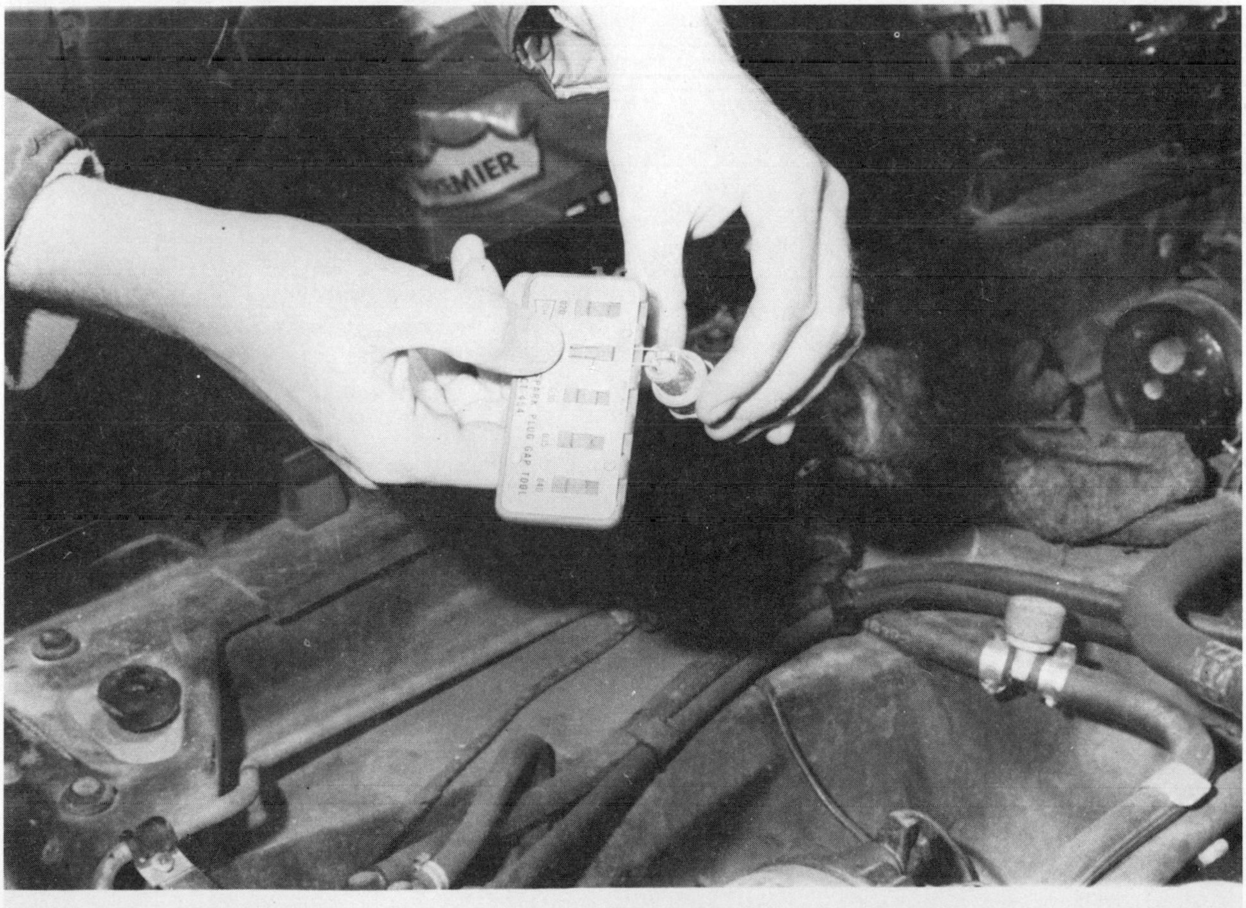

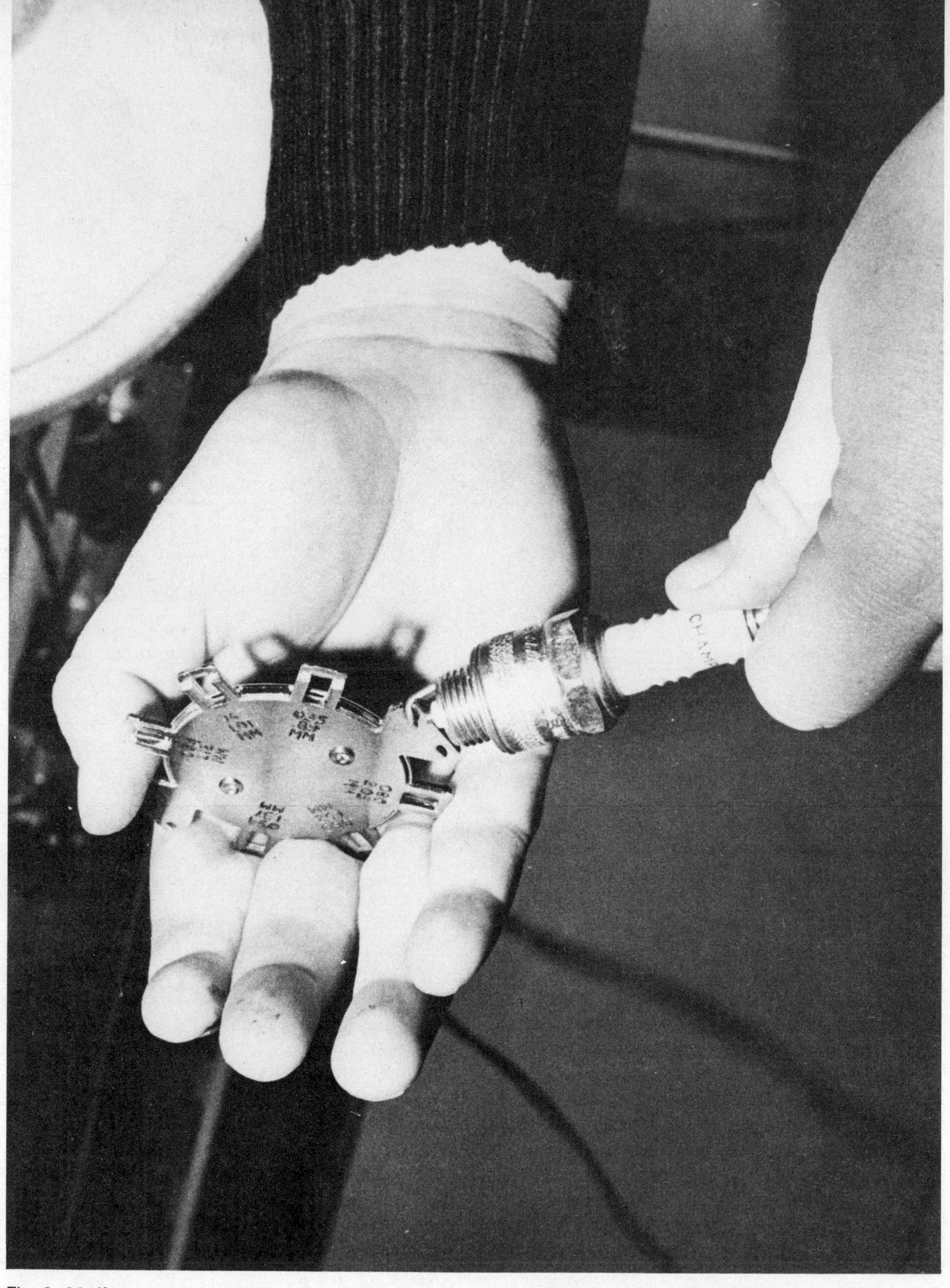

Fig. 2–83. If gap must be increased, use bending bracket found on all gauges. Only bend on side electrode.

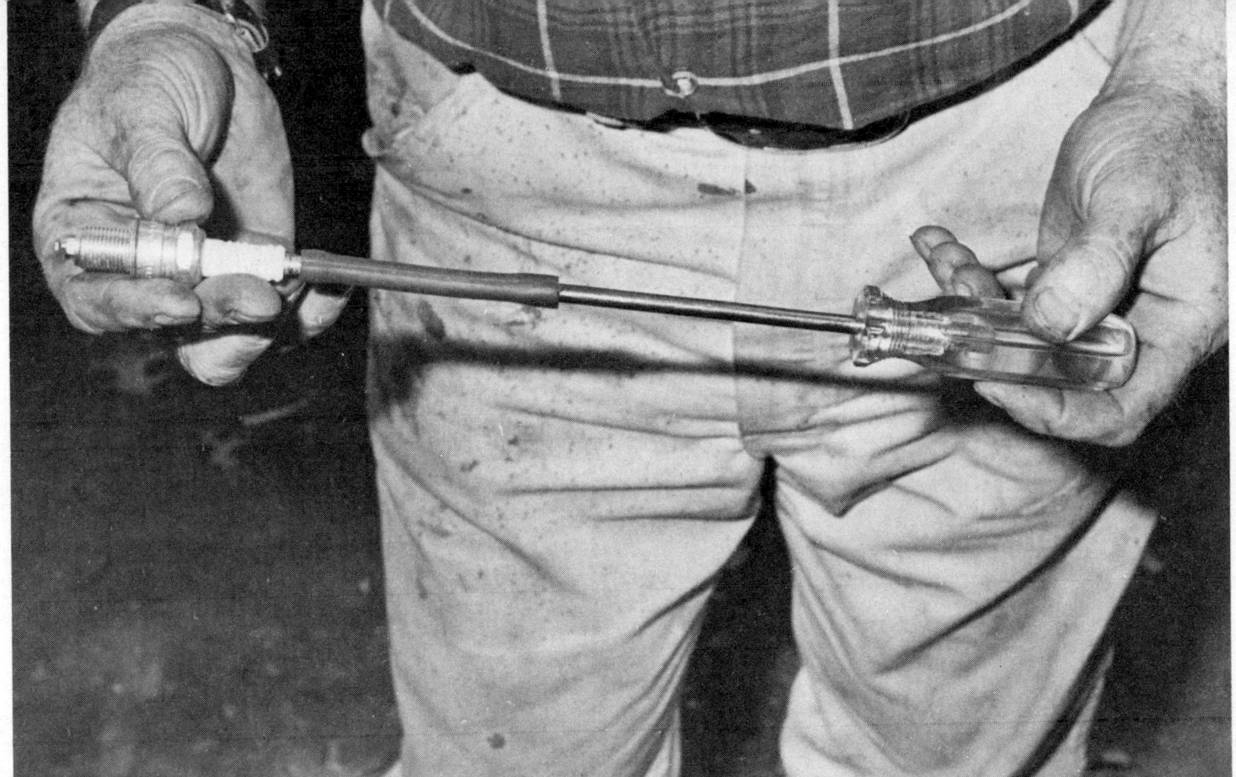

Fig. 2–84. If you have trouble starting a spark plug by hand because of limited clearance, a short piece of hose on a Phillips screwdriver can be used as a spark plug starter. Push open hose end onto spark plug as shown, then use screwdriver as a thread-in tool for the spark plug.

Fig. 2–85. A look at two spark plugs. The upper plug has a metal gasket at the end of the threads, a thread on cap for the outer end of the center electrode, and a 13/16-inch hex in the middle. The lower plug has a tapered surface at the end of the threads, so the plug wedges into place and seals without a gasket. The cap is a formed part of the center electrode, and the hex is 5/8-inch.

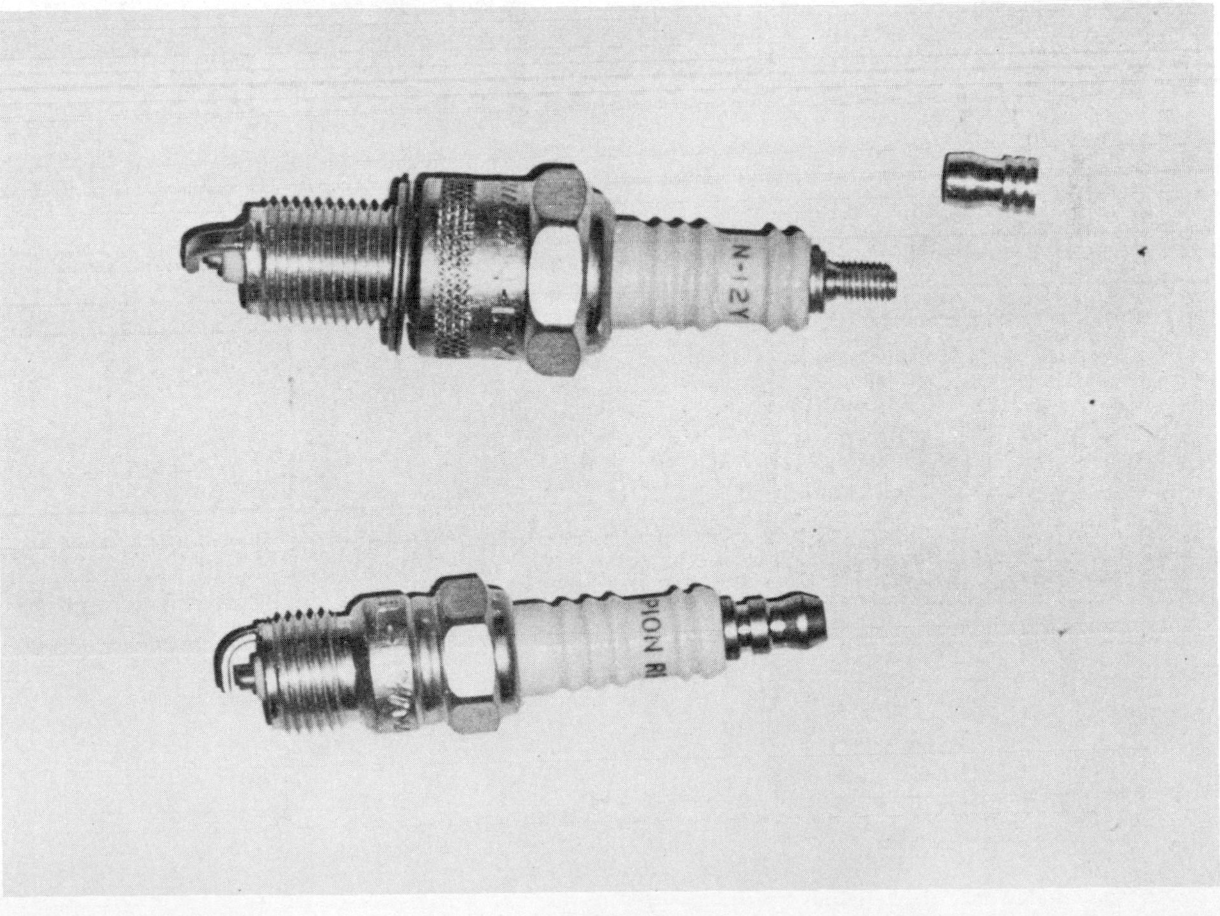

ohms per foot of length of the plug wire (for example, 22,500 ohms for a wire that is 1½ feet long). If it does, the wire must be replaced. If the reading is infinity, there is a break in the wire (if your engine is misfiring, this could be the cause). Note: On some electronic ignition systems, the acceptable resistance can run as high as 60,000 ohms per foot. Check specifications if resistance is high.

• *Voltmeter.* If you have, or can obtain, a voltmeter (a test instrument that measures voltage) you can check the wire this way: Connect the voltmeter's negative lead to the battery's negative post (the one to which the ground cable is attached) and the voltmeter's positive lead to one end of the plug wire. Touch the other end of the plug wire to the battery's positive post and you should get a reading of six volts or more. If the reading is less, the plug wire has excessive resistance. If you get a zero reading, there is a break in the plug wire.

Note: This test procedure may require disconnecting the plug wire from some of its retaining guides, so one end can reach the battery. In addition, this procedure should not be used on wires from electronic ignition systems.

Buying Replacement Plug Wires

Replacement spark plug wires are sold in various ways.

• *Rolls of wire.* These you cut to size and attach the terminals and nipples. This is the cheapest way, but even a professional using special tools may have difficulty making good connections for the terminals on resistance wire.

• *Semi-custom sets.* The plug wires are precut and a terminal (and perhaps a nipple) is factory-installed at the spark plug end. You shorten the wire if necessary by cutting with pliers and attach a terminal to the distributor cap end. If you're going to attach a terminal, this is the less-critical end for the connection will be

Fig. 2–86. Testing spark plug wire with leads from an ohmmeter. This spark plug wire has been removed completely from the engine. In most cases, however, you can simply disconnect at both ends and test by bringing the ohmmeter leads over.

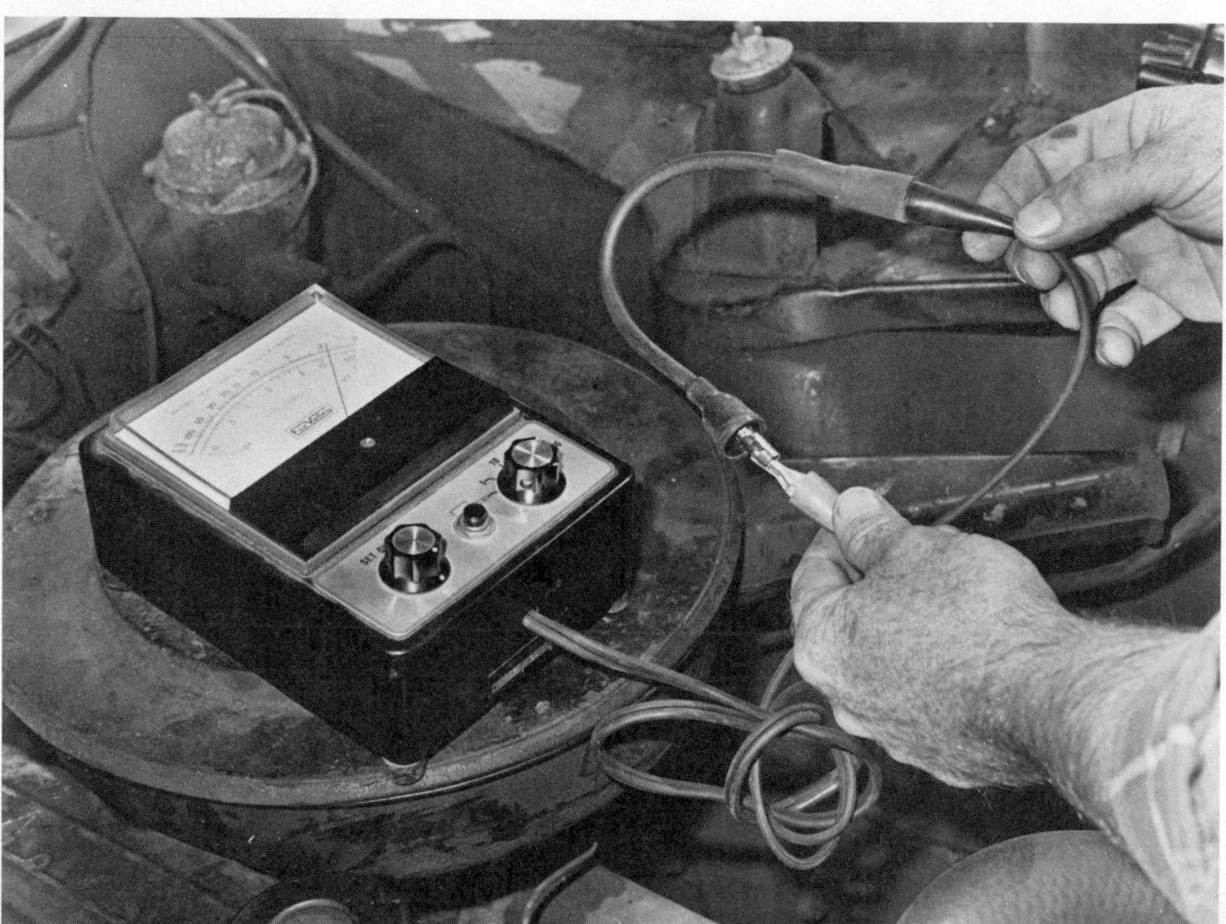

disturbed less frequently, and it is a type that can be more easily inspected.

• *Custom sets* (see Fig. 2–87). The plug wires pre-cut to length and the nipples and terminals factory-installed. This is the most expensive type, but it's worth it. Shop around for the best price.

Types of Wires

The late-model cars all use resistance wire as original equipment, and so should you for replacement purposes. It's true you can buy metal core wire in rolls, plus the terminals and nipples you need, and suffer little risk in assembly because of the metal-to-metal connections. However, the entire ignition system is based on the use of resistance spark plug wires (and in many cases resistance spark plugs as well), so experimentation is not recommended.

If you'd like something less fragile than ordinary resistance wire, there are two choices: 1) wire made with a core of aramid, a synthetic fiber also used in tires. In some respects it's stronger than steel; 2) an internally-shielded wire with a metal core. This type does contain a resistor, but the reading you should get is only 400–700 ohms per foot. Keep that in mind when you test.

The type of insulation jacket is your other consideration, and there are three basic kinds:

Standard. This is an inexpensive synthetic rubber, such as neoprene. It will last only in a much older car. Or if you plan to dispose of your car within a year, it may hold up on a later model. Just be sure you get it with silicone rubber nipples at the spark plug end, for maximum heat resistance where it's hottest.

Premium. This is usually hypalon, which was used on original equipment through to about 1974 on most cars. It is really the minimum you should consider, particularly if the car has electronic ignition. If this is what you get, be sure it has silicone rubber nipples at the spark plug end.

Super-premium. This is silicone rubber, designed for today's combination of high engine compartment temperatures and the high voltages of most electronic ignition systems. Both the wire jacket and the nipples are made of this material. Note: On some cars the conditions are

Fig. 2–87. If you need new plug wires, the completely custom set shown is the best choice. Nipples and terminals are already in place and the length of the plug wires is very close to the originals.

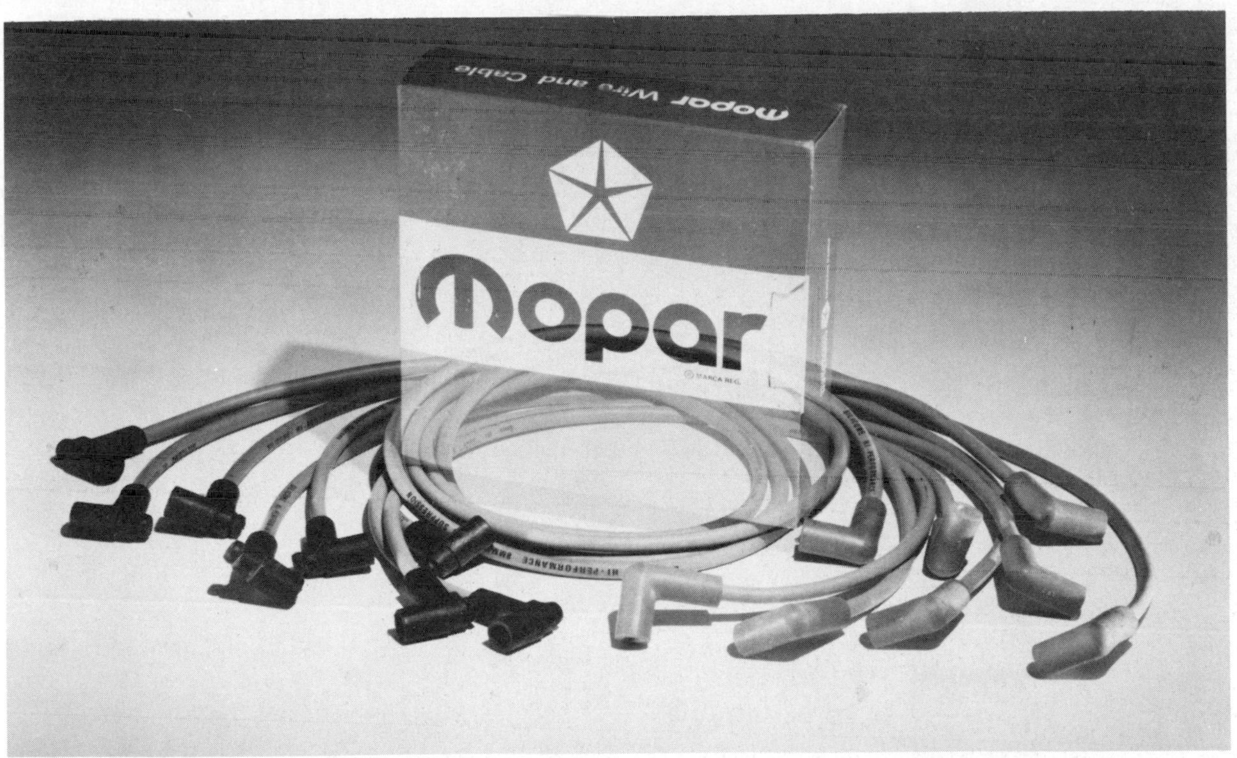

so severe that the manufacturer installs an extra-thick silicone rubber wire (8 mm instead of 7 mm), and if this is what the car maker uses, so should you.

Silicone rubber insulation wires are very expensive, but so long as they aren't cut they will really last, resisting not only high temperatures and voltages, but oil, grease, and road film deterioration. If you find just one bad wire, replace only that (see Fig. 2–88). Unlike other less-expensive types, silicone rubber wires can be purchased individually.

Changing Spark Plug Wires

If you must replace spark plug wires, do the job one wire at a time, to prevent an incorrect connection. Always duplicate the routing of the original wire, even if it seems somewhat unusual (see Fig. 2–89). The car makers attach the wires to little plastic holders and run the wiring so that the ignition system functions properly. Any deviation could cause the engine to miss.

Connecting the wires. The spark plug wire must make a good connection at each end. If you have disconnected it from the distributor cap, push it into or onto the cap terminal until it seats, then push down on the nipple.

At the spark plug end, first test the fit of the terminal by pushing it onto a used plug. If you feel any looseness, the terminal must be tightened. On a straight terminal, just push back the nipple carefully and squeeze the terminal together with pliers (see Fig. 2–90). On a right-angle terminal, the nipple can't be pushed back, so wrap the pliers jaws with tape, then squeeze on the nipple (the tape will protect the nipple). Once you have the terminal tight enough, you can install it on the plug.

Note: It's a good idea to first spray the inside of the nipple with silicone lubricant (see Fig. 2–91) or coat it lightly with silicone dielectric grease (a special grease sold in tubes in auto parts stores). The lubrication will not only help installation, but will keep the nipple from heat-sealing to the spark plug.

Fig. 2–88. The spark plug wires must be disconnected from both the cap and the plastic cage on General Motors High Energy Ignition distributors. This particular system does not have the coil built into the cap, but mounted remotely, slightly to the left (it's the rectangularly-shaped part that looks rather different from other ignition coils). Caution: If you also replace a distributor cap, be sure to align the new cap with the old (using the locking rods or spring clip bosses for reference) before you begin to transfer wires. When transferring wires, move them one at a time into a cap terminal that is exactly the same position as on the old cap.

Fig. 2–89. Notice the unusual routing of these spark plug wires. It was done to prevent a phenomenon called crossfire that results in engine missing.

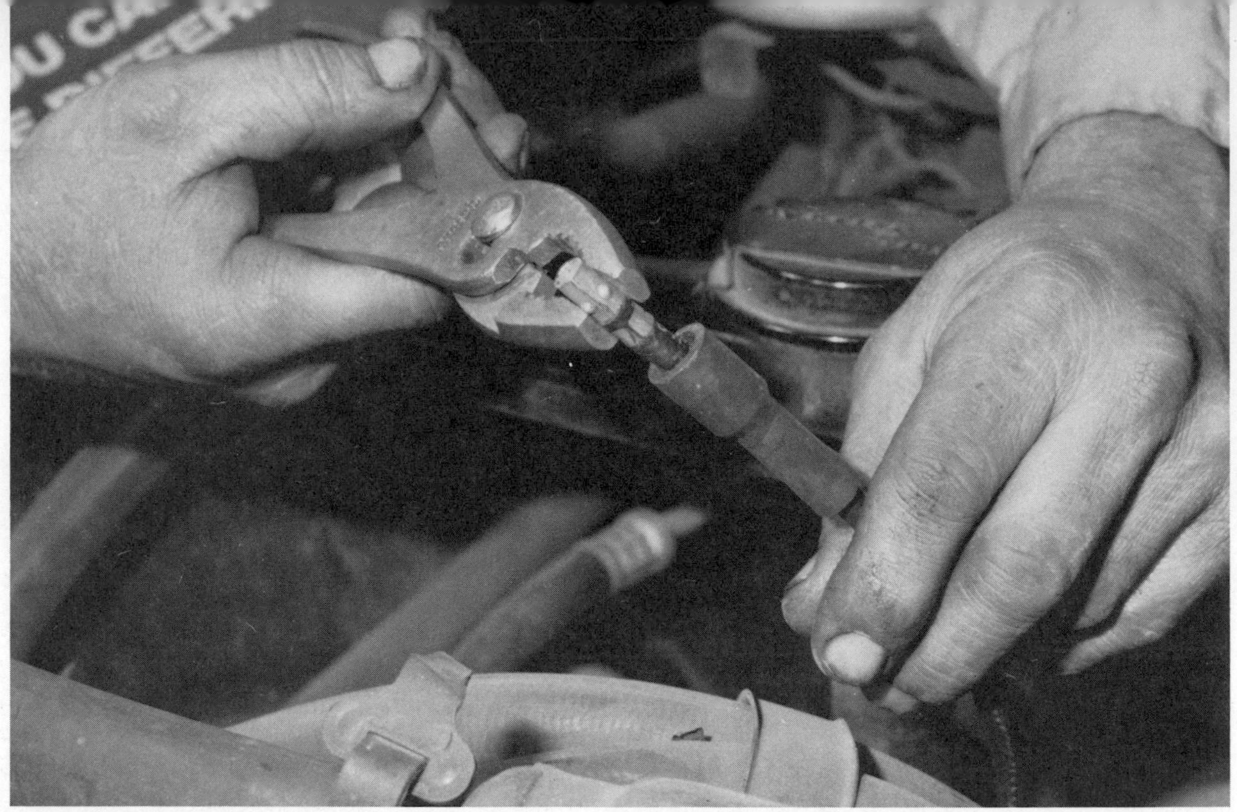

Fig. 2–90. If a plug wire terminal is a loose fit on the plug center electrode, push back nipple as shown and tighten terminal with pliers. If terminal is a right-angle type, cover pliers jaws with tape and squeeze carefully on the nipple, since you can't push it back.

Fig. 2–91. Spray silicone lube (or dab silicone grease) on inside of nipple at spark plug end before installation.

3 FUEL SYSTEMS FROM CARBURETORS TO TURBOCHARGERS

How the Carburetor Works

THE CARBURETOR (see Fig. 3–1) is the device that combines air and gasoline to produce a combustible mixture. The ideal air-fuel mixture depends on a number of factors, including engine temperature, so the carburetor must produce the correct mixture automatically. When the engine is cold, it requires what is called a rich mixture, one with more gasoline in it; as the engine warms up, a leaner mixture is desirable. Rich and lean are relative terms, for the mixture is always mostly air. Rich might be 10 parts of air to one of gasoline; lean would be 14 to 18 parts of air to one of gasoline.

The perfume atomizer, a gadget with which you may be familiar, illustrates one of the basics of carburetor operation. Squeeze the bulb and you push air quickly past the top of a tube that extends down into the bowl of perfume. The air rushing past the top of the tube creates a vacuum there that sucks up perfume droplets and expels them with the air as a spray.

In an engine, the movement of the pistons creates a vacuum in the engine. Nature abhors a vacuum, so air rushes in to fill it, through the only opening—a cylindrical barrel called the carburetor barrel (or air horn). As the air rushes in, it speeds past the top of a tube that extends from a bowl filled with gasoline. The vacuum created by the air rushing through draws out fuel droplets which mix with the air. The "spray" flows into a series of chambers—called the intake manifold—and through the manifold into the cylinders.

Now let's look at the controls the carburetor has which enable it to get the right amount of fuel, mix it correctly, and control the flow rate of the air-fuel mixture.

Venturi

If you look into a carburetor barrel you'll see it necks down at about the half-way mark. This restriction is called the venturi, and some carburetor barrels have multiple restrictions at the same point. The venturi restriction forces the air to speed up to get through to fill the vacuum in the cylinders. The higher speed creates a higher vacuum at the fuel bowl tube (called the main metering jet) so more fuel can flow. Don't confuse the vacuum in the cylinders and the little vacuum at the tip of the main metering jet. They're two different things. The air rushing in to fill the vacuum in the cylinders is being used to create a second, mini-vacuum at the tip of the main metering jet.

Throttle

There must be some way to regulate the flow of air-fuel mixture or the engine would, once started, run at top speed. The throttle, a circular hinged plate that blocks off the bottom of the carburetor barrel, does that job. The throttle is connected by a linkage to the gas pedal, so when you step on the pedal what you are doing is pivoting the throttle. The more the throttle plate pivots, the more air-fuel mixture that can rush through the barrel, speeding up the engine.

Idle Circuit

When the engine first starts, or if you take your foot off the gas pedal, there must be some way to keep it running. This is the job of the idle circuit, a passage with an opening above the throttle (a source of air) and an opening below the throttle (to permit it to bypass the throttle). The passage goes by the fuel bowl metering jet, and air flowing through picks up fuel droplets, with

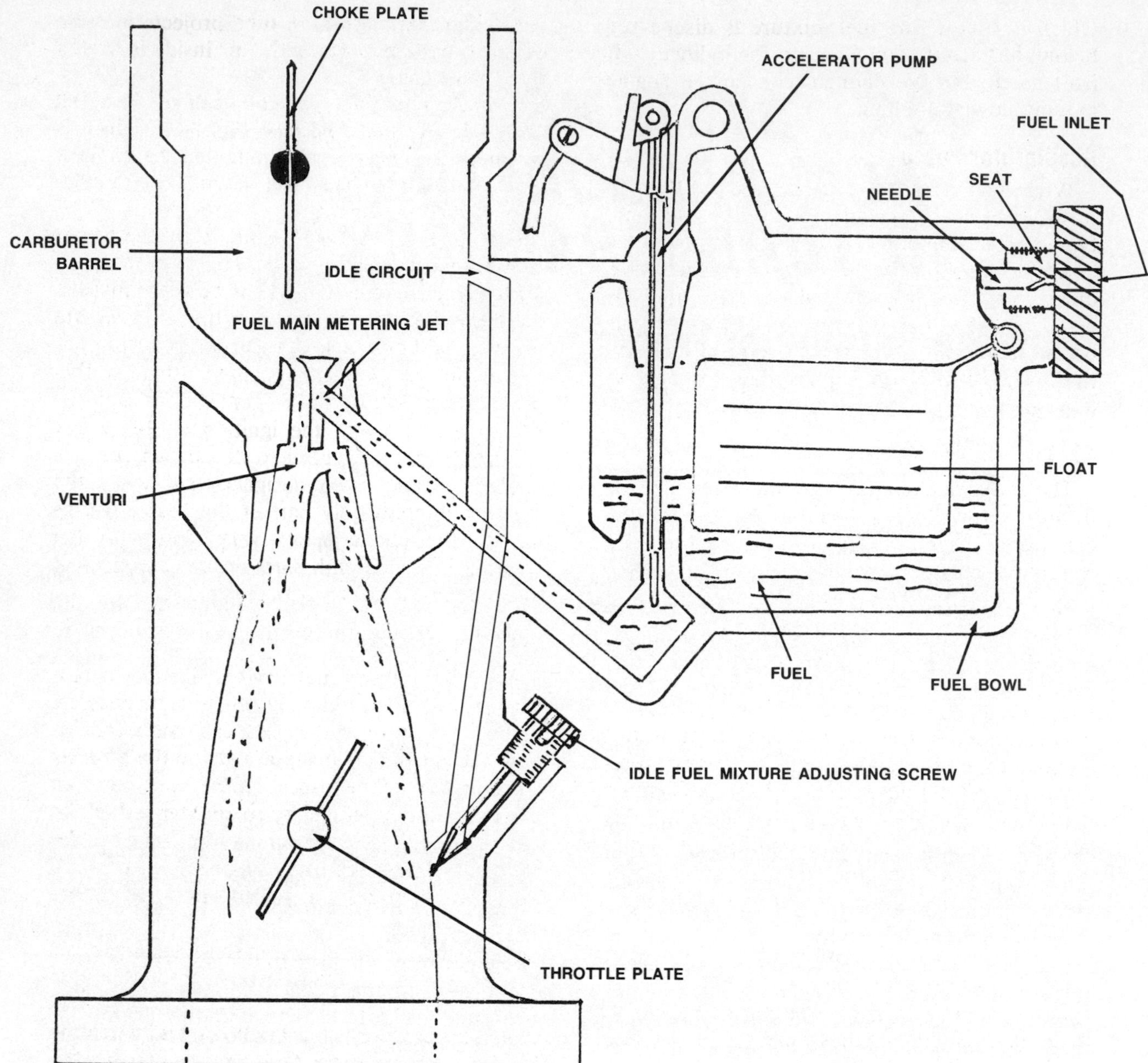

Fig. 3–1. This is a drawing of a simple one-barrel carburetor. Although no modern automobile has such a carburetor, it shows the basic principles of carburetor operation. Fuel is pumped into the fuel bowl and, when the level is correct, the float rises and pushes a tapered needle into a seat, shutting off the flow. As fuel is consumed the float drops and fuel under pressure pushes the needle off the seat, allowing the additional fuel to flow in. The fuel bowl contains a metering jet, a nozzle that projects into a tapered section of the carburetor barrel. When you step on the gas pedal, you pivot the throttle plate from horizontal toward a vertical position, allowing the engine to suck in air. The air creates a mini-vacuum at the end of the jet, drawing out droplets of fuel that mix with the air to form a combustible mixture that flows into the cylinders. To provide some fuel mixture when the throttle plate is closed, the idle circuit is used. It's a passage that runs from the top of the barrel to a point below the throttle. Even when your foot is off the gas pedal, the throttle isn't completely closed; it's open just enough to allow some air to flow past it. This air combines with the air-fuel mix that comes through the idle circuit to form a mixture that is both combustible and sufficient to keep the engine idling. Note the idle fuel mixture adjusting screw. Although preset and sealed on current models, on older cars the idle circuit permits adjustment of the idle fuel mixture for smooth idle. When you floor the gas pedal to accelerate, additional fuel is needed, beyond what the main metering jet will supply. At that time the accelerator pump is operated, permitting additional fuel flow.

which it mixes. The fuel mixture is discharged below the throttle and flows to the cylinders. It isn't much, but it's enough to keep the engine moving slowly, or idling.

Accelerator Pump

When you really floor the gas pedal, the fuel flow can't match the sudden need. To add extra fuel to the suddenly greater air flow, an accelerator pump is used. This is a mechanical device operated by the linkage from the gas pedal, and it squirts an extra bit of fuel into the barrel. The pump system will only work when the gas pedal is floored quickly, not when it is done gradually.

Choke

The choke is a hinged plate that resembles the throttle. However, it sits on the top of the carburetor barrel, well above the venturi, and can only regulate air volume into the barrel (see Figs. 3–1, 3–2). When it nearly covers the top of the barrel, there is little air flow and the overall air-fuel mixture is richer with fuel (perhaps a 10:1 air-fuel ratio). When it pivots to the vertical position (open), maximum air flow is allowed.

The choke plate position is controlled by two things: engine vacuum and a temperature-sensitive coil spring (see Fig. 3–3). Remember that the pistons are creating a vacuum and this vacuum exists in the intake manifold and up to the bottom of the throttle plate. Even when the throttle is opened, unless it's opened all the way there is some vacuum around.

A passage from the bottom of the carburetor goes up to either of two devices: a little piston in a little cylinder in the carburetor (see Fig. 3–4), or an external diaphragm connected to the vacuum passage by a hose (see Fig. 3–5). The piston or the diaphragm is connected by linkage to the choke plate, and because there's vacuum on the piston or diaphragm it's always trying to pull the choke plate open. The temperature-sensitive coil spring, however, keeps tightly wound up and holds the choke plate closed when the engine is cold. As the engine warms up, the coil unwinds and the vacuum piston or diaphragm pulls the choke plate open. Some carburetors have two choke diaphragms, designed to open the choke at different rates, for more precise operation.

Heat for the thermostatic coil typically comes from one of these:

- *Exhaust manifold.* A tube projects into the exhaust manifold, where the air inside is heated by exhaust gas.
- *Intake manifold.* The coil itself sits in a well on the intake manifold (see Fig. 3–6), which is warmed up by physically touching the exhaust manifold, or by exhaust gas which is run through a special passage in it.
- *An electric heater.* In some cars, the heater is a ceramic device warmed up by current tapped off the ignition circuit. Or it can be wired through both the ignition circuit and the oil pressure warning light circuit so that current is supplied to the choke only when the engine is running. This prevents premature choke opening if the motorist sits in the car with the ignition on for a few minutes before attempting to start the engine.

The ceramic is so designed that when the weather is cold, only part of the heater warms up, so the choke opens very slowly. If the weather is warm, both parts of the ceramic heat up, and the choke opens more rapidly.

Fuel Bowl

The carburetor's fuel bowl is filled as necessary by the fuel pump, normally a mechanical device bolted to the engine and connected to both the gas tank (for supply) and to the bowl by steel and/or rubber lines. Some cars have an electric pump, operated by current tapped off the ignition circuit. This pump may be placed in the gas tank or mounted in the engine compartment. Most fuel pumps are sealed, non-serviceable units.

The pump (covered later in this chapter) forces the fuel into the carburetor bowl through a little valve called the needle and seat. A toilet-bowl-type float is mounted in the bowl, and when the fuel level is correct the float rises and pushes the needle into the seat, shutting off the fuel flow. As the engine uses fuel the float drops and more gasoline is admitted.

The level of the fuel is important, for if too low, the engine will starve for gasoline; if too high, it will pour out the main metering jet and flood into the engine. Either condition can cause the engine to stall.

Other Controls

The modern carburetor has other control devices for precise operation. It is unlikely that you would ever service them, because that's work for

Fig. 3–2. Finger points to the choke plate on the top of the carburetor. When the engine is cold, this plate is closed as shown. When the engine starts, it cracks open a bit to allow some air flow; as the engine warms up, it pivots to the vertical position so air can flow through the barrel freely. The closed choke plate, by restricting air flow, richens the fuel mixture as is necessary for cold starting and cold engine driveaway.

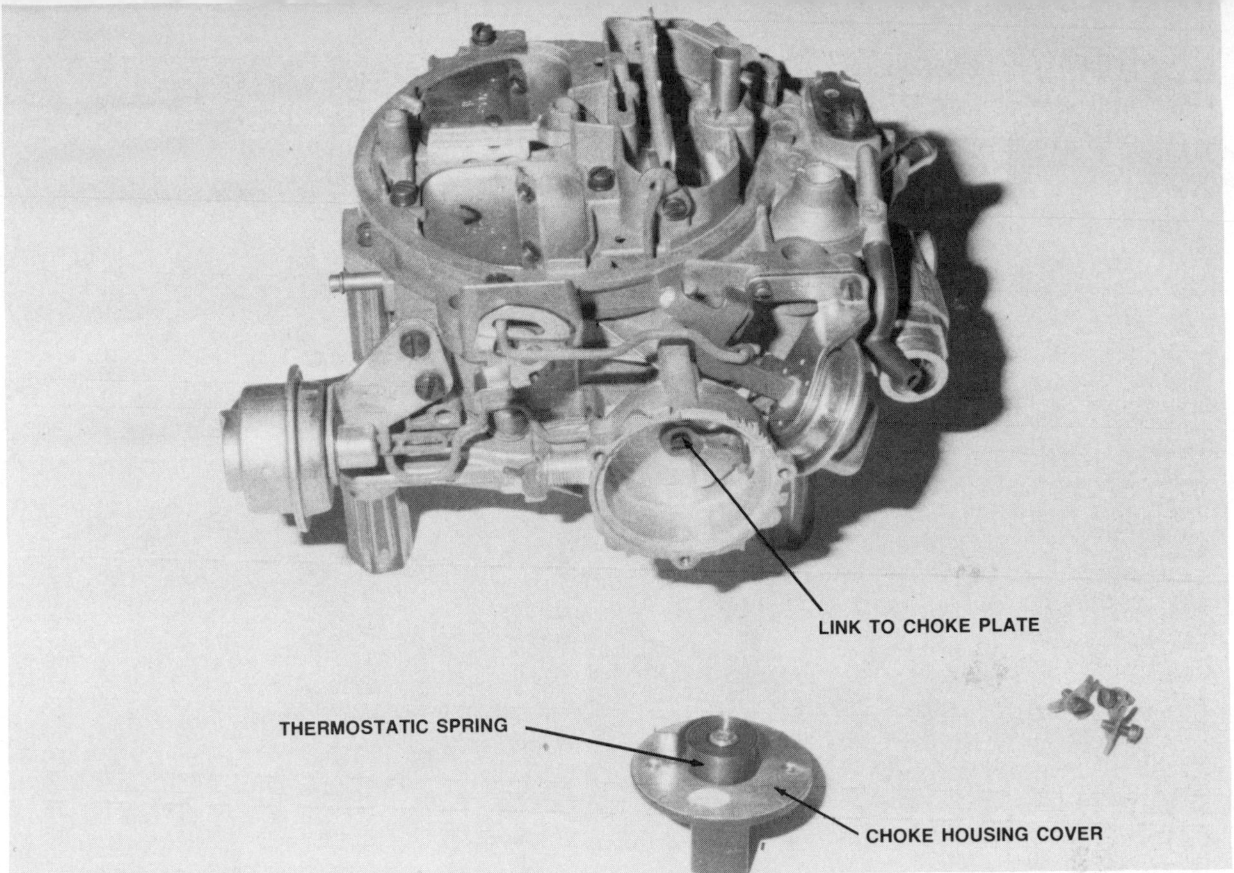

Fig. 3–3. The choke housing cover has been removed so you can see the thermostatic spring. When cold, the spring winds up tight and holds the choke plate closed by its link connection to the plate.

the carburetor rebuilder. If the carburetor does not respond to the service described later in this chapter, obtaining a rebuilt is the practical solution.

Variable Venturi Carburetors

The venturi in the typical carburetor is a fixed restriction. If we could change the size of that restriction in accordance with throttle opening, the air-fuel mixture could be kept more uniform and we wouldn't need such devices as the accelerator pump.

Some carburetors, called variable venturi designs, do exactly that. One wall of the cylinder barrel is movable, and to that wall is attached a tapered needle that fits into the main metering jet. As the wall is moved to enlarge the venturi it pulls the tapered needle out of the jet, allowing more fuel to be drawn out by the inrushing air. The shape of the needle is tailored to allow just the desirable amount of additional fuel flow (see Figs. 3–8, 3–9).

When you step on the gas pedal with this type of carburetor you pivot the throttle plate as in a conventional carburetor, but this action also allows engine vacuum to pass into a chamber with a diaphragm. The diaphragm is linked to the movable barrel wall, and it draws it open, increasing the size of the venturi.

Gasoline Filters

All carburetor systems have filters to remove dirt particles from the gasoline. The main filter, which is replaceable, is located in any one of three places: at the carburetor fuel bowl inlet, where the line from the pump is attached; spliced into the line between pump and carburetor bowl; or mounted on the fuel pump.

Replacement of this filter is a standard part of a tuneup. Some filters, primarily those at the fuel bowl inlet, are designed so that if they clog, fuel can bypass them. Others are not so designed, and if they fill with dirt gasoline cannot get through and the engine will starve and stall.

Electronically-Controlled Carburetors

Electronically-controlled carburetors are a product that results from strict emission regula-

Fig. 3–4. Finger points to little cylinder that contains a tiny piston connected by linkage to the choke plate. As the thermostatic coil heats and unwinds, the piston, which is open at the bottom to a source of engine vacuum, is sucked downward. The downward movement of the piston tends to tug the choke plate open.

tions. To meet them, the car makers must use an additional catalytic converter, which functions properly only when the air-fuel mixture going into the engine is very close to 14.7:1. To maintain this mixture under all operating conditions, a computer or electronic control unit is used to richen or lean out the fuel mixture as required. The computer is often a multi-function control, the same one used to regulate spark advance on some cars. See Computer-Controlled Spark Advance (*Chapter 2;* see also Fig. 3–10).

Key to the system is an oxygen sensor, an electronic device which checks for oxygen content in the exhaust, an accurate indicator of the air-fuel mixture that went into the cylinders. If the oxygen sensor indicates a rich mixture, it signals the computer to lean it out. If it indicates a lean mixture, it signals the computer to richen it. The electronically-controlled carburetor may also be triggered according to information from sensors that measure coolant temperature, engine speed, and throttle position.

The computer gets the carburetor to respond in one of many ways, of which the following are the most common.

- *A solenoid (electromagnetic switch) operates a vacuum valve* (see Fig. 3–10). The computer supplies or denies current to the solenoid so as to allow only the desired amount of engine vacuum to pass through this valve and through a hose to the carburetor. At the carburetor, the vacuum draws on a spring-loaded diaphragm connected to a fuel metering rod. If it pulls on the rod, it yanks it partly out of a fuel passage, and more fuel can flow through the passage to richen the mixture. When the computer wants to lean out the mixture, it denies current to the solenoid valve, no vacuum goes through to the diaphragm, and the spring pushes the metering rod back in, reducing the fuel flow through the passage.
- *The computer supplies current to a stepper motor* (see Fig. 3–9). The stepper motor is a little motor that can run forward or backward and stop in any position desired. The stepper motor operates a little rod in a carburetor vacuum passage up to the fuel bowl. When the computer wants to

Fig. 3–5. Many chokes have an external vacuum diaphragm connected by links to the choke plate instead of the little piston inside. An external vacuum diaphragm is shown. As soon as the engine starts, the diaphragm (connected by a hose to an engine vacuum source) tries to pull the choke open. As the thermostatic coil unwinds, it performs the same function as the little piston in Fig. 3–4. Also note the heat tube. It is run to a source of exhaust gas heat, such as the exhaust crossover passage in the intake manifold of a V-8, or to the exhaust manifold itself on an in-line engine. As the exhaust gas heats the air inside the tube, the air transfers the heat to the thermostatic coil inside the choke housing.

lean out the mixture it winds out the motor so the rod allows full vacuum to go to the fuel bowl and suck out most of the air. This lowers the air pressure below atmospheric, and with less air pressure on top of the fuel in the bowl, less can flow out in response to the mini-vacuum in the carburetor venturi. To richen the mixture the motor pushes the rod to close off the vacuum passage. Atmospheric pressure is instantly restored, and fuel flow increases.

Carburetor Service

The carburetor normally requires little attention, and a basic rule is: With the exception of certain external maintenance and minor adjustment, leave it alone. The exceptions are: cleaning, replacement of fuel and air filters, minor adjustment, and tightening. For those experienced weekend mechanics who wish to go further, this chapter covers replacement of the needle-and-seat valve and adjustment of float level, a job that frequently can be done without removing the carburetor from the car.

Cleaning

You could pour a can of gasoline additive into your tank periodically and hope that will keep the carburetor clean. However, it won't do the job that an at-the-carburetor cleaning will.

To a professional, at-the-carburetor cleaning means complete disassembly and a soaking in solvent. You can, however, get very close to the same results with this two-prong approach, performed every two years or 30,000 miles:

(1) Spray the carburetor barrel and the exterior of the carburetor, particularly the linkage, with an aerosol solvent (see Fig. 3–11). When spraying the barrel, block the choke plate open, perhaps with a rubber band on the linkage to hold it.

Fig. 3–6. This is a well-type choke. The thermostatic coil sits in a well in the exterior of the intake manifold, and is connected by a link to the choke plate. As the intake manifold warms up, it warms up the coil. Note that this choke system has a vacuum diaphragm.

(2) Run pure solvent through the internal passages of the carburetor with an on-the-car cleanout kit (see Fig. 3–12). Disconnect the fuel line at the carburetor and plug it with a tapered stopper, then attach a suitable fitting and a hose from the kit. Attach the can of solvent to the other end of the hose, hold the can upright, and run the engine at idle until the contents of the can are exhausted. If the carburetor has not been cleaned in four years or more, use two pint cans of solvent. Note: If the engine will not run (stalls out) on pure solvent, mix it 50–50 with gasoline.

Note: Either way this job will require removal of the air cleaner housing, which on the typical late-model car has several vacuum hoses connected to it. When you disconnect a hose, wrap a piece of masking tape on its end and another piece on the neck from which you disconnected it. Write an identification mark (such as the same letter) with a marking pen on each tape so you can reconnect accurately. Or you can obtain a set (six pair) of different–colored vacuum hose and neck plugs, sold in auto parts stores.

Fuel Filters

You will probably find one of the three types of replaceable fuel filters on your gasoline engine:

• *A little filter in the fitting that holds the fuel line to the carburetor bowl.* This filter, made of pleated paper or powdered bronze, should be replaced annually, or every 12,000 miles, as follows:

(1) Disconnect the fuel line (see Fig. 3–13).

(2) Loosen the fitting with a large wrench (an adjustable type normally will do).

(3) Unthread the fitting by hand and carefully remove it. The filter element will pop out (see Fig. 3–14). If a spring also pops out, note which way the spring faces. If the filter is powdered bronze, discard and obtain a pleated paper replacement for better filtration.

(4) Install a new filter element and spring, and thread in the fitting finger tight. Give it a final tightening with the wrench.

(5) Reinstall the fuel line and tighten.

• *An in-line filter, in the line between fuel pump and carburetor bowl.* See Fig. 3–15. It is a

Fig. 3–7. On this choke, electricity is used to warm the thermostatic coil. Note the electrical connector to the choke housing cover.

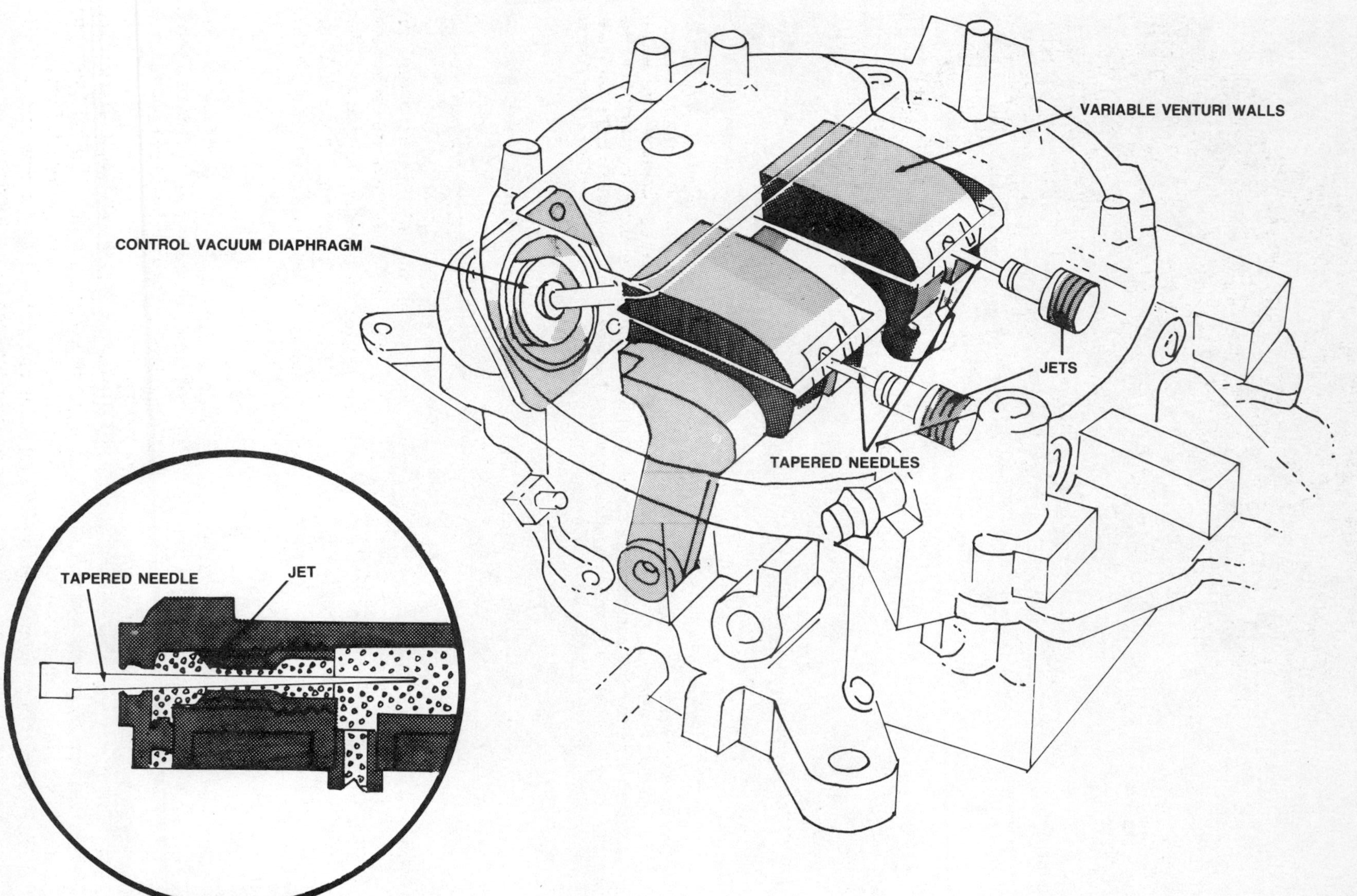

Fig. 3–8. This is a two-barrel variable-venturi carburetor. The venturi opening in each barrel is controlled by movable walls that are operated by a control vacuum diaphragm. As the venturi walls move to increase the venturi opening, they also pull tapered needles out of fuel jets, permitting increased fuel flow.

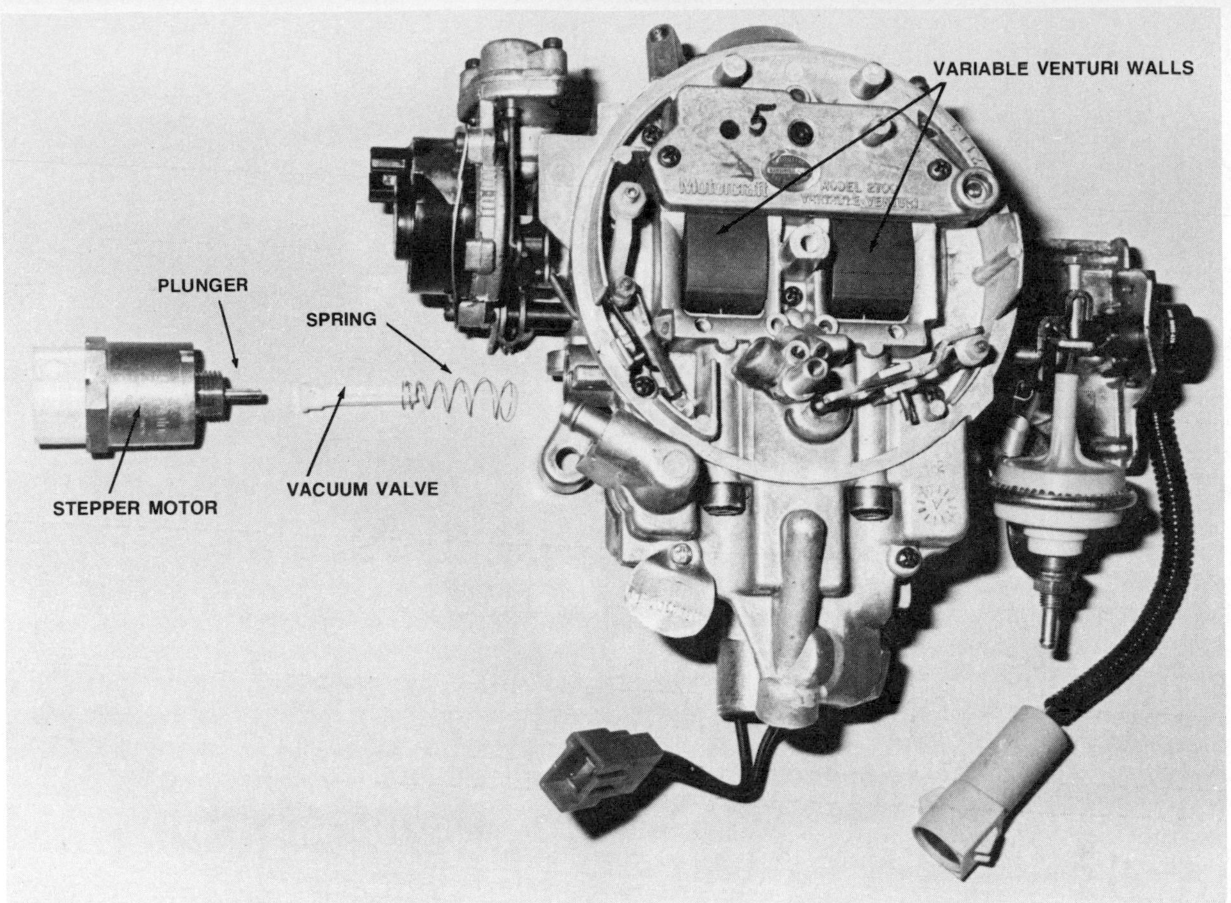

Fig. 3–9. Here you can see the tops of the movable venturi walls of a variable-venturi carburetor. This type of carburetor does not lend itself to the use of a choke, so instead there is a system that increases fuel flow when the engine is cold (this instead of restricting air flow as does the choke). This particular carburetor is used with a computer that regulates fuel mixture by operating the stepper motor and vacuum valve, shown removed at left. The position of the motor plunger (determined by the computer) regulates the flow of engine vacuum to the top of the fuel bowl. As vacuum is admitted, fuel flow is reduced; as the vacuum is reduced or cut off, the fuel flow is increased.

pleated-paper or nylon filter that should be replaced every two years or 24,000 miles as follows:

(1) Disengage the four hose clamps. If spring-type clamps are used, squeeze the tangs together with hose clamp pliers (see Fig. 3–16) and move them off the overlap section of hose and fuel line or hose and filter neck. If they are crimp-together type, pry them open and cut them apart with cutting pliers. If the filter is threaded into the carburetor fuel inlet (as on many Ford products), loosen the clamp and disconnect the fuel hose from the filter, then unthread the filter from the fuel inlet.

(2) Carefully work the short hoses off the filter necks or fuel line, whichever is easier. Be sure to hold the fuel line when you try to get a hose off so you don't damage the line's connection at the carburetor or fuel pump. Note: Some replacement filters come with new pieces of hose. In this case, remove the hoses from the fuel line and discard.

(3) Install the new spring clamps on the midpoint of each hose, then install the new filter so that the arrow on the filter points toward the carburetor. With the pliers, relocate each clamp so it secures the hose to the fuel line and filter necks. On Ford products with a hose neck that has two clamps, position each so it is not over the hose neck boss.

• *A filter element under a cover on the fuel pump*. Like the in-line type, it is made of pleated paper and is of a size that should last two years or 24,000 miles.

To replace (see Fig. 3–17) simply unscrew the cover (if it is tight, obtain a special band-type wrench, sold in auto supply stores). Remove the old filter element, install the new one, refit the cover, and tighten with the wrench.

COMBUSTION COMPUTER

COMBUSTION COMPUTER
AIR FILTER
COOLANT SENSOR
SOLENOID REGULATOR
CARBURETOR
CHRYSLER Plymouth
MANIFOLD VACUUM
OXYGEN SENSOR
EXHAUST MANIFOLD

ELECTRONIC FEEDBACK CARBURETOR CONTROL SYSTEM

Fig. 3–10. This is a schematic of a computer control for a carburetor. The computer housing contains circuit boards that control spark advance, the carburetor, and perhaps some other engine functions. An oxygen sensor in the exhaust manifold tells the computer whether the fuel mixture is too rich or too lean, and the computer makes an adjustment. In Fig. 3–9, the adjustment was to operate a stepper motor. In this case it controls a solenoid regulator, an electromagnetic switch that admits or denies engine vacuum to a diaphragm in the carburetor. When the diaphragm receives vacuum, it yanks on a tapered rod in a fuel passage to increase fuel flow. When vacuum is denied, a spring retracts the diaphragm and the rod moves back into the fuel passage, reducing fuel flow.

Fig. 3–11. Spray aerosol solvent into carburetor barrel and onto carburetor linkage to clean the parts.

Automatic Choke Service

A choke that sticks closed will cause poor gas mileage and reduced high-speed performance. A choke that sticks open will make starting, particularly in cold weather, very difficult.

• *Checking the choke.* The engine must be cold and ambient temperature must be 50°F. or lower to check the operation of the automatic choke. Remove the air cleaner cover (see Fig. 3–18) and look at the choke plate (see Fig. 3–19). The plate should cover the top of the carburetor barrel. If it does not, floor the gas pedal, hold it there for an instant, then release. The choke should close.

If the choke plate does not close completely, move it back and forth by hand. If you feel stickiness the choke should be cleaned, as will be explained.

If the choke plate closes, refit the air cleaner cover and drive the car for a couple of miles. Stop the engine, remove the air cleaner cover, and look at the choke again. The plate should be vertical. If it hasn't moved at all, or has moved very little, try to operate it by hand. If you feel stickiness, cleaning is required. If the choke has an external vacuum diaphragm (see Fig. 3–20), watch the link to that diaphragm as a helper starts the engine. If the diaphragm doesn't pull the link, check for a good hose connection, and if that's not the problem, unscrew and replace the diaphragm.

If the diaphragm moves the link, but not

Fig. 3–12. This is an on-car carburetor cleanout kit in use. Fuel line is disconnected from carburetor and plugged; kit is attached to carburetor inlet fitting as shown. Can of cleaning solvent is inverted as shown, and the engine is run at idle until the contents of the can are exhausted.

enough to fully open the choke on a warm engine, a choke vacuum break adjustment may do the job. The adjustment procedure is described later in this section. Note: Starting on 1981 models, the choke vacuum break adjustment is sealed by a cover plate riveted in place. To make an adjustment, therefore, the rivets first must be drilled out to permit removal of the cover. When reinstalling the cover, use self-tapping screws.

• *Cleaning and adjusting the choke.* Using aerosol spray solvent (see Fig. 3–11), spray the choke plate, its shaft, and all the external linkage that moves when you operate the choke plate by hand.

If the choke is the type without an external diaphragm, the same job is done by a little piston inside a housing built into the carburetor under a plastic cover (see Fig. 3–4). Take off the cover slowly so you can look inside and see how the tang on the thermostatic coil spring engages the choke linkage.

You'll see little piston and linkage. Squirt them with solvent and work the choke plate back and forth until the solvent has done its job and the choke plate operates absolutely freely (see Fig. 3–21).

Refit the cover disc, insert the screws (but do not tighten), then turn the disc until the mark on the disc aligns with the center mark on the carburetor (see Figs. 3–22, 3–23). The choke plate should close, but if it doesn't, continue turning the disc until its mark is no more than two marks past center. If the choke plate still isn't closed, replace the thermostatic coil. If the plate is closed, tighten the cover disc retaining screws. Reattach the electric connector if used.

• *Well chokes.* The choke with a thermostatic coil in the well in the intake manifold (see Fig. 3–6) may be adjustable. Remove it and if you see a locknut holding plates with marks (see Fig. 3–23), slacken the nut and align the single mark on one plate with the center of the marks on the other. If they're already aligned, reset up to two lines toward "R" (for rich mixture) to correct a hard-starting problem. Or replace the well assembly. Note: Well assemblies are not adjustable on cars built since the early 1970s.

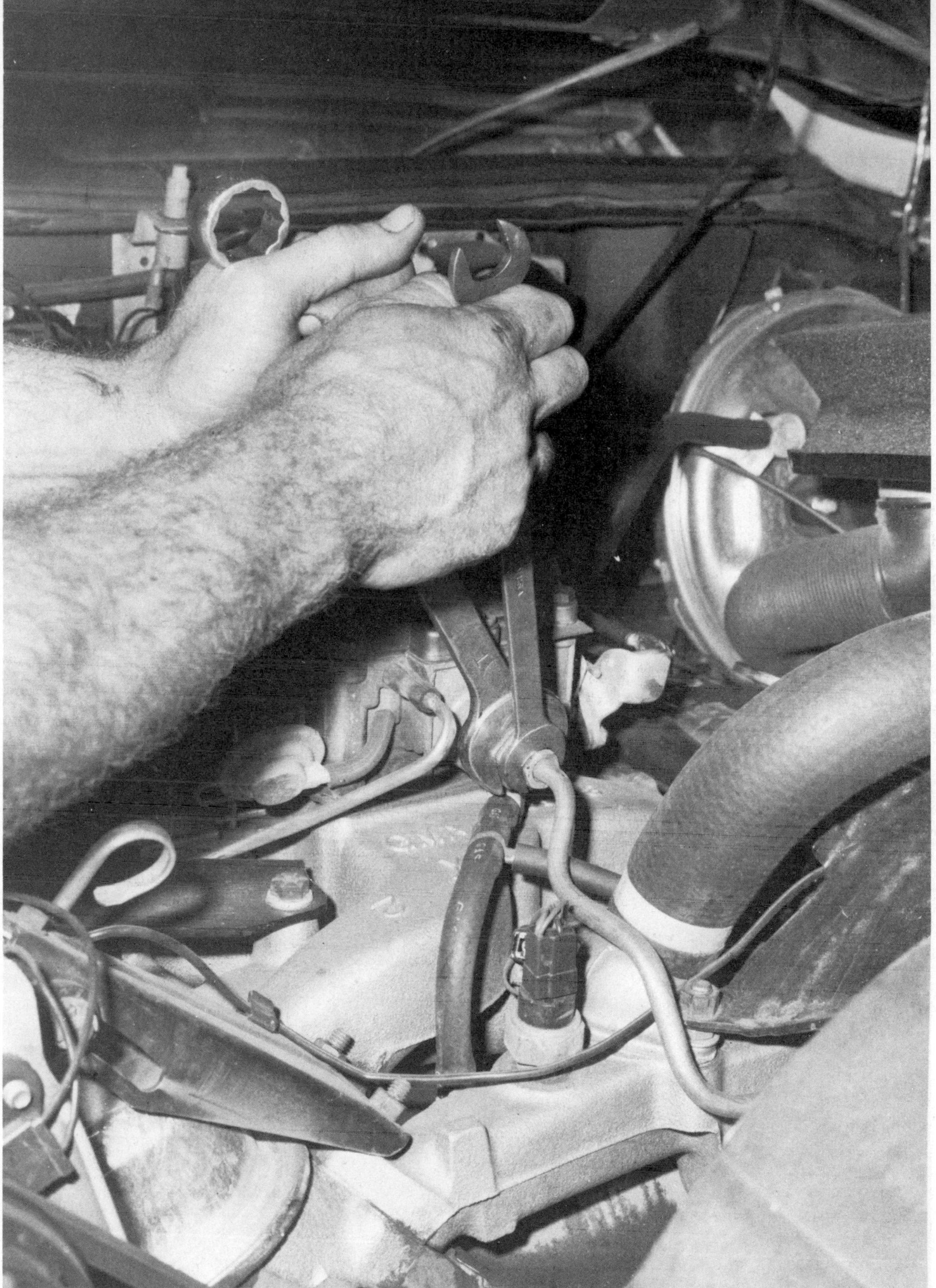

Fig. 3–13. Disconnecting the fuel line from the carburetor, using two wrenches—one on the carburetor fitting, a second on the fuel line nut. Apply tightening pressure on the fitting while you loosen the fuel line nut.

Fig. 3–14. After disconnecting the fuel line you can remove the carburetor fitting and take out the filter inside the fuel inlet. In this case it is a powdered bronze filter as shown.

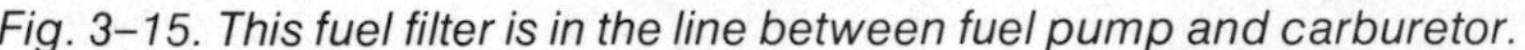

Fig. 3–15. This fuel filter is in the line between fuel pump and carburetor.

Fig. 3–16. If in-line fuel filter is held by spring clamps, move them off the overlap sections between flexible hose and steel iine, or hose and filter neck, using hose clamp pliers as shown. Then remove filter and hoses from fuel lines. When installing a new filter on a car that has two nipples at one end, the two-nipple end should face the carburetor side of the fuel line. Note: On some cars, primarily Ford products, the filter threads into the carburetor at one end and attaches to the fuel line with a short hose and clamp on the other.

Fig. 3–17. Unscrewing fuel pump fuel filter cover.

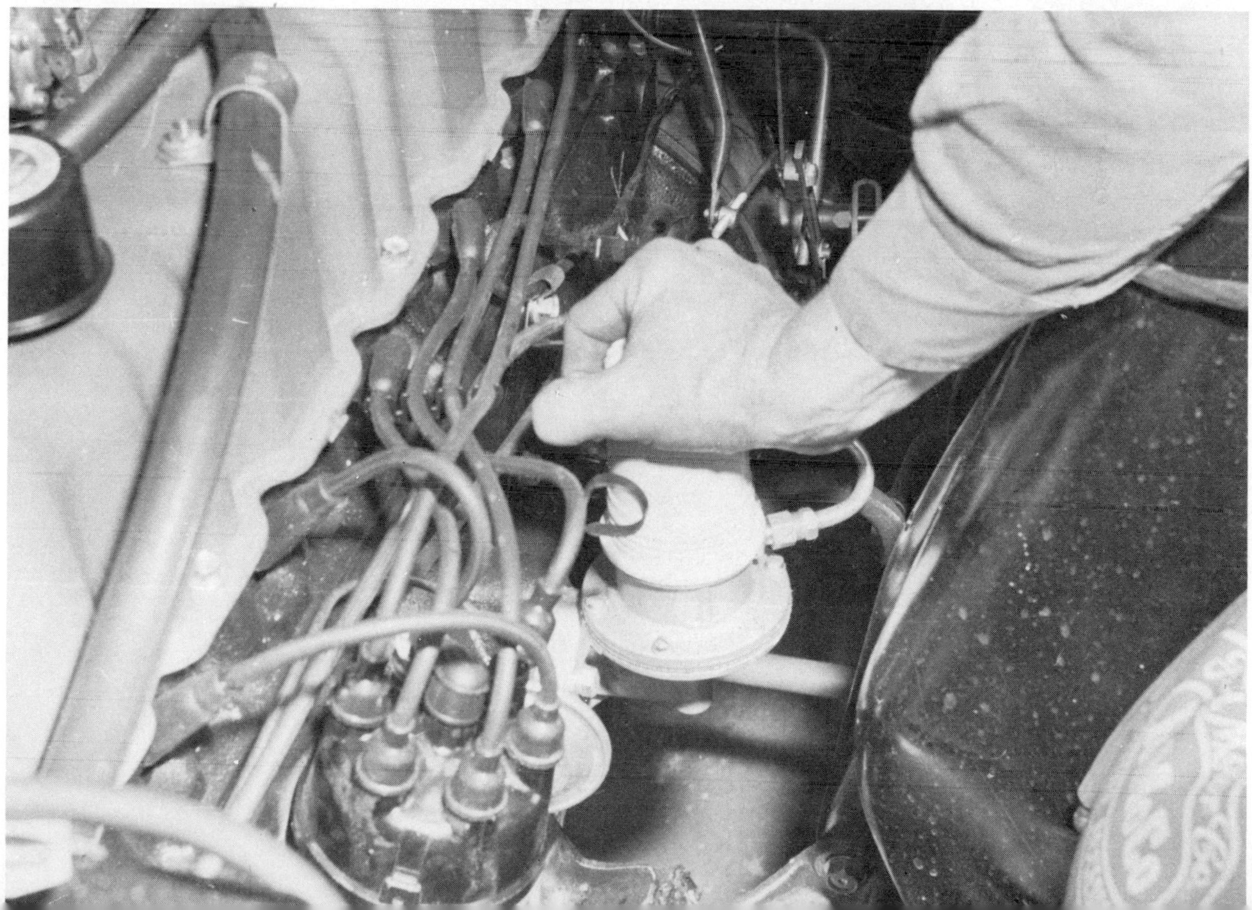

Fig. 3–18. Air cleaner cover is held by wingnut. Unscrew it and lift cover as shown. Air filter element can be lifted out for replacement if necessary at this time, but if it's clean just leave it in place.

Fig. 3–19. With air cleaner cover off you can see if choke plate is closed, covering the top of the carburetor barrel as shown.

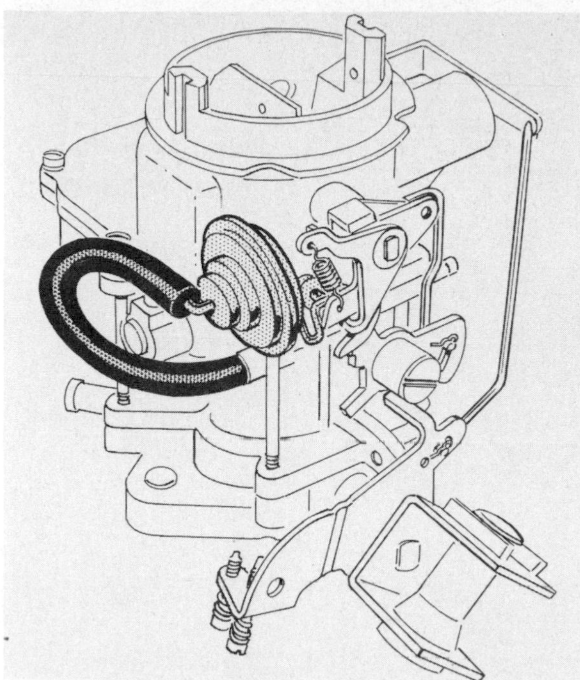

Fig. 3–20. This view of a carburetor clearly shows the vacuum diaphragm and the hose that connects from it to a source of engine vacuum (a neck that has a passage through the carburetor).

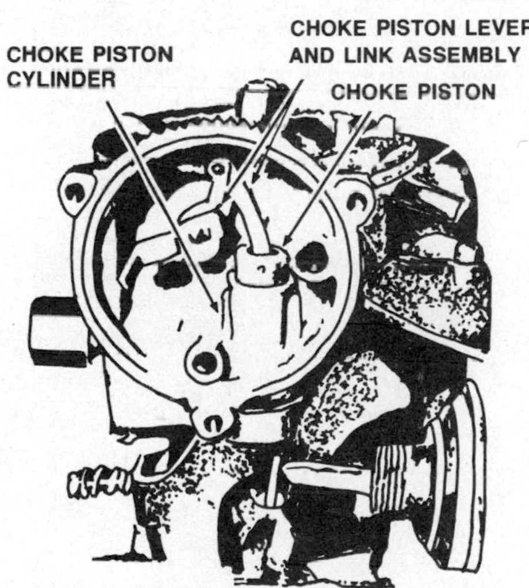

Fig. 3–21. This view inside choke housing clearly shows linkage and piston. Spray with aerosol solvent until choke plate, linkage, and piston all move absolutely freely.

• *Choke vacuum break.* When you first start a cold engine, the choke vacuum piston or diaphragm must pull the choke open just a little bit, ⅛ inch or so, to admit enough air to keep the engine running. If the choke opens too much, the fuel mixture will be leaned out and the engine will stall. If the choke isn't pulled open enough, the mixture will be too rich and, again, the engine will stall. If you have a cold engine stalling problem, check the opening.

To determine if the opening is correct, start a cold engine with the air cleaner cover off and insert a drill bit of the specified size (reference manuals will list the size for your car and carburetor) between choke plate and the wall of the carburetor barrel (see Fig. 3–24). If the drill bit can't fit, the opening is too small; if it slides in and out easily, the opening is too big. Shut the engine and make an adjustment.

The following adjustment arrangements are used:

A link with a U-bend. Remove the link and spread it at the U to decrease the opening; close it up to enlarge the opening. Make small changes in the U-bend and allow the engine to cool between adjustments.

An adjusting screw. Turn the screw to change the opening. On some carburetors the screw is secured with a sealer and won't turn. Apply heat with a soldering gun (see Fig. 3–25) to loosen the sealer before turning the screw.

• *Adjustment with engine off.* Because you have to make an adjustment and check, adjust and check, etc., it's better to do the job with the engine off and, on the majority of cars, which have external diaphragms, you can. Obtain a vacuum pump-gauge (see Fig. 3–26), an inexpensive tool with many other uses on the car. Disconnect the vacuum hose at the diaphragm, attach the hose from the pump-gauge, and operate the trigger handle to obtain 18 inches of vacuum. The choke plate will be drawn open and you can measure the gap between plate and carburetor barrel wall with the drill bit. If adjustments are necessary, you can make them over and over until you get the gap just right, without having to wait each time for the engine to cool off.

• *Choke heat tube.* On many chokes, the exhaust manifold may supply the heat for the

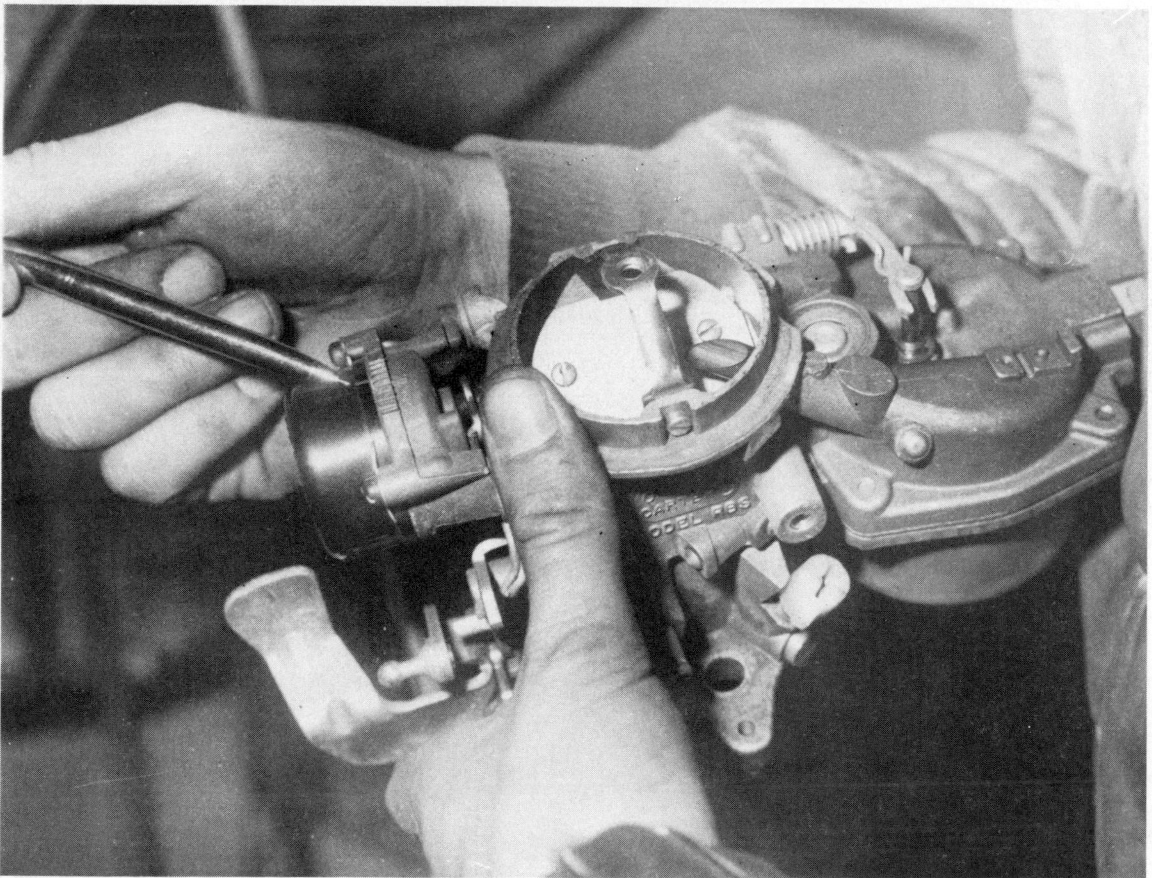

Fig. 3–22. With the cover screws loose you can turn the choke housing cover clockwise until the choke plate closes and the mark on cover aligns with the specified mark on metal body.

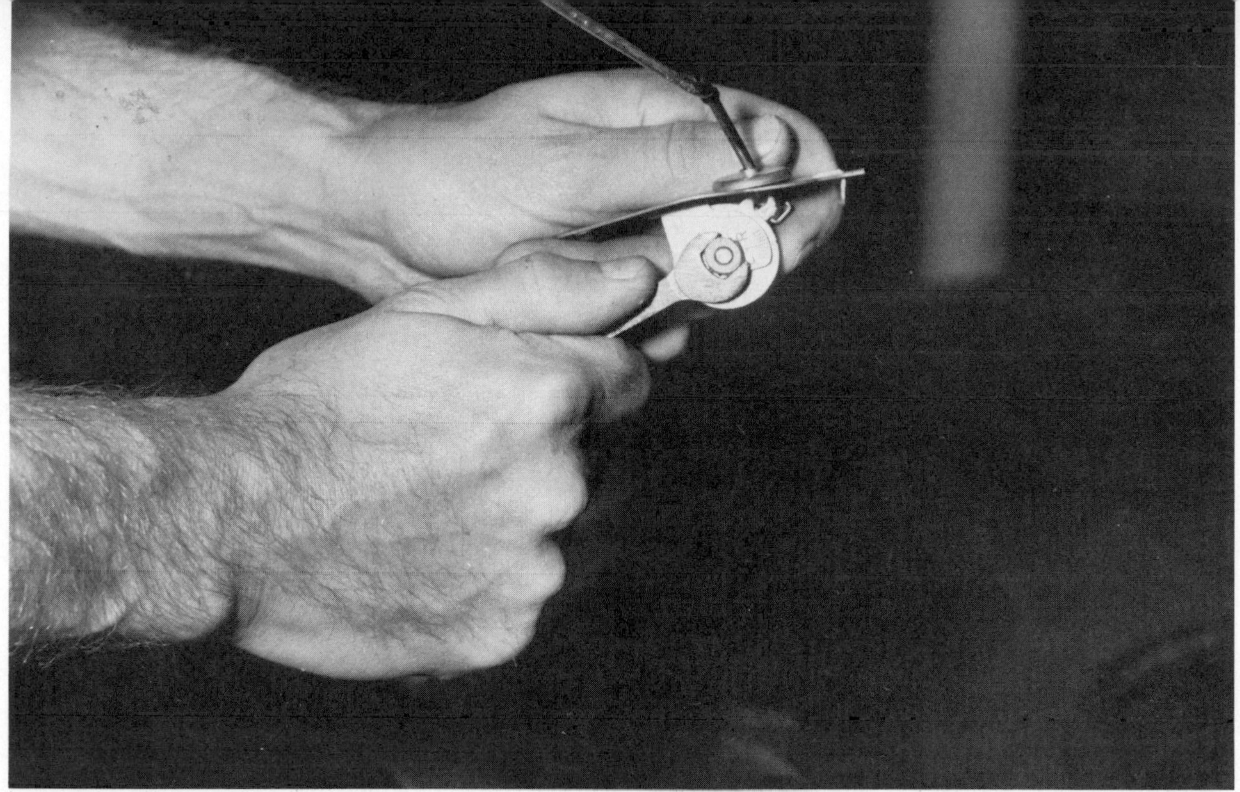

Fig. 3–23. On older cars, the well choke spring is adjustable by slackening nut and moving plate with single mark to align with correct mark on other plate.

Fig. 3–24. Inserting specified drill between choke plate and carburetor barrel when cold engine is first started. Drill should move in and out with some drag.

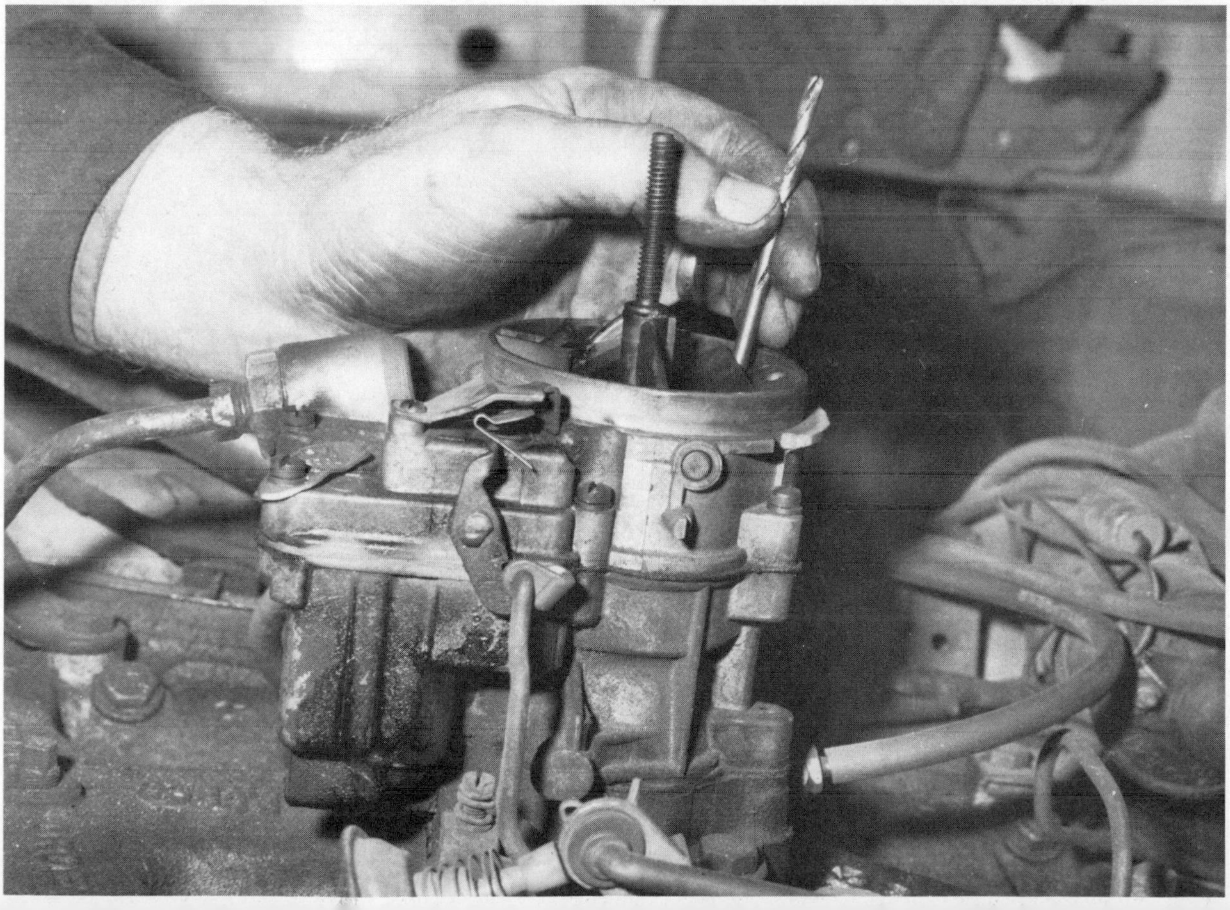

Fig. 3–25. Heat the adjustment screw with a soldering gun to loosen sealer.

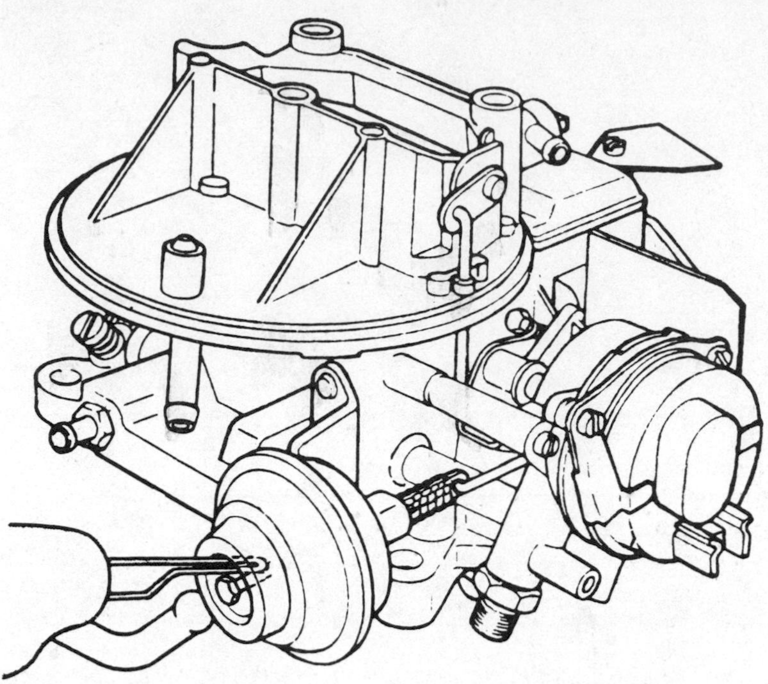

Fig. 3–26. Using vacuum pump gauge for choke vacuum break adjustment. Pump is applying pressure to vacuum diaphragm with engine cold, while break is checked with drill bit. This type has adjusting screw, which is being turned so that drill bit moves between choke plate and carburetor barrel with light-to-moderate drag.

thermostatic coil by means of an insulated tube from manifold to carburetor housing. If when cleaning the choke housing with solvent you notice a large accumulation of gummy deposits, you can't stop with the cleaning. The problem is that exhaust gas is getting into the housing and depositing there. Reason: The heat tube plugs into what seems like a hole in the exhaust manifold, but actually there's a sealed tube in the manifold that you can't see. If exhaust heat cracks that sealed tube, exhaust gases can leak into it, and then into the insulated tube to the carburetor housing.

The cure is a special service kit. There are several types, but the easiest for a weekend mechanic is a design that clamps to the exterior of the exhaust manifold. Just remove the old heat tube and drive a plug into the hole (the plug is furnished with service kit). Then attach a half-cylindrical retainer to a suitable location on the manifold, using a large worm-drive hose clamp (this clamp can be opened to install; it doesn't have to fit over an open neck, as do other clamps). Insert a replacement heat tube (covered with fiberglass cloth provided) into the edge of the retainer (where there's a suitable size hole), then carefully bend the heat tube and attach to the carburetor housing (see Fig. 3–27).

Note: On many cars the carburetor retaining nut and a tiny piece of hardware called a ferrule are all you need, and they're in the kit. If, however, your car has the tubing flared at the carburetor end you must install the nut and flare the end of the tube with a flaring kit (see Fig. 3–28). If you don't have a flaring kit, leave the job to a professional.

Idle Speed Adjustments

Whenever you perform a tuneup on a car you should set the idle speed to factory specifications. These are usually listed on a decal under the hood. The adjustment must be made with the engine fully warmed up. Note: On many 1980 Cadillacs and a wide variety of 1981 General Motors cars of other makes too, idle speed is controlled automatically by the engine computer, which drives a DC motor with a plunger against the throttle linkage. A similar arrangement also is used on the limited-production Chrysler Imperial. On these cars, do not disturb the idle speed adjustment. Few idle speed adjustments also should be needed on most 1980 and some 1981 VWs with a digital electronic idle stabilizer, and 1981 American Motors six-cylinder engines, which use an electronic control unit to operate a vacuum-electric device that regulates idle speed.

The factory specified idle speed is a compromise that provides maximum smoothness with resistance to stall, while at the same time minimizing fuel consumption and the creeping tendency of a car with an automatic transmission. To do the job you need a tachometer and a screwdriver or small wrench.

First, locate the idle speed adjuster and understand how it works. If it is an adjusting screw, it should bear against the throttle linkage on American cars (on some European carburetors it does not, and you should check your owner's manual for details). If there are two adjusting screws on the carburetor linkage (see Fig. 3–29), one is for fast-idle, a higher idle speed for when the engine is cold. That screw bears against a stepped part called the fast-idle cam, or against a link that bears against the fast-idle cam. It is the other screw you want.

If there is no other screw, there may be a solenoid, an electromagnetic switch with a plunger. When you turn on the ignition the solenoid is energized and ejects its plunger to bear against the throttle linkage. When you turn off the ignition the plunger retracts, allowing the throttle to close more completely. This minimizes a problem in many cars called "after-run" or "dieseling," in which the engine continues to run briefly after the ignition key is turned off. Many late-model cars have solenoids, and there are several types. Here's what you may find:

- *Solenoid with plunger that has a square or hex-shaped tip*. See Fig. 3–30. To increase idle speed you put a wrench on the hex or square and turn counterclockwise, to partly unscrew the plunger from the solenoid.
- *Solenoid on adjustable carriage*. See Fig. 3–31. The solenoid moves toward (increasing idle speed) or away (decreasing it) as you turn the carriage adjusting screw.
- *Solenoid plunger bearing against adjusting screw*. In this case (see Fig. 3–32), you turn the adjusting screw, which is on the throttle linkage.
- *Solenoid plunger adjuster on other side of solenoid*. See Fig. 3–33. The adjuster is a rod with a hex or square tip on the side opposite to the plunger.

Fig. 3-27. Here is one replacement for a heat tube that is broken inside the exhaust manifold. Above, the new heat tube retainer is held to the exhaust manifold by a clamp. The heat tube itself, covered with fiberglass cloth insulation, is run up to the choke housing. Below, you can see it being threaded onto the choke housing fitting.

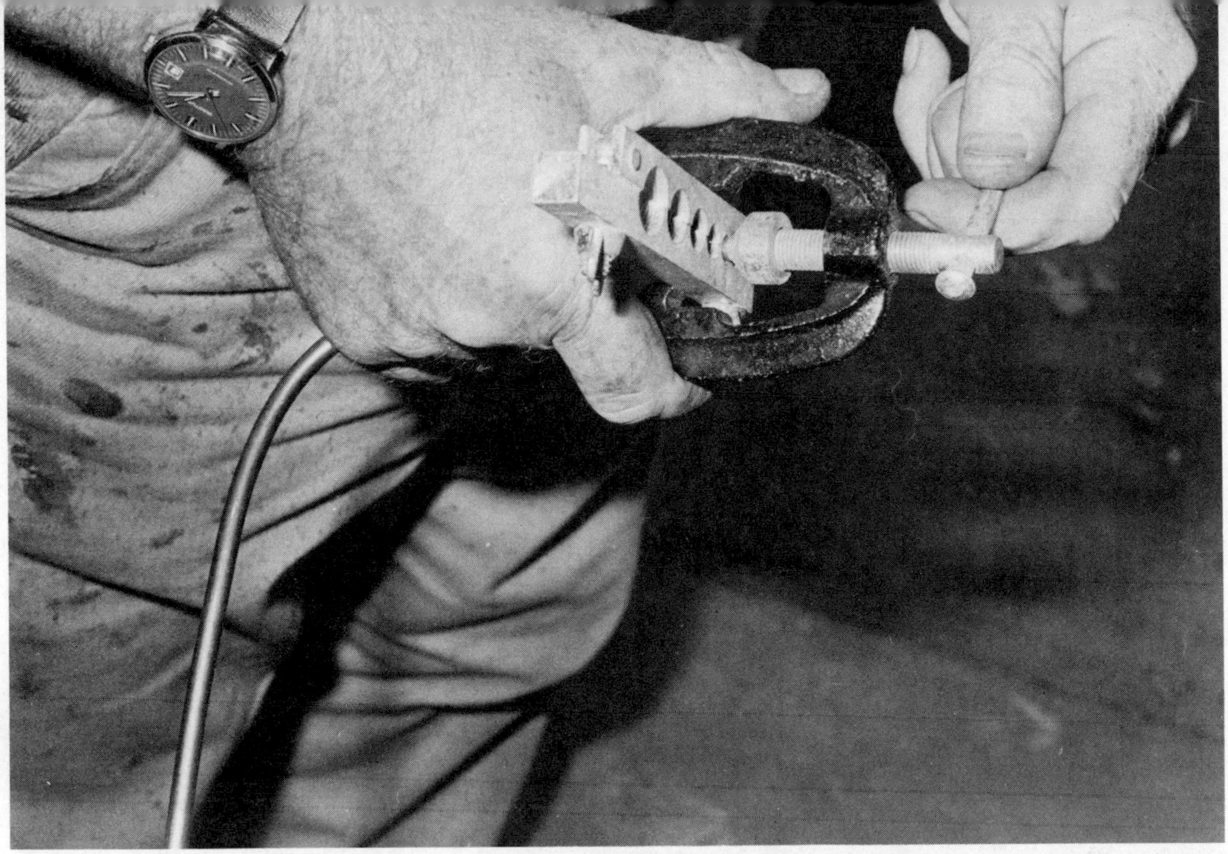

Fig. 3–28. If necessary, flare the choke housing end of the heat tube. The nut should be in place before this flaring tool is used, for once you flare the end you may not be able to get the nut on from the other end.

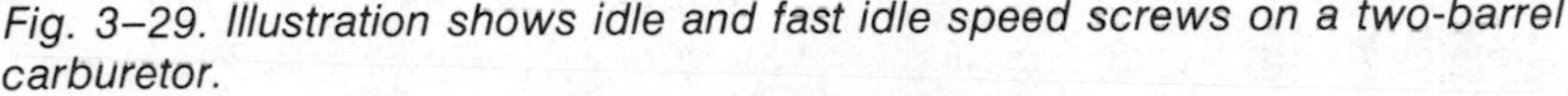

Fig. 3–29. Illustration shows idle and fast idle speed screws on a two-barrel carburetor.

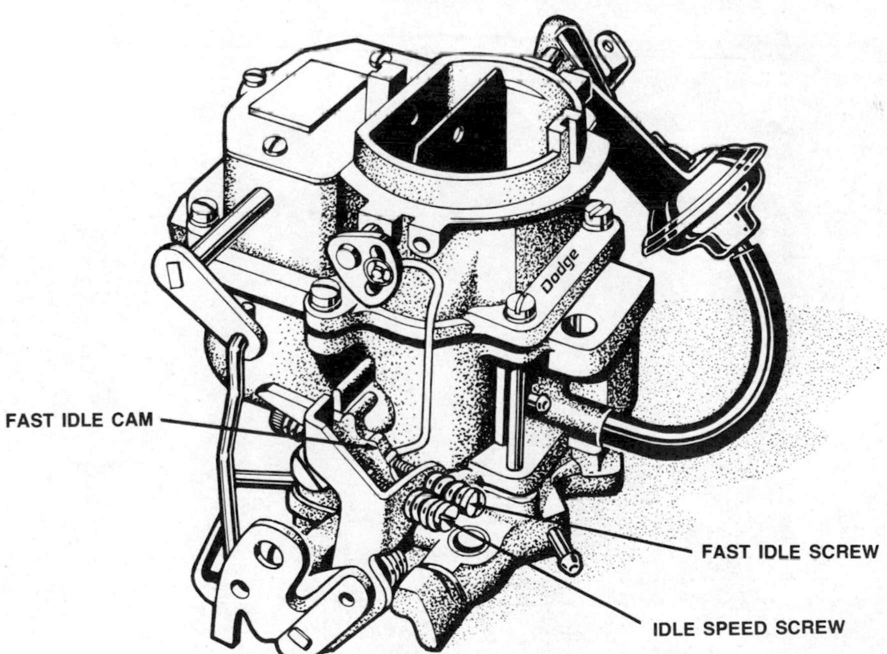

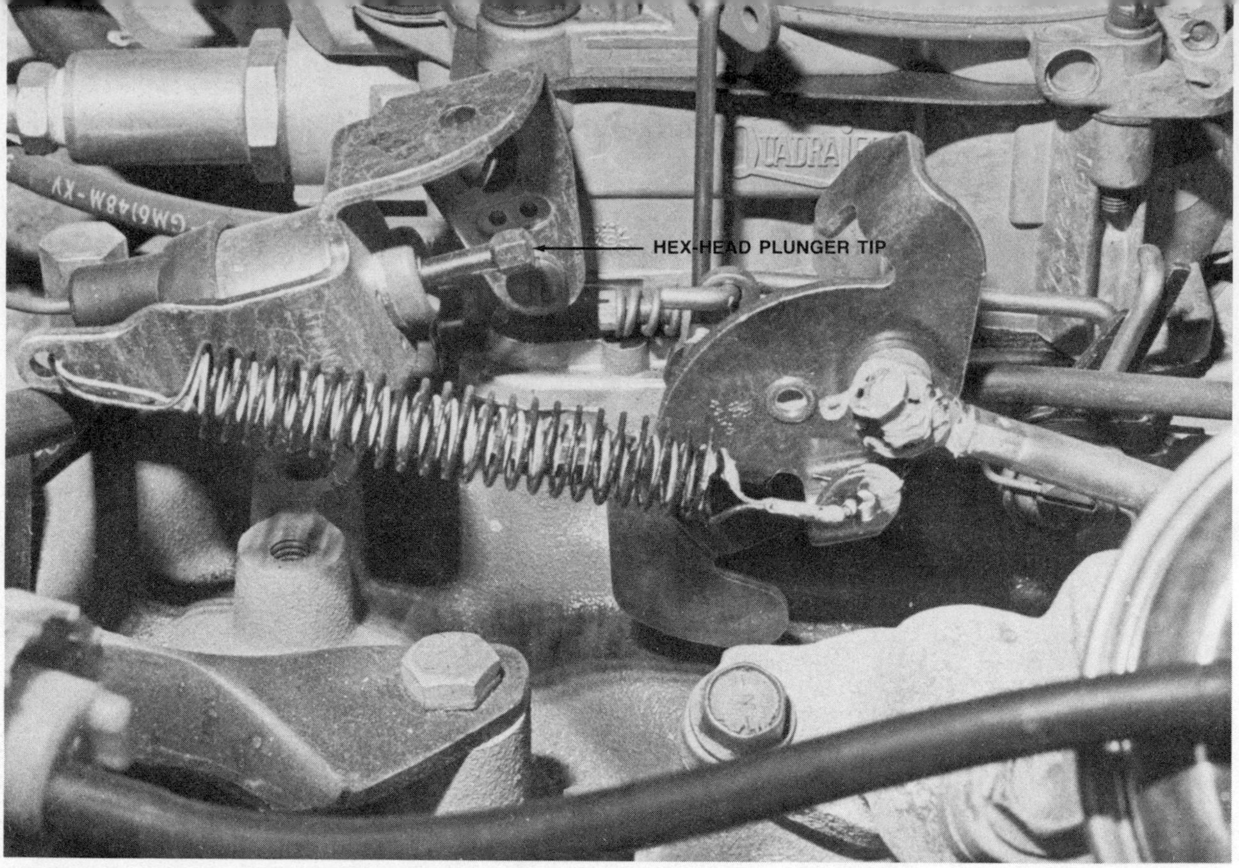

Fig. 3–30. This solenoid plunger has a hex-head tip. Put wrench on tip and turn to increase or decrease plunger length to adjust idle speed.

Fig. 3–31. This solenoid is on a movable carriage that is reset by turning the spring-loaded carriage adjusting screw as shown.

ADJUSTING SCREW

SOLENOID PLUNGER

Fig. 3–32. There's no adjustment on this solenoid. Instead, the solenoid plunger bears up against an adjusting screw on the throttle linkage.

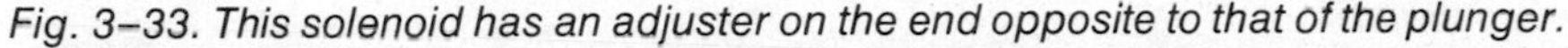

Fig. 3–33. This solenoid has an adjuster on the end opposite to that of the plunger.

Fig. 3–34. This solenoid is like a giant adjusting screw. To adjust your idle speed, turn the entire solenoid body in (against the spring pressure).

Fig. 3–35. Back view of solenoid shows the hex for wrench, to permit easy turning of solenoid body. Wire has been disconnected from solenoid to avoid twisting it.

• *Entire solenoid body is adjustable, plus internal adjuster in back of solenoid.* This design is found on many General Motors cars. The basic adjustment is made by turning the entire solenoid body (you may have to disconnect a wire from the solenoid to avoid twisting it) and the adjustment is fine-tuned by turning an internal adjuster. See Figs. 3–34, 3–35, and 3–36.

You may find more than one likely adjuster, such as both a solenoid and a screw. In this case you must find out if the solenoid has another purpose, namely to raise the idle speed to prevent stalling when the air conditioning is turned on.

To check, have a helper start the engine, hold down the gas pedal about an inch, and turn on the air conditioning. If the solenoid is wired into the air conditioning circuit you'll see the plunger extend only after the AC is turned on. Note: The gas pedal must be held down a bit or the plunger may not extend. The solenoid has only enough energy to hold the plunger in the extended position, not to push the throttle open.

If the solenoid plunger does respond only to the air conditioning switch, it is not used to set the basic idle speed. You should, however, check its operation after setting idle speed to make sure that idle speed is maintained (or rises slightly) when the air conditioning is turned on.

• *Setting idle speed.* Connect your tachometer to an electrical ground and to the negative terminal (marked with a minus sign or "CB") of the ignition coil. On cars with electronic ignition, see Figs. 2–31 and 2–32.

Start the engine and compare the tachometer reading with specifications, which usually appear on an underhood tuneup decal.

If the reading is lower or higher than specifications, do one of the following:

Turn the adjustment screw on the linkage; turn the carriage screw, plunger, or rear adjuster on solenoids so equipped; or, on solenoids of the type shown in Figs. 3–34, 3–35, and 3–36, turn the internal Allen adjuster with a ⅛-inch Allen wrench clockwise until the adjuster comes to a stop (don't force it past the stop). Then turn the

Fig. 3–36. Allen wrench is inserted through hole into back of solenoid to turn internal adjuster in order to fine-tune idle speed.

Fig. 3–37. Screwdriver is turning plastic limiter on idle mixture screw. The limiter has a tab that permits only a modest amount of mixture adjustment.

entire solenoid body (disconnect wire if necessary) with a wrench on the hex in back (see Fig. 3–35) until idle speed is 100–150 r.p.m. above specifications. Finally, back out the internal Allen adjuster to reduce idle speed to specifications.

Note: The idle speed specifications given, even if it doesn't say so, are plus or minus 50 r.p.m., so don't struggle beyond what is necessary.

After each turn of the idle speed adjuster, have a helper step on the gas pedal for an instant. This will pull the throttle linkage away from the solenoid and allow the plunger to extend. Wait 10–15 seconds for the engine speed reading to stabilize before attempting another adjustment.

• *Low idle.* On some tuneup decals there will be a specification for an engine speed called "low" or "slow" idle, in addition to "curb idle." In this case, the solenoid controls curb idle, rather than just acting as a speedup for the air conditioning, and there is a "low idle" screw somewhere at the throttle linkage (except for the type with internal Allen adjuster, in which case the solenoid body itself serves this purpose). To check low idle, disconnect the wire from the solenoid, and if low idle is not within specifications, turn the throttle linkage screw (or readjust the solenoid with the internal adjuster) to set it.

• *Idle fuel mixture adjustment.* The fuel mixture is designed into the carburetor, and in some cars is controlled electronically. On older cars, however, there still is one adjustment remaining: the idle mixture. This is the mixture that bypasses the throttle to keep the engine fueled when you have your foot off the gas pedal. Years ago you could turn the mixture screws (one for a one-barrel, two for a two- or four-barrel carburetor) a fair amount. With emission controls came the limiter, a plastic cap (see Fig. 3–37) that restricted the amount you could turn the screw. On the newest cars (since 1979 on most cars) there is no mixture adjustment (see Fig. 3–38).

In any case, don't look for a mixture adjustment to cure a real performance problem, for once you step on the gas pedal the idle passages have little effect, and above 25 mph usually none at all.

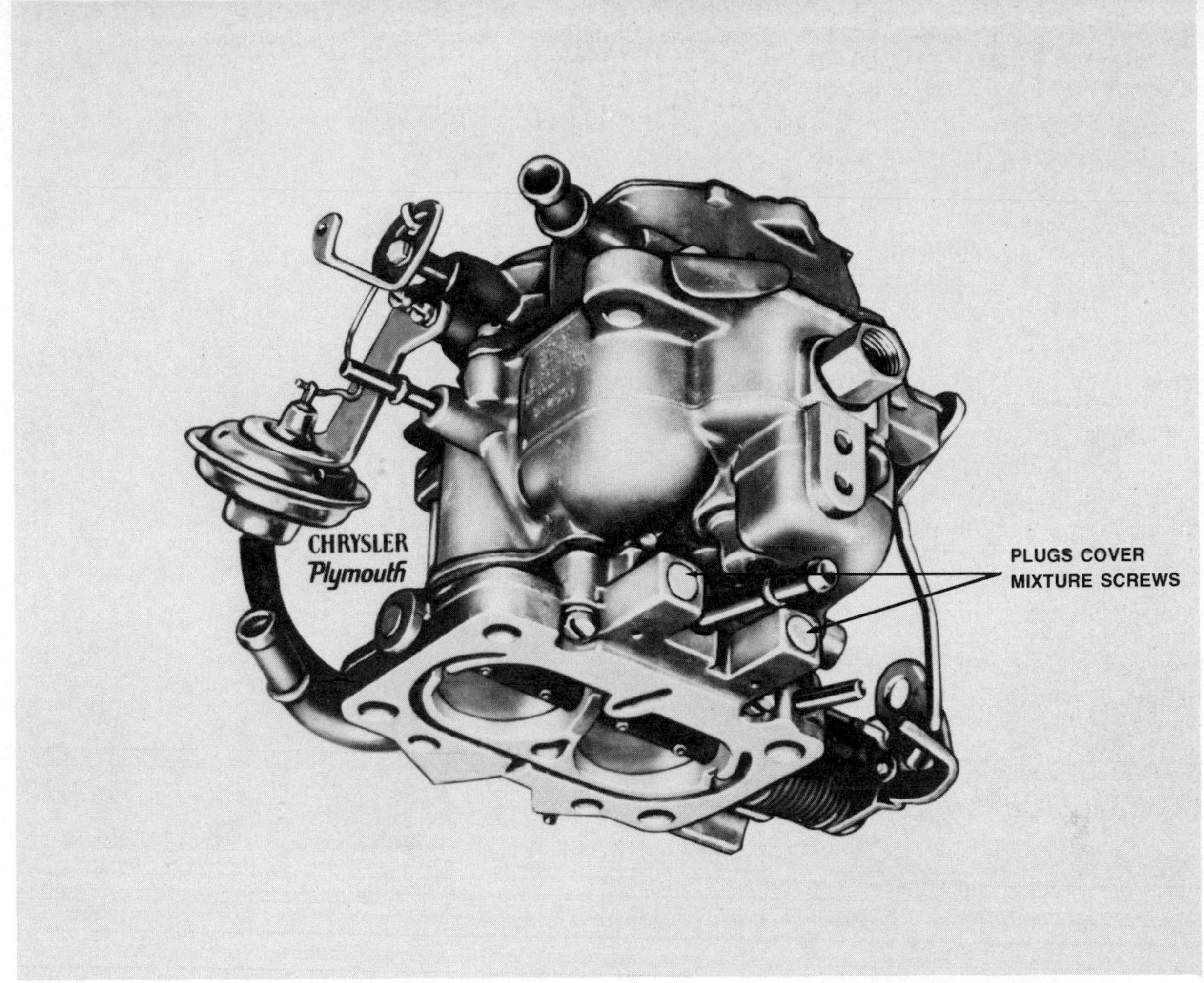

Fig. 3–38. Current-model cars have mixture screws preset at factory and sealed by plugs as shown.

If the idle is rough, however, a mixture adjustment may help. If your state has periodic motor vehicle inspection with an emissions check, however, rough idle may be the price of passing inspection.

To determine if your car has adjustable mixture screws, look for them at the base of the carburetor, and if you see little screws with springs around them, not threaded into the linkage but into the carburetor itself, they're the mixture screws. Refer to Fig. 3–37, which shows the typical screw with its limiter cap.

Turning clockwise leans out the mixture; turning counterclockwise richens it.

Fuel Mixture Adjustments

The most accurate way to set the fuel mixture is with a method called "Artificial Enrichment" or "Propane Enrichment." Propane is fuel and if engine speed increases beyond an acceptable level with propane added to the carburetor, the fuel mixture the engine is getting is too lean. If the speed does not increase at all (or below specifications), the mixture is too rich.

Check the underhood tuneup decal. If it has some reference to artificial or propane enrichment, you will use it. If it doesn't, or if the car has no decal, you should set the mixture so there is a 0–10 r.p.m. gain with propane flowing.

On some decals you will see the words "RPM Gain" and a number, usually in the range of 0–50 r.p.m. In this case you should get a minimum engine speed gain of that number with the propane flowing. Or you may find two idle speed numbers, one with propane flowing, one with propane off.

• *What you need.* To set idle mixture with propane you need a household propane torch, a rubber or plastic hose at least three feet long that fits over the valve outlet on the torch, plus a

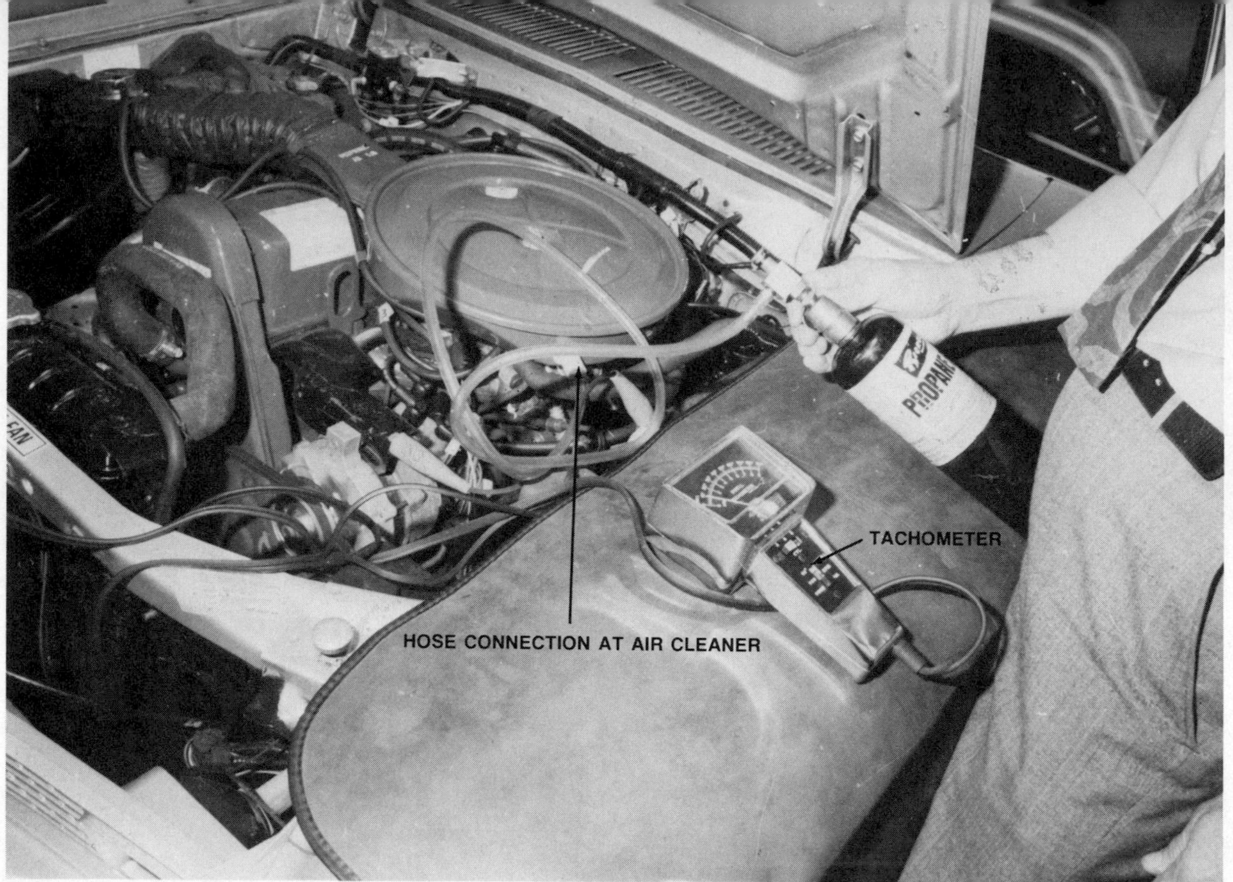

Fig. 3–39. Propane tank hose is connected to neck on air cleaner. The neck on this particular car is for a hose from the fuel vapor control cannister. Hose was disconnected for propane test.

connector for the other end. A universal-type tapered cone connector and hose is available at many auto supply stores.

First, warm up the engine. On Ford products, disconnect the air pump drive belt on cars so equipped, and the fuel vapor cannister hose at the air cleaner. On GM cars, disconnect the vacuum hose, and plug the end with a pencil, at the parking brake of those models with automatic (vacuum-operated) parking brakes; also disconnect the hose from the fuel vapor cannister at the air cleaner. To both Ford and GM cars, connect the propane tank hose to the fuel vapor hose neck on the air cleaner (see Fig. 3–39).

On Chrysler products, pull the PCV hose out of its valve cover grommet and let it sit (hose attached) on top of the engine. Disconnect the fuel vapor cannister hose at the air cleaner or carburetor. On models with Lean Burn computers, ground the idle stop carburetor switch with a jumper wire (see Fig. 3–40). On models with an electric radiator fan (such as Omni and Horizon), disconnect the bullet wiring connector at the fan motor, ground the female part of the terminal and wire the male part to the battery's starter terminal, using jumper wires (see Fig. 3–41). This will actuate the fan, necessary for the adjustment procedure. Finally, disconnect the vacuum hose from the thermostatic air cleaner sensor at the carburetor (the hose goes from the underside of the air cleaner to a neck on the carburetor base) and attach the propane tank hose to the carburetor base neck.

Connect a tachometer and start the engine. Allow 60-90 seconds for the reading to stabilize, then note the reading.

With the engine idling, crack open the propane torch valve to get the propane flowing (keep the tank vertical) and very gradually open the valve to increase propane flow until you get the maximum speed on the tach. If the engine speed suddenly drops (from too much propane), repeat the procedure. When you have a stabilized maximum reading with the propane flowing, proceed as follows: On Ford products, the underhood decal provides a specification gain for propane. If the r.p.m. gain is too high, richen the mixture by turning the screw(s) evenly. If it's too

Fig. 3–40. Before making propane test on Chrysler products with Lean Burn computer-controlled spark advance system, connect a jumper wire from the carburetor switch to an electrical ground.

low (or nonexistent), lean out the mixture. This must be done within the range of the mixture limiters. If a greater adjustment is necessary, the job should be given to a professional.

On Chrysler and General Motors automobiles, set the idle speed to specifications by turning the throttle linkage screw or solenoid, as explained earlier in this chapter, with the propane flowing. Close the propane tank valve and again read idle speed on your tachometer.

For General Motors cars there are two specifications, one with propane flowing and a second, curb idle speed, which is lower than the speed for the propane-flowing situation. If the reading with propane off is too low, richen the mixture; if too high, lean it out by turning the mixture screw(s) evenly. As with Ford products, see a professional if the mixture screw limiters prevent proper adjustment.

On Chrysler products, shut the propane tank valve, wait 10-15 seconds, and if idle speed drops 25 r.p.m. or less from the curb idle speed specification (it may or may not refer to propane), the mixture adjustment should be considered acceptable. If the drop is greater, turn the mixture screw(s) evenly until the idle speed is within 25 r.p.m. of the figure reached with propane flowing. As with Ford and GM cars, do not attempt to exceed the range of the mixture screw limiters.

After the adjustment is complete, remove the propane tank hose and restore all hose and wiring connections, and on Ford products, refit and adjust the air pump drive belt.

Note: If the drop without propane is beyond what you can correct with a mixture screw adjustment, there is a possibility that you have an air leak either at the gasket joint between carburetor and intake manifold or between intake manifold and engine.

First, connect a vacuum gauge (refer to *Chapter 8*). If the gauge reading at idle is steady but low, the odds of an air leak are great. Try to tighten all carburetor and intake manifold nuts and/or bolts, and if the vacuum gauge reading rises to normal, you've found and cured the problem. If it doesn't, double-check by squirting oil around each gasket joint. The oil acts as a temporary seal, so if the engine speeds up and/or if the vacuum gauge reading rises, you've confirmed the problem. Refer to Chapter 5 for gasket

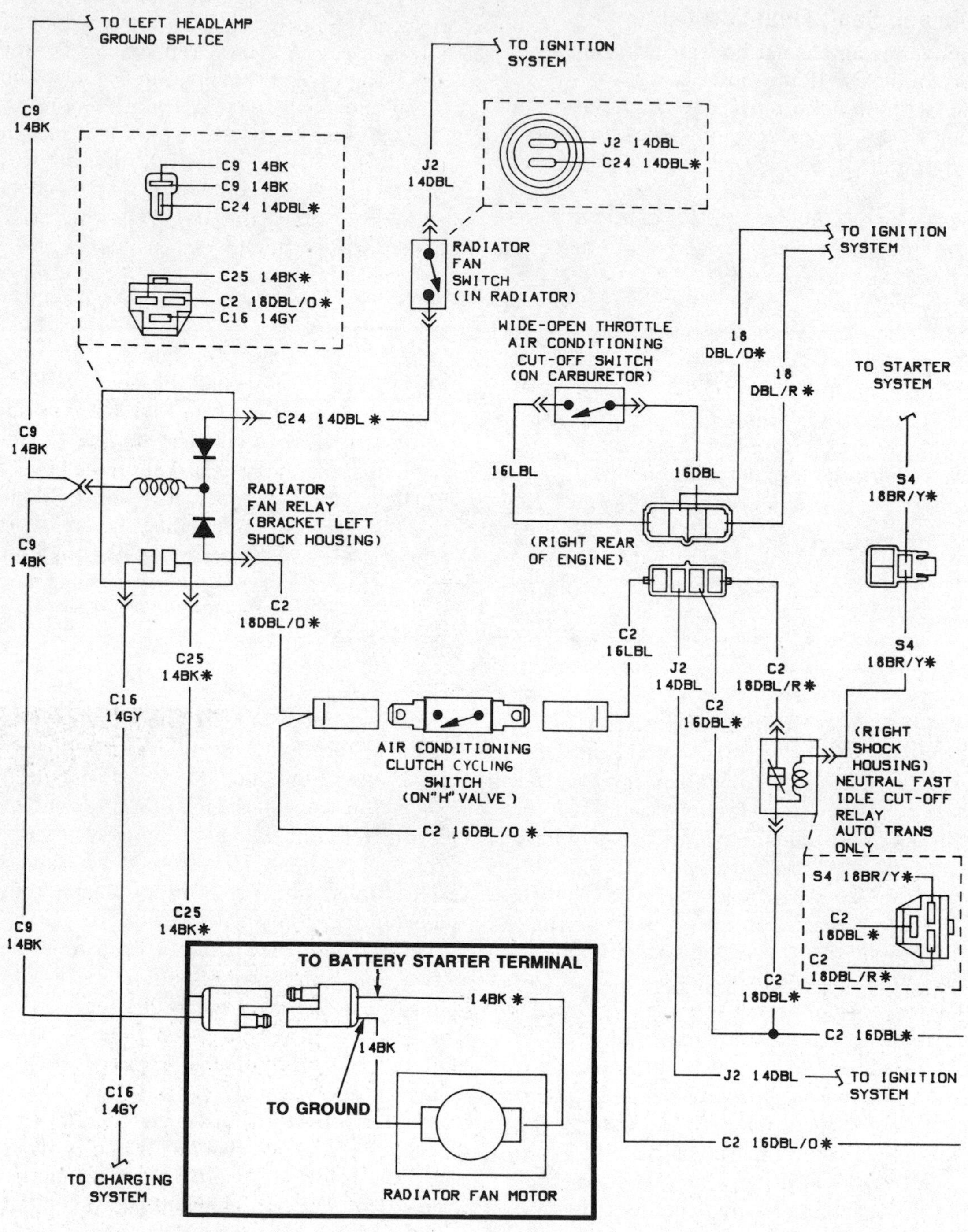

Fig. 3–41. Before making propane test on Chrysler products with electric radiator fan, have fan operating. Undo bullet connector near fan motor, ground female terminal, and connect male terminal to battery starter terminal using jumper wires.

service procedures.

Needle and Seat, Float Level

Remove the air cleaner housing and look at the carburetor bowl. If the bowl cover can be removed without taking off the carburetor and without extensive carburetor disassembly you can service the needle and seat valve and adjust float level. This operation is indicated if the engine floods out repeatedly (stalls and produces a strong smell of gasoline).

First, obtain a carburetor kit, one that includes a new needle and seat valve, a new gasket for the carburetor bowl, and a float level gauge.

Remove the carburetor bowl cover. If you have to disconnect a link or remove a part, make a note and/or sketch so you can reconnect correctly. If you must disconnect the fuel line, refer to earlier instructions under *Fuel Filters*.

On most carburetors, the float is hinged to the bowl cover, which also contains the fuel inlet fitting and the needle and seat valve (see Fig. 3–42).

Remove the hinge pin with pliers (see Fig. 3–43) and you can take off the float. The needle will then come out of the needle and seat valve. Inspect the needle tip with a magnifying glass; if you see gouge marks, replace needle and seat. The seat itself normally threads into the bowl cover and can be removed with a wrench.

Invert the new seat on a hard, flat surface, then place the needle inside. Hold the seat with your fingers and tap the outside end of the needle with a hammer (see Fig. 3–44). This will seat the new needle and it should seal properly. This step is unnecessary if you have a needle with a synthetic rubber tip or if the needle-and-seat is preassembled.

Install the seat and tighten, then drop the needle into it and install the float. Refit the float gauge over the float (see Fig. 3–45) as specified in the instructions with the service kit (you may have to place the new gasket on the bowl cover). If the gauge does not just touch the float, reset the float position by bending the tab on the float arm (see Fig. 3–46). Caution: Although this book does not cover Japanese cars, you are warned

Fig. 3–42. On most cars the float is hinged to the bowl cover.

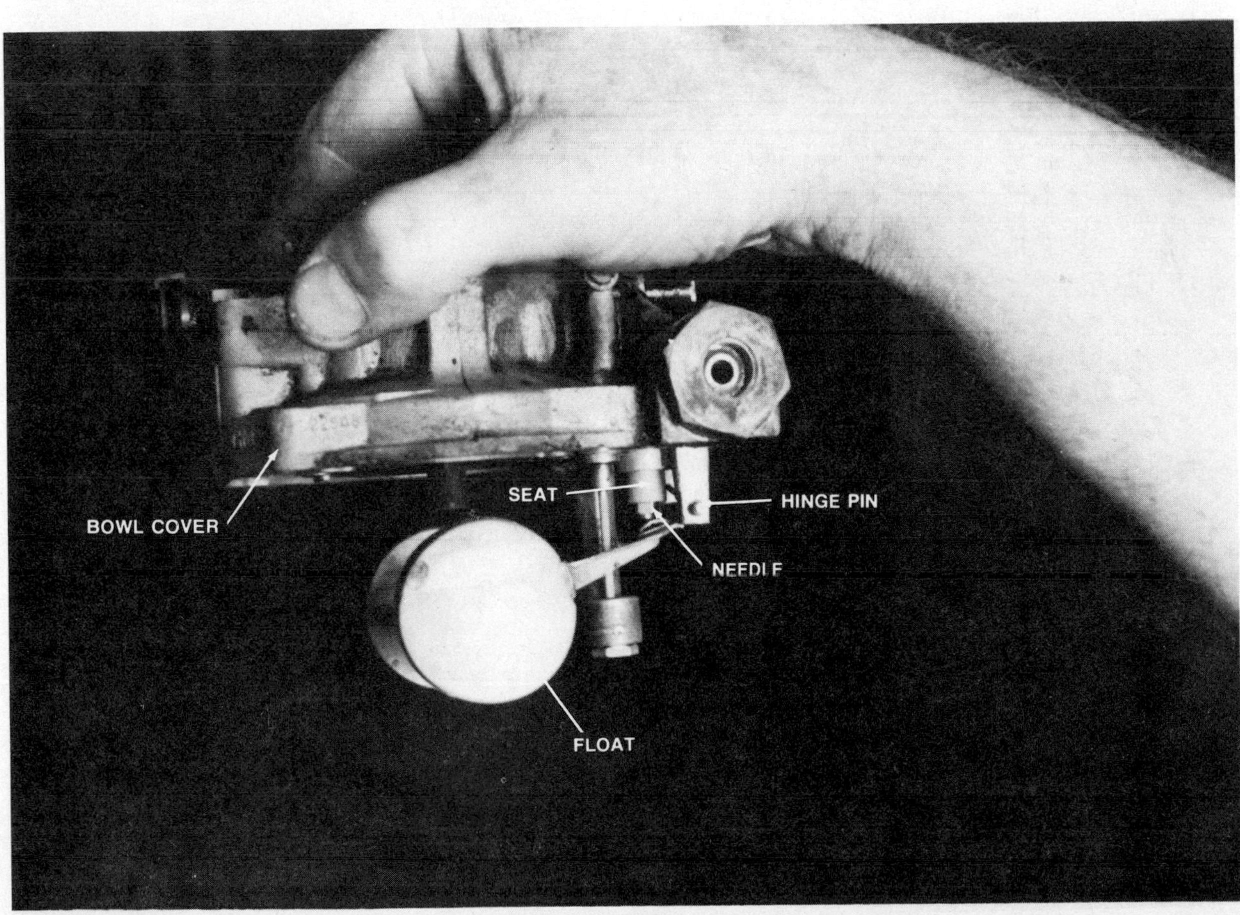

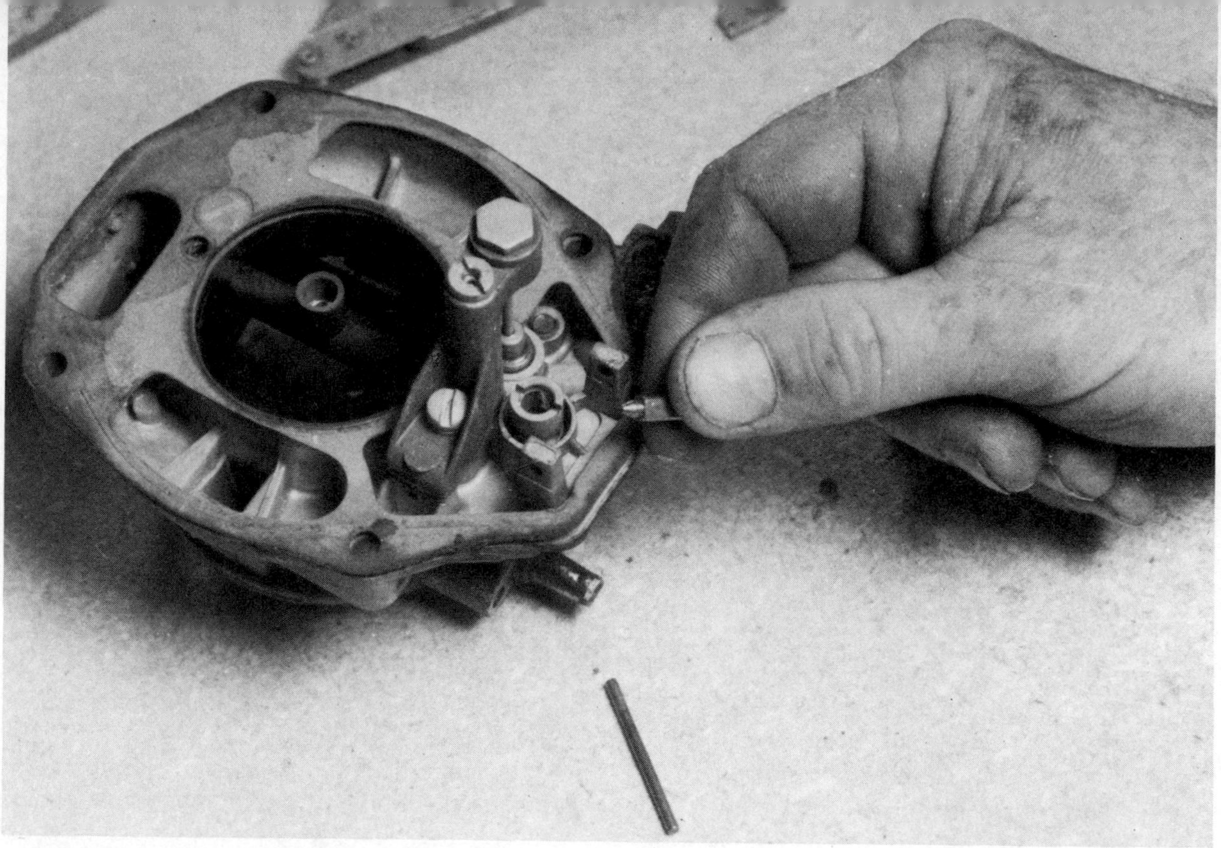

Fig. 3–43. With hinge pin out and float removed, you can take needle valve from seat and inspect both parts.

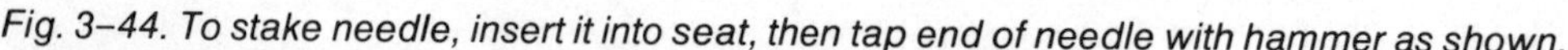

Fig. 3–44. To stake needle, insert it into seat, then tap end of needle with hammer as shown.

Fig. 3–45. If needle and seat kit includes a gauge, you can measure float level easily. With the gauge shown, you invert cover, position gasket, and place gauge as shown. The gauge should just touch both the gasket and the center bottom of the floats.

Fig. 3–46. If adjustment is necessary on this float, pry on tab as shown.

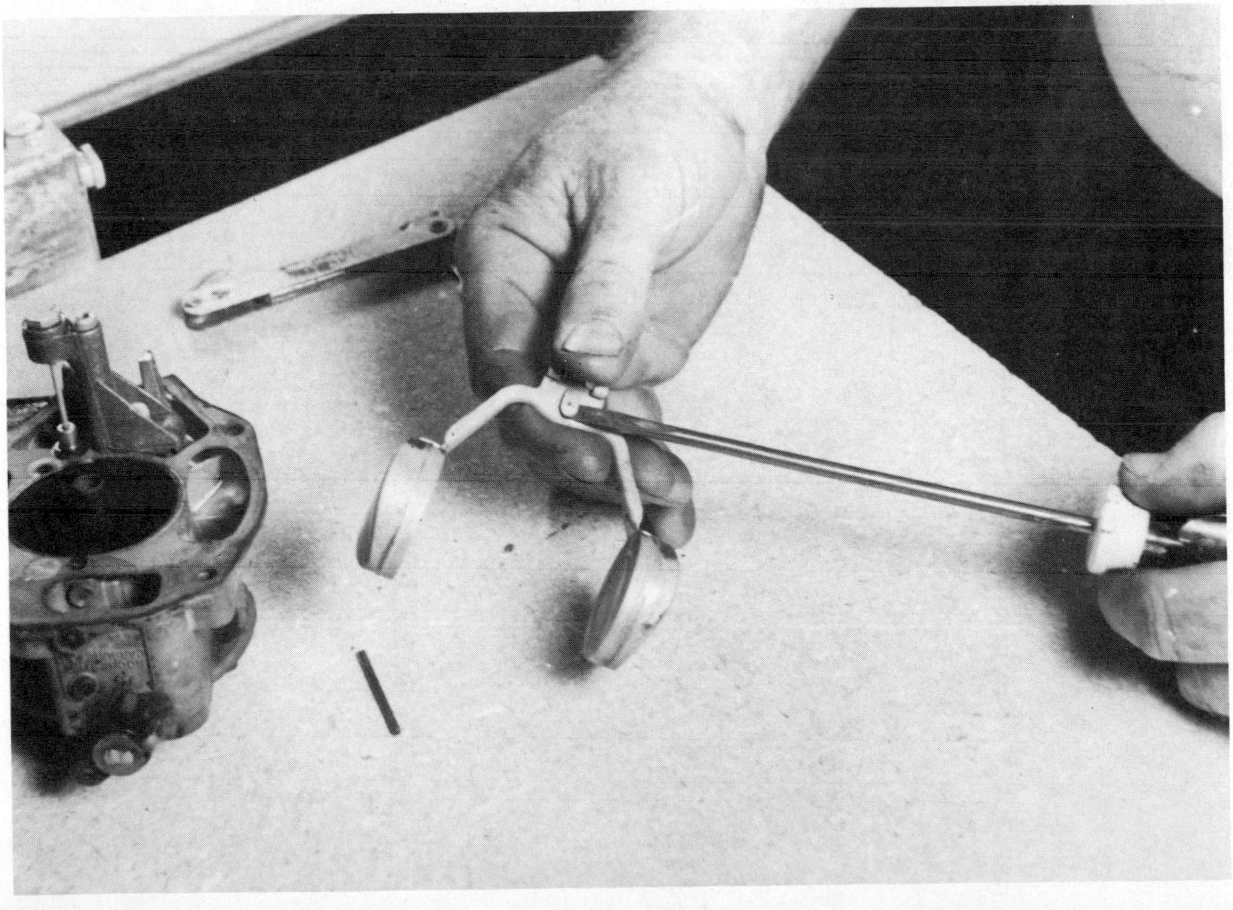

that on some Dodge Colt and Plymouth Arrow and Champ Models, special shims between the seat and carburetor inlet are used to change float level.

Note: If there is no float gauge with the kit you can set float level with a metal ruler as shown (see Fig. 3–47). Check a reference manual for the measurements and whether or not the bowl cover gasket must be in place. With the ruler, measure the distance from the bowl cover gasket or gasket surface to the top of the float (remember the float is upside down with the cover inverted). The measurement may be specified to be taken at the center of the float or at the heel and toe (heel is the end to which the arm is attached; toe is the opposite end).

Once the float level is set accurately, reinstall with the new gasket, install and tighten the cover screws, and reconnect the fuel line.

• *Float in the bowl.* If the float is in the bowl, the adjustment procedure is the same, except that you measure from the top of the float to the top of the bowl with a ruler or gauge (see Figs. 3–48 to 3–55). If the reference manual specifies a "wet float level" or a "fuel level" instead of simply the float level, the check is made with the bowl filled with fuel until the float has moved up and pushed the needle all the way into the seat.

If the check is made "dry" you must hold up the float as you make the measurement with the ruler. This can be done with a piece of stiff wire such as a paper clip.

Electronically-Controlled Carburetors

Although the electronically-controlled carburetor requires no regular service, you should check all vacuum hose and electric wire connections from the carburetor to the solenoid vacuum valve (if used) and to the computer and oxygen sensor as part of a tuneup.

The oxygen sensor itself (see Fig. 3–56) should be replaced in accordance with manufacturer's recommendations. Remove the electrical connector and take out the sensor with a wrench.

Although the electronic system controls the fuel mixture, the range in which it operates is relatively narrow. If it malfunctions it will affect exhaust emissions and performance, but there should be no effect on starting and basic running of the engine.

Fig. 3–47. If you don't have a special gauge, check specifications manual for measurement with ruler. Specifications chart will tell you what the distance should be from gasket to bottom of float at center.

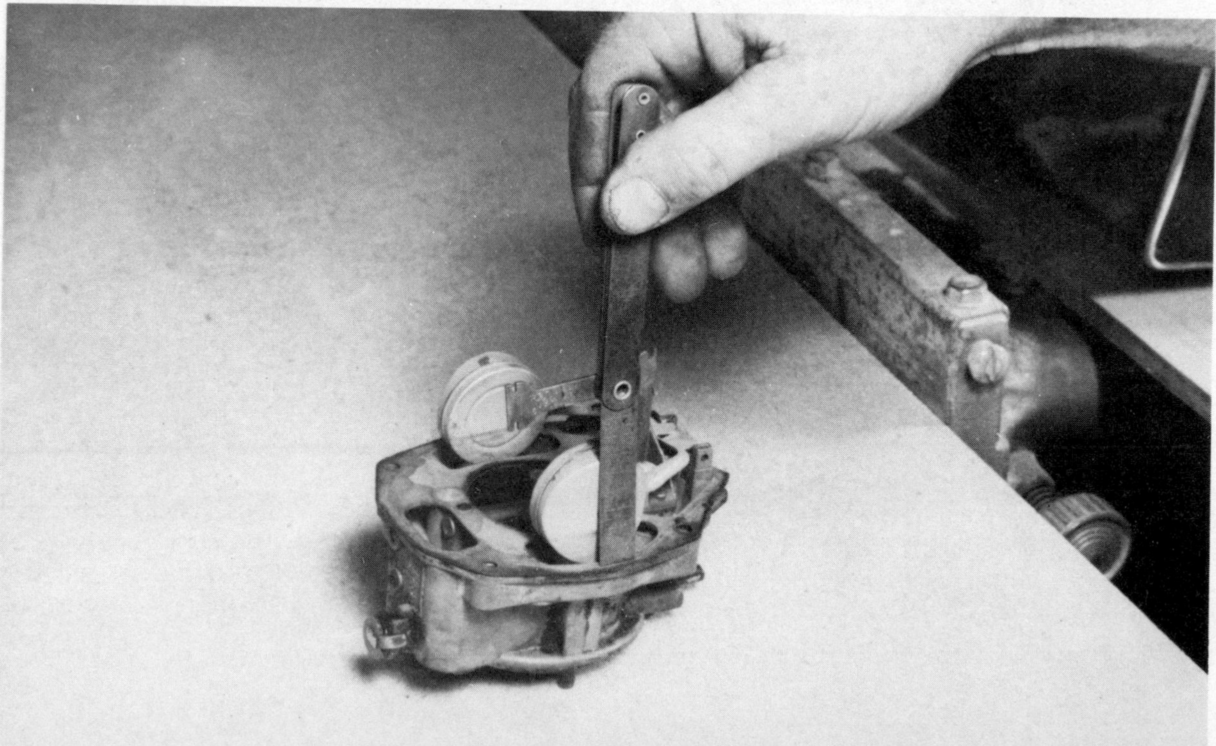

Fig. 3–48. To set float level on this car's Holley 1945 one-barrel carburetor, begin by removing the air cleaner assembly and the air cleaner gasket. Disconnect the choke assembly from the thermostatic coil in the manifold well. Remove the fuel bowl vent hose, then the fast-idle cam retaining clip, and then the cam itself and its link. Remove the vacuum break diaphragm unit, the vacuum hose, and the link to the choke. Then, as illustrated, remove the throttle return spring, the accelerator pump arm, and linkage assembly.

Troubleshooting

Although the basic system is warranteed for five years or 50,000 miles, at some point you may have to troubleshoot it. GM's computer–controlled system, introduced in 1979 on a few cars built for sale in California, includes an on-board diagnostic system to enable you to find the most common problems easily. The system was called C-4 (Computer–Controlled Catalytic Converter) because the carburetor mixture is controlled, as explained earlier, to permit a special catalytic converter to function properly. In 1981, when the computer was installed on all models (except diesels and Cadillacs with throttle body fuel injection), the name was changed to CCC (Computer Command Control) and the black box itself was modified. The modifications primarily are intended to enable the computer to handle new functions, such as ignition timing advance, operation of many emission controls and idle speed.

If a "Check Ignition" light on the dash goes on, you head for the underdash computer near the glovebox (see Fig. 3–60) and connect a jumper wire from the trouble code connector—a white and black wire hanging down from the computer—to an electrical ground on the car body. The Check Engine light then will flash a trouble code number three times, pause, then repeat, as long as the jumper is connected and the ignition is on. Just look at a chart and you'll see the number corresponds to a particular

Fig. 3–49. Remove the seven bowl cover screws and lift the bowl cover straight up until the vacuum piston stem, accelerator pump, and main well tube are clear of the carburetor main body. Depress the float as shown to allow fuel (under residual pressure in the fuel line) to overfill the bowl to within 1/8 to 1/4 inch below the top of the bowl. If the residual line pressure isn't enough to fill the bowl, pour in fuel from a can. With two wrenches, loosen the fuel line nut (leave the carburetor fitting tight). Place the bowl cover gasket on top of the fuel bowl.

problem. Example: 14 is a shorted coolant-temperature sensor circuit; 15 is an open in the coolant-temperature sensor circuit.

The built-in diagnostic system also can find an intermittent problem. Once the "Check Engine" light goes on, the fault goes into the computer memory. With an intermittent problem, the fault code is lost if your turn off the ignition. Therefore, a second wire hangs down from the computer to handle this situation. The second wire is orange, and if you connect it with a jumper wire to the car battery, the computer won't forget. Once you get the information and correct the problem, you must disconnect the jumper to the battery, or in time the battery could run down.

As of 1981, the orange wire is wired from the computer to a battery circuit, so there is no need to connect it. Therefore, all you have to do to probe the computer memory is ground the black/white wire (Fig. 3–57), and that's easier as of 1981 too. Locate a five-wire connector (Fig. 3–57) under the dashboard, perhaps held by screws to the base of the dash. At one end you'll find a black wire (which is to an electrical ground) and right next to it is the black/white wire. To ground the black/white wire, therefore, just shove an opened paper clip into the adjacent terminals of the connector.

If you get any trouble code but 12 on the check engine light (12 is light pulse, pause, then two more light pulses), there is a problem somewhere in the computer control system, its sensors or operating devices. Actually 12 is a trouble code too. It means that the computer is not receiving an engine r.p.m. signal from the electronic ignition unit. However, when you made the test (grounding the black/white wire) with the engine off, naturally there is no engine r.p.m. signal to send. If you make the test with the engine running (the engine will start and run despite many problems in the system that may occur), when you ground the black/white wire you should not get the 12 code.

Problems not pinpointed by the on-board diagnostic system can be checked out with ordinary test equipment (ohmmeter, test lamp, dwellmeter, vacuum gauge, etc.), following a procedure in the service manual for the car. Note: On cars with this system, the TACH terminal in the distributor cap flange is wired. To connect your tach, trace the wire or wires from the terminal. One may go to an unconnected terminal, to which you can connect your tachometer. If it doesn't, trace the wire to the nearest connector, separate the connector and insert a metal rod, such as a nail, to rejoin the terminals without pushing them back together. Then attach the tachometer lead to the nail.

Early Ford computer systems (Electronic Engine Control I and II) required special test equipment for precise checkout. As of 1980, however, diagnostic connectors were installed on EEC III and the new system Microprocessor Control Unit, to permit troubleshooting with ordinary test equipment. The self-test procedure for EEC III is given in Chapter 2, under *Checking Computer-Controlled Timing*. If your car is equipped with Microprocessor Control Unit (MCU), the computer presently controls fuel mixture, not ignition timing. MCU is used on all computer-equipped cars except the big Fords, Mercurys and Lincolns (Escort and Lynx have no computer). MCU has been made with two self-test procedures, one (on 1980 models) a long procedure using a voltmeter, tachometer, vacuum pump and similar, routine equipment. The length of the procedure precludes inclusion in this book. In 1981 the second procedure was introduced, and it is more automatic. To determine if your 1981-on MCU system is apparently sound, activate the self-test, as follows:

(1) Locate a six-wire connector on the passenger's side of the engine compartment (see Fig. 3–58).

(2) Connect a voltmeter positive lead to the battery starter terminal, and the voltmeter negative lead into No. 4 terminal. Connect a jumper wire from No. 5 to No. 2 terminal.

(3) Turn on the ignition and you should get a code 11 (voltmeter needle pulses to the middle of the scale, pause, then another pulse). If you get this, the system passes the engine-off test.

(4) Run the engine, four-cylinders at 2500–2800 r.p.m., V-8s at idle. When you get two needle pulses on a four, four pulses on a V-8, the self-diagnosis is underway.

(5) On V-8s there is a device called a knock sensor (Fig. 3–59) that tells the electronic ignition unit to retard the ignition timing when the engine has started to knock (ping, etc.). You must trigger this sensor on V-8s for proper operation of the self-diagnosis feature. Immediately tap vigorously on the intake manifold very close to the knock sensor (not on it), using a long bar and perhaps a hammer. If you don't do this promptly, or if the knock sensor is bad, you'll get a code 25 (two needle pulses, pause, then five needle pulses).

The self-test then will continue automatically. After about a minute, you should see a code 11 if the computer system is operating properly. If you see other codes, the system needs attention (it's warranteed for five years or 50,000 miles, so leave the work to the dealer).

The four-cylinder computer will give out only one trouble code. If there are more faults, you must fix the one shown, then repeat the test in order to get any other, again just one at a time. On the V-8, all fault codes will be shown, and the sequence will be repeated once.

Because the MCU self-test relies on needle pulses, you can only use a voltmeter with a needle and dial, not a digital type.

American Motors cars also use the MCU computer, but it's a bit different than the Ford setup. There was no self-test procedure in 1980. There is a self-test in 1981, wired to a dashboard check engine light. If a failure occurs, the light goes on and pulses out a trouble code (it stops pulsing after the code has been repeated four times). Any number of trouble codes could be pulsed when you are driving the car.

Trouble codes on all the engine control computers are not always indicative of precise problems. In many cases, the code only indicates a trouble area, and it is necessary to follow a manual checkout procedure to pinpoint the failure. The manual checkout is always based on the use

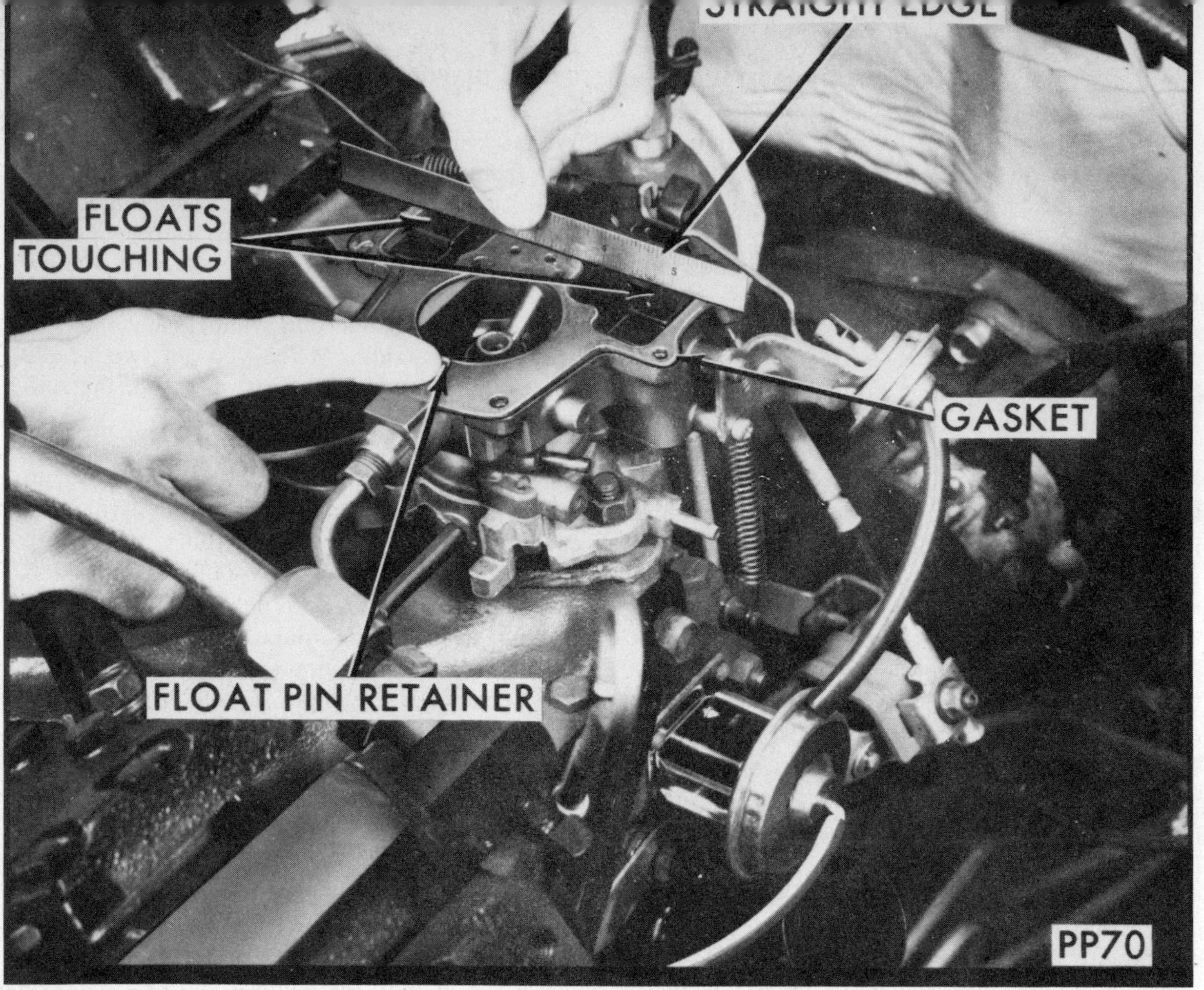

Fig. 3–50. Firmly seat the float pin retainer by hand and place a straight edge across the gasket surface as illustrated. The portions of the float furthest from the fuel line inlet should just touch the straight edge as shown. If adjustment is necessary, bend the float tab. Remove the bowl cover gasket.

of routine service instruments, such as voltmeter, tachometer, test lamp, vacuum gauge, etc.

Chrysler products computers do not have a self-test at this time. However, you can make a rough check of computer control of ignition timing by starting the engine cold and letting it warm up. As the engine gets warm, the timing should retard. This test will require an adjustable timing light, as the initial readings will be off the timing marks indicator scale.

If you still have the car after the five-year, 50,000-mile warranty on the computer system has expired, obtain a service ("shop") manual for your car to get the manual checkout procedure, both for cars with self-diagnosis and those Chrysler products without it. Checkout without special equipment also is possible on Chrysler and AMC systems, following a service manual procedure.

Chrysler's electronic system is also designed for special shop equipment, but a manual checkout procedure, with ordinary test equipment, is available in the factory service manual.

Air Cleaner Service

Unless the engine has an adequate source of clean air it will not run very well for very long. Virtually all modern cars have a pleated paper air filter under a cover in the air cleaner housing. Just remove the cover wingnut or spring clips, lift off the cover, and there's the filter element. If it's dirty on the outside, or has gone two years or 25,000 miles, replace it (see Fig. 3–18).

Some cars have a filter element enclosed in a sealed cannister, so you can't see the condition of the paper. The recommended replacement interval is five years or 50,000 miles, but few make it that long. If the engine runs smoother

Fig. 3–51. Drain fuel from the accelerator pump well with a syringe, as shown, to prevent internal parts still there (a check ball and weight) from being moved out of position when the bowl cover is reinstalled. Put the gasket back and carefully install the bowl cover, making sure the leading edge of the accelerator pump sealing cup isn't damaged as it enters the pump bore. Be careful not to damage the main well tube. Install the cover screws and tighten alternately and in stages. Install the throttle return spring, accelerator pump arm and linkage, fast idle clip, cam and link, and bowl vent hose; reconnect the choke, install the air cleaner and gasket, and set idle speed to specifications.

with the filter off (test very briefly, for a minute or two, to prevent excessive dirt from getting in), replace the sealed cannister (see Fig. 3–61). Or replace every two years or 25,000 miles to be sure.

Fuel Pumps

The mechanical pump contains a diaphragm with spring-loaded discs that cover and uncover ports inside. One port is connected to a line from the gas tank; the other is connected to the line to the carburetor bowl. The diaphragm is flexed up and down to produce the pumping action. Operating the diaphragm is a spring-loaded arm that is pushed by a lobe on a shaft (called the camshaft) inside the engine. When the arm is pushed in, the diaphragm is flexed downward, drawing in fuel from the tank. The spring then retracts the arm and the diaphragm and the fuel is pushed up to the carburetor (see Figs. 3–62, 3–63).

The electric fuel pump is either a diaphragm device operated by a set of contact points (much like the breaker points in an ignition system) or a motorized device with the motor operating a turbine, piston, etc.

The exact method of internal operation of a fuel pump is not important, for the typical fuel pump is sealed. If it doesn't work you replace it.

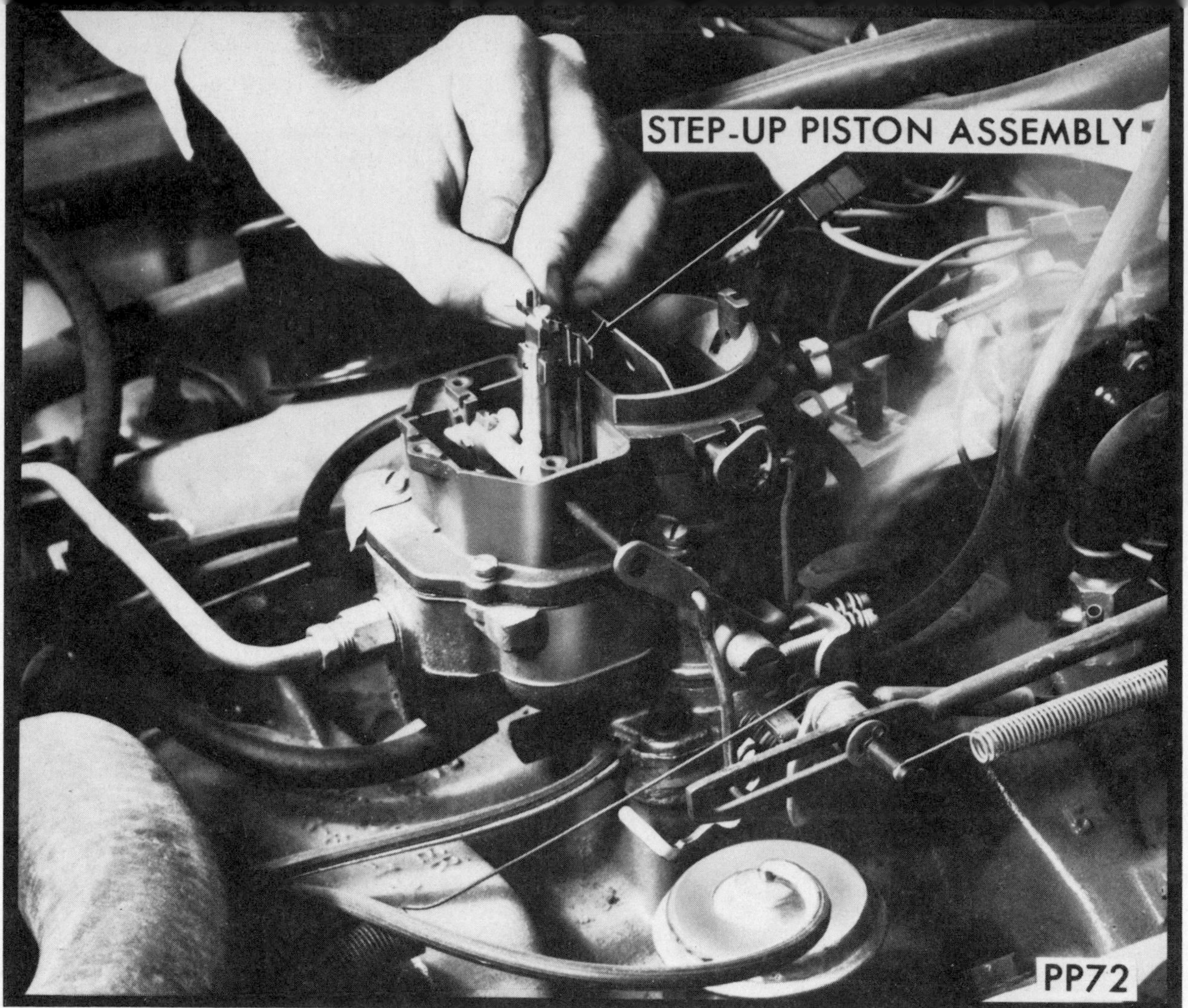

Fig. 3–52. To set the float level on this car's two-barrel Carter BBD, begin by removing the air cleaner and gasket and the air cleaner mounting bolt assembly. Disconnect the choke and the hoses from the choke vacuum diaphragm unit and the idle enrichment diaphragm (a triangular part held to the carburetor by three screws). Remove the retaining clip from the accelerator pump arm link and take out the link. Remove the fast idle cam retaining clip and take off the link. Remove the step-up vacuum piston cover plate and gasket from the top of the carburetor. Remove the metering rod lifter lock screw, then lift the step-up vacuum piston and metering rod straight up and out, as illustrated.

A defective pump may be responsible for starting failure, or engine misfire or weak performance.

Testing the Pump

First check to see if fuel is getting to the carburetor. Remove the air cleaner cover, push open the choke plate with your finger, have a helper pump the gas pedal, and look for squirts into the carburetor barrel (from the carburetor's accelerator pump). If there is no fuel squirting from the carburetor, the problem likely is that no fuel is being delivered.

To test the fuel pump itself, disconnect the fuel line from the carburetor and aim the end into a quart container (see Fig. 3–64). Have a helper start the engine and run it on the fuel in the carburetor bowl. A healthy pump should deliver a stream of fuel amounting to a pint in 30 seconds on a V–8, 45 seconds on a four- or six-cylinder engine. If the engine does not start, just crank it for 30 seconds. The amount of fuel collected in the container should be close to what you'd get with the engine running at idle.

If the pump is not delivering, or delivering much too little, there are four possibilities:

• *Clogged fuel filter.* To check, remove the filter as explained in the previous section and repeat the test. With the in-line type, place the

Fig. 3–53. Remove the baffle in the fuel bowl, then depress the float manually (or pour in fuel) to bring the fuel to within 1/4 to 3/8 inch of the top of the bowl. Loosen the fuel line nut, but keep the carburetor fitting tight.

filter aside and position the container at the end of the section of fuel line from the pump.

• *Restricted fuel line.* Visually inspect the fuel line from pump to carburetor, looking for kinks in flexible hose sections and dents in steel sections. Next, jack up the car, support it on stands (see *Chapter 5*), and follow the line back to the tank to check for kinks or dents. A kink may be eliminated by repositioning the line slightly, particularly in the area of a retaining clip. A dent can only be corrected either by replacing the line or by cutting out the dented section with a tubing cutter and installing a piece of flexible fuel line hose which is secured with hose clamps. If you attempt this job on a line between a mechanical pump and rear gas tank, you should jack up the front of the car as high as possible, so as little fuel as possible will flow out of the cut fuel line while you're working. Install a pan underneath to catch the fuel after you cut the line.

• *On electric pumps, a problem in the electrical circuitry.* The problem could be a bad fuse, a defective electrical ground at the pump, a poor connection at the pump or, on some cars, a defective oil pressure sensor (the sensor feeds current to the pump so that the pump operates only when the engine is running, not just when the ignition is on).

First check for a defective fuse. On some cars there is a second fused circuit that operates the pump when the engine is cranking. If your car starts easily you might never know it failed unless the car runs out of gas, in which case it won't restart unless the circuit is repaired or the engine is started by pouring some gasoline directly into the carburetor.

If the problem is not a defective fuse, go to the pump and look for a black wire that is simply screwed down to the car body. The screw-down point could be anywhere, including in the trunk

Fig. 3–54. Seat the float pin retainer by hand while you measure the distance from the top of the float at its center to the top of the bowl (gasket off), and compare with specifications. If you have a gauge such as the one illustrated, the crown of the floats should just touch the gauge as you lay the gauge across the bowl's top surface. If necessary to adjust, hold the floats in the bowl and bend the float tab toward or away from the needle. When bending, don't allow the tab to push against the needle because it has a rubber tip and this will be compressed enough to provide a false reading.

compartment. This is the electrical ground, and if it's loose or corroded the pump won't work. Clean and tighten.

Other problems, including replacement of a defective electric pump, require experience to check and should be left to a professional mechanic.

• *A defective fuel pump.* If this final possibility seems to be it, you can double-check using a vacuum gauge with a pressure section on the dial (most vacuum gauges except the pump-gauge described under *Vacuum Choke Break* have a pressure section). Fit the gauge hose over the carburetor end of the fuel line and have a helper crank or run the engine (see Fig. 3–64). The reading you get should be compared with factory specifications. In the absence of specifications, look for a reading of at least 2½ p.s.i. with the engine running at idle.

Replacing a Mechanical Pump

Most electric fuel pumps, unless they are located in the engine compartment (and the typical one is in the gas tank), require special tools to replace. Therefore, this section is limited to mechanical pumps bolted to the engine.

First, disconnect the fuel lines from the pump. If they are metal, disconnect them with two wrenches, as for a fuel line at the carburetor (see Fig. 3–65).

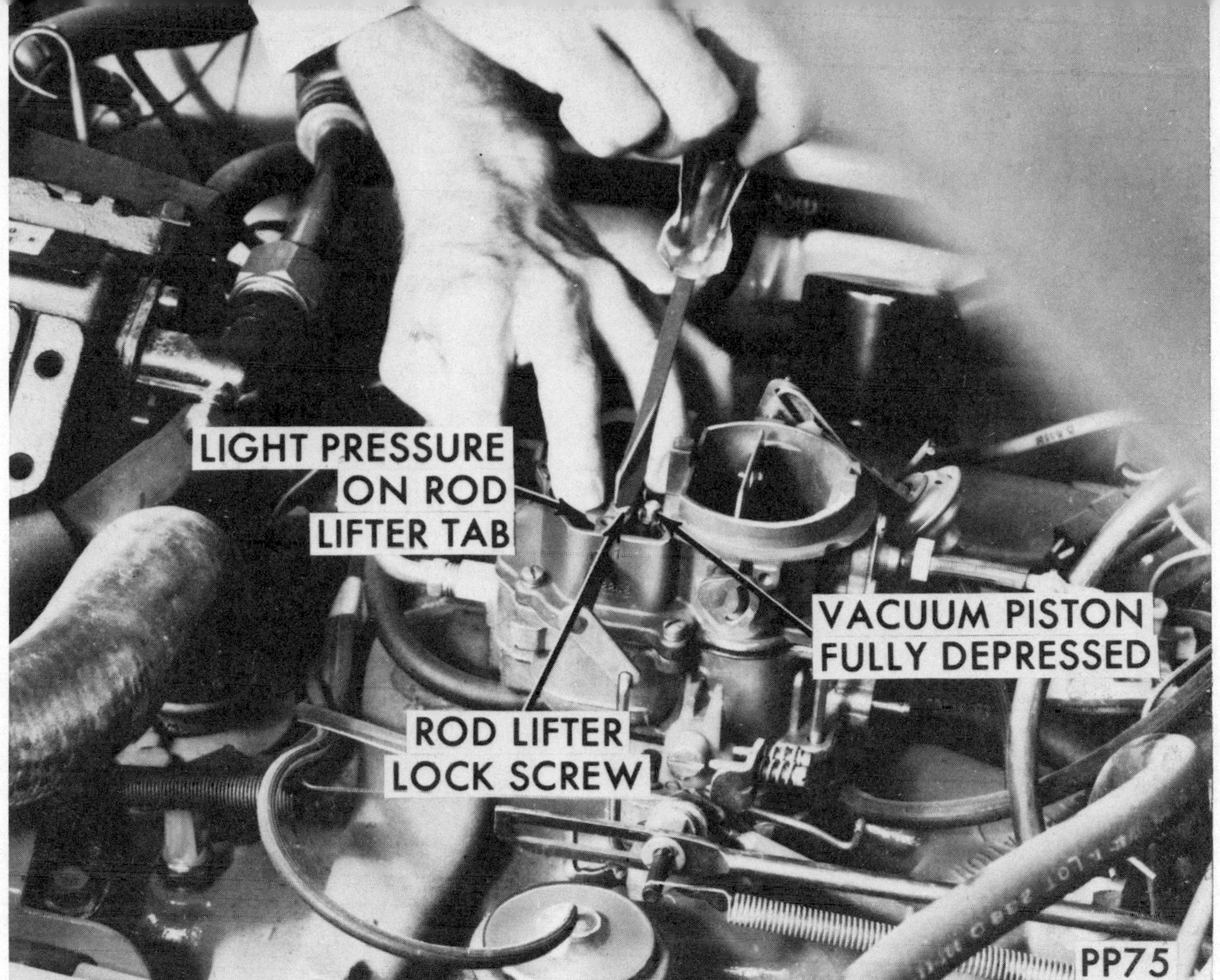

Fig. 3–55. Install the step-up vacuum piston and metering rod assembly, and then the metering rod lifter lockscrew. Slacken the idle speed screw until the throttle plates are fully closed (look inside the carburetor barrel). Count the number of turns of the screw so you can return it to the original setting. Fully depress the step-up vacuum piston while holding light pressure on the metering rod lifter tab, then tighten the lockscrew, as illustrated. Install the step-up vacuum piston cover plate and gasket, return idle speed screw to original setting, install fast idle cam link and retaining clip. Connect the choke vacuum diaphragm unit and idle enrichment diaphragm hoses, install choke, air cleaner mounting bolt assembly, air cleaner gasket, and air cleaner itself. Adjust idle speed.

Next, remove the pump mounting bolts or nuts with a ratchet and socket, or with whatever wrench will fit in if a socket won't (see Fig. 3–66).

Pull the pump out (see Fig. 3–67) and scrape any gasket residue from the engine.

Apply a thin film of non-hardening gasket sealer to the replacement pump's gasket surface only. The sealer is sold in small tubes in auto supply stores. Position the gasket on the pump, and the sealer will hold it in place.

Refit the pump to the engine. Note: On many General Motors engines, particularly Chevrolet, the camshaft does not push on the arm directly. Instead, there is a plunger between them. The camshaft pushes on the plunger which pushes on the rod. When you take out the old pump, the plunger drops down. You have to push it back up and hold it there so you can install the new pump with the diaphragm arm under it. It doesn't sound hard, but there's no room for your fingers. Special tools are available, but here are two other solutions:

(1) First, reach in with your fingers and smear some grease on the plunger; then push it up. The grease will hold it there long enough to get the pump in if you work fast.

(2) Push the plunger up with your finger and hold it up with a piece of stiff wire (such as a paper clip) bent into an L-shape. As you hold it up with the wire, slip in the pump and, just as

Fig. 3–56. Oxygen sensor is threaded into exhaust pipe. To replace it, remove electrical connector and unscrew it with a wrench, such as a deep socket and ratchet.

you're ready to seat the pump into bolt-on position, yank out the wire.

Install the bolts or nuts, starting them by hand, then tightening with a wrench. Reconnect the fuel lines and the job is done. Note: On some cars the location of one of the fuel line fittings on the pump is such that with the pump in place you will struggle to reconnect. If you see this problem (perhaps you encountered a similar one when disconnecting), make the reconnection before you install the pump.

Fuel Injection

In a carburetor, inrushing air draws fuel droplets out of a nozzle and carries them into the intake manifold, and from there into the cylinders. Fuel injection is a much more precise system, compared even with electronically-regulated carburetors, because the amount of fuel can be more precisely controlled and, in most designs, it can be sprayed in even amounts very close to the cylinders.

If the control is electronic it often is a computer, the same one that regulates spark timing on some cars. If the control is hydraulic it may be modified to include some solenoids (electromagnetic switches) for further regulation by a master computer. See *Computer-controlled Spark Advance* in Chapter 2.

Three types of fuel injection systems are in use:

- *An electronically-controlled system with electrically-operated injectors,* one for each cylinder, is on some Cadillacs and a few imports. The injectors are located just outside the intake ports of the cylinders. Electronic, electrical, and mechanical sensors for engine and ambient temperature, engine speed, throttle position, etc., tell a computer how much fuel the engine needs, and the computer then supplies current to the injectors for an appropriate length of time. The longer the computer energizes the injectors, the more fuel they spray into the intake ports (see Fig. 3–68).

- *A hydraulically-operated system with mechanical injectors,* one for each cylinder, is used on many imported cars. A flap in the air intake (see Fig. 3–69) moves inward as airflow into the engine increases. As it does it pushes up

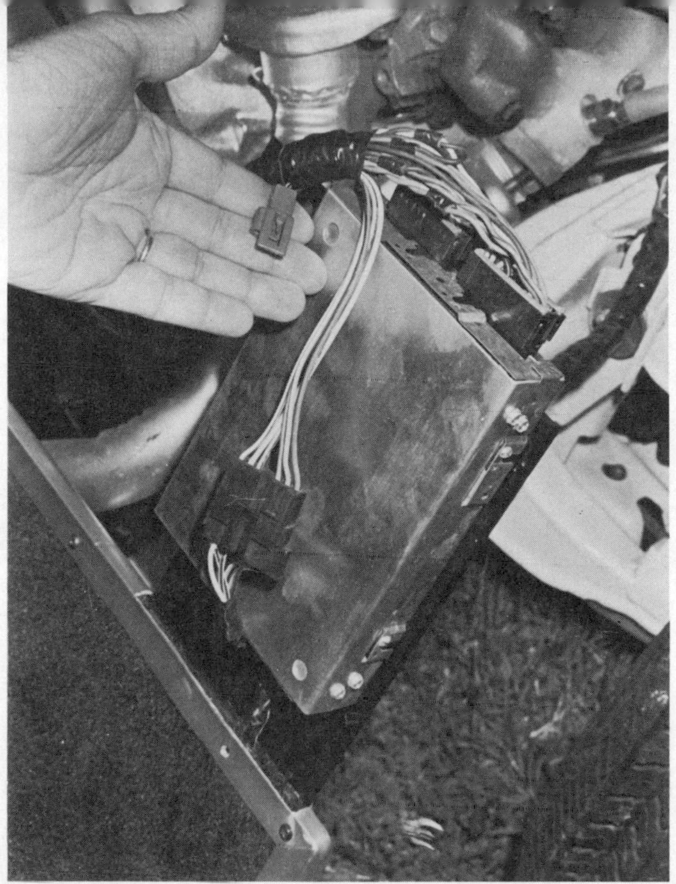

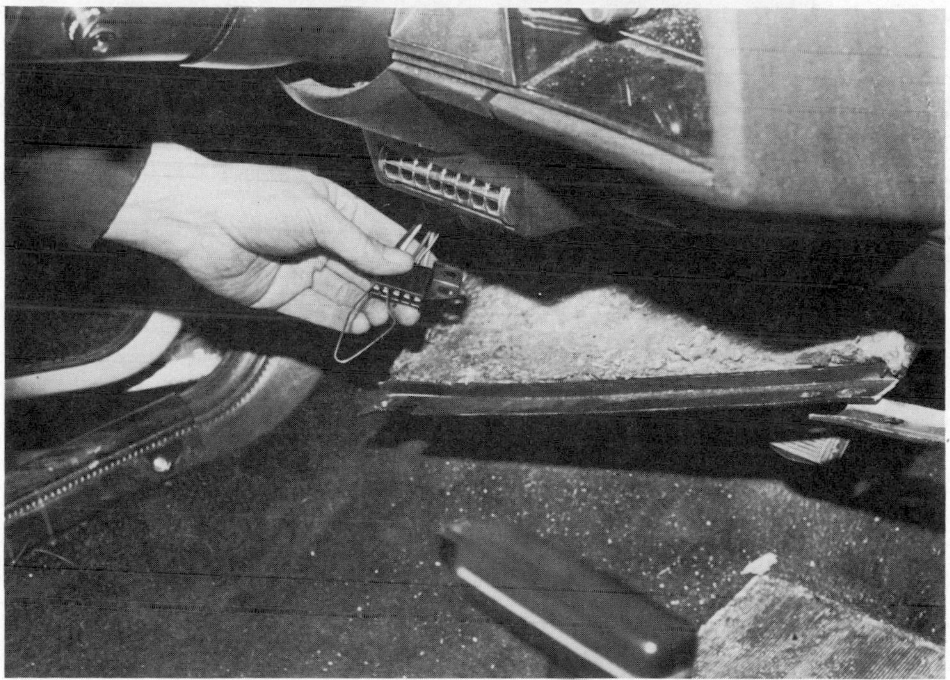

Fig. 3–57. To probe the General Motors computer (all cars except Cadillac with throttle body fuel injection and all diesels), locate the special black/white wire hanging down under the dashboard. On pre-1981 models, it is a single wire hanging down on the passenger's side from the computer, as shown in the top photo. On 1981 models, it is part of a five-wire connector just toward the driver's side of the dashboard (you may have to remove a trim panel underneath to find it). With the single wire, connect it with a jumper wire to an electrical ground. With the five-wire connector, insert a paper clip as shown between the terminals for the black/white wire and the black wire next to it at one end.

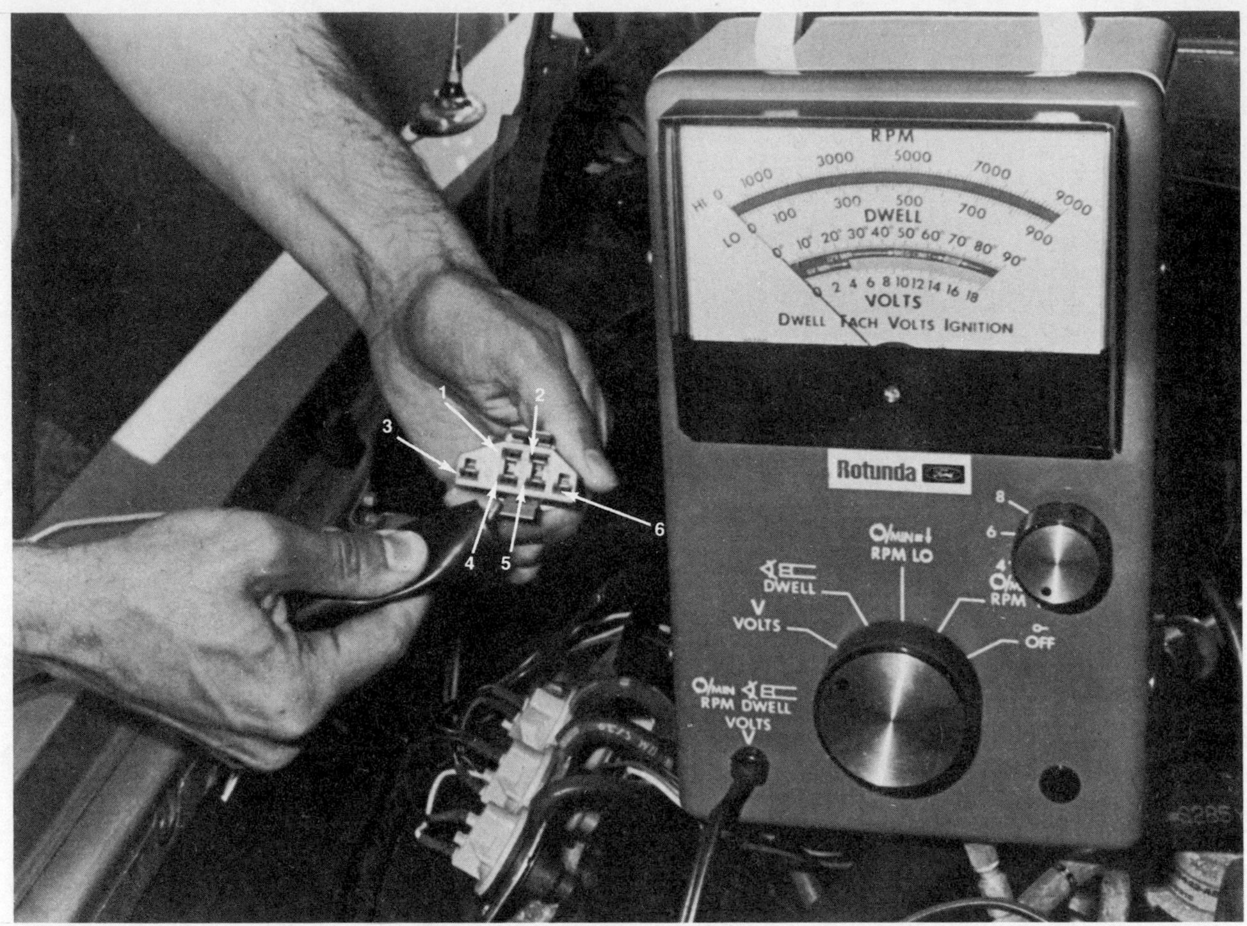

Fig. 3–58. If you have a 1981 Ford product with the Microprocessor Control Unit (MCU) computer, locate the six-wire connector on the passenger's side of the engine compartment. Connect the voltmeter positive lead to the battery starter terminal and the negative lead to the No. 4 terminal of the connector (numbering applies to the unwired side). Then connect a jumper wire from No. 5 to No. 2 (or to an electrical ground).

on a valve that regulates fuel flow to the injectors. The higher the valve rises, the greater is the fuel flow. As the engine warms up, however, it needs less fuel. To regulate the flow, fuel under pressure is also fed to the top of this valve. When the engine is cold and maximum fuel flow is required, this fuel pressure is bled off, so the valve rises. As the engine warms up the bleeding of the fuel pressure is stopped, so it pushes down harder on the valve, preventing it from rising so much, thereby reducing the fuel flow to the injectors. The injectors themselves are like shower heads with a tiny check valve that prevents spraying until there is a minimum amount of pressure.

• *A throttle body fuel injection design* has a provision to inject fuel only at one point—the center of the intake manifold. It is built into a module that replaces the carburetor. However, it has a barrel or two barrels with throttle plates, just like a carburetor. Hence the name.

In the most popular design, used by Ford (see Fig. 3–70) and General Motors, there are one or two electrical fuel injectors in the throttle body, and they are activated and deactivated by the engine control computer. This is far less expensive than an injector at each cylinder, and although perhaps not as effective, it is simpler and more reliable than the complex carburetors needed on modern cars.

A second throttle body design is that introduced on 1981 Chrysler Imperials. In this type, the fuel is delivered to the throttle body module by a low-pressure pump in the gas tank. At the

Fig. 3–59. This is a Ford V-8 knock sensor on intake manifold.

throttle body the fuel is picked up by another, high-pressure pump controlled by the computer. The fuel flows through this high-pressure pump into bars across the throttle body barrels, just above the throttle plates. These bars have calibrated holes in them, and the fuel sprays out into the air stream. If the computer senses that more fuel is needed, it supplies more current to the throttle body pump, which turns faster as a result and pushes more fuel into the spray bars. (See Fig. 3–71.)

Precise Mixture Control

Whatever the system, it provides very precise fuel mixture control, which is important if the car is to meet exhaust emission standards.

In many of these systems, the computer receives signals from an oxygen sensor threaded into the exhaust manifold. Oxygen in the exhaust is an indicator of the fuel mixture going through the injection system, so if necessary the computer can richen or lean out the mixture as required for low emissions. The oxygen sensor arrangement was introduced on cars built for sale in California, which has very strict emissions regulations of its own. With this device, the mixture can be controlled so precisely that a special catalytic converter can be used, one that controls all major air pollutants. As explained earlier in this chapter, some carburetor fuel systems also have oxygen sensor electronic controls. The electronic control may be a computer or a simple on-off switching unit. In most cases beginning in 1981, the control is a computer.

Fuel Injection Service

Although major fuel injection service is a job for a professional, a weekend mechanic can

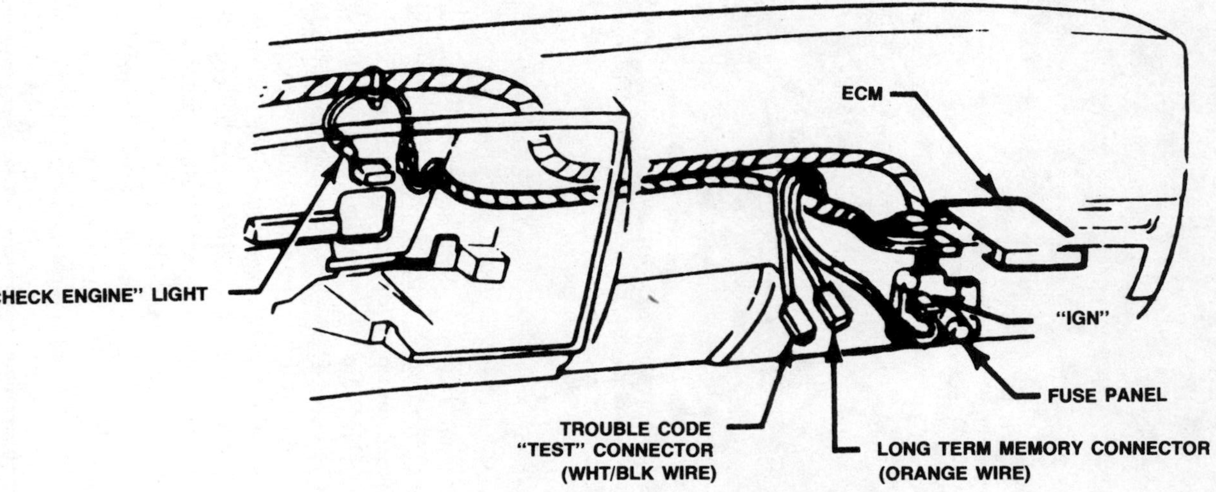

Fig. 3–60. This cutaway drawing of a GM dashboard shows the computer with on-board diagnosis. The white and black wire at left is grounded to produce a trouble code on the Check Engine light. The orange wire at right is connected to the car battery to keep any trouble code in the computer's memory if the ignition is shut off.

Fig. 3–61. This is a sealed cannister air cleaner, viewed from the underside. You can peer in from the intake at left or through the circular opening in the center, but you won't be able to tell if the element is clean or dirty.

perform the only required maintenance—replacement of the fuel filter—and do some basic troubleshooting.

Before you try, however, you should understand that the injection system is basically reliable, and you first should look to other items, such as the ignition system, for a solution to a performance or starting problem. If the engine cranks well and there is a good spark, then you can look at the fuel system, and even then, consider first these routine inspection items.

Routine Inspection

On all fuel injection systems, you should check to make sure that no lines are kinked or dented, and that all electrical connections are secure. With electronic systems, the computer ground is important. It may be a wire secured to the engine, or through the computer mounting screws. A loose ground wire, or loose and/or corroded computer mounting screws can cause a variety of problems.

• *Checking for fuel.* There are various fuel delivery checks you can make if the engine won't start, depending on the system. With the electronic system as used on Cadillac gasoline engines with an injector at each cylinder, you can remove a cap from the fuel tubing loop on the engine and expose a tire-type valve (see Fig. 3–72). With the engine cranking, just depress the valve tip (keep a rag handy to catch spillage), and if fuel squirts out you know there's fuel coming up to the system. On imports that use a similar system, there is no tire-type valve, so for the same check you must disconnect a flexible fuel hose.

If there is no fuel up to the tubing, look for a kinked fuel line between the engine and the gas tank, and check the in-tank electric pump. General procedures for this service are covered in the previous section of this chapter on fuel pumps. Also note that on some models the fuel filter is located between an electric fuel pump near (not in) the tank and the tank itself. In this case, check

Fig. 3–62. Exterior view of a mechanical fuel pump. Pump's inlet fitting is at lower right, outlet is at left. At upper right you can see part of the spring-loaded arm, also called an actuating lever.

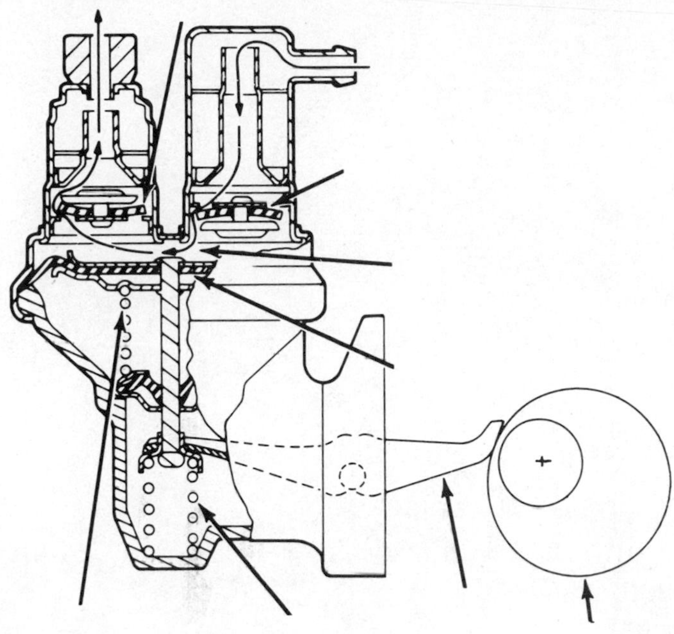

Fig. 3–63. This drawing shows how the pump works. As eccentric lobe on camshaft turns, it pushes on actuating lever that flexes the diaphragm. When the diaphragm flexes down, it creates a vacuum that draws open the inlet valve and pulls fuel from the gas tank. The camshaft turns, relaxing pressure on the lever, so the diaphragm spring is pushed back. This creates pressure against the fuel, which pushes open the outlet valve, and flows up a piece of tubing to the carburetor.

for a plugged fuel filter, and in very cold weather, for one in which moisture has frozen.

On the hydraulic fuel injection (called K-Jetronic) you can check both for fuel going into the fuel distributor (disconnect the line at the distributor as in Fig. 3–73) and either connect a gauge or see if fuel spurts out. To test for fuel at the injectors, proceed as follows:

(1) pull an injector out of its bore and aim it into a jar;

(2) gain access to the airflow sensor by removing the duct (see Fig. 3–74);

(3) disconnect any wiring at the airflow sensor housing (see Fig. 3–75) and the back of the alternator.

Locate the fuse box under the dashboard. On pre-1981 models, you will see a line of metal boxes (switching devices called relays) above the fuses. One has a fuse attached to its exterior; this is the fuel pump relay. Unplug it. With a jumper wire, bridge terminals L-13 and L-14 (they are at 12 and 6 o'clock and numbered) in the underdash panel. Starting in 1981, the cars have the small spade-type fuses and the relays are installed in the fuse box. The relay terminals in the fuse box, however, are not numbered, but the ones you must bridge with a jumper wire again are those at 12 and 6 o'clock (see Fig. 3–76).

(4) with pliers, grasp the center of the airflow sensor and move it. You should see fuel spray from the injector into the jar (see Fig. 3–77).

Note: Stop quickly, or the other cylinders may be filled with fuel and hydraulically lock up. If fuel is going into the fuel distributor, but not going to the injectors, check injector tubing for

Fig. 3–64. Fuel line is disconnected from carburetor, and hose is installed over the end of the line. Other end of hose then is aimed into quart container (a clean, empty oil can in this case) or fitted to pressure gauge, as shown above, and the engine is cranked. Amount of fuel discharged into container is measured and compared with specifications.

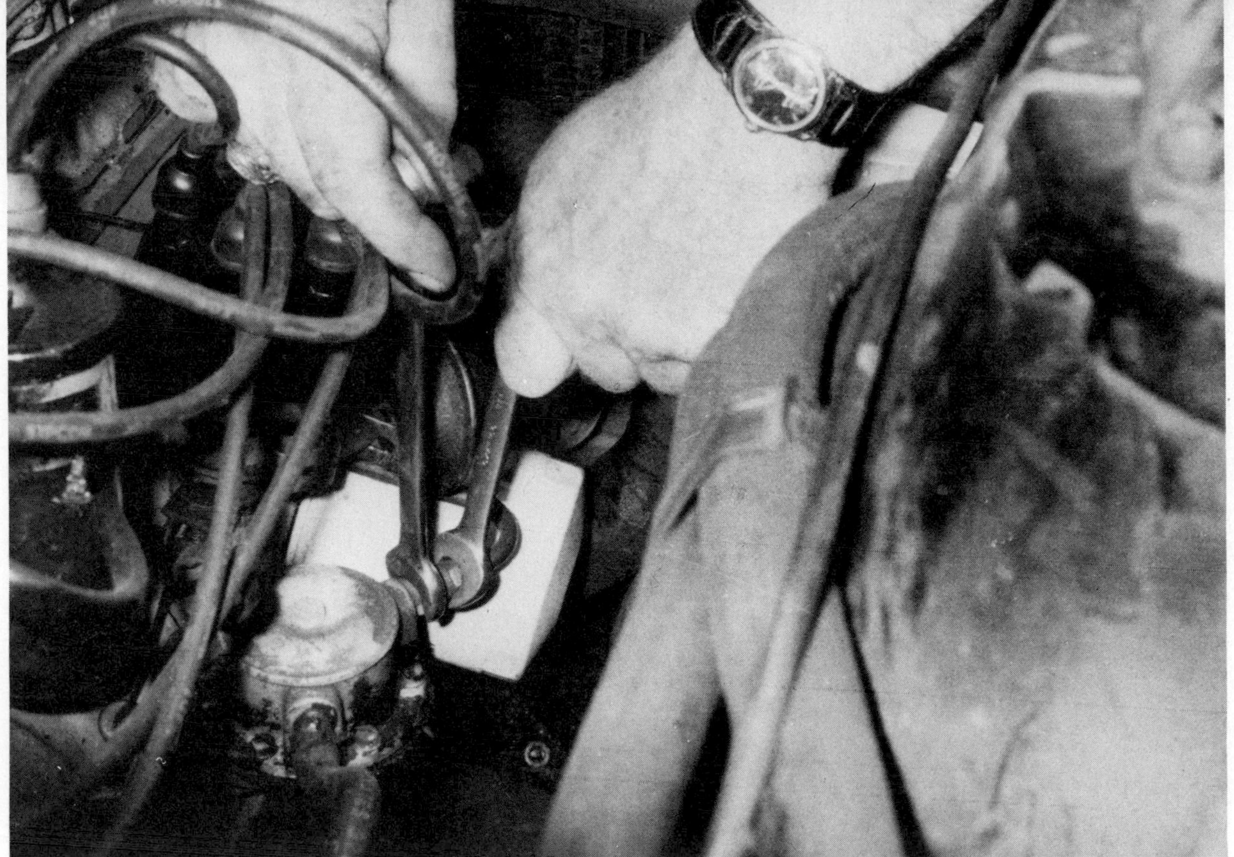

Fig. 3–65. Using two wrenches to disconnect fuel line from pump. One wrench holds the pump fitting while other slackens fuel line nut.

kinks, and if not kinked, the problem is in the fuel distributor, which is replaced on an exchange basis.

With injector-type throttle body fuel injection, begin by looking for a fuel spray from the injectors as the engine is cranked (air cleaner cover must be off, and in some cases the entire air cleaner housing should be removed for easier viewing). On the Chrysler Imperial system, have a helper turn on the ignition as you look for a single squirt from the spray bars.

If you don't see the spray as the engine is cranked, or the single squirt from the Chrysler spray bars, locate a Schrader (tire-type) valve in the fuel line near the throttle body on GM and Chrysler cars, on the throttle body of Ford products (see Figs. 3–9 and 3–10). Press the Schrader valve pin while the engine is being cranked, and you should get a strong squirt of fuel (keep a rag handy to catch the fuel). If you get a strong squirt of fuel from the Schrader valve, check the fuel filter (remove and install a piece of fuel line hose in its place, then repeat the Schrader valve test). If the filter passes, the problem is in the throttle body, a job for a professional. If there still is no fuel to the Schrader valve, check for a kinked fuel hose or dented fuel line. Also inspect for loose or corroded electrical connections at the fuel pump, which project from the in-tank pump.

On '82 GM cars with throttle body injection, there's no Schrader. If fuel doesn't spray from the injector (engine cranking), disconnect the fuel line to check pump action.

Fuel Filter

Change the fuel filter at the intervals specified by the car manufacturer. On cars in which the filter is located in an inconvenient place, use fuel-line anti-freeze regularly in cold weather to minimize the possibility of a moisture freezeup that might be physically difficult to correct.

Air Intake

Even more so than on carburetor systems, air leaks can cause performance and starting problems. Check tightness of the nuts or bolts holding the intake air distributor to the engine as part of a tuneup.

A clean air filter, as with a carburetor system, also is important. Refer to *Carburetor Service*, earlier in this chapter, for details.

Fig. 3–66. Once fuel lines are disconnected, remove the pump mounting bolts, as shown.

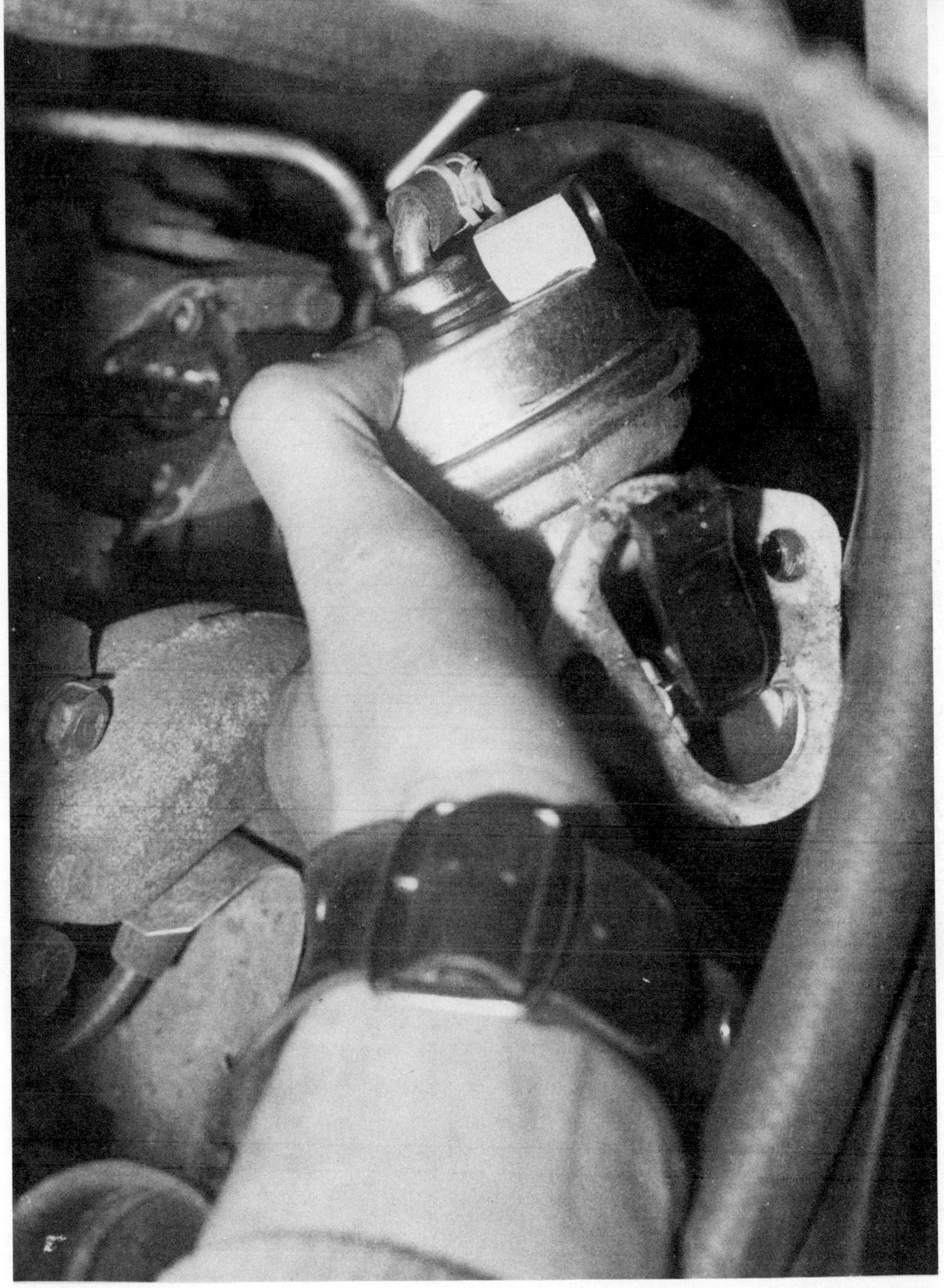

Fig. 3–67. Here's the fuel pump being removed from the engine. Notice the actuating lever. In this case, the flexible fuel hose was left in place. The pump is being brought to a convenient working position, where the hose will be disconnected.

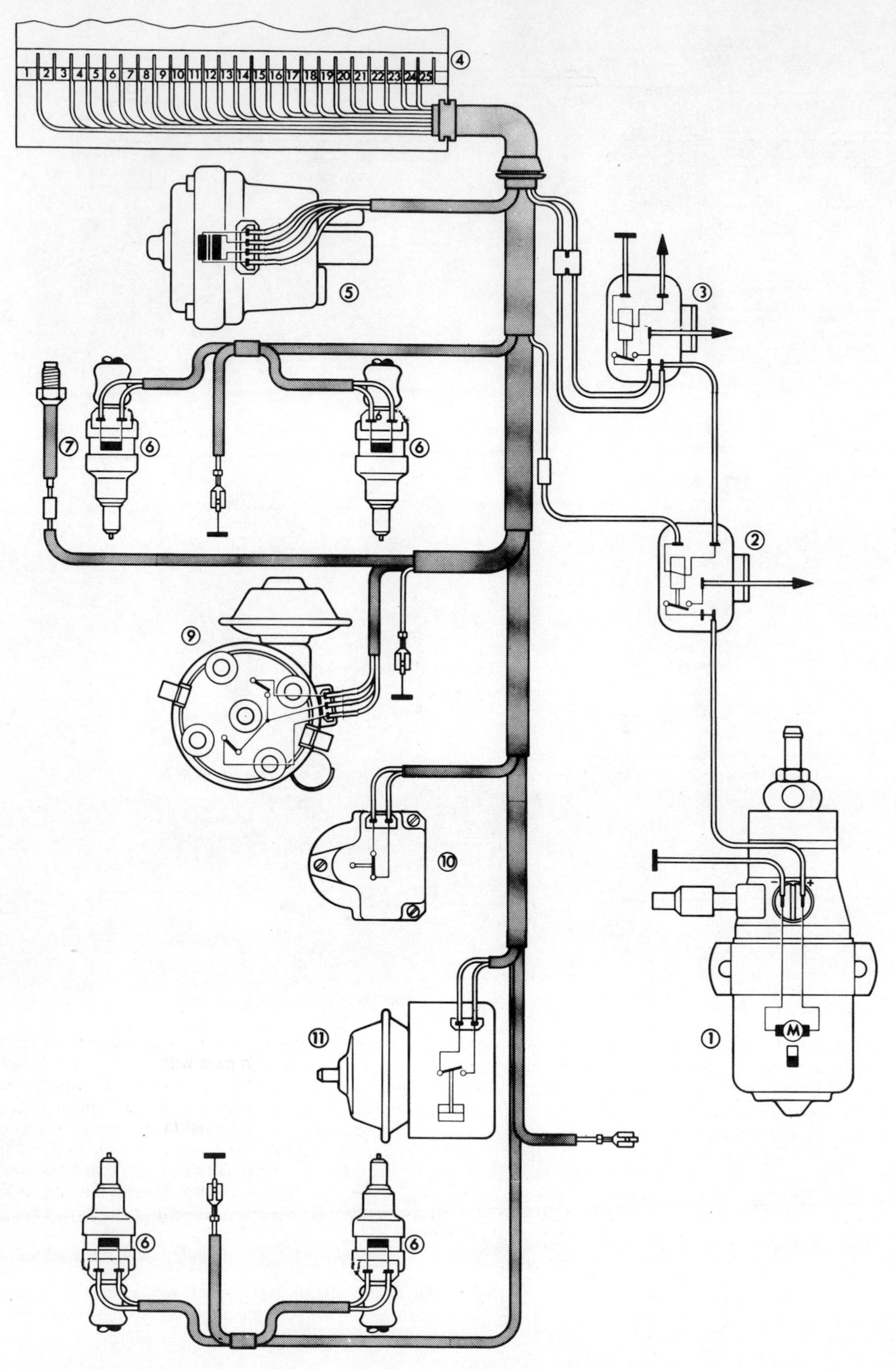
1
2
3
4
5
6
7
8
9
10
11
12
13
14
15
16
17
18
19
20
21
22
23
24
25
④
⑤
③
⑦
⑥
⑥
②
⑨
⑩
⑪
M
①
⑥
⑥

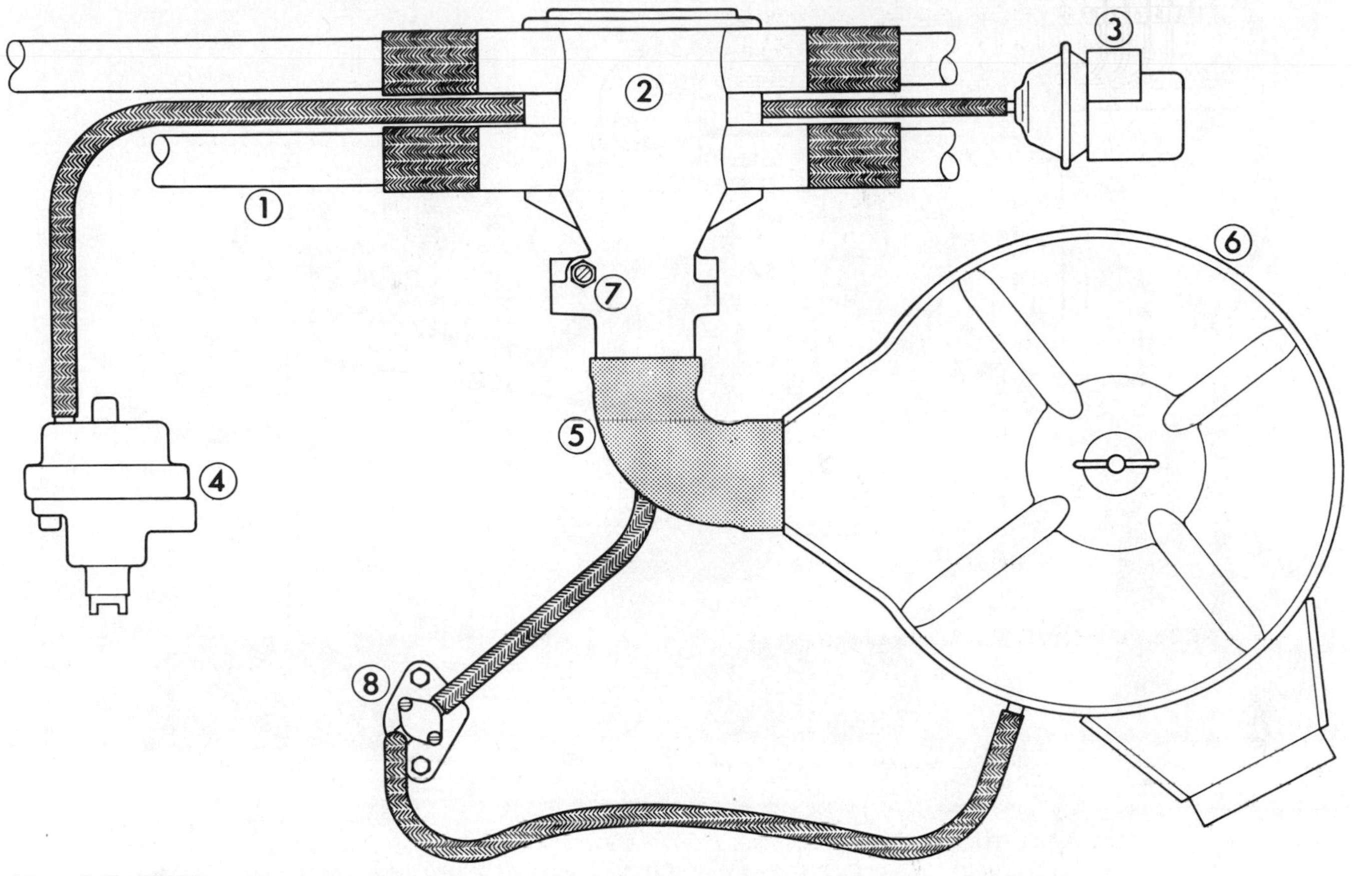

Fig. 3–68. At left is a schematic drawing of an electronic fuel injection system. The parts are:

(1) electric fuel pump
(2) pump relay
(3) ignition relay
(4) computer
(5) pressure sensor
(6) fuel injectors
(7) temperature sensor
(9) distributor with trigger contact points
(11) pressure switch

Above is a drawing of the air distribution system:

(1) air pipes
(2) intake air distributor
(3) the pressure switch
(4) pressure sensor
(6) air cleaner
(7) idle speed adjusting screw

Here's how it works: The computer receives current from the battery via the ignition relay and information from pressure and temperature sensors and from the distributor trigger contacts (which in cars with breaker points is a second set of points that provide an engine speed signal to the computer). The computer uses this information to decide how long to hold open the injectors—electromagnetic switches that operate little fuel valves. The longer the injection time, the more fuel supplied to the cylinders. Fuel is injected into ports just outside the engine's intake valves. The air flow is through pipes and the intake air distributor, controlled by a throttle plate connected to the accelerator pedal. Note No. 8, the auxiliary air regulator, a valve that opens when necessary to allow an increased air flow for such things as cold driveaway. In addition to this design, many foreign cars have a fuel injection system that draws information from a flap in the intake air distributor. As that flap moves in response to increasing air flow, it signals the computer to also increase the fuel injection.

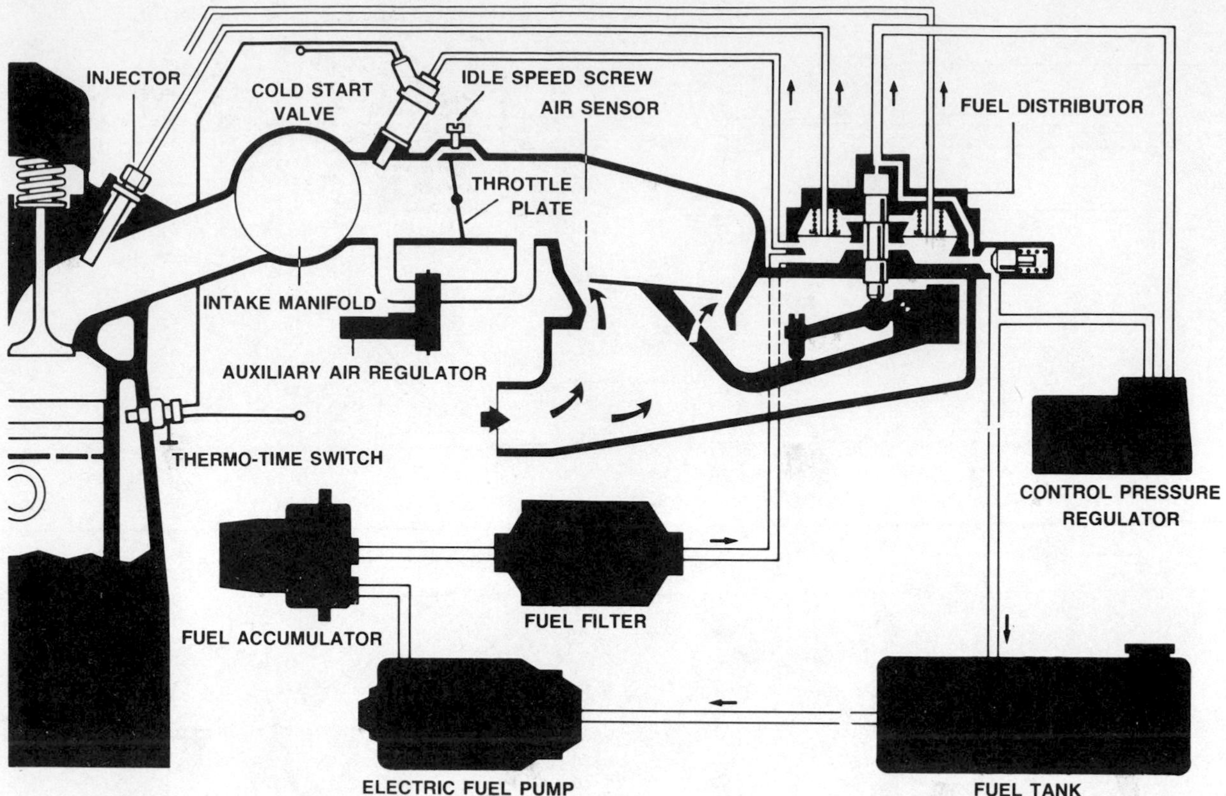

Fig. 3–69. Here's a hydraulically-operated system with an intake air flap (called an air sensor). Instead of signalling a computer, however, it has an arm that pushes on a cylindrical valve in a component called the fuel distributor. The greater the air flow, the further the arm pushes the valve, the greater the fuel flow to the injectors. A control pressure regulator controls the fuel mixture, leaning it out as the engine warms up. Note that it is connected by tubing to the fuel distributor on top of the valve. When the engine is cold, the regulator is open, allowing fuel pressure to bleed off back to the gas tank. The valve, therefore, moves in accordance with the push from the intake air flap arm. As the engine warms up, however, the regulator closes partially, so that much of the fuel pressure remains on top of the valve, tending to push it down. This downward pressure (opposing the upward push of the air flap arm) reduces the fuel flow through the cylindrical valve, leaning out the fuel mixture. The injectors are very similar to shower heads in that they spray continuously so long as there is adequate pressure in the fuel line. As in the electronic system, the auxiliary air regulator admits additional air for cold driveaway. The fuel accumulator is a spring-loaded valve device that insures a solid column of fuel under pressure in the system at all times. The thermo-time switch is a temperature-sensitive switch that operates the cold-start valve—an injector much like the type in the electronic system that supplies extra fuel when the engine is started at very low temperatures.

Turbochargers

The turbocharger, or turbine-driven supercharger, is a common method used to provide power on demand for a small engine. With it, the small engine develops normal fuel economy in easy driving. When you mash the gas pedal, the turbocharger cuts in and gives you the extra power you want.

The turbocharger provides the extra power because it is a form of compressor. Its paddlewheel-like design draws in extra air-fuel mixture from the carburetor and forces it into the intake manifold, and through that into the cylinders. More air-fuel mixture in the cylinders means more to burn, and so more performance.

The power to drive the turbocharger is provided by a turbine-like fan on the same shaft, but separated from the intake air-fuel system (see Fig. 3–78). This turbine fan is in a chamber that is part of the exhaust system. The harder you step on the gas pedal, the faster the engine runs, so the greater is the exhaust gas flow. Exhaust gases flowing through the turbine chamber spin the

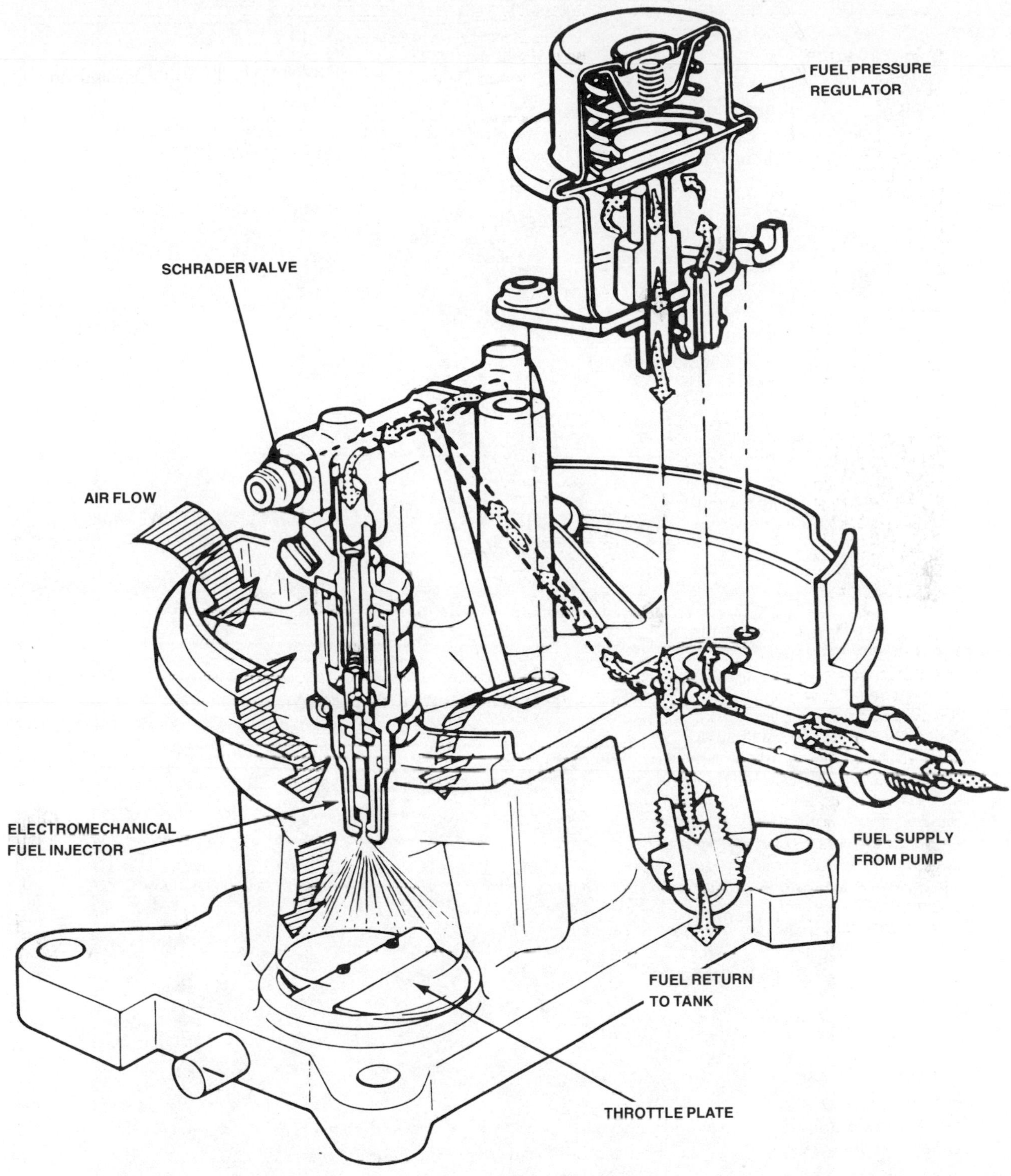

Fig. 3–70. This is the Ford throttle body fuel injection unit. The Schrader valve is on the top of the unit itself. The Cadillac system is similar, but has the Schrader valve in the fuel line to the throttle body unit and operates at much lower pressure (10 p.s.i. vs. 39 p.s.i.).

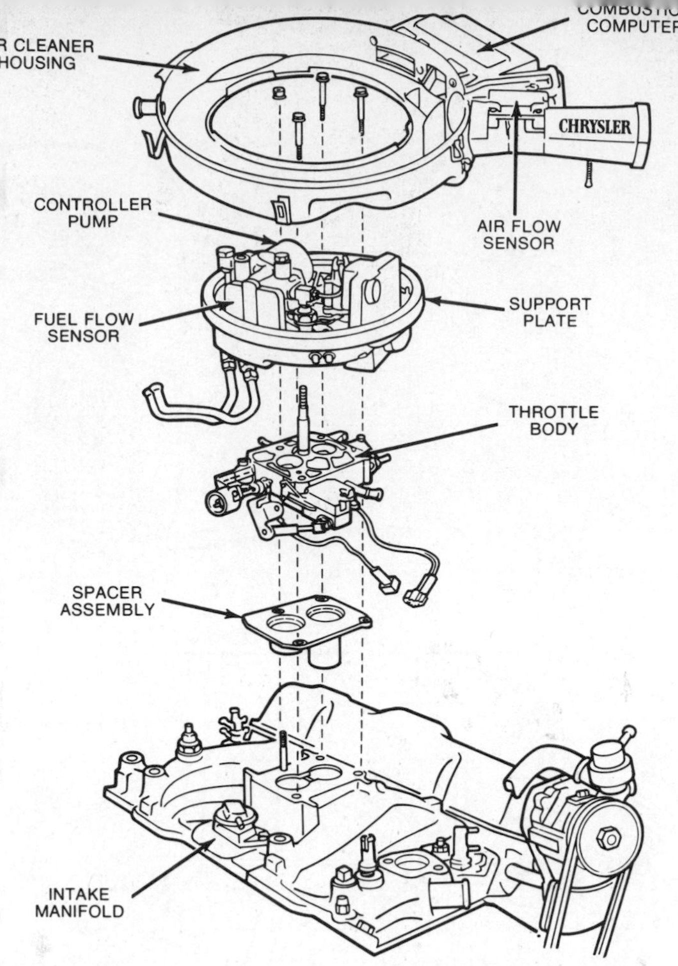

ELECTRONIC FUEL INJECTION

Fig. 3–71. This is the Chrysler throttle body fuel injection system. Instead of injectors it uses spray bars. There is a pair of spray bars in each barrel—one for normal operation and a second to which fuel is supplied for extra enrichment during high-speed operation.

Fig. 3–72. Remove cap as shown to expose tire-type valve on fuel tubing of Cadillac electronic fuel injection system. If valve needle is depressed with engine cranking, fuel under pressure should squirt out.

turbine before flowing into the exhaust piping. The exhaust gases were heading that way anyway, so the power to operate the turbocharger is essentially free.

There is a limit to how much air-fuel mixture the cylinders can accept before engine loads increase beyond the safe limit. When the engine starts to knock, severe damage can result.

To keep the power boost within safe limits, two controls are used: a waste gate and a spark timing retard system.

• *Waste gate on the turbocharger.* See Fig. 3–79. When the pressure in the intake manifold reaches a certain level, it operates a diaphragm connected by a rod to a part called the waste gate, which is a type of valve. The diaphragm rod pulls the waste gate open and exhaust gases bypass the turbine chamber and go straight to the exhaust piping. In 1981 General Motors went to a new type of waste gate control diaphragm, using vacuum in the plenum (a chamber between the carburetor and the compressor wheel of the turbocharger, as shown in Fig. 3–78). The vacuum holds the diaphragm and therefore keeps the waste gate open during idle and part throttle, to reduce exhaust back pressure and thereby improve performance and fuel economy. At medium throttle, plenum vacuum drops and the diaphragm springs force the waste gate closed, allowing turbocharger power boost to build. At full throttle, the exhaust back pressure builds up and forces the waste gate open to prevent excessive power boost. The waste gate opens only enough to keep the intake manifold pressures within specifications; the engine still might knock if ambient temperatures and humidity levels are unfavorable. To prevent knock, the spark timing is retarded by an electronic control unit. Two popular ways to trigger this control unit are:

With a knock sensor threaded into the intake manifold. See Fig. 3–80. When the engine just starts to knock, this electronic sensor "reads" a change in engine vibration and signals the electronic control unit. Depending on the strength of the knock signal, the control unit retards the spark a certain number of degrees, up to a maximum of 12. General Motors uses this system.

Fig. 3–73. Pressure gauge can be used to check hydraulic fuel injection system. You can connect a standard gauge into fuel distributor inlet line (from pump, accumulator, and filter) and so check pressure into fuel distributor, then connect the gauge between fuel distributor and line to control pressure regulator, warm up the engine, and see if control pressure drops to specifications. Or you can use the special gauge with shutoff valve, as shown, connected between the fuel distributor and the line to the control pressure regulator, and with this one connection make both tests.

Fig. 3–74. To check the injectors, begin by removing the air duct over the intake air flap as shown.

Fig. 3–75. Disconnect the wiring connector on the air distributor next to the air flap as shown. Also disconnect wiring from the alternator.

Fig. 3–76. As of 1981 the fuel pump relay is plugged into the new fusebox, shown in this photo. When bypassing, remove it and connect across the terminals at 12 and 6 o'clock with a jumper wire. VW recommends use of a jumper wire with an 8-amp fuse, which can be made from an in-line fuse holder—sold at auto parts stores—and a pair of alligator clips.

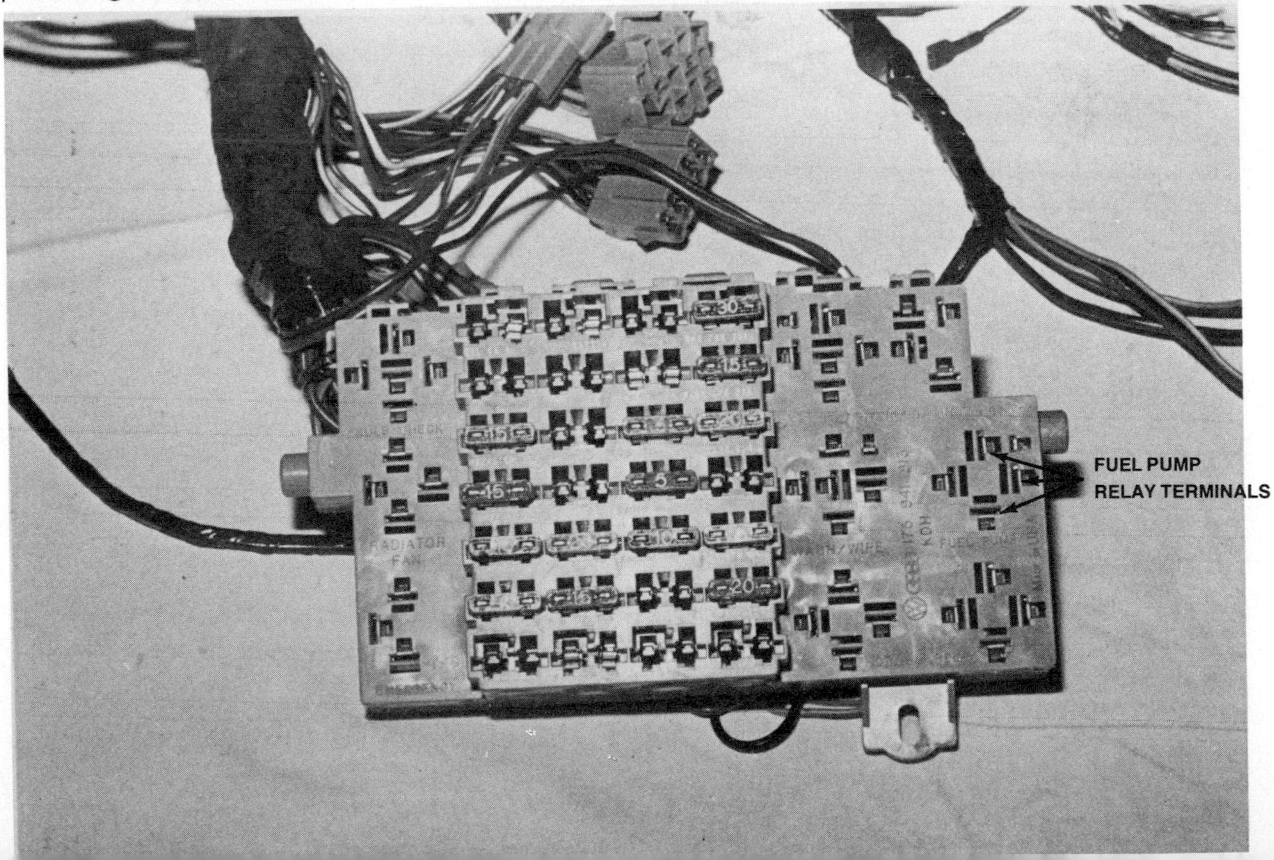

Fig. 3–77. Pull an injector out of its bore, aim it into a beaker as shown and, with fuel pump relay bypasses as explained in this chapter, pull on the intake air flap with needle-nose pliers as shown (or with a magnet). Fuel should spray from injector into the beaker. Pull on flap only long enough to confirm fuel spray, for other injectors also are spraying, and if you do this too long the cylinder could fill with fuel. That could result in engine damage if you try to crank the engine under such conditions.

With a two-stage pressure switch mounted on a fender apron and connected by a hose to the intake manifold. When manifold pressure reaches a certain level, the switch triggers the electronic control unit and a small amount of spark timing retard is provided. When the pressure reaches the second level, the switch again triggers the control unit and the spark is retarded further. Ford uses this arrangement, and in early models installed two switches, side by side, so that in case one failed the other would be enough to do the job.

On one car, the Swedish Saab Turbo (not included in this book), only a waste gate control is used. The knock problem is handled by an engine modification (a low compression ratio) and a switch that cuts out the electric fuel pump to stop the engine if pressure in the system builds up to a certain level.

Turbocharger Service

To check out the turbocharger, first go to the waste gate diaphragm and apply calibrated air pressure (see Fig. 3–81). When you apply 7–10 p.s.i., the diaphragm rod should move. If it doesn't, have the system checked immediately, for the engine might be severely damaged in high-speed use.

If you have the plenum-vacuum-controlled waste gate diaphragm used only on '81 General Motors cars (you can tell by the fact that there's just one hose on the diaphragm and it goes to the plenum), do not make this test. Instead, apply 20 inches of vacuum with a manual vacuum pump (engine off) and you should see the diaphragm link move.

Testing the Ford system also calls for a calibrated source of compressed air (three and six p.s.i.). Disconnect the hoses from the intake manifold at the switches and plug the hose ends. Then connect a timing light, run the engine at idle, and apply the low-pressure air to the switches (see Fig. 3–82). When you apply the low-pressure air, you should see the spark retard at the timing marks.

To test the spark timing retard on a GM car, connect a tachometer or a timing light (aim the light at the marks). Run the engine at 2000 r.p.m.

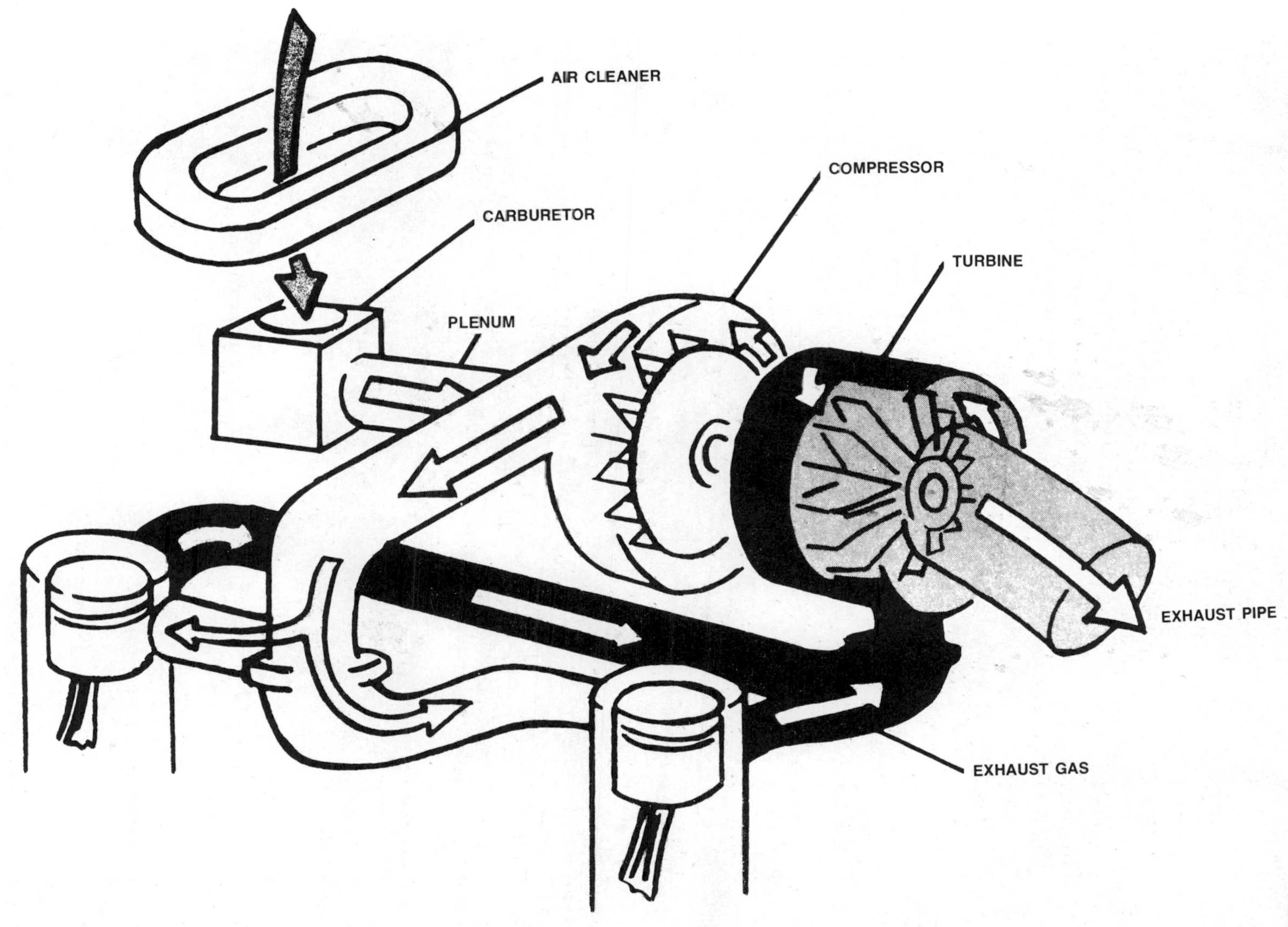

Fig. 3–78. Here's how the exhaust turbocharger works. Exhaust gas flows from cylinders into the turbine chamber, spinning the turbine, and then flows out to the exhaust pipe. The turbine is on the same shaft as a compressor wheel in a chamber between the carburetor and intake manifold. As the compressor spins it pulls an extra amount of air-fuel mixture and forces it into the cylinders, so they produce more power.

and have a helper repeatedly rap the intake manifold next to the knock sensor with a rod (see Fig. 3–83). You should see engine speed drop 300–500 r.p.m. or spark timing retard (perhaps 20 seconds after you stop rapping, the timing and r.p.m. will return to normal). If you don't get the timing retard or engine r.p.m. drop, shut off the engine and disconnect the wire from the knock sensor and hook up an ohmmeter (one lead to the sensor terminal, the other to electrical ground). Compare the reading with specifications (typically 175 to 375 ohms). If it is outside these limits, replace the knock sensor. If the reading is acceptable, have a professional check out the system.

Oil And Filter Change Interval

The turbocharger turbine wheel spins very fast, and so it must be lubricated well to survive. Oil is supplied by a line from the engine. This handles the lubrication problem, but some exhaust heat contamination of the engine oil results. The useful life of the engine oil in a turbocharged engine, therefore, is shorter than in a conventional—called a normally aspirated—engine. The factory-recommended oil and filter change interval is much more frequent with turbocharging, and you should observe that interval carefully. A 2000 to 2500 mile oil and filter change interval is a good practice.

Fig. 3–79. The waste gate is a valve that opens automatically to prevent the exhaust-driven turbine from turning so fast that it causes excessive power boost from the compressor. In this design, when the pressure of the air-fuel mixture in the intake manifold exceeds specifications, the pressure (transmitted by a hose) pushes on a diaphragm unit that operates a rod that opens the waste gate, allowing the exhaust gases to bypass the turbine.

Fig. 3–80. The knock sensor, an electro-mechanical accelerometer, is threaded into intake manifold as shown.

Fig. 3–81. To check manifold-pressure-controlled waste gate operation, disconnect diaphragm unit hose from intake manifold and connect to a hose tee (available from auto parts stores). To the other two necks of the tee, connect a vacuum-pressure gauge and a bicycle pump. Apply specified pressure with bicycle pump and you should see diaphragm rod move. If you can't apply sufficient pressure with a single stroke of the pump, install a tire valve in the hose between the tee and the pump, which will permit you to pump as necessary to build up pressure.

Fig 3–82. To check ignition timing retard on General Motors V-6 or V-8 turbo, connect tachometer or timing light, run engine at fast idle (2000 r.p.m.), and repeatedly whack the intake manifold next to the knock sensor with a socket extension rod or similar bar. You should see engine speed drop 300 to 500 r.p.m., or the ignition timing retard.

Fig. 3–83. Ford system uses pressure switches to signal an electronic control unit to retard the spark timing. To check them, disconnect the pressure hose from the intake manifold and apply calibrated compressed air, such as with bicycle pump and vacuum-pressure gauge, as shown in Fig. 3–81. At one p.s.i., the spark timing should retard six degrees; at four p.s.i. it should retard an additional six degrees. If the retard does not work, replace the appropriate switches. Note: On some cars the three spark retard switches may be under the warning light switches instead of above them as shown.

4 AIR POLLUTION CONTROLS

THE MODERN AUTOMOBILE is equipped with a wide variety of systems designed to reduce air pollution. Many are built into the engine, but others are external and (except for the catalytic converter) should be checked and serviced as part of tuneup. Although it is widely believed that all pollution controls reduce performance, many of the external ones either have no effect, or their malfunction can hurt performance. In any case, they are the law and good citizenship calls for keeping them functional.

Although some checks and services are explained in this chapter, you should know that emission control systems (except Positive Crankcase Ventilation) are warranted for five years or 50,000 miles, whichever comes first. Therefore, if a system seems to be malfunctioning and is still under this warranty, take the car to the dealer for free service.

There are six areas that a weekend mechanic can check and service.

Positive Crankcase Ventilation

The crankcase of your engine, where the crankshaft churns through a bath of oil, is contaminated by fumes from unburned gasoline that slip past the sealing rings on the pistons. These fumes can build up pressures that could ruin oil seals, so they must be purged. On older cars they were allowed to escape into the atmosphere, but on all cars since the early 1960s they are recirculated back into the cylinders for burning with a system (see Fig. 4–1) called Positive Crankcase Ventilation (PCV). If this system plugs, not only can oil seals be damaged, but the engine may idle poorly and stall.

The key to the system is a flow control valve—called the PCV valve—that sits in a rubber grommet in the cylinder head cover on most cars (see Fig. 4–2). On others it is spliced into a hose or threaded into a hole that goes through the intake manifold into the engine.

The PCV valve is attached by a hose to the base of the carburetor, where engine vacuum draws up fumes from the crankcase to the top of the engine, through the valve, and into the intake manifold. Here it combines with the fuel mixture already there and flows into the cylinders for burning.

The PCV valve regulates the flow. When the engine is idling and can accommodate very little additional fuel, the valve is nearly closed. As the engine speed increases, engine vacuum drops and a spring inside the valve pushes a plunger, allowing the increased flow that the engine can now tolerate.

The drawing out of crankcase fumes creates a light vacuum in the crankcase. This is filled by fresh air. The air may go into the rocker cover through an oil fill cap with a filter built in, or come from the carburetor air cleaner. In the latter case, the air cleaner housing has a small additional filter for crankcase air built in.

Checking the System

Unlike other pollution control systems, maintenance of the PCV is the owner's responsibility, so here's how.

To check the operation of the PCV system, disconnect the PCV valve from the engine top cover or from the hose that connects it to the cover or other part of the engine. Shake the valve and you should hear a jiggling noise, indicating that the valve parts inside are not stuck.

Run the engine at idle and feel the end of the valve with your finger (see Fig. 4–3). If you feel a vacuum, the valve and the hose that connect it to the carburetor base are functional.

If the valve fails the jiggle test it should be replaced. If you don't feel vacuum on the second test, check the hose to which the valve is connected for kinks or plugging, and if it is okay, replace the valve. In any case, change the valve at least every two years or 24,000 miles, but never exceed the car maker's recommendations.

Finally, check the fresh air filter. If it is the type built into the oil fill cap, clean it in solvent. It may be so dirty, however, that a solvent soaking won't help, in which case it should be replaced. If the filter is built into the air cleaner housing, inspect it, and if it's dirty remove the retainer clip and change it (see Fig. 4–4).

Thermostatic Air Cleaner

The thermostatic air cleaner is a temperature-controlled flap device that provides preheated air for the carburetor during engine warmup. If the valve does not provide the warm air, the cold

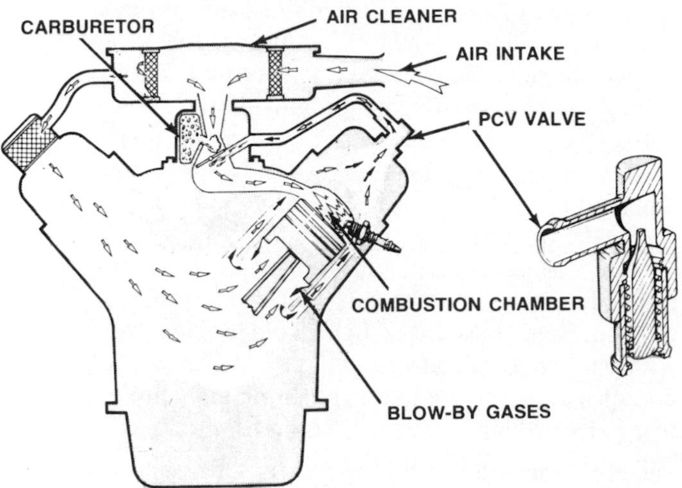

Fig. 4–1. Schematic of positive crankcase ventilation system. PCV valve draws crankcase fumes up into engine top chamber, through valve, and into carburetor. Fresh air enters through air filter, then flows into crankcase to help purge fumes.

Fig. 4–2. The PCV valve sits in a rubber grommet in the engine top cover. It has been pulled out so you can see what it looks like.

engine will stumble and stall. If it fails to shut off when the engine is warm, the air to the carburetor will be overheated, and warm engine performance will be affected. Refer to Fig. 4–5.

How It Works

The air cleaner housing has an inlet neck—called the snorkel—with a duct that connects to the exterior of the exhaust manifold. This is the source of warm air. When the engine is warm, the duct is covered and cool air comes in through the opening in the front of the snorkel.

The control device is a simple flap valve. When the engine is cold, it covers the front of the snorkel, blocking the cold air and allowing the engine to draw in warm air from the exhaust manifold duct. As the engine warms up, the flap pivots and starts to block off the air duct and permit airflow through the front of the snorkel. Finally, when the engine is fully warm, the flap closes off the warm air duct completely and only cool air can enter, through the front of the snorkel.

Operating the flap on the most common—vacuum controlled—system is a spring-loaded diaphragm device connected through a temperature sensor (inside the air cleaner) to a source of engine vacuum at the carburetor base.
sensor (inside the air cleaner) to a source of engine vacuum at the carburetor base.

The temperature sensor (see Fig. 4–6) is a bleeding device. When the engine is cold, it passes all the vacuum from the engine through to the diaphragm device on the snorkel. As the engine warms up, the sensor starts to open and allows some of the vacuum to bleed off into the atmosphere. Less vacuum stays at the diaphragm device and so it exerts less pull on the flap. A spring on the diaphragm moves a link connected to the flap to the warm air duct. Finally, when the engine is warm, all vacuum supplied to the diaphragm is bled off and the flap closes the warm air duct completely.

Note: Some cars have different systems. Typically, they have thermostats that directly move the flap. Some GM cars (such as older Chevettes with one-barrel carburetors) have a thermostatic coil spring, much like that in an automatic choke, that coils up tight when it's cold and pivots the flap to block off the intake air snorkel. As the

Fig. 4–3. Feeling end of PCV valve for vacuum.

Fig. 4–4. Removing fresh air filter from air cleaner housing. When the clip is removed as shown, the entire filter assembly can be removed. Or you may be able to simply remove filter from the assembly and install a replacement.

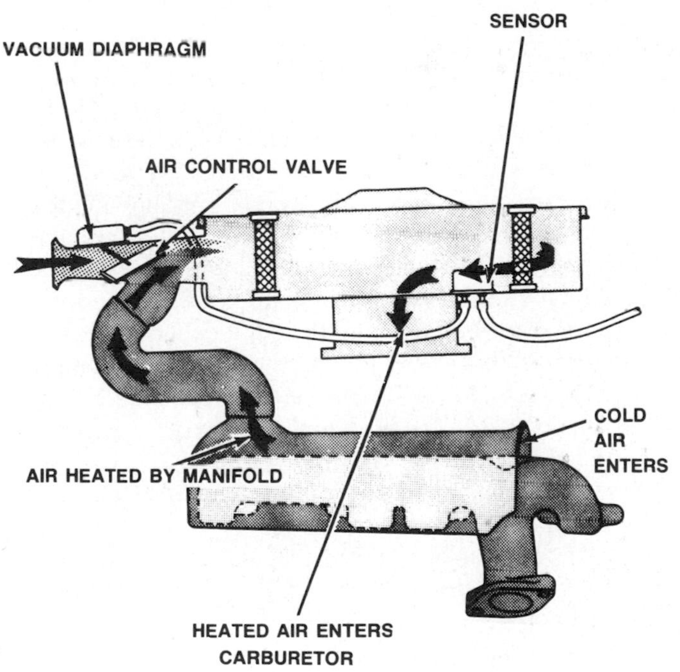

Fig. 4–5. Schematic of thermostatic air cleaner. The vacuum diaphragm operates flap in air cleaner. The diaphragm is connected by hose to the temperature sensor, which is connected by another hose to vacuum source at the base of the carburetor or on intake manifold. When the engine is cold, the sensor passes vacuum from carburetor or manifold to diaphragm to operate flap. When engine is warm, sensor bleeds off vacuum to atmosphere, and spring in diaphragm assembly pivots flap to close off hot air duct.

Fig. 4–6. With air cleaner cover off and filter element out, you can see the thermostatic sensor.

engine warms up, the heat causes the coil to unwind and pivots the flap to open the intake air snorkel. Many Ford products have a thermostat with a link that connects to the flap, which does basically the same thing.

Testing the System

To check out the system, start the engine when cold and probe the snorkel with your finger (if the snorkel is a short one) and with a screwdriver if your finger can't reach. The opening should be blocked by the flap. Recheck when the engine is warm and the snorkel should be open. On some cars (see Fig. 4–7) you may have to disconnect a cold air duct for access to the flap.

If the flap is stuck in the heat-off position with the engine cold, disconnect the vacuum hose from the diaphragm device and feel for vacuum. If there is vacuum, the diaphragm is defective and must be replaced. If there is no vacuum, check the hose connections at the carburetor base and the underside of the air cleaner (see Fig. 4–8) and if they are good, and the hoses are not kinked, check the sensor by placing a finger over the vacuum bleed neck (see Fig. 4–9). If the system now works, the sensor is defective.

If the flap is stuck in the heat-on position with the engine warm, disconnect the vacuum hose from the diaphragm device and check to see if the flap now closes. If it does, replace the temperature sensor, which is apparently not allowing the vacuum to bleed off. Note: Some new cars have a second, solenoid–operated flap in the snorkel which closes when the engine is off to keep vapors from getting into the atmosphere. Do not confuse this with the thermostatic air cleaner. This second flap should be partly opened by the solenoid when the ignition is turned on (otherwise the engine will not breathe).

Replacement of parts in the vacuum-operated system is an obvious procedure.

If your car has one of the self-contained thermostatic controls also used, you may have to replace the entire snorkel assembly. If a Ford thermostat is available as a detail part, thread it into its mounting bracket to the same depth as the old thermostat. Check operation of the flap after installation, and readjust the thermostat on its bracket if necessary. (See Fig. 4–10.)

Fig. 4–7. With part of air cleaner snorkel disconnected from cold air duct, you can see this flap clearly.

Fig. 4–8. Checking hose connections to sensor on the underside of the air cleaner.

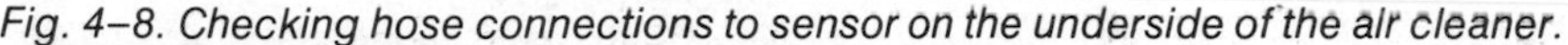

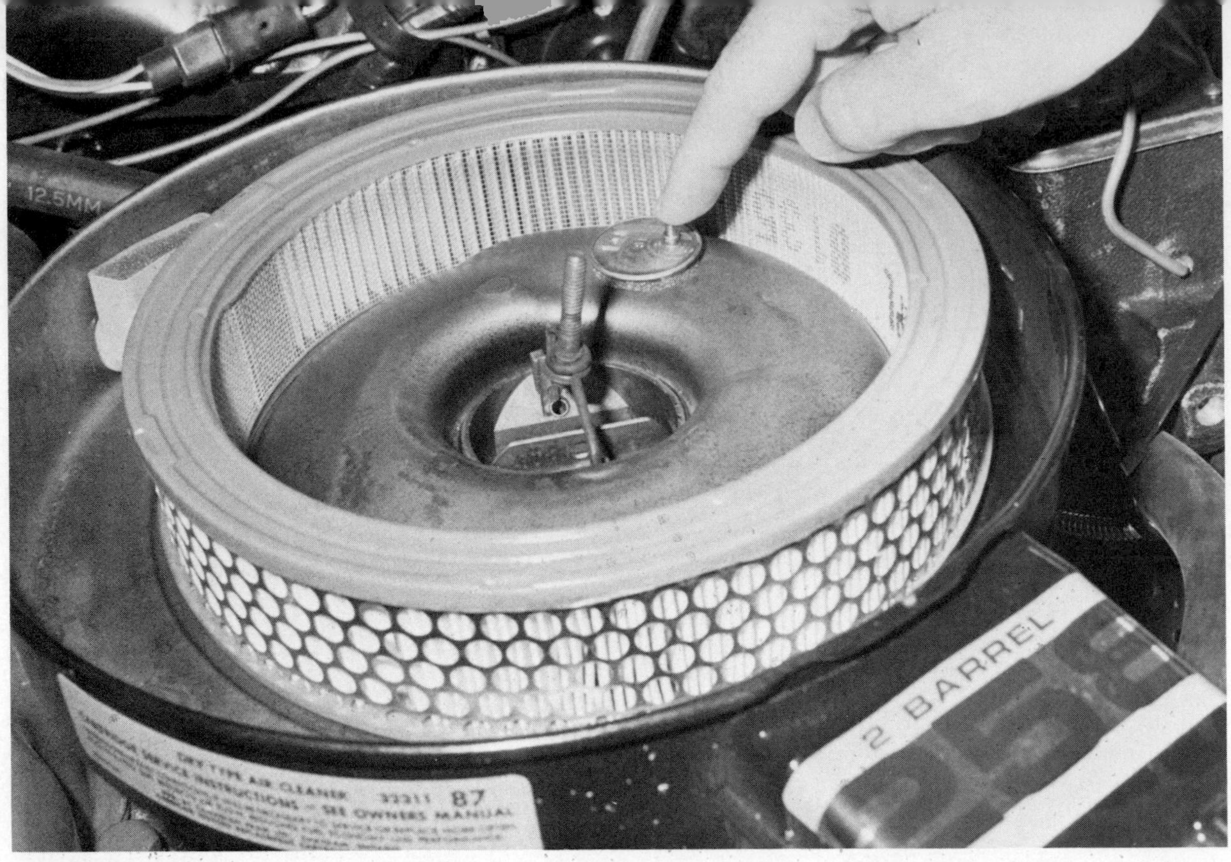

Fig. 4–9. Placing finger over vacuum bleed neck. If system now works, sensor is defective. Some systems may not have this type of bleed, but in any case there will be some type of opening through which vacuum can bleed off.

Fig. 4–10. Many Fords have used this type of thermostatic air cleaner in which all movement of the hot air flap is by the thermostat. There is a vacuum diaphragm attachment unit in the bottom but its function is to pull down the flap, opening the cold air snorkel for acceleration. If the thermostat is defective, note the depth it is threaded into the bracket and install a replacement to the same depth. You may find, however, that only a replacement snorkel with the thermostat already installed is available, in which case the precision adjustment isn't necessary.

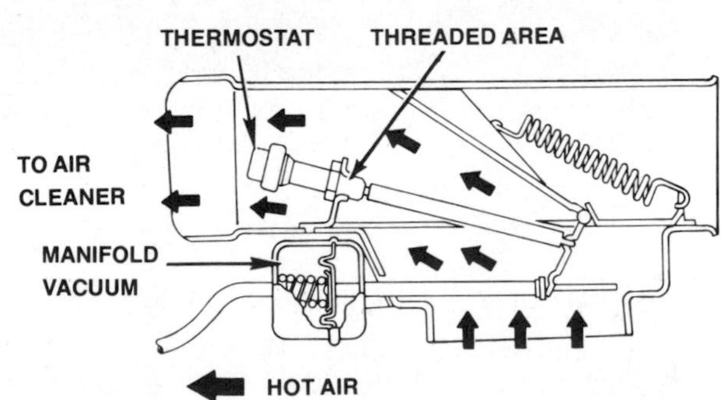

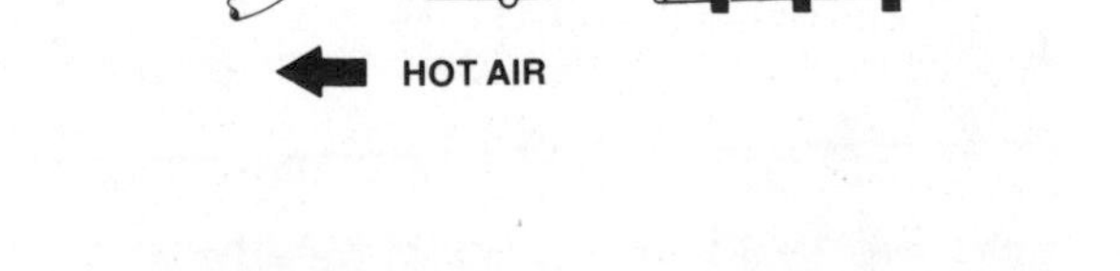

Exhaust Gas Recirculation

The exhaust gas recirculation system (EGR) allows a calibrated amount of exhaust gas to enter the intake manifold, where it combines with the air-fuel mixture from the carburetor. The exhaust gas is a contaminant, and it lowers the temperature of the fuel mixture when it burns in the cylinders. This doesn't help gas mileage, but the lower temperatures do reduce the air pollutant called oxides of nitrogen (see Fig. 4–11).

The heart of the system is the exhaust gas recirculation valve, a diaphragm unit controlled by engine vacuum or air pressure supplied by the air pump, another pollution control system. The valve is a flow control device that regulates the amount of exhaust gas that can get into the engine. Most cars with EGR have additional controls to modulate and limit the operation of the valve, but these are so complex that their service is best left to the car dealer.

You should check the EGR to make sure it is functioning, for these reasons: a malfunctioning system contributes to air pollution; on some cars, the vacuum systems are so carefully balanced that loss of EGR under some circumstances could affect engine performance; and, if the valve operates when it shouldn't (such as engine

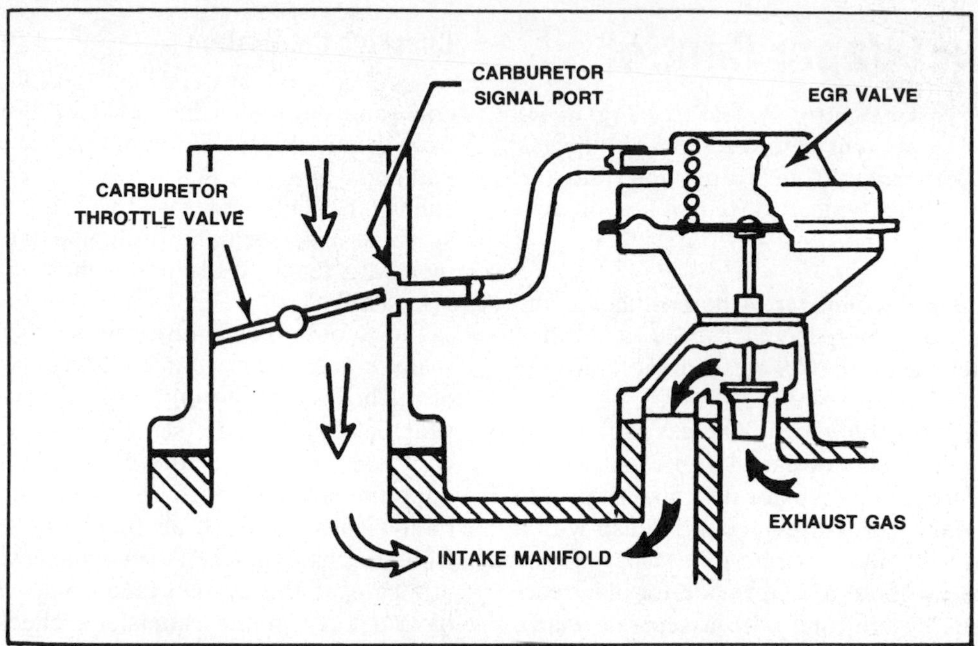

Fig. 4–11. Schematic of exhaust gas recirculation system. When carburetor vacuum is applied to the diaphragm inside the valve, it pulls up on the valve, allowing exhaust gas to flow into intake manifold. On many cars there is more than the simple hose shown between carburetor signal port (the vacuum source) and the EGR valve. Special sensors and time-temperature delay hardware may be used to tailor exhaust gas flow more precisely.

cold and low ambient temperature) the engine will perform poorly. Too little exhaust gas recirculation can cause the engine to knock. Too much at idle can cause rough idle.

On many engines with computer control, the EGR system is controlled by the computer. If you do not get the EGR motion that is described next, the problem may be in one of the solenoid controlled valves that regulates it. The computer-controlled system should be left to the car dealer. It is like other emission controls, covered by a five-year, 50,000-mile warranty.

Checking EGR

First, start the engine from cold, on a cold day, and immediately look at the EGR valve, which is at or near the base of the carburetor. Have a helper hit the gas pedal. The EGR valve stem (see Fig. 4–12) should not move. If it does, and you've had cold stalling and other cold performance problems, this could be the reason. Have a mechanic check out the system, for most cars incorporate a cold start delay in the EGR system.

Repeat the test with the engine warm—after the engine has idled for at least a minute. This time the stem should move. You must wait the minute, for many cars have warm-restart delays also built into the system.

If you can't see the valve stem on your car, disconnect the hose at the valve. Repeat the test conditions, but this time have your helper hold the gas pedal at part throttle (about one-third down) as you hold your finger over the hose end. You should feel vacuum (or on some cars air pressure). Reconnect the hose and the engine should slow down just a bit. If you're not sure, connect a tachometer to the engine (see *Chapter 2*).

If the EGR valve does not appear to be functioning, unbolt it from the engine and inspect it. If you see an accumulation of carbon deposits on the inside, wire-brush off the deposits or knock them off with a soft hammer (see Fig. 4–13), then check by trying to move the stem with a screwdriver. If the valve seems free, reinstall and check it. If it still doesn't work, the diaphragm is apparently defective and the valve must be replaced.

Fuel Vapor Control

The fuel vapor control system (see Fig. 4–14) is designed to prevent gasoline vapors in the gas tank and carburetor from evaporating into the atmosphere. The typical system relies on two components:

- *A charcoal cannister.* Charcoal holds the vapors that otherwise could escape, and when the engine starts they're discharged into the carburetor.
- *A special gasoline cap.* The gas cap on cars with vapor control is designed to prevent vapors from escaping. If you replace the cap, install only this type, called pressure-vacuum. A cap with a plain vent will allow vapor to escape. If you install a cap without a vent (made for older cars without this system and with a separate vent), the engine will not draw fuel from the tank and will quickly stop. In some cases, the cap without a vent could result in a vacuum in the tank that might cause the tank to collapse.

Checking the System

To check this system, just look for the charcoal cannister under the hood by following hoses from the air cleaner or top of the carburetor until you come to it (see Fig. 4–15). On some cars the cannister is in a fender well with a plastic cover, so if the hose seems to disappear into the sheet metal at a fender well, that's where the cannister is located.

Check only the cannister under the hood for a foam or fiber filter pad in the end opposite to that of the hoses, and if you find it, replace it once a year or 12,000 miles (see Fig. 4–16). On many cars you can feel for the filter, pull it out with your fingers, and push in a replacement. That filter cleans the fresh air that purges the fumes from the charcoal when the engine is started.

Whatever the system (and some cars do not have a filter in the cannister), check for good hose connections at the cannister and the carburetor or air cleaner. If any of the hoses are oil-soaked or cracked, replace them. Only use a hose made for the cannister system.

Fig. 4–12. The EGR valve stem, to which the screwdriver points, can be visually inspected for movement to check valve operation.

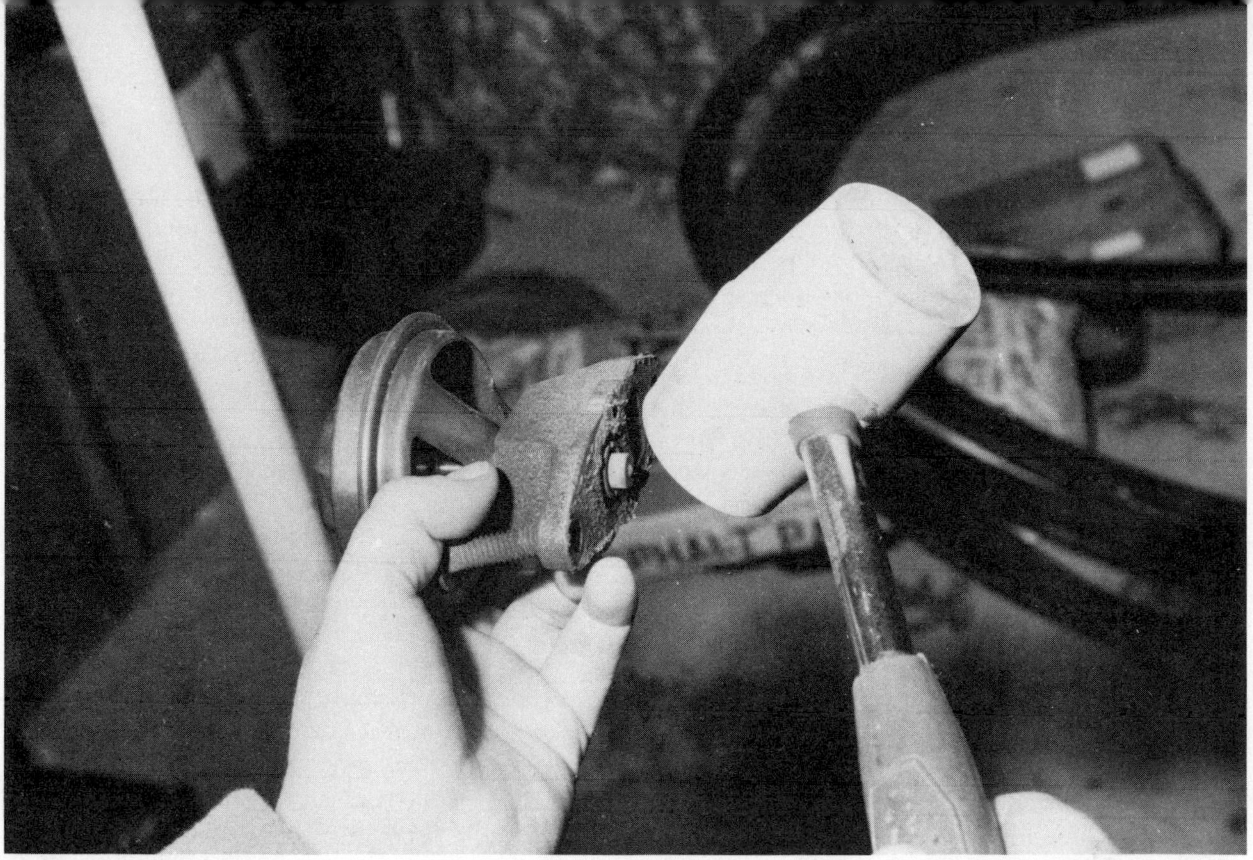

Fig. 4–13. Use a soft hammer to knock deposits off EGR valve stem.

Fig. 4–14. This schematic shows the fuel vapor control system and the hoses and lines which, in the layout shown, connect in some cases to the ignition distributor vacuum control unit and the Positive Crankcase Ventilation system.

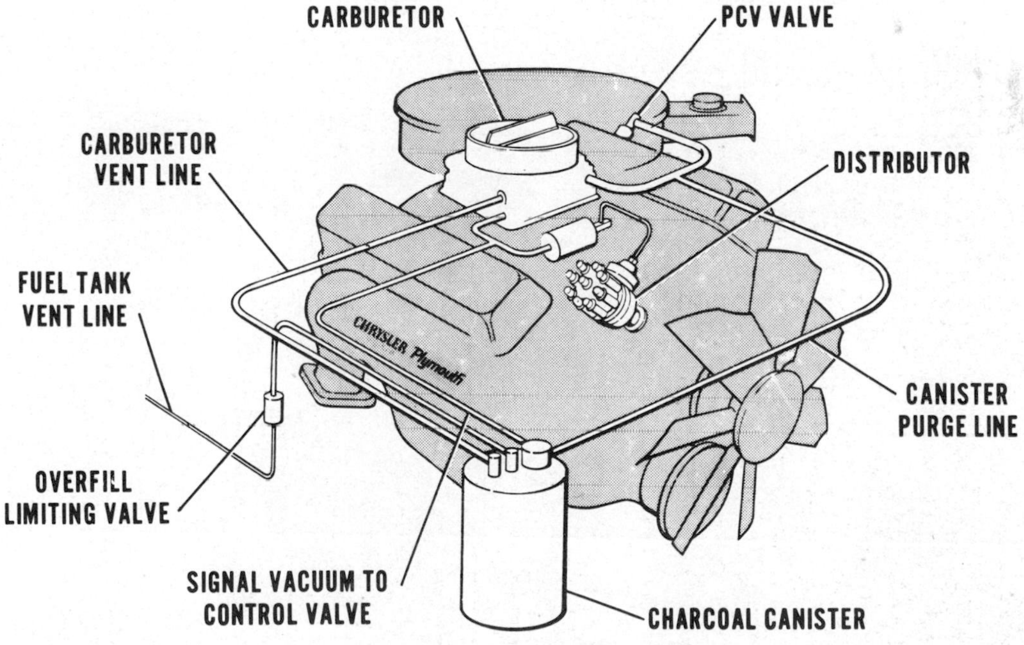

Maintenance services and the gas cap in the fuel vapor system may not be covered under warranty.

Air Pump

The air pump has been used on American cars as a pollution control system since 1967. For the last several years (pre-1981), it was dropped on many cars in favor of the simpler Pulsair, a valve device. It made a comeback in 1981 as most American car makers went to special catalytic converters and needed the higher air output of the pump to meet more stringent emission controls. In these new cars, it may be combined with solenoids wired to the engine control computer, so the air pump output is effectively regulated by the computer. Such systems should be left to the car dealer for service.

The object of the air pump is to supply pressurized air to the exhaust system, where it connects to the engine. This air contains extra oxygen to burn off any gasoline that was not consumed in the cylinders, before it could get into the atmosphere.

On most cars that are equipped with the system, the pump's pressurized air flows through external tubing (called an injection manifold) to the exhaust manifold where it connects to the cylinders (see Fig. 4–17). On a few General Motors engines, there is a single connection to the cylinder head, which has internal passages for the air.

On many late-model cars with the catalytic converter, the air is injected into a single point in the exhaust system, just before the converter. In this case, the extra air does its job only in the converter.

The pump system also includes a device called an air bypass (or anti-backfire) valve. During engine deceleration, the throttle is closed and engine vacuum is very high. This tends to pull a very rich fuel mixture into the engine, and as a result the exhaust has a higher-than-normal content of unburned gasoline. If air were injected into the exhaust during deceleration, that gasoline would explode and the engine would backfire. Letting that gas go into the atmosphere is the only solution, for backfire is unacceptable.

This means that air injection must be stopped, and this is the job of the air bypass valve. When

Fig. 4–15. This photo shows a typical location for the charcoal cannister under the hood. On some cars, however, the cannister is hidden behind a fender skirt or built into the air cleaner housing.

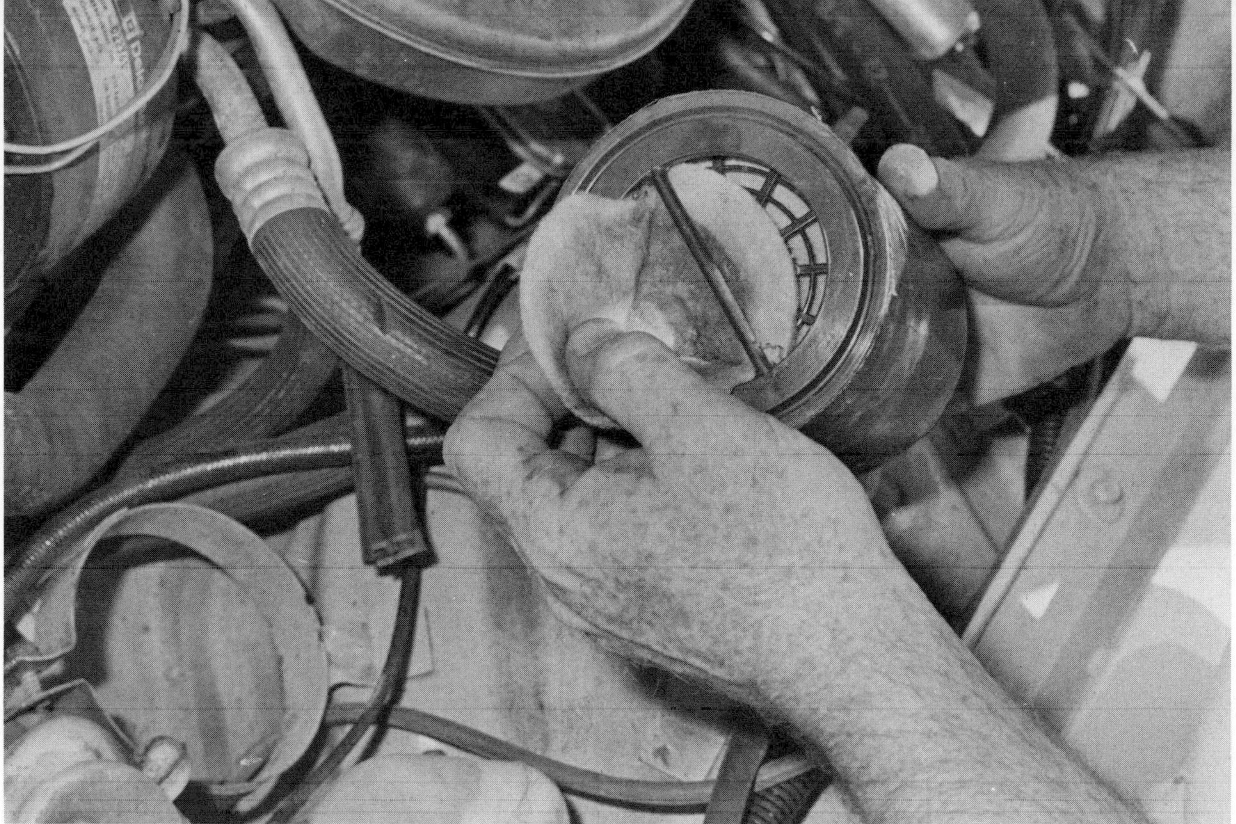

Fig. 4–16. Removing the fiber filter from bottom of charcoal cannister is easily accomplished with the cannister removed from the car and inverted.

vacuum is very high, signalling deceleration, it operates the air bypass valve and sends the air from the pump to one of three locations:

(1) to the atmosphere
(2) to the air cleaner
(3) into the carburetor, where it helps lean out the fuel mixture.

Because the air pump system, when it malfunctions, can cause backfire (as well as higher emissions), checking it and correcting any defects is a part of tuneup.

The Hardware

The air pump system checkout starts with the belt-driven pump at the front of the engine. Some pumps have a paper air filter connected to a hose. Others have a fan-like centrifugal filter just behind the belt pulley. You can replace the paper air filter yourself, but the fan type is a job for a professional mechanic. The work should be done every two years or 24,000 miles.

Always check the pump drive belt and replace or tighten as required (see Fig. 4–18). Refer to Chapter 7, *Cooling System,* which covers drive belt inspection and service. Drive belt and filter service are not covered under warranty.

To determine if the pump is working, disconnect the hose from the pump to the air bypass valve and run the engine at fast idle. You should be able to feel air under pressure at the pump's hose neck. If the valve is bolted directly to the pump, unbolt it for the test.

If you can't, and the belt is adequately tight, the pump is defective and should be replaced.

Follow the thick hose from the pump to the air bypass valve, and then the thick hose from that to another valve, just before the injection manifold. This is the check valve (see Fig. 4–19) and a V-type engine may have one or two (for each bank). Disconnect the hose from the air bypass valve at the check valve and run the engine at part-throttle (warmed up). You should feel air flow at the hose end (check at both check valves that may be found on a V-type engine).

If there is no air flow, and the air pump checked out satisfactorily, check all hoses that connect to the air bypass valve (at both ends) and, if they're not kinked or loose at their ends, the air bypass valve probably is defective. A test procedure is described later.

While the hose is disconnected look at the

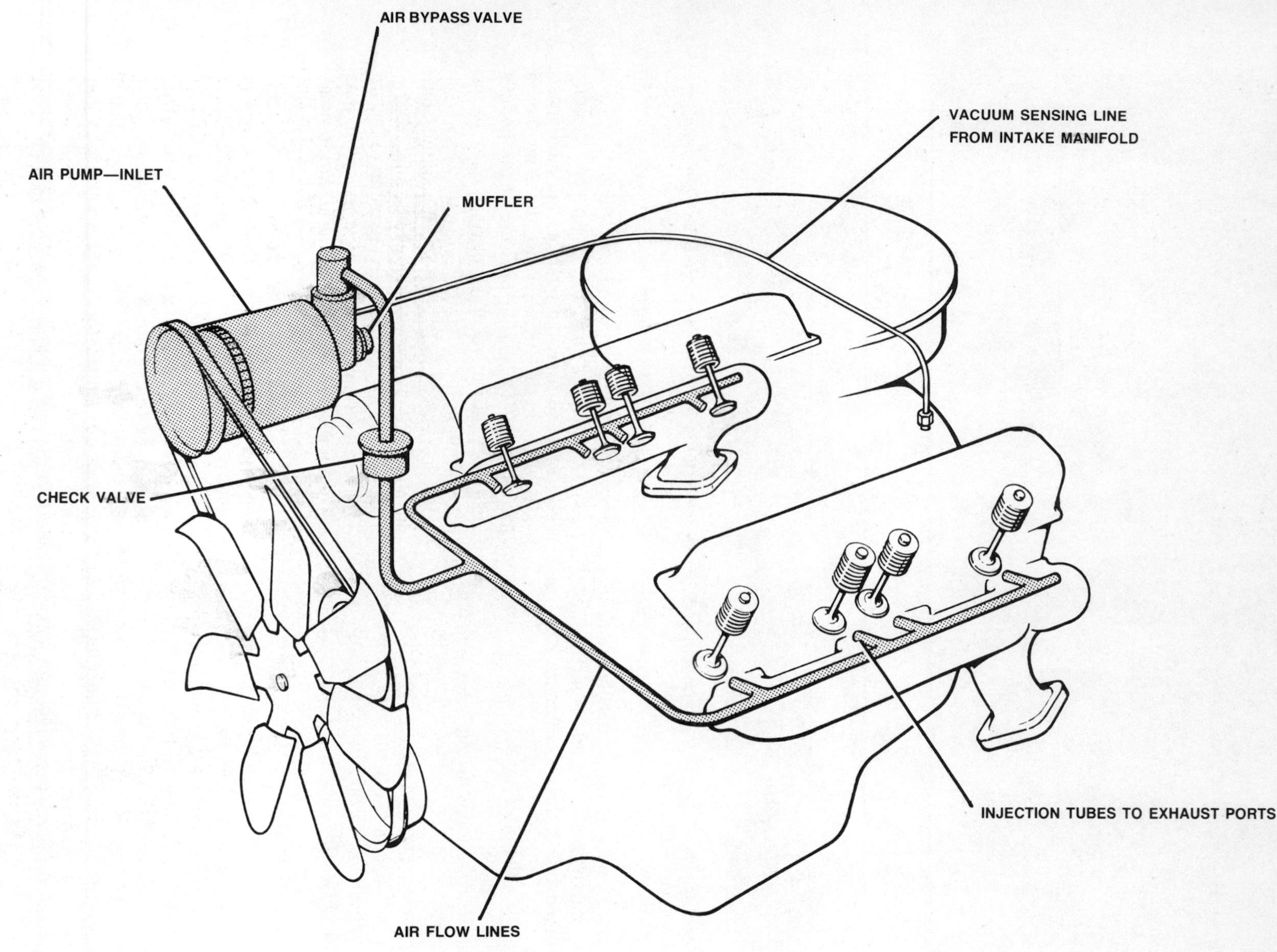

Fig. 4–17. This schematic of an air pump system shows only a single check valve for a V-8; on many cars there is a separate check valve for each bank of the V-type engine.

check valve, and if you see any exhaust gas leakage replace the check valve. The leakage can contaminate the air bypass valve and pump.

Now test the air bypass valve. Begin by locating the manifold vacuum hose—a narrow diameter hose that goes (perhaps through little control valves) to the base of the carburetor. (There may be a second narrow diameter hose on Ford products' air bypass valve that goes through little control valves, including one on the air cleaner, to a neck on the middle of the carburetor. This is not the one.) With the engine at part throttle, feel around the air bypass valve to determine if air is coming out (it should not be). With a Mity-Vac manual vacuum pump (tool sold in auto parts stores), apply 22 inches of vacuum or more to the manifold vacuum hose neck on the air bypass valve, and air should now vent from the air bypass valve. If the air bypass valve fails this test, it is apparently defective. The air typically vents through a perforated opening in the valve called a muffler. By placing your hand over the valve, you'll feel the air flow.

Finally, shut the engine and manually test the check valves by inserting a Phillips screwdriver and pushing against the valve part inside. You should feel a part move, and then spring back into position when you remove the screwdriver. If you can feel no movement the air will not get into the injection manifold, and the engine may backfire. Replace the check valve.

Upstream-Downstream Controls

The air pump controls on the newest cars are more sophisticated than the type just described. There are two valves, one called the air bypass, the other called the diverter. You should understand these two terms have previously often been used interchangeably on systems with only one valve.

In the newest systems, called "upstream-downstream," the air goes from the pump to the air bypass valve, which either dumps it to the atmosphere or passes it straight through to the diverter valve. The diverter valve has two possible positions: in one it will pass the air into the exhaust manifold, called upstream; in the second it will shift the flow to a position between two catalysts in the modern converter, called downstream. One of these catalysts (the second in the series) can use the extra air to perform better, but only after it's warmed up. Therefore, when the engine is cold the air goes to the exhaust manifold, where it helps consume unburned gasoline droplets. When the engine and the catalytic converter are warm, the airflow shifts. Under certain

Fig. 4–18. Slackening pump bolts prior to adjusting belt tension.

conditions, such as wide-open-throttle, deceleration and extended idling, the air bypass valve will vent the air from the pump to the atmosphere.

The air pump system valves may be computer-controlled electrically, with solenoid valves that regulate vacuum flow from the engine, or directly by electricity to solenoids in the two air valves. Or the air pump system may be controlled more simply, through a thermostatic valve that regulates vacuum flow to the air control valves.

A simple check of the system can be made with the engine cold and later with the engine warm. Trace the hoses from the control valves and determine which goes to the exhaust manifold, which to the mid-point of the two catalysts in the converter. When the engine is cold, air should go to the exhaust manifold; when it's warm, it should go to the converter when the throttle is cracked open. If you can't feel the airflow by squeezing on the hoses, disconnect the one through which air should be flowing to make sure.

Pulsair

Pulsair is a diaphragm device connected by hoses and tubing to the exhaust system at one end (just before the converter) and the air cleaner or carburetor at the other. The flow of exhaust gases is constantly pulsating, and half the pulses are a light vacuum that draws in air through the valve into the exhaust. There isn't as much as with a pump, but on some cars the catalyst doesn't need as much, and the very inexpensive valve does the job. To check it, disconnect it from the carb or air cleaner with the engine running and feel the end for the vacuum pulsations (see Fig. 4–20). If you can't feel them, replace the valve.

Fig. 4–19. The arrow points to check valve, located at the start of the air injection manifold for a V-8 engine.

Fig. 4–20. Feeling for vacuum pulsations at end of Pulsair valve.

Fig. 4–21. Feeling vacuum hoses for flexibility.

Vacuum Hoses

Look under the hood of any car and you'll see what looks like a plate of black spaghetti—hoses running from the carburetor all over the place. These are vacuum hoses, and they control many systems, including such non-emission devices as the heater, air conditioning, power brakes, and automatic transmission. You may have trouble telling which hose controls what, but so does a professional mechanic.

Feel each hose for flexibility (see Fig. 4–21), and if any is heat-hardened, replace it (really heat-hardened hoses may crack apart as you feel them).

If only the hose's very end is hardened, and there is some slack in the hose, break off the hard end and reconnect. Flexible vacuum hose will make a more secure, leak-free connection. If a hose is merely loose, push it back on.

5 BASIC GAS ENGINE SERVICE

CHANGING THE ENGINE OIL and replacing the oil filter are basic maintenance services on any gasoline engine (or for that matter, on the diesel). Although service intervals have been extended by the car makers, the more conservative motorist will change both oil and filter at least twice a year on gasoline engines, every 3000 miles on a diesel. On some gasoline engines with turbochargers, a 2500-mile interval also is advisable, according to the car makers.

Changing Engine Oil and Filter

Oil and filter service normally is performed underneath the car, so you need a provision for jacking it. There are three possibilities a weekend mechanic may consider:

• *Drive-on ramps*. These are relatively inexpensive and fairly easy to use. You simply drive up on them or, if working at the rear, back the rear wheels up onto them. However, they do not leave the wheels hanging free, a necessity for many repair jobs, such as brakes.

• *Scissors jack and safety stands*. The scissors jack expands like a pair of scissors when you turn a giant screw with a special handle. It takes some time and effort to raise a car with this jack, but its low price makes it suitable for the person who will be doing a limited amount of work on his car. The safety stands are absolutely mandatory with any kind of jack, for a jack itself is not a reliable device for holding up the car, only for raising it (see Fig. 5–1).

• *Hydraulic floor jack and safety stands*. The hydraulic floor jack (see Fig. 5–2) is the type many professionals use. It has wheels or casters for easy positioning, and with just a few pumps of the handle it has the car up at working height, ready for positioning of the safety stands. Note: Do not confuse this with a plain hydraulic jack. Although there are moderately priced models for weekend mechanics, the hydraulic floor jack still is much more expensive than the plain version, but does a job the plain hydraulic cannot. The plain hydraulic jack has a small jacking pad and its closed height is much too great for underbody jacking.

Where to jack. On some late-model cars, particularly front-wheel-drive, there are reinforced pads built into the lightweight bodies, and you should jack only against these pads. Consult your owner's manual for details.

If you have a conventional front-engine rear-drive car, the most suitable jacking points are in the center of the lower control arms of the front suspension (see Fig. 5–3) and under the rear axle or the attachment points for the rear springs if they are the leaf type. A coil spring rear suspension has control arms (bars from frame to axle); you may jack under them, or at their attachments, too.

If you wish a front central jacking point on a front-engine rear-drive car, use the cross-member, a boxed section that goes from side to side of the chassis, just under the engine. In the rear, jack under the banjo-shaped part of the rear axle.

Oil Change

Engine oil is drained by removing a plug in the engine's oil pan with a wrench (see Fig. 5–4). You should use a wrench that is a perfect fit, rather than an adjustable wrench. If you have a wrench assortment, one will be the correct fit (see Fig. 5–5). Or, you can purchase a combination oil-filter wrench that combines a wrench

Fig. 5–1. To use scissors jack and safety stands, place the jack under a front crossmember. Safety stands are placed under the suspension lower control arms. Once stands are positioned, jack can be lowered and removed if desired.

Fig. 5–2. Hydraulic floor jack is under front crossmember and safety stands are under the suspension lower control arms. As with scissors jack, the hydraulic floor type may be lowered and removed after stands are in place, if desired.

Fig. 5–3. These two photos show underbody jacking and safety stand points. Above is front of vehicle, below is rear. Points marked "S" are for side jacking or support with safety stands. Points marked "J" are for central jacking.

Fig. 5–4. The oil drain plug looks like a bolt threaded into the oil pan at a low point.

Fig. 5–5. Removing oil drain plug with a socket wrench.

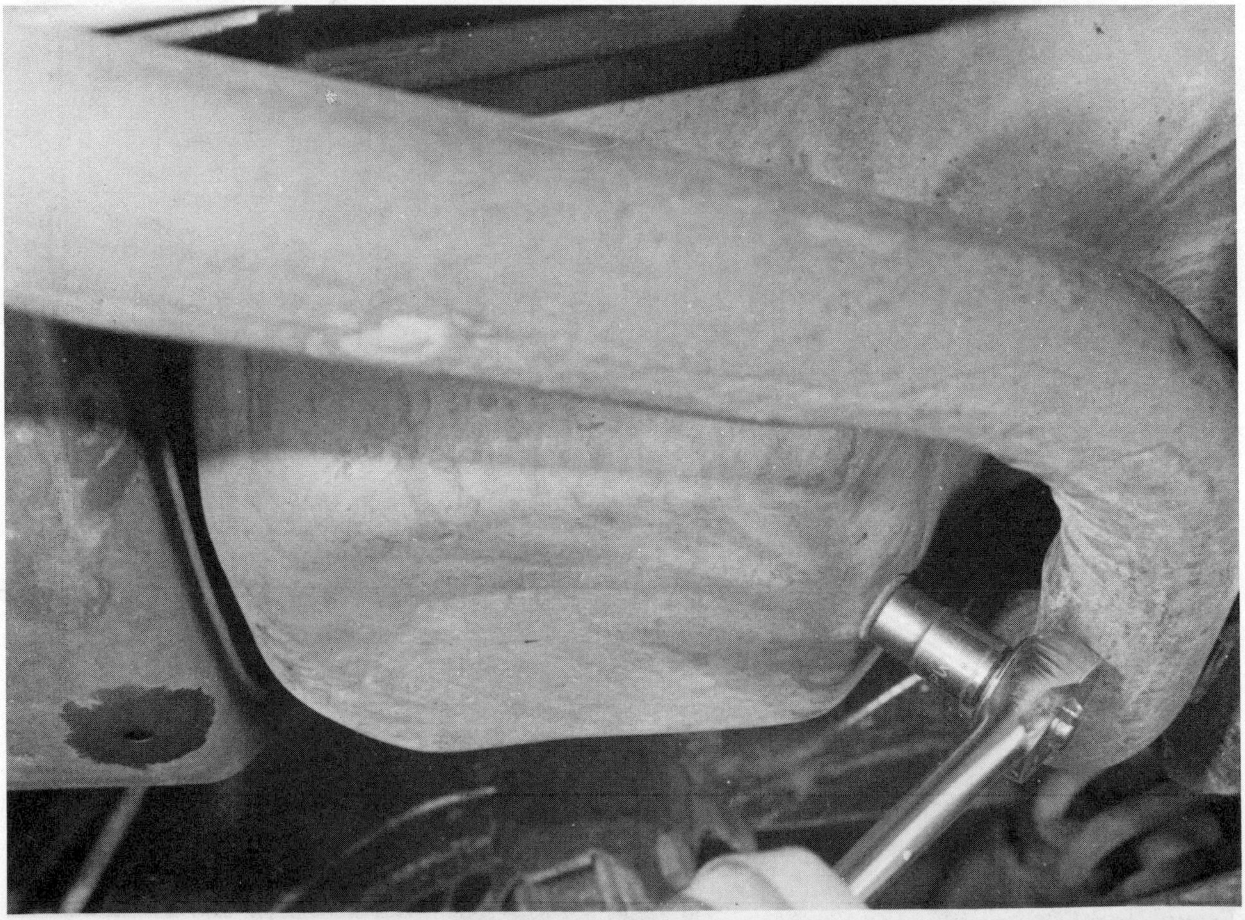

with all popular sizes for the drain plug and a band wrench for filter removal (see Fig. 5–6). Note: Many new General Motors cars have small-diameter filters and the combination filter wrench-drain plug wrench will not clamp tightly around them. Obtain a wrench made specifically for the smaller-diameter filter, or get a filter wrench with a cloth strap that will handle any size filter.

Select an oil of high quality, by a reputable refiner, for although there are standards, there is no policing of them. A name brand oil that is labeled for Service "SE" or "SF" is recommended for gasoline engines, one for "Service CC" is for diesel engines. Use a multi-grade oil (10W–40, 10W–50, etc.) for season-to-season suitability, easier cold starting, and high-temperature suitability. In extremely cold weather (consistently below 0° F.), use a lighter oil, such as 5W–30 or 5W–40, and avoid high speeds (above 50 mph). A synthetic oil of 5W–20 grade is an exception, for it can be used year-round everywhere on late-model cars. Note: If your engine leaks oil, do not use this lightweight synthetic, or any very lightweight oil, or it may leak out very quickly. A synthetic also may not be approved for all diesels; check your owner's manual.

Obtain enough oil to refill the engine—usually about five quarts.

Doing the Job

Start with the engine warm, so the oil will be hot and drain readily. Jack up the car and support on safety stands, or drive onto ramps. Place a drain pan of at least six quarts capacity under the engine's oil pan, that metal pan at the bottom of the engine. The drain plug looks like a bolt threaded into a low point in the pan and should be obvious. Note: On Volkswagen Beetles, there is no oil pan as such, and the drain plug is threaded into a small circular metal plate at the center bottom of the engine. If that plate has no drain plug, the entire plate must be unbolted from the engine for the oil to drain out. Procure a new

Fig. 5–6. Using combination wrench to remove oil drain plug. Notice that drain plug wrench end has three sizes to fit virtually all cars. Oil filter wrench end will fit most (but not all) cars, so check your filter before you buy.

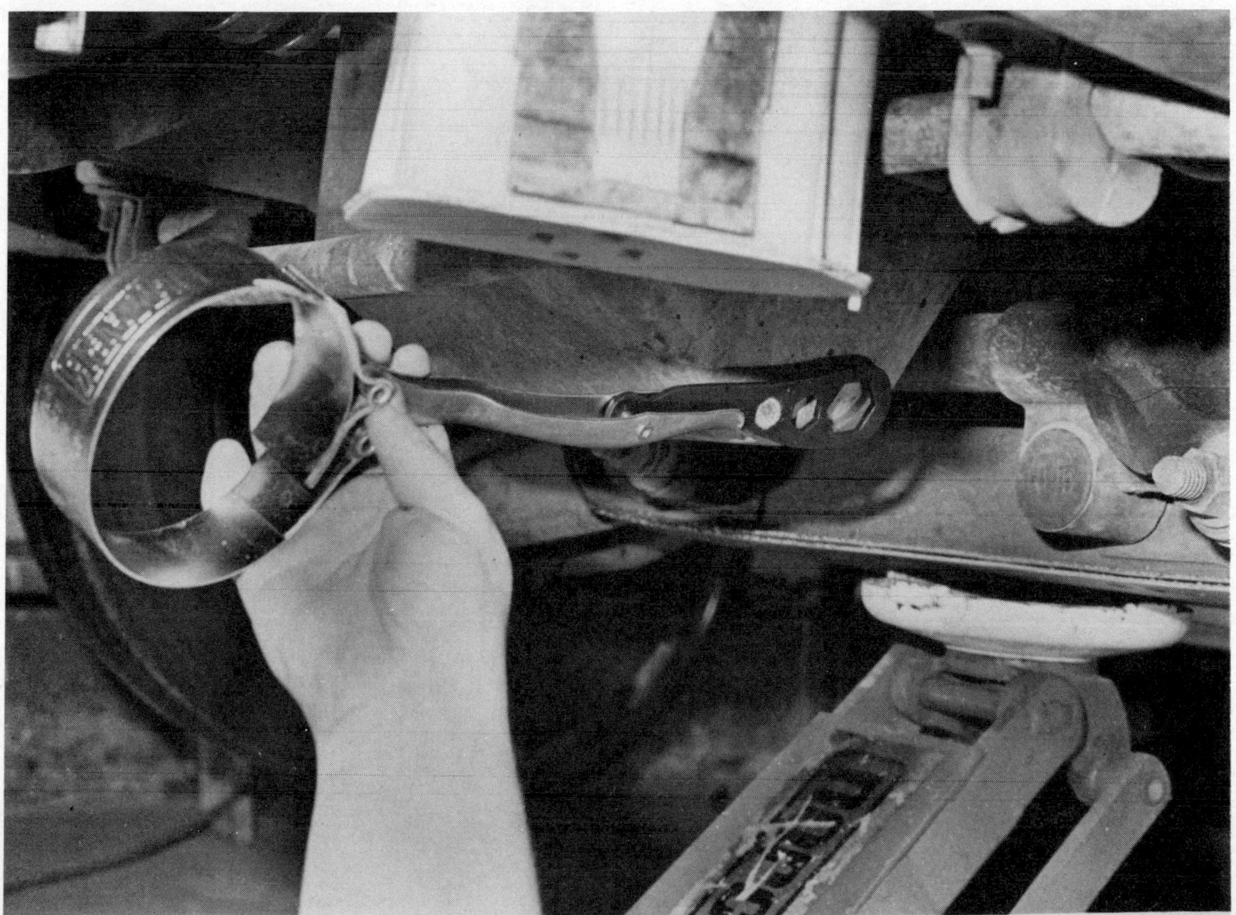

gasket for this plate and replace the old one when you change the oil.

Undo the drain plug. When you're underneath a car, you may become disoriented and attempt to turn the plug the wrong way, so if it seems to be tightening rethink your position. As soon as the plug is completely loose, remove the wrench, move the drain pan directly underneath, and complete the drain plug removal by hand.

Allow five minutes more for all the oil to drain out, then thread back the plug by hand, and complete the tightening with the wrench.

Caution: When you thread back the plug by hand, be sure it is threading normally, not crossthreading. Feel binding? Unthread and start over. If any threads are dirty, clean with a wire brush. Also, don't apply excessive tightening force with the wrench, or you could ruin the threads in the pan or on the drain plug.

Finally, if the worst occurs and the plug will not thread back properly, don't panic. There are several types of repair kits for this situation; they're inexpensive and they're sold in most auto parts stores. They come with complete instructions and they're easy to use.

The Oil Filter

When you select an oil filter for your car, be sure to buy one that is both a reputable brand and designed for the service you intend. If you change your oil and filter moderately often (three or four times a year and not exceeding 6,000 miles) you do not need the so-called 15,000-mile filters. The standard filter will actually do a better job for you. All modern cars have a spin-on oil filter, except VW Beetle. Though the Beetle does not have an oil filter as standard equipment, it may be retrofitted with a kit. If so equipped, the filter also may be a spin-on.

To remove the old filter, just position the filter wrench on it so the wrench band closes up and tightens around the filter as you turn the wrench handle counterclockwise (see Fig. 5–7). When the band is tight, continue turning until you loosen the filter. This procedure sounds simpler than it often is, however, because other parts in the engine compartment may prevent easy removal. You may have to reposition the wrench so that there is swing room for the handle with the band tight. Or you may have to move the band along the length of the filter cannister to find a location where there is swing room for the wrench handle.

Once the filter is loose, move the drain pan underneath it and complete the removal by hand. Coat the rubber O-ring of the new filter with a film of clean engine oil (see Fig. 5–8), and thread on the filter until you can feel that the rubber O-ring is just touching the engine, then move ¾ to one full turn more. In most cases you will have to use the filter wrench (installed the opposite way so the band closes when you move the handle clockwise) to complete the tightening.

Finishing up. Add fresh oil to the engine, through the oil filler neck, until the engine dipstick reads FULL. Start the engine, run it at idle and check underneath for leaks. If you see any from the oil drain plug or the filter, stop the engine, get underneath and check tightness of the leaking part. If there are no leaks, let the engine idle for a couple of minutes (the oil filter will fill with oil) and recheck the dipstick. Top up if necessary.

Gasket Service

In mass production no two flat surfaces mate perfectly. If air, oil, or water must flow between two mating parts, it is customary to use a gasket between them. When the parts are bolted together the gasket compresses and compensates for minor irregularities in the surface finishes.

Gaskets may be made of cork, fiber, treated paper (actually a form of fiber), synthetic rubber, and even soft metal.

Although there are many gaskets on an automobile, this chapter will cover only those that a typical weekend mechanic might attempt to replace. Some weekend mechanics might replace the cylinder head gasket, the one between cylinder head and engine block, but this is a complex job beyond the scope of this book.

It should be understood that the actual replacement of a gasket is usually a simple job. The problem is to get the part off for access to the gasket. No general book (indeed, even many specific make-of-car factory manuals) can assess the degree of difficulty in removing the part for access. Under the specific gasket headings, however, general comments regarding procedures for handling the most common problems are given.

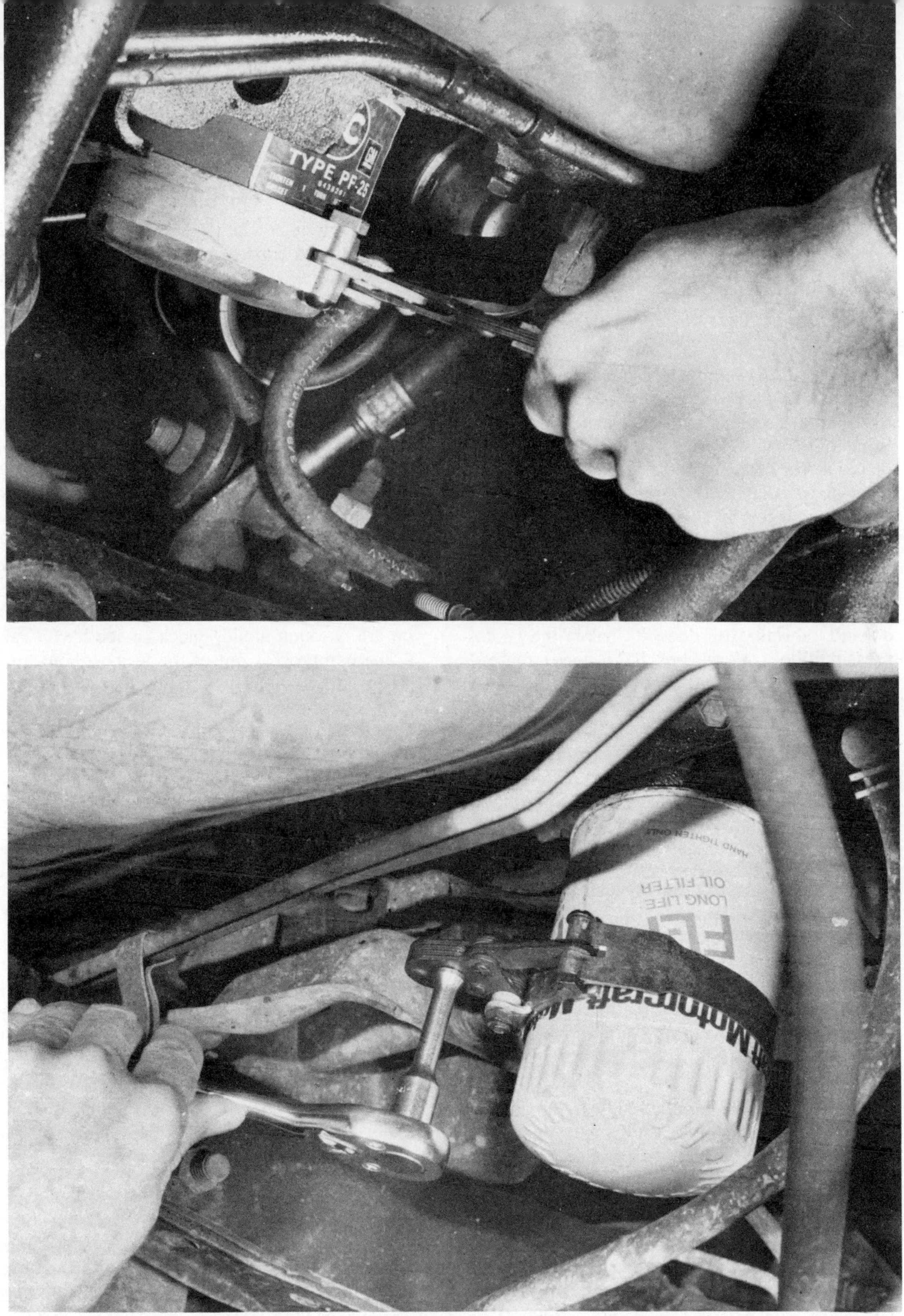

Fig. 5-7. The top photo shows oil filter wrench with built-in handle. The bottom photo shows a filter wrench that accepts an extension and rachet.

Oil Sealing

Gaskets that must seal against loss of oil are used on the engine's metal top cover (one on each bank of a V-type engine), engine oil pan, tappet (valve lifter) chamber cover, fuel pump, transmission oil pan, a rear axle cover on some cars, and, on engines with a chain or gear-driven camshaft, the front cover.

Types of gaskets for oil-sealing and their advantages and disadvantages are:

• *Cork.* It's very compressible, so if mating surfaces are warped it does the best job. However, it takes a set and you may have to retighten frequently. It also shrinks when stored in a dry place, is very fragile, and could break if you're trying to install it in tight quarters.

• *Fiber.* It has less compressibility than cork, but doesn't shrink and is somewhat less fragile. Cost is very little more than cork.

• *Synthetic rubber.* It's much more expensive than cork or fiber, but it won't break and it doesn't shrink. It doesn't have the compressibility of cork or fiber, but it does a reasonable job in this area. Neoprene is the most popular material, and in some cases silicone may be used.

• *Synthetic rubber paste.* This is usually silicone, it extrudes from a toothpaste-like tube, and you use it to form a gasket in place. It does a good job of sealing and has great heat resistance. It cures in place, and if it ever breaks when you remove the part for other service, you just have to squeeze out a ribbon of new paste to repair the break. Note: If you find the tube difficult to use, silicone paste is available in a cartridge that fits into a caulking gun. Silicone rubber paste relies on moisture in the air to cure, and it starts curing the minute it is squeezed from the container.

• *Anerobic sealer.* This is a fluid sealer, but not nearly as thick as silicone rubber paste. Unlike silicone, it cures in the absence of air, so after you squeeze it from its tube, you can let it stand without the problem of it curing while you attend to other things. However, it cannot seal between surfaces that are not both machined (such as an engine top cover). Anerobic sealer is suitable for water pumps and thermostat housings, for example.

• *Synthetic rubber-cork or synthetic rubber-fiber.* This type combines cork or fiber with rubber to make a gasket that has the good compressibility of the fiber or cork and the unbreakability of synthetic rubber. Price is slightly higher than the plain synthetic rubber type.

Getting to the Gasket

• *Engine top cover (also called rocker or valve cover).* At the least, you will have to disconnect a couple of hoses from a pollution control system to get at this, and usually the air cleaner assembly must come off. When removing the air cleaner, note the hose disconnections you must make, and label them with masking tape at both ends so you can correctly reconnect. In other cases, the fuel line to the carburetor also must be disconnected if it passes over the cover. If the air conditioning compressor bracket interferes check to see if you can get the cover off without disturbing it. If this isn't possible, work with a helper so one person can hold up the compressor while the other removes the cover. If you are working alone, check to see if you can lay the compressor out of the way without putting any strain on the air conditioning hoses.

If a cover or part sticks after all screws are out, carefully work a putty knife into the gasket joint and gently pry all around.

• *Engine oil pan.* On some cars you can simply remove the bolts or screws and lower the pan. In many cases, however, you must unbolt the engine mounts and jack up the engine to gain clearance so the oil pan can slip past a chassis crossmember.

• *Engine front cover (also called timing gear or chain cover).* Access to this usually requires draining and removing the radiator, and removing the crankshaft pulley assembly (a job that may require a puller). The bottom of the cover normally seals against the oil pan gasket and when you remove the cover you may tear the protruding segment of oil pan gasket. If this happens, get a timing cover gasket kit that includes a replacement for the torn segment. You will not have to drop the oil pan and change its gasket.

• *Fuel pump.* To remove the fuel pump, refer to Chapter 3.

• *Transmission oil pan.* This is usually wide open; just remove the retaining screws and carefully lower the pan. Caution: It's filled with transmission fluid. Also refer to Chapter 13.

Fig. 5–8. Apply some clean engine oil to filter gasket as shown, then spread it over gasket with your finger.

• *Rear axle.* On some rear axles there is a cover plate at the rear. To remove, just undo the screws and lift it away. There is seldom any interference of any significance, although it may be necessary to push on a parking brake cable to gain clearance to a particular screw.

• *Tappet chamber.* The tappet (valve lifter) chamber cover is on the side of an in-line engine, between the banks of a V-type engine. It is rarely easily accessible on the in-line because it's usually blocked by the exhaust system, which must be dropped to get to it. On a V-8, you must remove the intake manifold (see reference later to intake manifold gasket).

Servicing the Gasket

Scrape off all residue of the old gasket from both the cover and the mating surface on engine or transmission with a putty knife or scraper. Wire brush a metal pan or cover if necessary for final cleaning.

Check a cover for warpage by placing against a flat surface and looking at the cover edge. Sheet metal covers, if necessary, can be straightened with a soft hammer, while you hold the gasket surface against a flat block of wood (see Fig. 5–9). Check a machined surface, such as a fuel pump or thermostat housing, by laying a machinist's ruler across at various points. Try to slip a .005-inch (.12-mm) feeler gauge underneath (see Fig. 5–10). If it doesn't fit, warpage is minimal. If it does, try two gaskets and coats of pliable gasket sealer (one on the pump surface, another between the gaskets).

If you're using a pre-formed gasket, make sure it's a correct fit. Never try to stretch a gasket that seems to be a bit small—it could be the wrong part. If the gasket is cork and seemingly dried out, soak it in a pan of water for half an hour or so and that may restore it to size.

Apply a thin film of pliable gasket sealer to the gasket surface (see Fig. 5–11) of the part you removed (never to the engine or transmission), then position the gasket on the coated surface (see Fig. 5–12). If you are working with a gasket that has tabs, lock them into the slots in the cover.

Position the plate or part on the engine or transmission gently (to insure the gasket does not shift), then start threading in the screws, preferably by hand. If clearances are limited, put the screw into a socket with an extension, and start

Fig. 5–9. To straighten cocked engine top cover, brace against a flat piece of wood and gently pound back into shape with hammer and wood.

the screw in by hand-turning the extension. Either of these techniques is designed to prevent cross-threading.

Once all screws are finger tight, give them a final tightening with a wrench. If you have a torque wrench, check manufacturer's specifications and tighten the screws accordingly; if not, you must guess, and guessing really is for the experienced, not the beginner.

Always tighten in stages, first finger-tight, then moderately tight, and last, securely tight. With a rectangular part that has many screws, start in the center and work alternately toward each end, as indicated in Fig. 5–13.

Formed-in-Place Gaskets

The use of silicone paste to make a formed-in-place gasket is basically similar to installing a pre-formed gasket. After cleaning the surfaces, extrude a continuous ⅛-inch diameter bead on the gasket surface of the part you removed and, in addition, encircle each bolt hole with a bead about ⅛-inch or more (if the width of the gasket surface permits) from the edge of the hole (see Fig. 5–14).

Install the part immediately, and allow a few hours for the silicone to cure.

The procedure with anerobic sealer is basically the same. After you have cleaned the gasket surfaces, apply a coat of primer (from an aerosol can) to one surface and allow a couple of minutes for it to dry. Then apply a bead of anerobic sealer as you would the silicone paste. However, there's no rush to install the part. Allow a half hour after bolting down the part for the sealer to cure.

Water-Sealing Gaskets

Water-sealing gaskets are used at the joint between the water pump and the engine, and the upper radiator hose neck (actually the thermostat housing) and the engine. The radiator cap also has gaskets, but they are not serviceable; if they leak the cap must be replaced.

The general rules regarding installation of oil-sealing gaskets also apply to those on the water pump and thermostat housing. It is common practice, however, to coat both sides of a water pump gasket with pliable sealer.

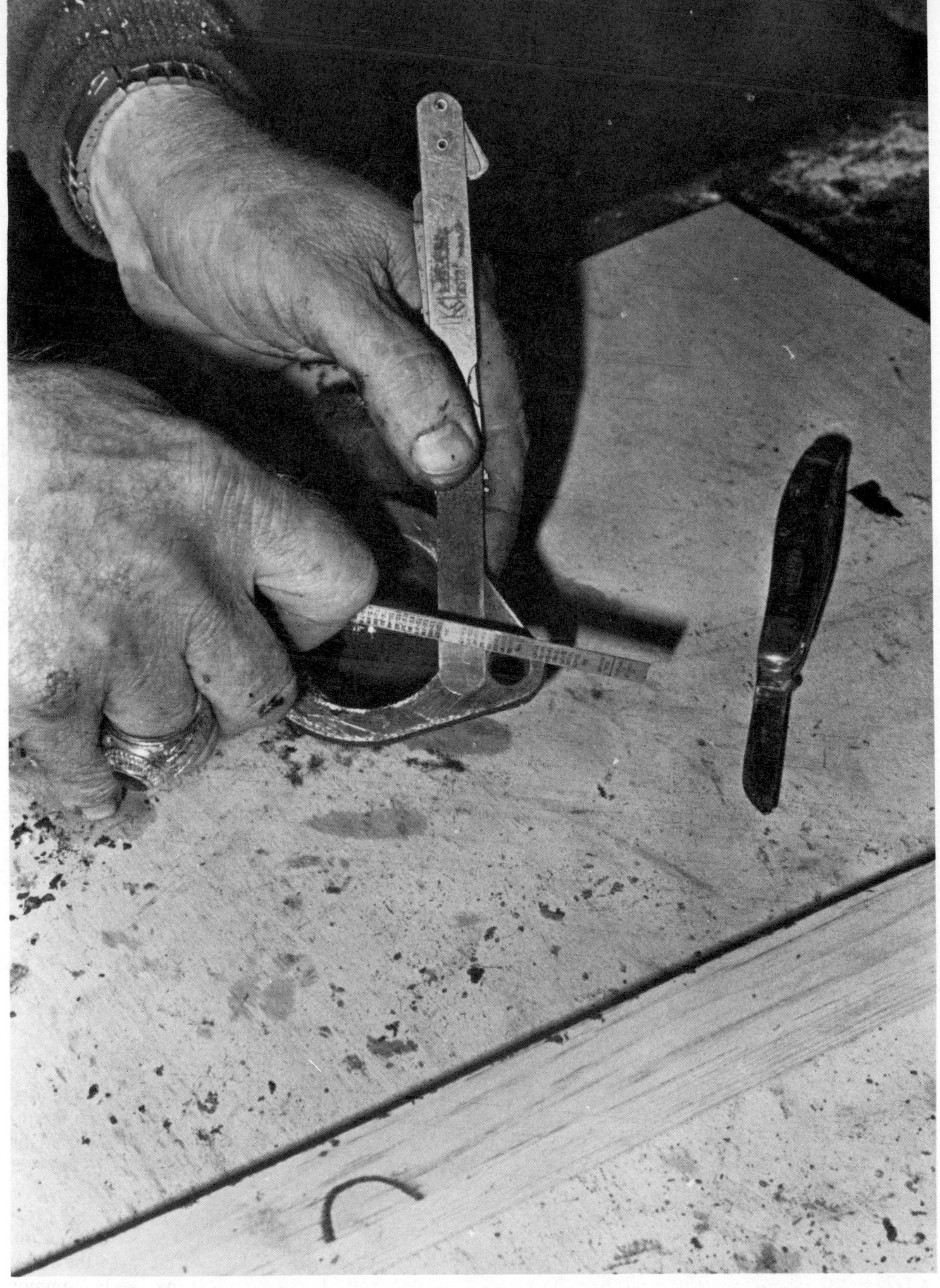

Fig. 5–10. Try to slip .005-inch feeler gauge under machinist's ruler pressed down on gasket surface. If it fits under, gasket surface is warped.

Fig. 5–11. Applying ribbon of gasket sealer to cover's gasket surface. Spread ribbon with finger until it is a thin coat all around.

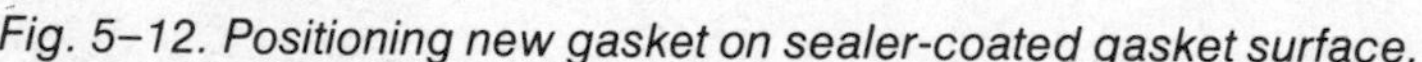

Fig. 5–12. Positioning new gasket on sealer-coated gasket surface.

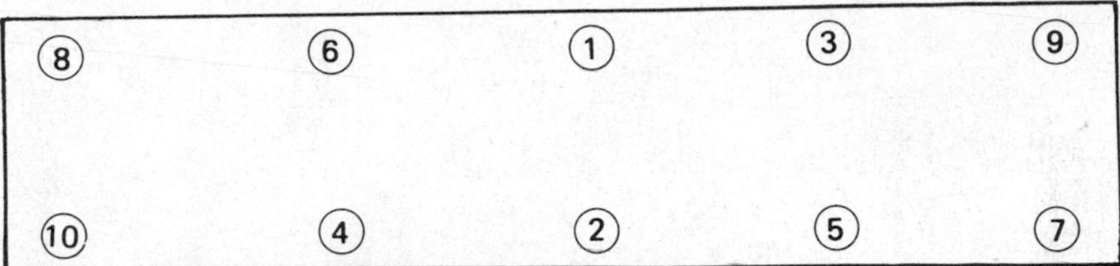

Fig. 5–13. Typical tightening sequence for bolts or nuts.

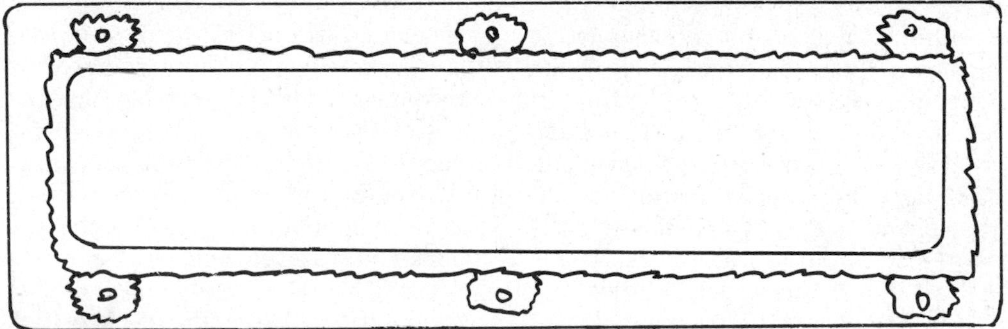

Fig. 5-14. The upper photo demonstrates use of formed-in-place gasket. The proper pattern for spreading is shown below.

Fig. 5–15. Using fine draw file to correct for warpage. This procedure may be used instead of two gaskets or, in severe cases of warpage, in combination.

Removal Techniques

• *Upper radiator hose neck (thermostat housing).* This neck is removed whenever it is necessary to test and/or replace the thermostat. In most cases, just undo two nuts or bolts and lift up the part. If it sticks, pry up gently with a putty knife. Be sure to check this part for warpage, as for a fuel pump (described earlier). If warpage of this part is excessive, you may be able to salvage it by resurfacing with a fine draw file (see Fig. 5–15).

• *Water pump.* The radiator almost always must come out, to gain working room, if the pump is conventionally mounted on the front of the engine. In addition, the fan belt and fan should be removed. Note: When reinstalling, you should tighten the bolts or nuts to factory specifications with a torque wrench if at all possible. As with the thermostat housing, check the gasket surface for warpage. If excessively warped, but still functional, try two gaskets and coats of pliable gasket sealer as described earlier for the fuel pump.

Air-Holding Gaskets

There are three important gaskets that seal in air: between air cleaner housing and carburetor; between carburetor and intake manifold; and between intake manifold and cylinder head. If the gasket between air cleaner and carb leaks, dirty air can get into the engine; if either or both of the other two leak, extra air can get in and lean out the air/fuel mixture, causing engine stalling, stumble, etc.

• *Air cleaner housing.* Just take off the air cleaner cover and lift the filter housing to check the gasket (if any hoses come off, be sure to reconnect). If it's hardened, or dented, replace it.

• *Carburetor base and intake manifold.* Connect a vacuum gauge to a vacuum neck at the base of the carburetor and run the engine at idle. If vacuum is normal (16–22 inches), the gaskets apparently are sealing. If the reading is low, squirt some oil along the entire gasket joint as a temporary seal (see Fig. 5–16). If the vacuum gauge reading rises, the gasket is leaking. First, try tightening carburetor or manifold nuts or

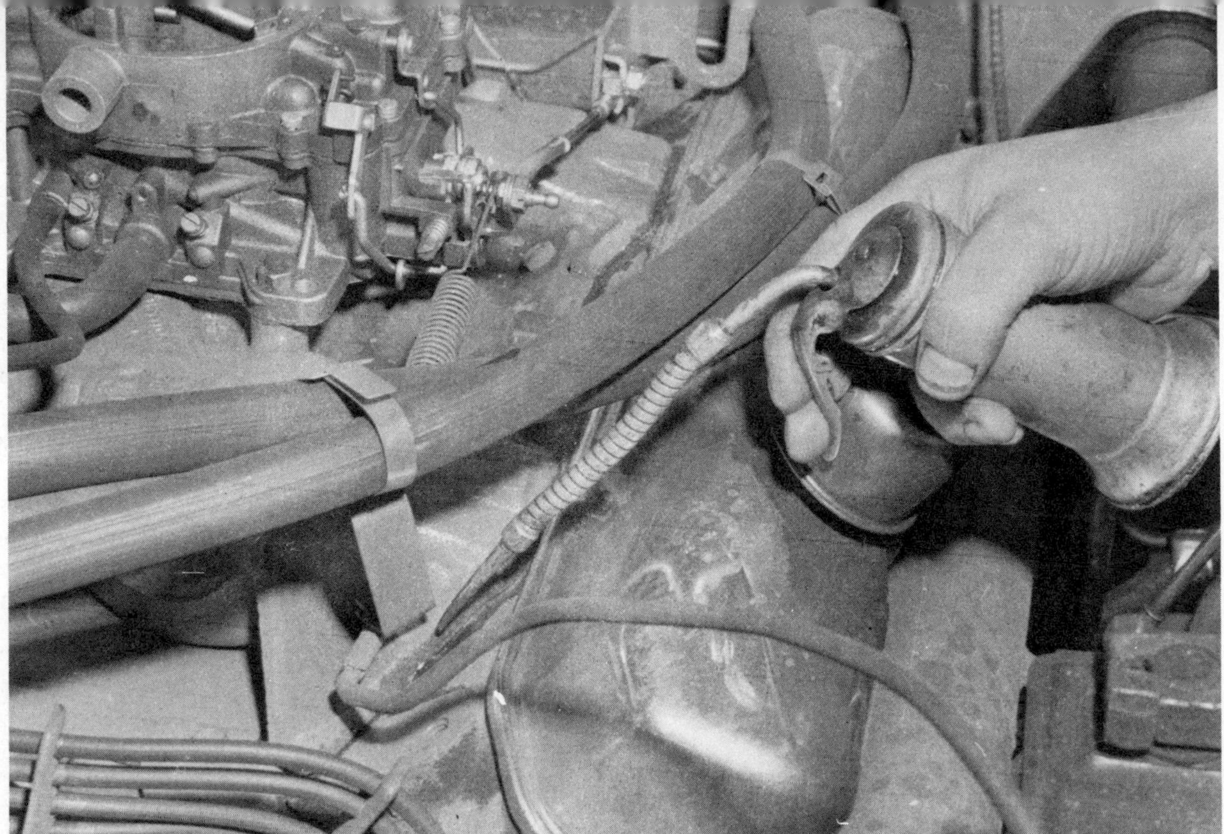

Fig. 5–16. Squirting oil along intake manifold gasket joint as a temporary seal. If vacuum gauge reading rises or if engine idle smoothes out or increases, the gasket is leaking.

Fig. 5–17. Before replacing a leaking intake manifold or carburetor base gasket, try tightening the nuts or bolts as shown.

bolts (see Fig. 5–17), but if that doesn't cure the vacuum leak, replace the gasket.

To replace a carburetor base gasket, remove the nuts that hold the carb to the intake manifold and, if you're lucky, you will be able to lift the carburetor high enough to remove the old gasket and install a new one. Note: The carburetor base gasket is specially made to resist heat and often is very thick, to help isolate the carburetor from heat. Do not fabricate a gasket from sheet cork or other material. If you can't raise the carburetor enough, you must remove it. Look for electrical connectors and vacuum hoses, and disconnect as required, taping each connection (and marking the tape) so you can reconnect correctly. It should not normally be necessary to disconnect the linkage from the gas pedal, but if it is, the detachment (which varies from car to car) should be obvious.

• *Intake manifold.* Once the carburetor is off, remove the intake manifold bolts. On many cars with air conditioning, the AC compressor bracket is connected to the manifold, and this must come off too. In addition, you may find other brackets attached to the engine by the intake manifold bolts.

Important: On some cars the ignition distributor must come out in order to take off the intake manifold. This must be done in a special way, so you can refit the distributor without

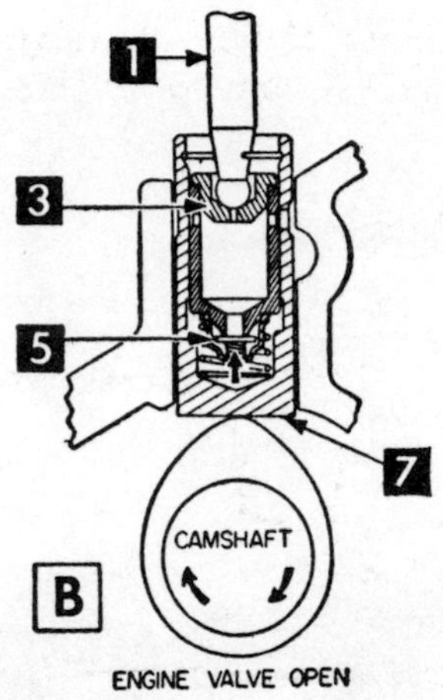

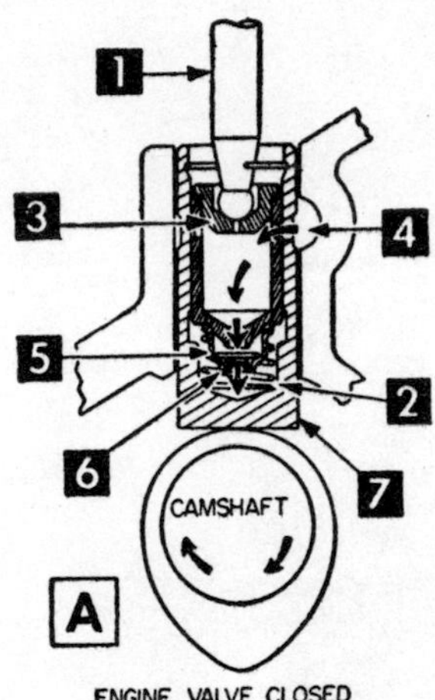

Hydraulic Lifters

How They Work

The hydraulic lifter has two interrelated purposes: (1) to eliminate shock in the valve train by keeping all parts in constant contact; (2) to compensate for wear in the parts.

When the valve is closed (camshaft lobe is not pressing on the lifter) there is no pressure on the pushrod (1) so that the lifter plunger spring (2) pushes up the plunger (3); engine oil under pressure from the oil pump enters the lifter through port holes (4) in the lifter body (7), then through a check valve (5) in the lifter into the space (6) below the plunger.

When the camshaft turns so the cam lobe is pushing up on the lifter, the check valve (5) closes, making the lifter and plunger (and the oil between them) into one solid part. The lifter then pushes up on the pushrod and rocker to open the valve.

The lifter is designed so there is a slight leakage—called leakdown—between the plunger and the body. This controlled leakage allows the oil to escape when the engine is shut off for an extended period. Thus, when you first start the engine, the lifters are noisy until they fill up with oil. The plunger spring is a rather mild one, yet if functioning properly it will take up the slack in the valve train during the leakdown process. If there is any sticking of the plunger, either from gums or dirt in the oil, the mild spring will not be able to push it up to compensate for clearances in the valve train.

Frequent oil changes, using the best quality oil, is your best insurance against any and all lifter malfunctions.

disturbing ignition timing. Remove the distributor cap and note the exact position of the rotor pointer, as if it were the hour hand on a clock. Also mark the location of the distributor body with a dab of nail polish on the distributor base at the location of the lockbolt bracket.

Remove the lockbolt (see *Chapter 2*) and lift out the distributor. If it sticks, twist it back and forth with an upward motion to free it. When you refit the distributor with a gear on the bottom of the shaft, you will have to "lead" the rotor, for once the gear meshes the rotor position will change slightly (but slightly off is unacceptable). You will understand this if you do it wrong, for the rotor will not be in exactly the same position as when the distributor was removed. If the bottom of the distributor shaft has a female hex and a gear, proper refitting may take repeated tries, so that both gear and hex engage properly. Caution: Never crank the engine with the distributor removed, or you'll never get it back in correctly without professional help.

Once the intake manifold is off, you have access to both the manifold gasket(s)—one on each side of a V-type engine, and the V-8 valve lifter chamber and its cover gasket. Or a single gasket may be used for both.

Like carburetor base gaskets, intake manifold gaskets are specially made for the applications, so never use a home-made. Tighten manifold bolts to factory specifications in stages with a torque wrench if possible. Also, use the criss-cross tightening pattern for rectangular parts described earlier.

Hydraulic Lifter Service

The rhythmic clatter of valve lifters (also called tappets) is a common engine noise. On the

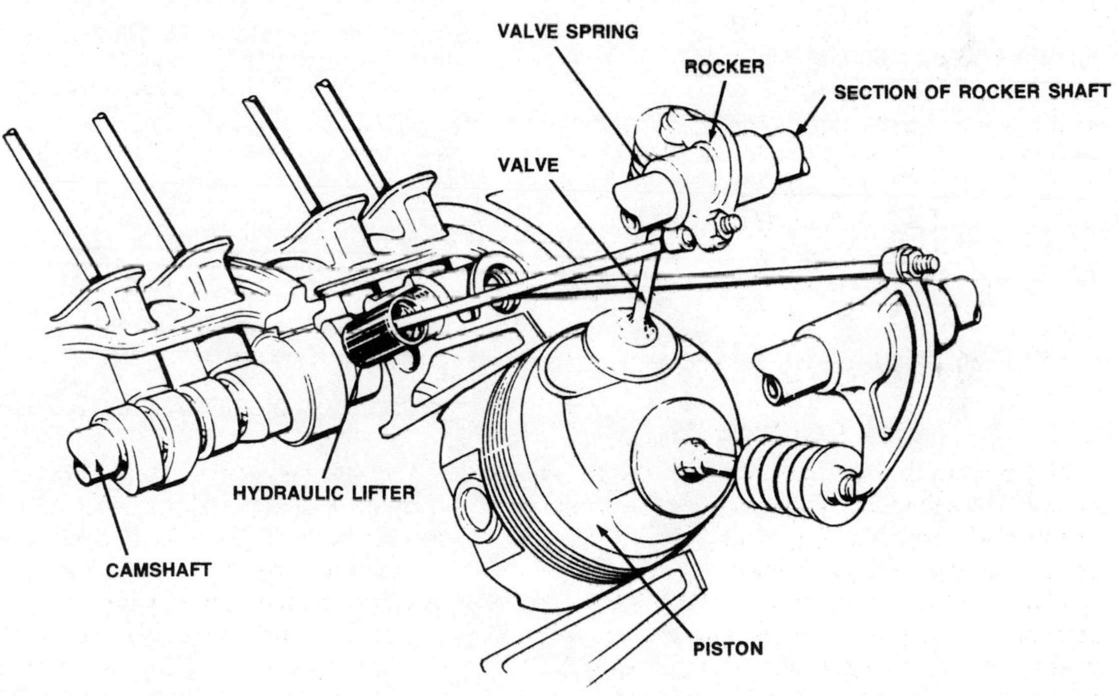

Fig. 5–18. This is a typical pushrod-type valve train with the engine block and cylinder head removed for clarity. Hydraulic lifters rest on camshaft lobes, and when the camshaft spins the lobes force the lifters up. The lifter pushes up the pushrod, which causes the rocker to pivot on its shaft (or on a ball-shaped stud). When the rocker pivots, it pushes down on the valve stem against the pressure of the valve spring. When the camshaft lobe no longer exerts upward pressure on the lifter, the valve spring closes the valve and pushes all the parts back toward the camshaft.

modern car with hydraulic lifters the noise should stop after a half minute or so as they fill with engine oil. If they don't fill up, the engine is not only noisy, but there is loss of power and premature wear of the valve train—the system that opens and closes the valves (see Fig. 5–18).

Once the engine is warmed up, hydraulic lifter-caused tapping is a matter for concern, and here's how to check it out.

First, inspect the oil level on the dipstick. If it's too high or too low, air can get into the oil and then into the valve lifters. Adjust the oil level and, if the noise continues, determine if the lifters are really the problem.

Obtain a piece of heater hose about thirty inches long and hold one end to your ear. With the engine idling, slide the other end along the engine top cover (also called the valve or rocker cover)—both covers on a V-type engine. If you hear a loud clicking noise through the heater hose the problem is in the valve system.

Remove the engine top cover (see *Gasket Service* earlier in this chapter). This time place the end of the hose on the pushrod side of the rocker arm. A noisy lifter will sound much louder at its rocker arm than another. Note: On overhead camshaft engines (which have no pushrods) place the hose end on the lifter side of the rocker. If there is no rocker, place the hose as close to the lifter as possible.

When you find what seems to be the noisy lifter, make these additional tests before condemning it:

• *For worn valve stem and/or valve guide.* With the end of a hammer handle, apply pressure at the side of the valve spring (trying to push against the coils). Repeat at the opposite side. If the noise is reduced the problem is a worn valve stem and/or valve guide, a repair job for a professional. Note: Valve guide service can be expensive. If the problem is accompanied by high oil consumption, you may solve the oil consumption problem yourself, as described later in this chapter, with new valve stem seals.

• *Check for a loose rocker shaft.* On some engines the rockers all are on a shaft, and if it is loose, the tapping noise can occur.

Fig. 5–19. If oil squirts through the hollow pushrod, then through the hole in the pushrod end of a rocker on a ball-stud, you can minimize the flow by inserting a paper clip as shown. Pull out the open end of the clip, insert into hole, then hook remainder of the clip on the underside of the rocker.

Fig. 5–20. With clips in place, oil merely oozes out of rocker hole as you turn the ball-stud nut to adjust clearance in the valve train.

• *Where adjustment can be made.* There is provision for an adjustment on some Ford and Chevrolet engines. On these engines there is no rocker shaft; rather, the rocker is on a ball stud and is held by a nut. With the engine idling back off the nut until there is severe clatter, then turn it down until the clatter is just eliminated, plus one full turn of the wrench (¾ turn on 1976–77 Chevrolets). This often cures lifter tapping if it is caused by wear in the valve system rather than a defective lifter. If you exhaust the adjustment range, the wear in the system is excessive or the lifter is defective. Note: During this adjustment oil may squirt out of the rockers and make a mess. You can prevent this by first inserting the end of a large paper clip into the hole (on the pushrod side) and bending the clip over so it holds against the bottom end of the rocker (see Figs. 5–19, 5–20).

• *On cars without a provision for adjustment,* insert the thickest feeler gauge possible between the rocker arm and the valve stem tip with the engine at idle. If the noise goes away and does not come back the problem is either wear in the valve system, probably in the rocker or valve stem end, or a lifter that is stuck internally. If it's in the rocker, you probably can replace the rocker yourself. If the valve stem tip is the problem, it's a job for a professional.

If the noise comes back almost immediately, the problem is in the valve lifter, which is leaking down excessively (see *How Hydraulic Lifters Work,* elsewhere in this section).

If the problem is an internally sticking lifter you may be able to get it working again. Drain the oil and change the filter, then refill the crankcase with four quarts of kerosene (or a gum solvent made for lifters that contains kerosene and special chemicals) and one quart of SAE 10W–30 oil.

Run the car for two hours at very low speed—under 25 mph. Then drain the oil-kerosene mixture, change the filter once more, and fill the crankcase with the normal quantity of oil. If you're lucky the stuck lifter will be free. If it is not, or if the problem is a lifter leaking down excessively, you must replace the lifter. In a professional shop it is standard practice to replace all lifters when one has failed, on the theory that the others soon will follow and so will

Fig. 5–21. With intake manifold and engine valley cover removed, you can see pushrods and lifters.

customer complaints. When you do the job yourself, however, you have the option of replacing only the defective one.

Replacing Lifters—Pushrod Engines

On a V-type engine, the access to the lifters is through a cover that is under the intake manifold in the center of the engine (see Fig. 5–21). You must disconnect the throttle linkage and remove the carburetor and intake manifold (see *Gasket Service* earlier in this chapter).

On an in-line pushrod engine, there are three possible ways to remove the lifters. Only one will apply to any particular engine. The cylinder head may have to be removed, a job for a professional. The lifters may come out through the pushrod holes, after removing the pushrods. This is a most convenient way, although a special lifter pulling tool is commonly needed. The 1981 Chrysler slant six-cylinder and the 1982 General Motors "J" car (Chevrolet Cavalier, etc.) four-cylinder have hydraulic lifters that can be removed through pushrod holes. Or there are covers on the side of the engine block for access to the lifters. In this case, which is more common, you must remove parts that interfere with simple unbolting of the covers, perhaps dropping the exhaust system. The covers have sealing gaskets that probably will require replacement.

Before you go to the trouble of removing these parts, make sure you can get the defective lifters(s) out.

It is possible for the base of a lifter to mushroom, preventing you from removing the lifter even with a special puller. Although this is not the common situation, it does mean that it must be pushed out through the bottom. To provide clearance, the camshaft must be removed first, and this is a job for a professional.

There is no simple way to know if a lifter has mushroomed. If you encounter the problem, you'll have to give up and reassemble. Because the possibility is so remote, don't let it stop you from trying the job.

After gaining visual access to the lifters by removing the carburetor and intake manifold, or in-line engine side covers, you must take out the pushrods. On cars with rocker shafts you can sufficiently slacken the bolts that hold the shafts to provide clearance to pivot the rocker, disengage it from the pushrod, and move it aside. If you can do this, lift out the pushrod.

On engines without rocker shafts, slacken the

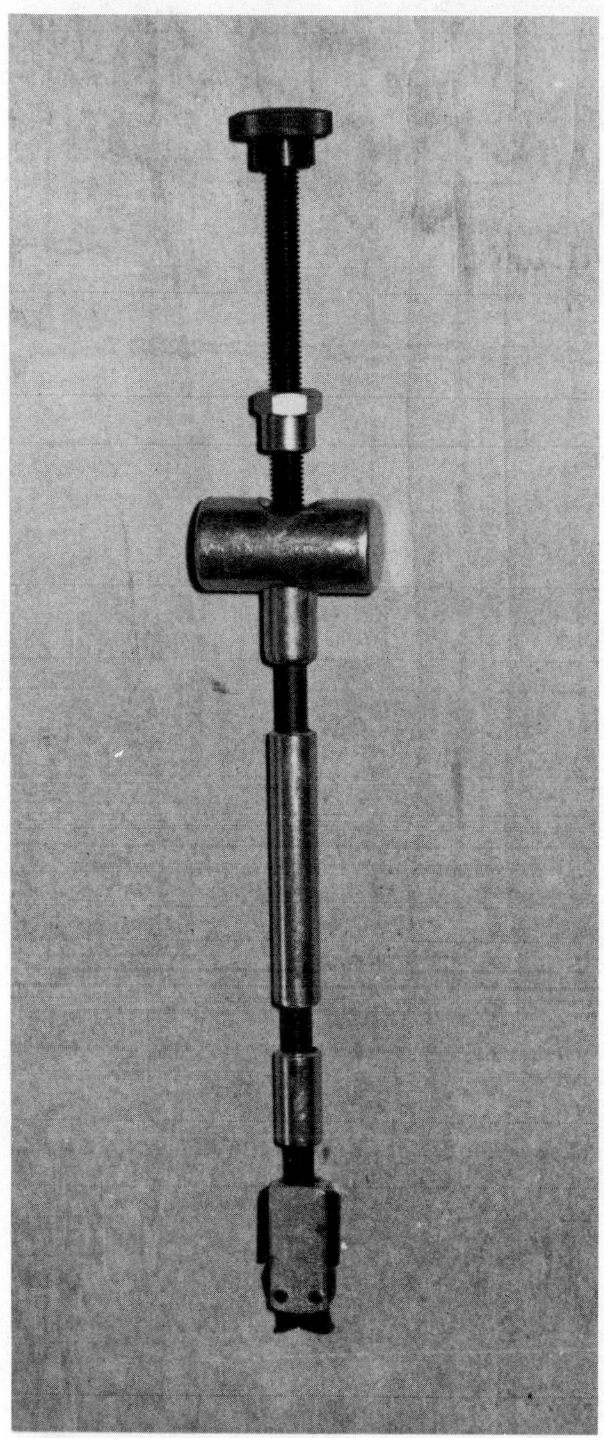

Fig. 5–22. Special tools, such as the slide hammer puller shown in this photo, simplify removal of stuck hydraulic lifters. After the pushrod is out, insert the bottom of this tool into the top of the stuck lifter. Turn knob at top of tool and jaws at bottom lock against inside wall of lifter.

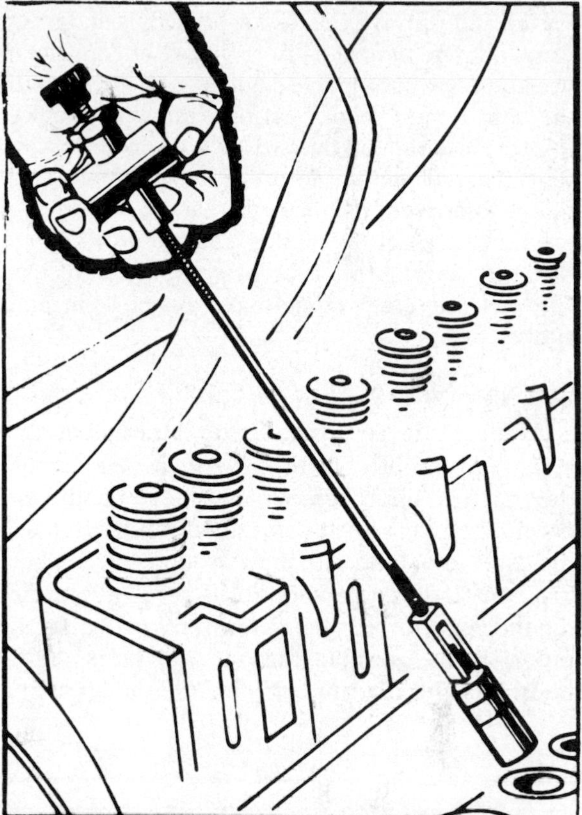

Fig. 5–23. Illustration of the lifter pulling tool in use. The T-shaped slide hammer is yanked upward to shock lifter free.

rocker retaining nut sufficiently, or remove it entirely, lift up the rocker, and pull out the pushrod. If you remove the rockers completely, keep them in order, so if serviceable they can be refitted to the same position. If you're lucky you can just lift the lifter from its bore. If not, squirt some penetrating solvent between the lifter and the engine block bore and try again. If a lifter is really stuck, obtain a special valve lifter pulling tool, an inexpensive item available at auto supply stores (see Figs. 5–22, 5–23).

Replacing Lifters— Overhead Camshaft Engines

The typical American overhead camshaft engine has rockers, and the lifters are not under the camshaft. Therefore the camshaft does not have to be removed, which would make lifter replacement a job for a professional.

You will, however, need a special compressor for the valve spring so you can compress the spring and remove the rocker, which is under the camshaft and over the lifter. Before attempting to use the tool, crank the engine until the peak of the cam is no longer bearing against the rocker (Note: Disconnect thin wires from coil or pink wire from General Motors HEI distributor cap flange to prevent engine from starting.)

The next section of this chapter, *Curing Oil Burning,* shows this type of compressor in use. Various universal models are available in auto supply stores.

Buy New Lifters

Although it is possible to disassemble and clean a stuck lifter with solvent, it is not really the practical approach. Unless you are both very careful and very fortunate, the cleaned lifter will not last very long, so shop around for the best discount and buy new lifters. Before installation, oil the exterior of the new lifter and squirt as much oil as possible through the holes in the body into the interior, or immerse the lifter in a pan of clean engine oil and operate the plunger by pressing down with a pushrod or dowel.

Curing Oil Burning

When your engine starts to burn oil you may psychologically be prepared for a major engine repair. In many cases, however, the problem is caused by something relatively simple.

The first step is to be sure you are burning engine oil, for on many cars with automatic transmissions it is possible for transmission oil to be sucked into the engine through a defective diaphragm device called a modulator (see *Chapter 13*). If the power brakes have failed, it's possible brake fluid is being sucked in. Or, it is possible that the exhaust smoke is caused by a defective fuel system, or leakage of coolant. If the smoke is blue-gray, it's oil; black, it's fuel; white, it's coolant. Check the tip of the tailpipe for oily deposits to help confirm the problem.

Fig. 5–24. Squirt a tablespoon of oil into the cylinder through the spark plug hole as shown. Even if rings are worn, this will seal the piston temporarily, and if you perform another compression test, the reading should show a significant rise if the rings are the problem.

If you know it's oil, the next step is to trace the problem. It could be piston rings, a major repair, but much more likely it is one of the following:

• *Positive Crankcase Ventilation System.* See also Chapter 4, *Pollution Controls*. If you see oil contaminating the carburetor air filter element, oil may be blowing out of the crankcase because of a defective PCV valve. Also pull the valve out of the engine top cover on cars so equipped and look for a baffle in the top cover. If there is a baffle and it's loose, oil droplets in the chamber below may be bypassing this baffle. In this case they will be drawn into the engine through the PCV valve. Either replace the top cover or have a shop weld the baffle back into position so it prevents oil loss.

• *Intake manifold.* On many cars with V-type engines, the design of the intake manifold is such that a loose or defective intake manifold gasket can readily allow oil from the engine to be drawn into the intake manifold. There is one check you can make: Squirt oil all around the exterior joint of the intake manifold and cylinder head, and if the engine speeds up a bit the gasket is defective (the oil acts as a temporary seal). This test is not a positive check for a defect in the gasket on the oil-sealing inner joint, so the gasket could pass this check and still be defective. You still must keep the gasket under suspicion. One helpful thing to keep in mind is that if the oil consumption has suddenly become severe, the gasket is a likely cause.

• *Piston rings.* Take a compression test (see *Chapter 2*). If the compression readings are lower than specifications, squirt about a tablespoon of oil into a cylinder (see Fig. 5–24) and immediately repeat the test. The oil acts as a temporary seal, and if compression readings suddenly jump to normal or above (showing an increase of more than 35–40 p.s.i.), the piston rings are suspect. You should understand that piston rings can be worn and still not be responsible for much of the oil consumption, for there is one ring (called the oil ring) that has little to do with sealing compression pressures but does most of the job of sealing against oil consumption. However, if compression rings are badly worn, there is a good chance the oil ring also is deteriorated.

Fig. 5–25. Oil seals (also called shields) are placed over end of valve stem in engine top chamber. The valve spring has been removed for illustrative purposes.

Valve Stem Guides and Seals

On most late-model engines, defective valve stem seals (shields) or worn valve guides are a major cause of oil consumption (see Fig. 5–25). The stem of the valve slides in a machined hole—called the guide—in the cylinder head. The seals are little rubber or teflon rings, sometimes shaped like miniature umbrellas, that are placed on the part of the valve stem that protrudes into the chamber under the engine top cover.

If there is excessive wear between valve and guide, accompanied by deterioration of the seal, a lot of oil can flow down the guide into the cylinder, where it is burned. The seal is the more critical item because even if there is a fair amount of wear between stem and guide the oil loss may remain within normal limits if the seal is good.

To narrow down the possibilities, take this approach: If compression is acceptable and the engine is an in-line, go straight for the stem seals and valve guides. If the compression is acceptable and the engine is a V-type, go for the possible cause—stem seals or intake manifold gaskets—that is easier to do. If the intake manifold is extremely difficult to remove, do the stem seals first, particularly on a car with 35,000 miles or more on the odometer. At that mileage, there

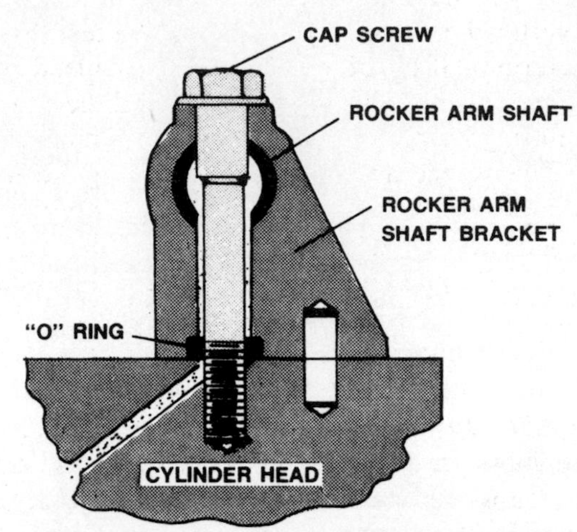

Fig. 5–27. To shut off oil flow around the bracket screw shank, insert an O-ring that fits tightly around the shank of the screw, as shown.

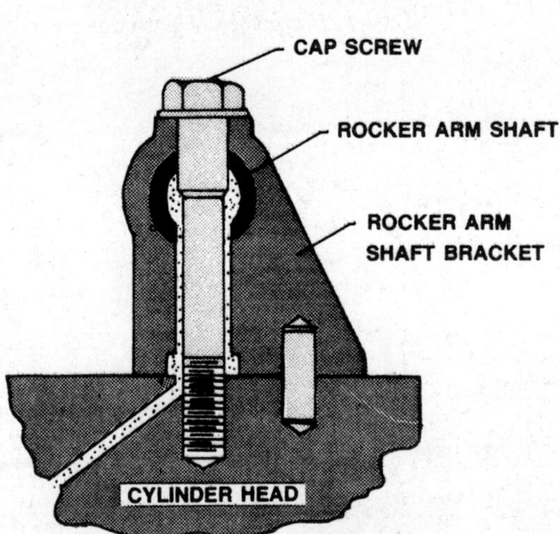

Fig. 5–26. Engine top chamber is lubricated in this design by a passage in the cylinder head. Oil flows up and around the rocker shaft bracket screw shank.

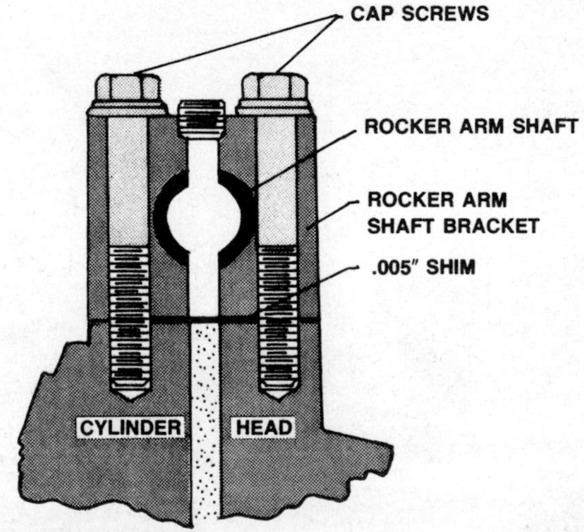

Fig. 5–28. If oil flows up to rocker shaft through passage, as shown, use a thin shim (about .005 inch) between bracket and cylinder head to shut off the oil flow.

is a reasonably good chance that new stem seals will help, even if the manifold gasket is the major problem.

Pushrod Engines

If you have an engine with pushrods (not an overhead camshaft) you can make a test that will determine if the oil consumption is caused by the stem seals or valve guides, provided the engine is emitting visible smoke through the exhaust.

The procedure consists of removing the engine top covers, shutting off the oil supply to the engine top chamber, and seeing if the smoking stops. An engine can safely be run for about 45 minutes or so with the oil cut off to that chamber.

Here's how: If there is no rocker shaft (that is, if the rockers are on ball-stud pedestals), the oil comes up to the rocker chamber through passages in the pushrods. Loosen the rocker retaining nut (count the number of turns) until there is enough clearance between the rocker and pushrod to insert a little piece of rubber (tire inner-tube thickness) between pushrod and rocker. Tighten down the nut the number of turns you loosened, less two (to compensate for the thickness of the rubber). Do this to all the pushrods and rockers and the oil flow will be stopped.

On Chrysler and some Ford powerplants, the rockers are on a shaft that is lubricated through passages in the shaft and a shaft retaining stand. Remove the rocker shaft, find the oil supply hole, plug it with a piece of rubber (perhaps a pencil eraser), then reinstall the shaft and tighten the bolts that hold the stand, using a torque wrench, to factory specifications (see Figs. 5–26 to 5–28).

Run the engine, and if the smoking stops, the valve guides and/or seals are the problem. Remember to remove the plugs blocking oil flow after the test is completed.

• *Servicing stem seals.* To replace valve stem seals you need a special valve spring compressor that permits compressing the springs without removing the cylinder head from the engine. There are many universal designs on the market, most of which are relatively inexpensive.

You also need a method of holding up the valves when you operate the spring compressor.

Fig. 5–29. To change stem seals, you must hold up the valves. Remove spark plug and use adapter at left, into which you thread a coupler (shown installed), then attach the air line at right.

Fig. 5–30. With adapter installed and air line connected as shown, compressed air flow will keep the valve up.

Fig. 5–31. With the valve held up, you can use a special valve spring compressor to press down on the spring. The spring compressor shown here is a universal type being used on an overhead camshaft engine. There also are types designed for pushrod engines only; they may be less expensive.

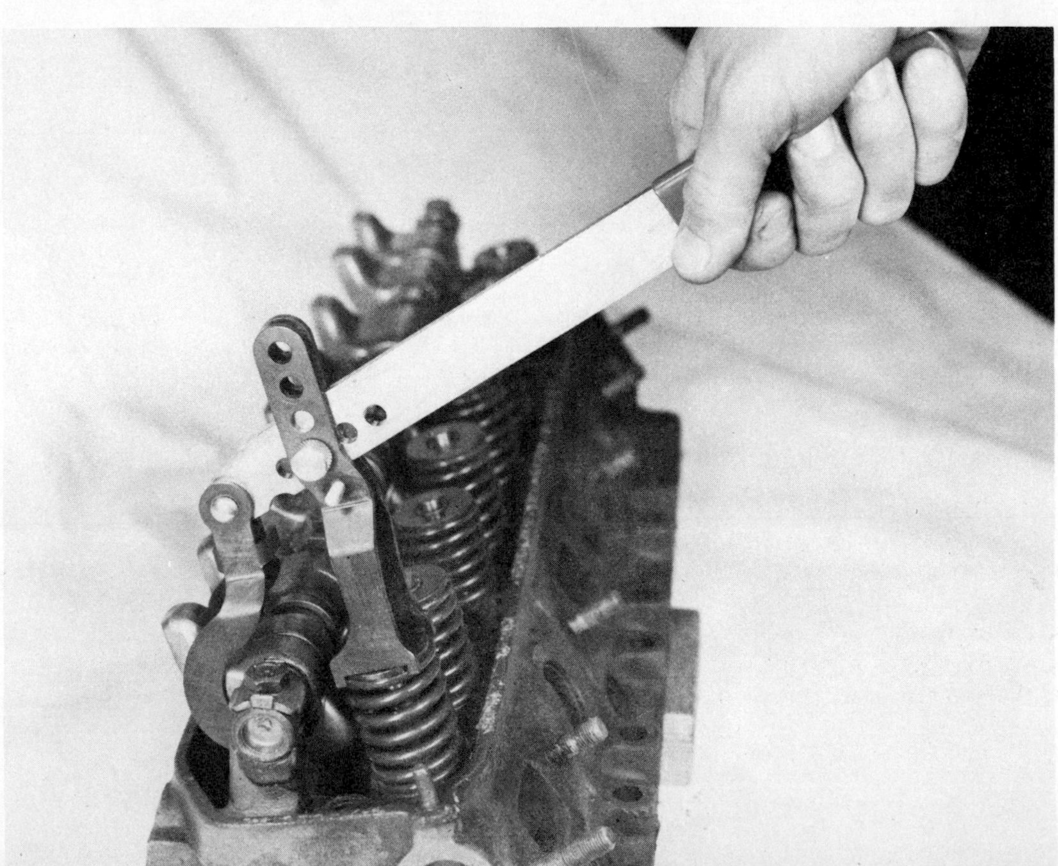

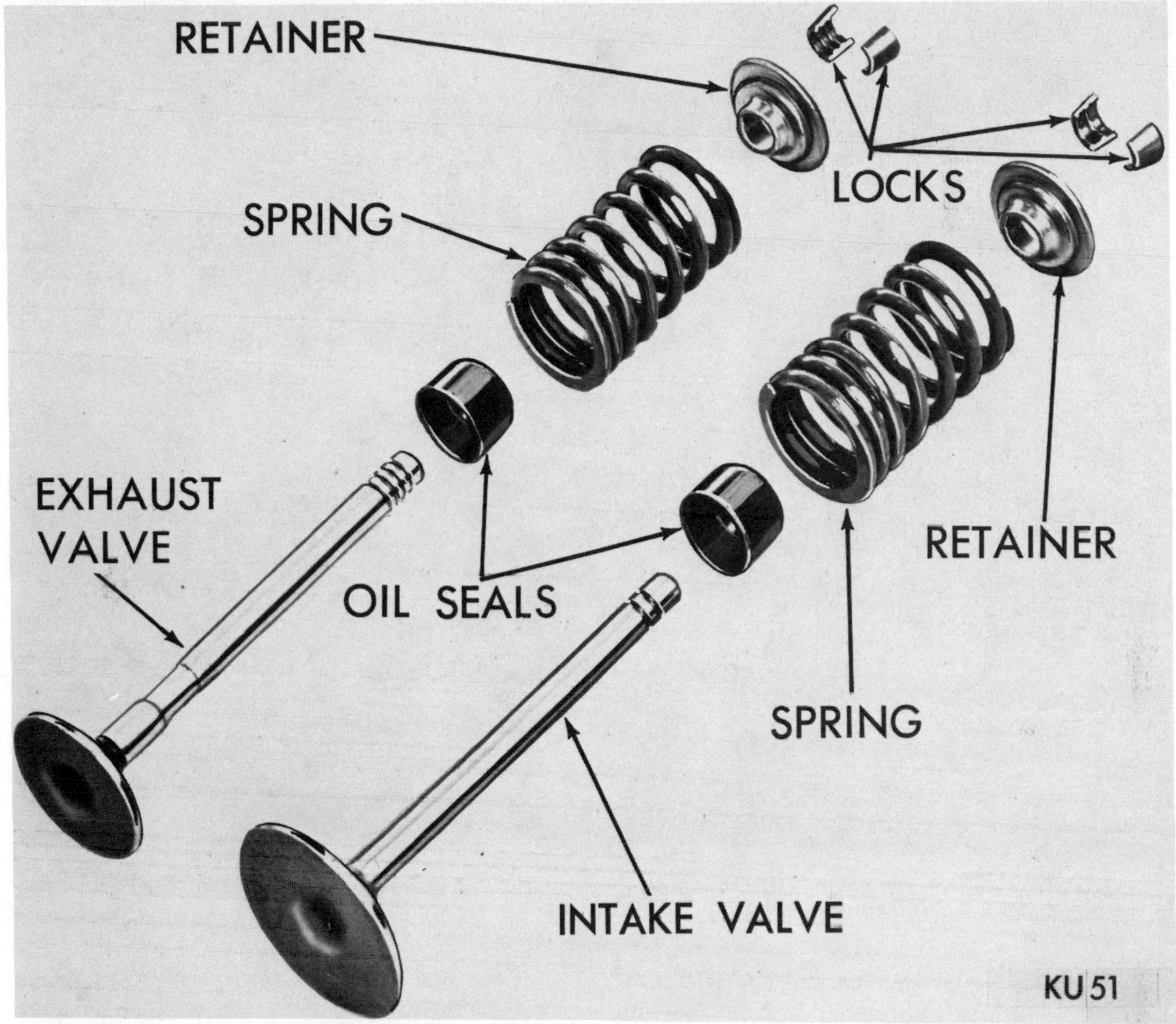

Fig. 5–32. Here's an exploded view of the parts. When you press down on the retainer with the compressor, you expose two half-moon locks that fit against grooves in the top of the valve stem.

It is possible to do this by stuffing the cylinder with clothes line (through the spark plug hole), but this is a tedious procedure. A better way is to use compressed air through the spark plug hole. Available in any auto supply store are special adapters (see Fig. 5–29) that thread into the spark plug hole and attach to a compressed air line (see Fig. 5–30). If you have a paint-spraying compressor or a compressed air tank (an inexpensive item you refill at service stations), you can use the air method.

Remove the rocker arm shaft of the individual rockers (keep them in order so you can refit them to the places from which they were removed). Thread in the adapter and apply compressed air to hold up the valve. Note: If you can feel a considerable flow of compressed air escaping through the tailpipe, an exhaust valve is leaking badly; if you can feel the airflow through the carburetor air cleaner, an intake valve is leaking. If engine performance is poor, the engine needs major valve service, a job for a professional.

Follow the instructions with the tool to compress the valve spring (see Fig. 5–31). Lift out the two semi-circular spring retainer locks (see Fig. 5–32). A magnetic tweezers can be used, as

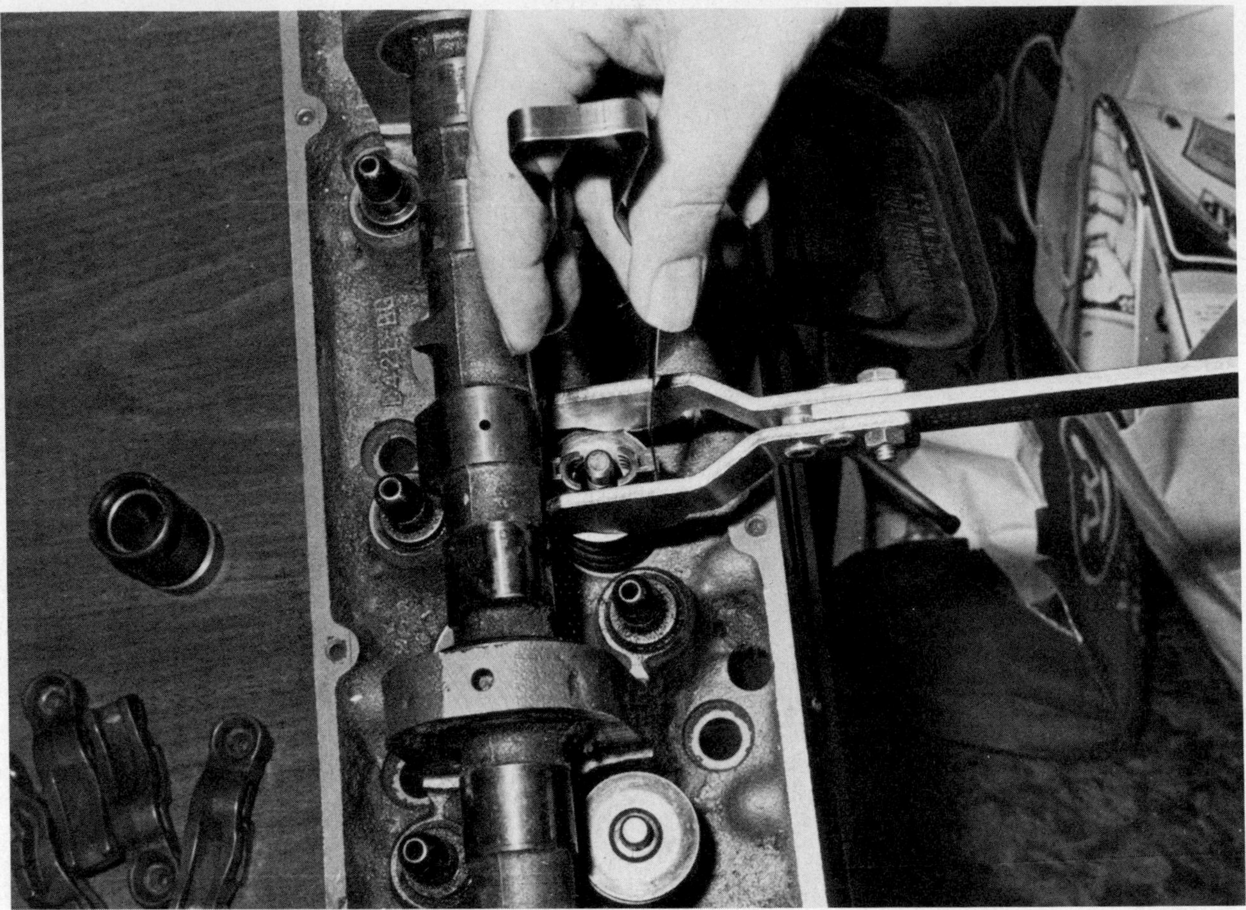

Fig. 5–33. Top view of a universal valve spring compressor in use, plus magnetic tip tweezers to withdraw the half-moon locks.

shown in Fig. 5–33. Then carefully release pressure on the spring and remove it. Note: As with rockers, springs must be refitted to the same place.

Try to rock the valve stem side to side, and if you feel more than just barely perceptible movement the clearance between the stem and the guide is excessive, and professional service is required. If there is just slight play, the replacement of the stem seals should be sufficient.

When you buy a set of new stem seals, ask the store for the best available, such as the umbrella design or one of the new teflon and rubber seals. Note: If the top of the valve guide has a machined surface, you can install a teflon and rubber seal that will retain positively and do a superior job. If it doesn't and you want to have the job done professionally, a machine shop can cut the top of the guide and install this type (see Fig. 5–34).

If the seal kit comes with plastic caps, place them on the top of the valve stems, then push the seal onto the stem as far down as it will go (see Figs. 5–35, 5–36). The cap will protect the seal from damage; remove it after the seal is installed. Some seals are very flexible, will not be damaged by the valve stem tip, and no cap is supplied.

Check the valve spring for straightness by placing it against a right angle. If there is more than $^{1}/_{16}$-inch clearance between any coil and the side of the right angle, the spring should be replaced. For balanced performance, change all valve springs.

Position the valve springs (or both springs on some cars) over the stem, then compress. Apply some chassis grease to the half circle locks, then put them into place on the end of the valve stem (the grease will help keep them in position). Gradually release pressure on the spring and it will rise, securing the locks to the spring's

Fig. 5–34. If you wish, a machine shop can cut a machined surface on the top of the valve guide as shown, so the guide will accept a premium seal. Or you may be able to rent the tool and do the job yourself. Or even better, your engine may already have the machined surfaces on the guides.

Fig. 5–35. A plastic cap installed on valve stems protect new seals.

Fig. 5–36. A new seal is pushed down over the plastic cap into place on the machined surface of the guide. Then plastic cap is removed and discarded.

washer–like retainer.

Repeat this procedure for all valve stems, then refit the rocker shaft and tighten to specifications with a torque wrench, in three relatively even stages (if the specification is 25 lbs.-ft., for example, first tighten all nuts or bolts to about 10, then 20, finally 25). If the rockers are on ball-stud pedestals, tighten to specifications with a torque wrench, unless they are the Ford or Chevrolet adjustable type, in which case you should follow the adjustment procedure described under *Hydraulic Lifter Service,* earlier in this chapter.

Overhead Camshaft Engines with Rockers

The basic procedure for overhead camshaft engines with rockers is similar to the pushrod type. First, however, crank the engine until the valve is closed (the cam peak will not be bearing against the valve rocker). Compress the valve spring and remove the rocker, then take out the half-circle retainers and the valve spring, and you'll have access to the seal.

On overhead camshaft engines without rockers, the job requires that the camshaft be removed, so leave it to a professional.

Manifold Heat Control Valve Service

As the fuel mixture passes through the carburetor, it enters the intake manifold, a casting with a series of chambers that carries it to the cylinders. When the engine is cold, the fuel in the mixture could condense inside the manifold, and either not get to the cylinders or be improperly distributed to them.

To prevent this, most engines have a provision for heating the intake manifold when the engine is cold. In a few cases, engine coolant runs through passages in the intake manifold, but on most cars there is a device called a manifold heat control valve, or a "heat riser."

The heat riser is actually located in the exhaust manifold. On in-line four- and six-cylinder engines it is a valve that pivots open to allow hot

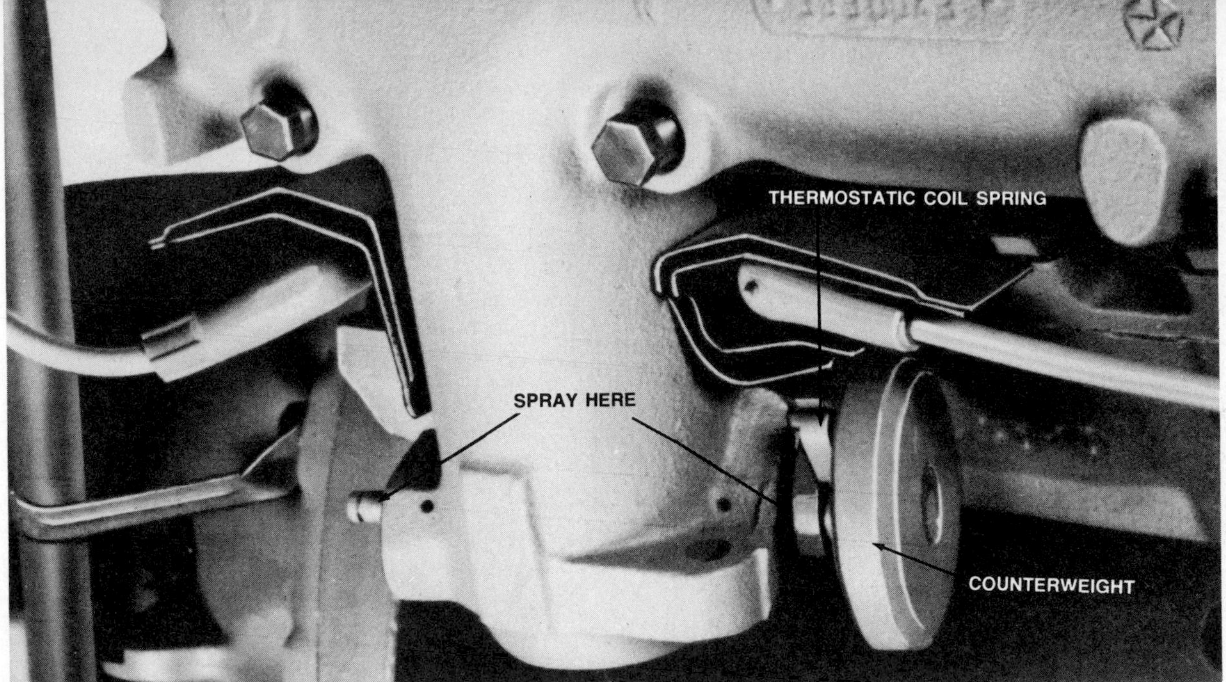

Fig. 5–37. Self-contained manifold heat control valve on a six-cylinder engine.

exhaust gas to flow up against the top of the exhaust manifold (the casting that carries the exhaust from the engine to the exhaust system). The top of the exhaust manifold butts up against the bottom of the intake manifold at one point, so the exhaust flow heats the intake manifold.

When the engine is warm the valve closes, stopping the exhaust flow.

On V-type engines there is an exhaust gas passage in the intake manifold (separate from the fuel mixture passages, but close enough to them to transfer heat). In this case, the heat riser opens and exhaust gas is allowed to flow through the intake manifold, warming it up. When the engine is warm, the valve closes and stops the exhaust flow.

There are two types of heat risers, and both should be checked periodically and serviced if necessary. If they stick in the heat-off position, the engine will perform poorly until it warms up. If they stick in the heat-on position, the fuel mixture will be overheated so badly that the engine will starve for fuel, stall out, and have poor high-speed performance.

Valve Locations

The locations of heat risers can vary from car to car, but if you look around the exhaust manifold you should find yours. The one on an in-line engine, if it is used, is always near the point where intake and exhaust manifolds come close together. On a V-type engine you will have to look at the exhaust manifold on each side, and in some cases you will have to jack up the car, support it on safety stands, and look from underneath.

Self-Contained Heat Riser

The self-contained heat riser has a thermostatic coil attached (see Fig. 5–37). With the engine off and cool, just try to flip it back and forth, holding it at the external counterweight. If it isn't absolutely free in its movement, spray with penetrating solvent, aiming the spray into the joint between the shaft and the bore in the exhaust manifold. If the valve is badly stuck, tap the counterweight with a hammer to start the freeing-up process. Do not stop working and spraying until the valve flip-flops freely.

Vacuum-Operated Type

The vacuum-operated heat riser (see Figs. 5–38, 5–39) has an external link connected to a vacuum diaphragm. The diaphragm is connected by a hose to a thermostatic switch threaded into the engine's water jacket, and this switch has a hose to a source of engine vacuum at the base of the carburetor.

When the engine is cold, the thermostatic vacuum switch is open, and engine vacuum flows from the carburetor base, through the switch, to the diaphragm, yanking on it. The movement of

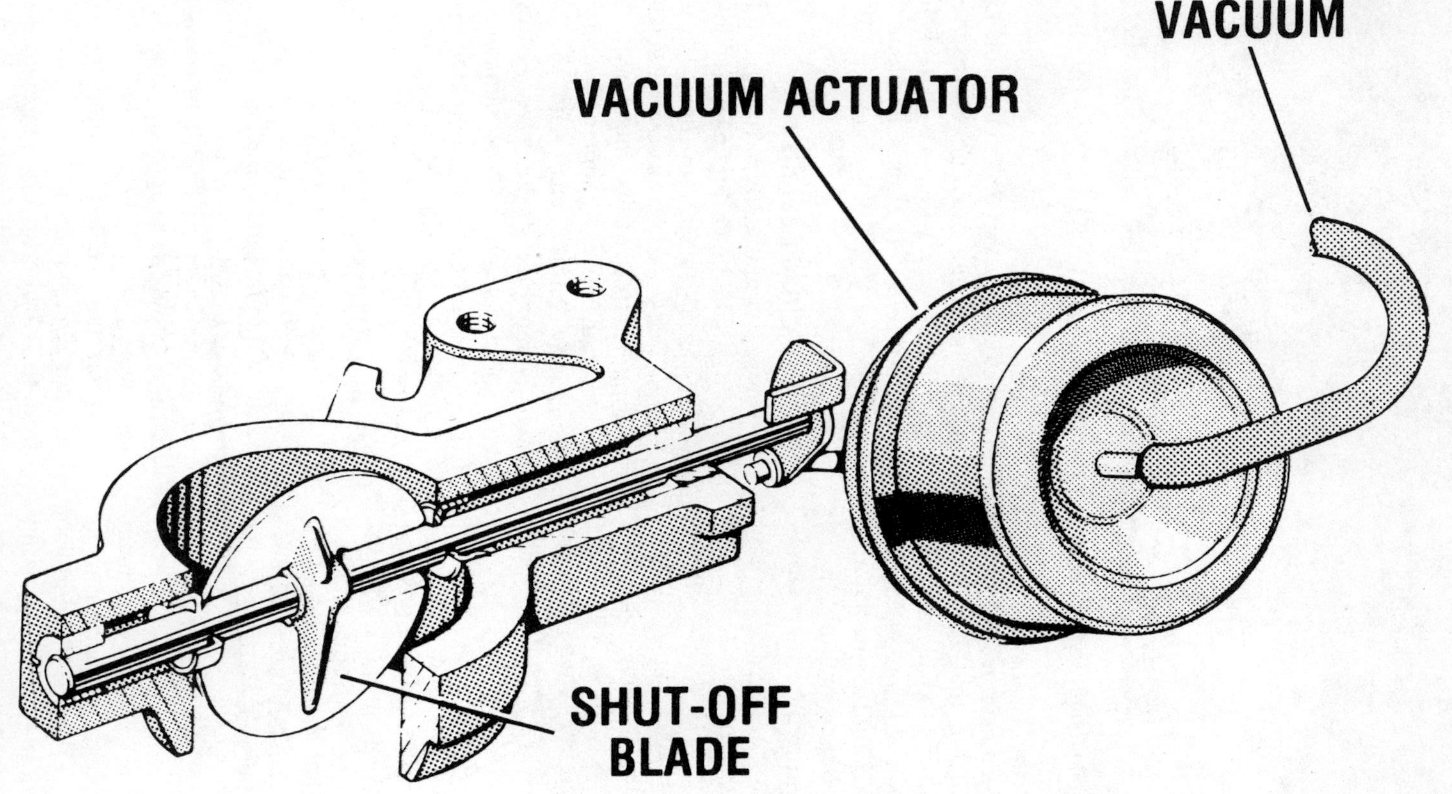

Fig. 5–38. This heat riser is controlled by a vacuum actuator (diaphragm unit). The vacuum hose runs to a coolant-temperature-controlled vacuum switch threaded into the engine, and from there to a source of vacuum at the carburetor base or intake manifold. When the coolant is cold, vacuum can pass through the switch to operate the heat riser. When the engine warms up, the coolant temperature switch closes, shutting off the vacuum and allowing a spring in the vacuum actuator to close the shut-off blade.

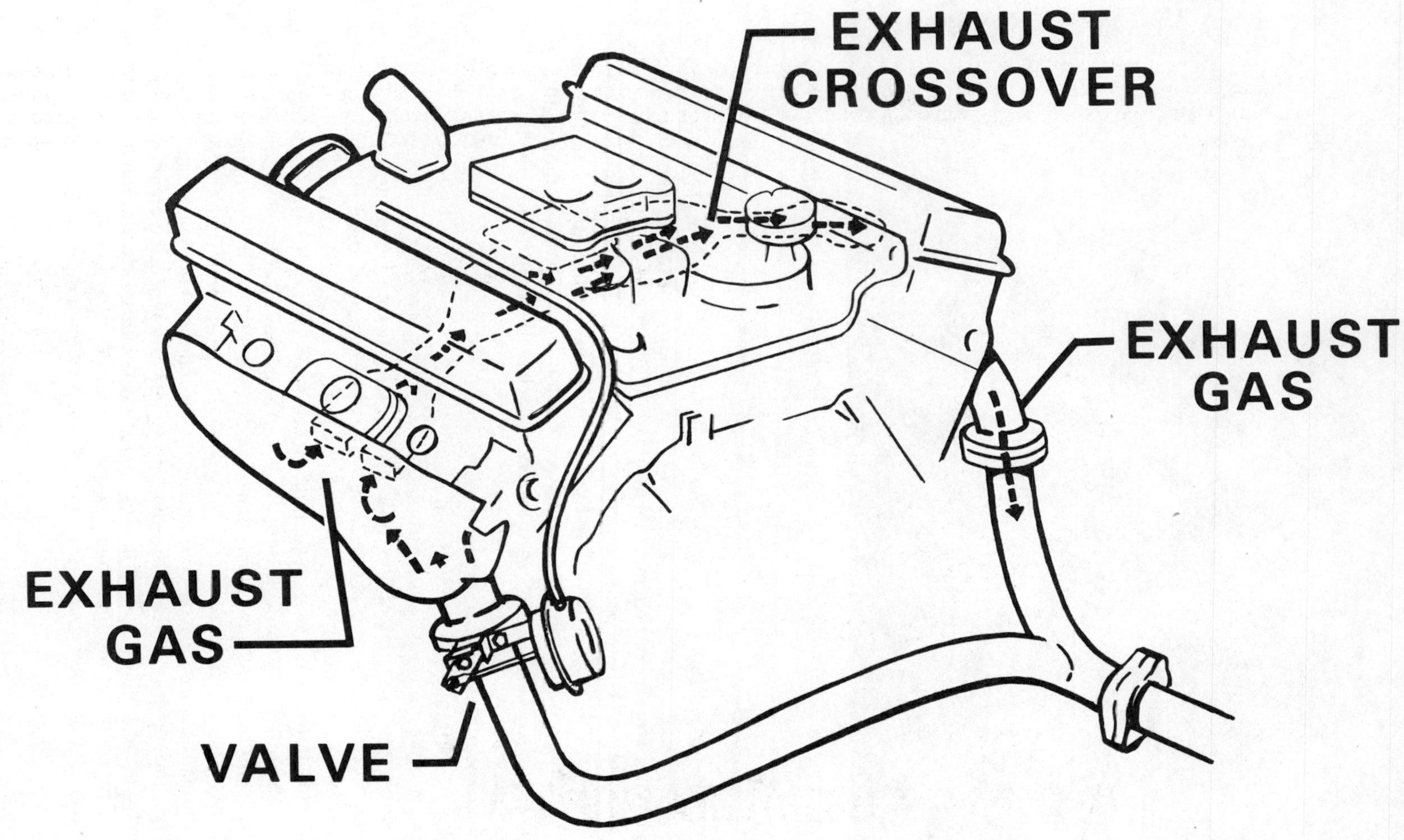

Fig. 5–39. Vacuum-operated manifold heat control valve as actually installed on a V-type engine. In this design, the valve permits exhaust gas flow through a crossover passage in the intake manifold.

Fig. 5–40. Pulling off the vacuum hose at the diaphragm unit. Feel the end for vacuum when engine is cold and idling.

the diaphragm pulls on the linkage to pivot the valve.

To check this type, start a cold engine as you watch the valve linkage. If it is pulled up, the system is functioning correctly when cold. Recheck when the engine is warmed up; the linkage should not move.

If the linkage fails to move when the engine is cold-started, disconnect the hose from the diaphragm (see Fig. 5–40) and feel the end for vacuum. If there is none, follow the hose to the thermostatic switch and disconnect the other end of that hose. Feel for vacuum at the hose neck. None? Disconnect the other hose and feel its end for vacuum. If there is vacuum at this hose but none at the other hose's neck, replace the thermostatic switch, which is simply unthreaded from the engine with a wrench.

If you find vacuum at the diaphragm hose end, but no diaphragm movement, pull on the diaphragm rod. If it can be pulled (against moderately stiff spring pressure), the diaphragm assembly is defective. Unbolt and replace it. If the rod can't be pulled, the thermostatic valve in the exhaust manifold is sticking. Apply penetrating solvent spray and free it up in the same basic manner as the self-contained type.

If the linkage moves to the heat–on position when you restart a thoroughly warm engine, replace the thermostatic switch.

Electric Heater

On a few cars starting in 1981, the heat control valve has been dropped in favor of an electric heater. General Motors builds the heater (a ceramic grid) into the carburetor base gasket. American Motors has an electric heater that screws into the intake manifold from the underside of the manifold. With both these designs, you should check for current to the electrical connector when the engine is cold and the engine first is started. If you must unplug the electrical connector for any service, be sure to reconnect or cold engine operation will be poor.

6 THE DIESEL ENGINE

How it Works

WITH FUEL ECONOMY STANDARDS TIGHTENING every year, the car manufacturers, in some cases, have turned to the diesel engine. The diesel differs from the gasoline engine in that there is no spark ignition system. Instead, the air/fuel mixture is squeezed even more tightly than in the gasoline engine, building up so much heat that the fuel self-ignites. The compression ratio of a diesel is typically 22–1 or higher, compared with perhaps 8–1 for the gasoline engine today.

As in the gasoline engine, the pistons go up and down in a four-stroke cycle: down on intake, up on compression, down on power, and up on exhaust. During the intake and compression strokes, however, only air is in the cylinder. As the piston approaches the completion of the compression stroke, fuel is sprayed into the cylinder by an injector, and it ignites in the hot air (see Figs. 6–1 to 6–3).

The diesel engine does not have a throttle plate. When you step on the "gas" pedal, you operate linkage to the pump that supplies fuel to the injectors—one injector for each cylinder. The linkage moves mechanical parts in the pump to increase the amount of fuel injected as the pedal is depressed.

Unlike gasoline, which is rated according to octane, the diesel fuel is rated according to cetane, which is basically the opposite. Octane is resistance to burning (so the gasoline doesn't self-ignite and cause engine knock).

There are two diesel fuels, No. 1 for cold weather, No. 2 for warm. If No. 2 is used in cold weather, it can form wax crystals that can plug fuel lines and prevent fuel flow.

Because the diesel fuel must self-ignite from the heat of compression, starting in cold weather can pose problems because the air entering the engine is cold and there is no spark to get ignition going.

Glow Plugs and Car Battery

All passenger car diesels, therefore, are made with pre-chambers; that is, miniature chambers next to the cylinder that contain glow plugs. It is into these chambers that the fuel is sprayed (see Figs. 6–4, 6–5). The glow plug is a heating element that warms the air so that it will ignite the fuel in cold weather. It takes a lot of current to operate the glow plugs, so in very cold weather when batteries do not perform at their best there is quite a drain on the car battery. Further, it takes a lot more electrical power to crank a diesel because of higher compression. Since the tiny pre-chamber is a lot easier to heat than the complete combustion chamber it extends the life of the battery besides contributing to smoother performance in the diesel.

The heavy current requirements for cranking and glow plugs make it important to be sure the battery terminals are clean and tight and, if the battery is a conventional type, topped up with water. Note: The GM diesel car is equipped with two batteries to provide extra current; the VW has only one battery, but it's a heavy-duty model. If a diesel battery fails, make sure you install a replacement of equal or greater power. In extremely cold climates, install a battery warmer, a device that plugs into household current and keeps the battery at a temperature that will enable it to develop full power. Battery warmers are available at auto parts stores and are easily installed.

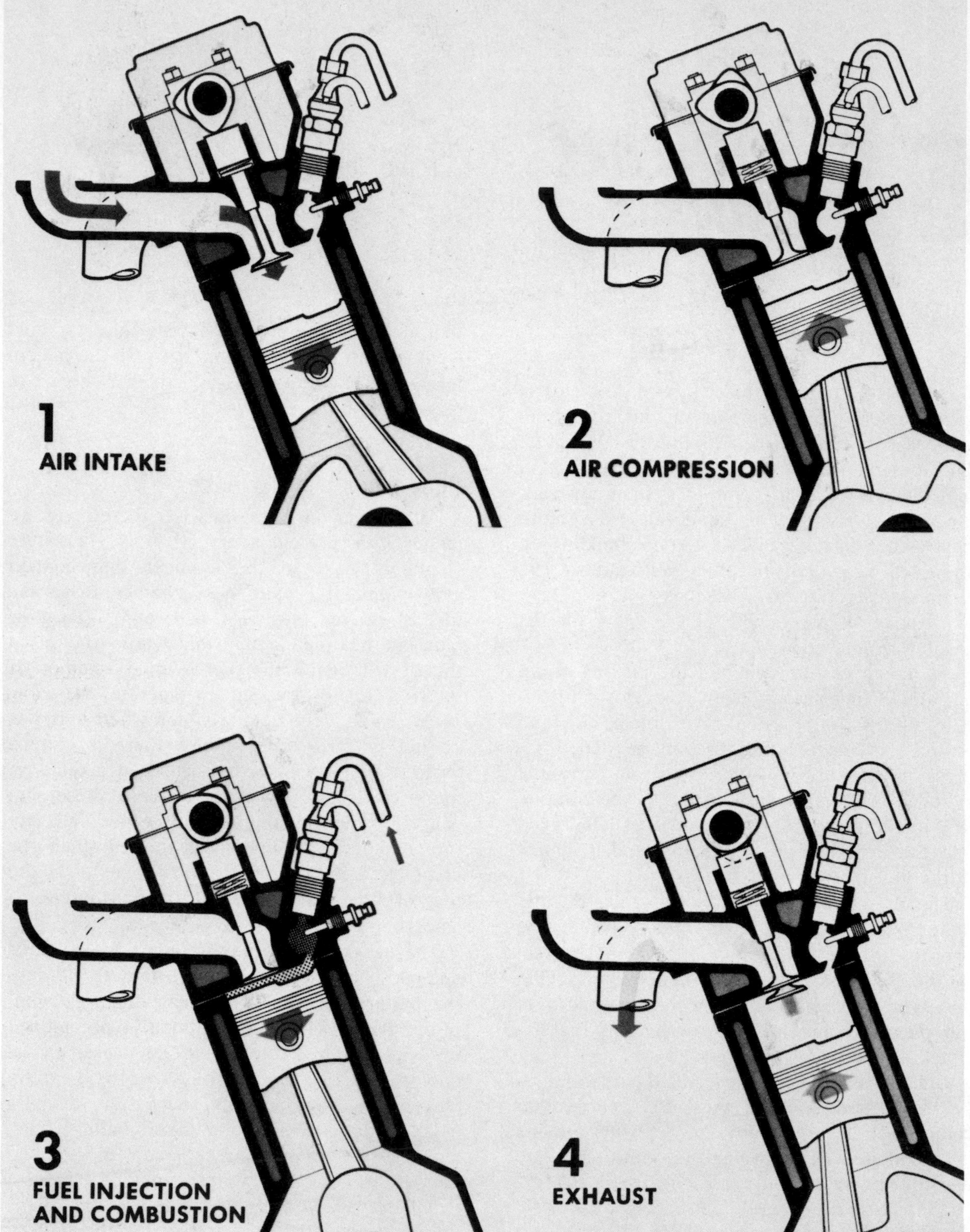

Fig. 6–1. The diesel engine cycle. *Note that the fuel is sprayed into a tiny pre-chamber just above and to the right of the top of the cylinder. This is common design practice for passenger car diesel engines, so that the glow plug, which threads into the pre-chamber, also can be used, to ease cold-weather starting.*

Fig. 6–2. Oldsmobile V-8 diesel, used on many GM cars and some trucks.

Coolant Heaters

Many diesels come with engine coolant heaters as standard or optional equipment to keep the engine warm and help prevent starting problems. If your diesel is not so equipped, install a retrofit kit, available from most auto parts stores. Like the battery warmer, it plugs into household current. Also useful is a fuel heater powered by battery current.

Although the diesel does not have spark plugs, it does require periodic maintenance in addition to battery care. Here are other aspects of diesel maintenance you should know.

Engine Oil and Filter

Use only an oil recommended for diesels, one rated for Service CC (see markings on the can). Many gasoline engine oils also are rated for diesel service. Change oil and filter every 3000 miles or three months. The diesel combustion process results in very rapid contamination of engine oil, so the importance of frequent oil change cannot be overemphasized.

Fuel Filter

The diesel fuel injection system is composed of parts made to very precise tolerances. Tiny dirt particles can cause malfunctioning, so fuel filtration is extremely important. The size of the filter used with different diesel engines varies and, therefore, so does the replacement interval. Do not exceed the manufacturer's fuel filter replacement interval. The filter may be a spin-on cannister, serviced much like an engine oil filter (see Fig. 6–6), or one with fuel lines threaded on. Use tubing-type open-end wrenches to slacken the fittings on this second type.

Idle Speed

The idle speed adjustment is made to the linkage at the fuel injection pump (see Figs. 6–7 and 6–8). Although you simply turn the screw (or slacken the locknut and turn the screw on the VW diesel), there is one problem: You cannot use a standard tachometer because it works off an electric ignition system, and there is none on the diesel. You have these choices:

(1) Buy a vibration sensor, which VW sells (see Fig. 6–9). It clamps magnetically onto the engine top cover and connects to the car battery and a conventional tachometer. Engine vibration produces a signal proportional to engine speed, and the sensor converts this to an engine speed pulse.

Fig. 6–3. Volkswagen four-cylinder diesel. The gear at the front right center is the one that drives the fuel injection pump.

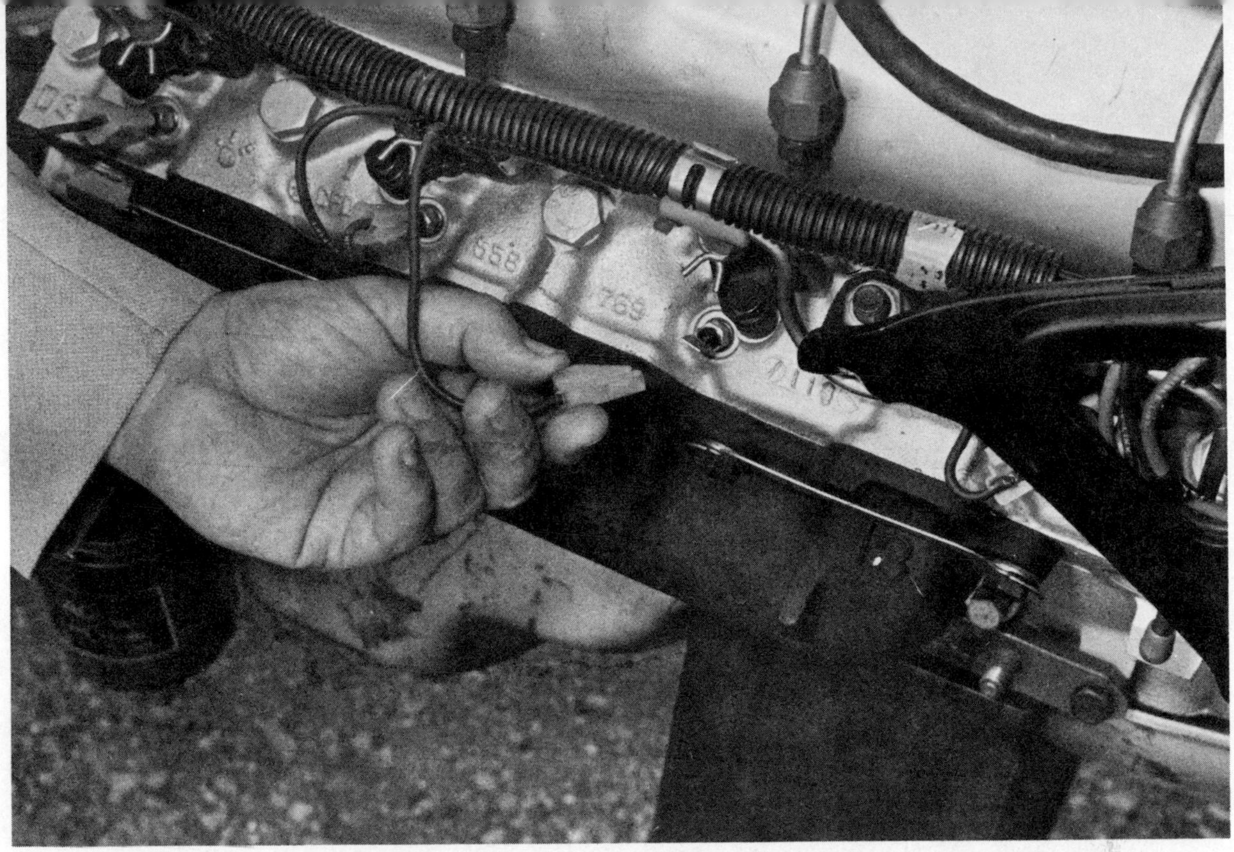

Fig. 6–4. When the wiring connector is removed, you can see end of glow plug projecting from cylinder head.

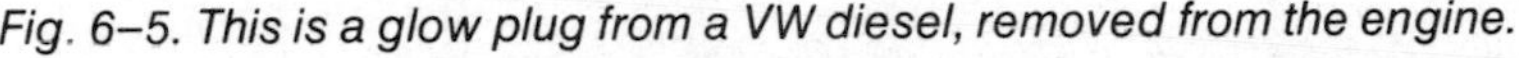

Fig. 6–5. This is a glow plug from a VW diesel, removed from the engine.

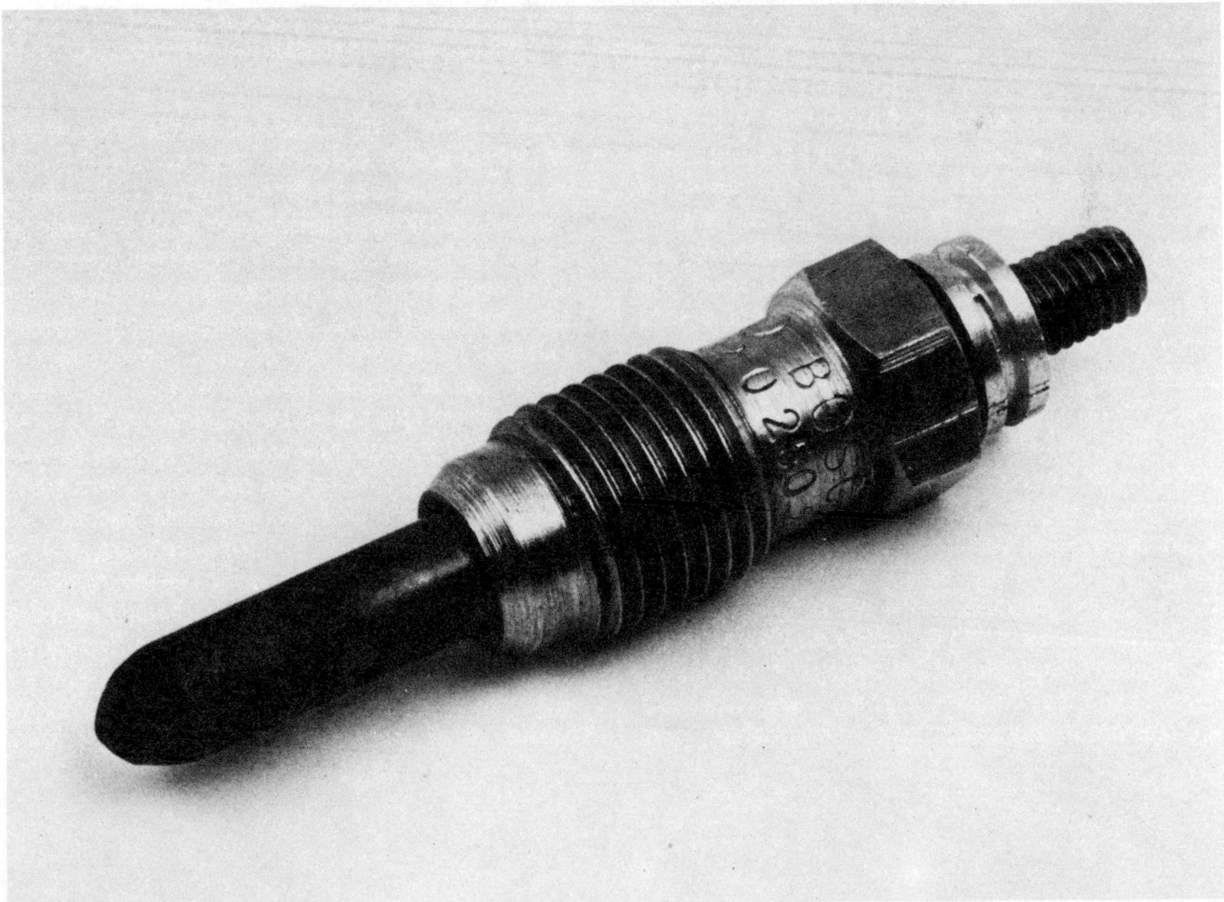

Fig. 6–6. Removing spin-on fuel filter from VW diesel.

Fig. 6–7. Adjusting idle speed on Olds (GM) diesel. Air cleaner was removed for photograph, but need not be disturbed for adjustment.

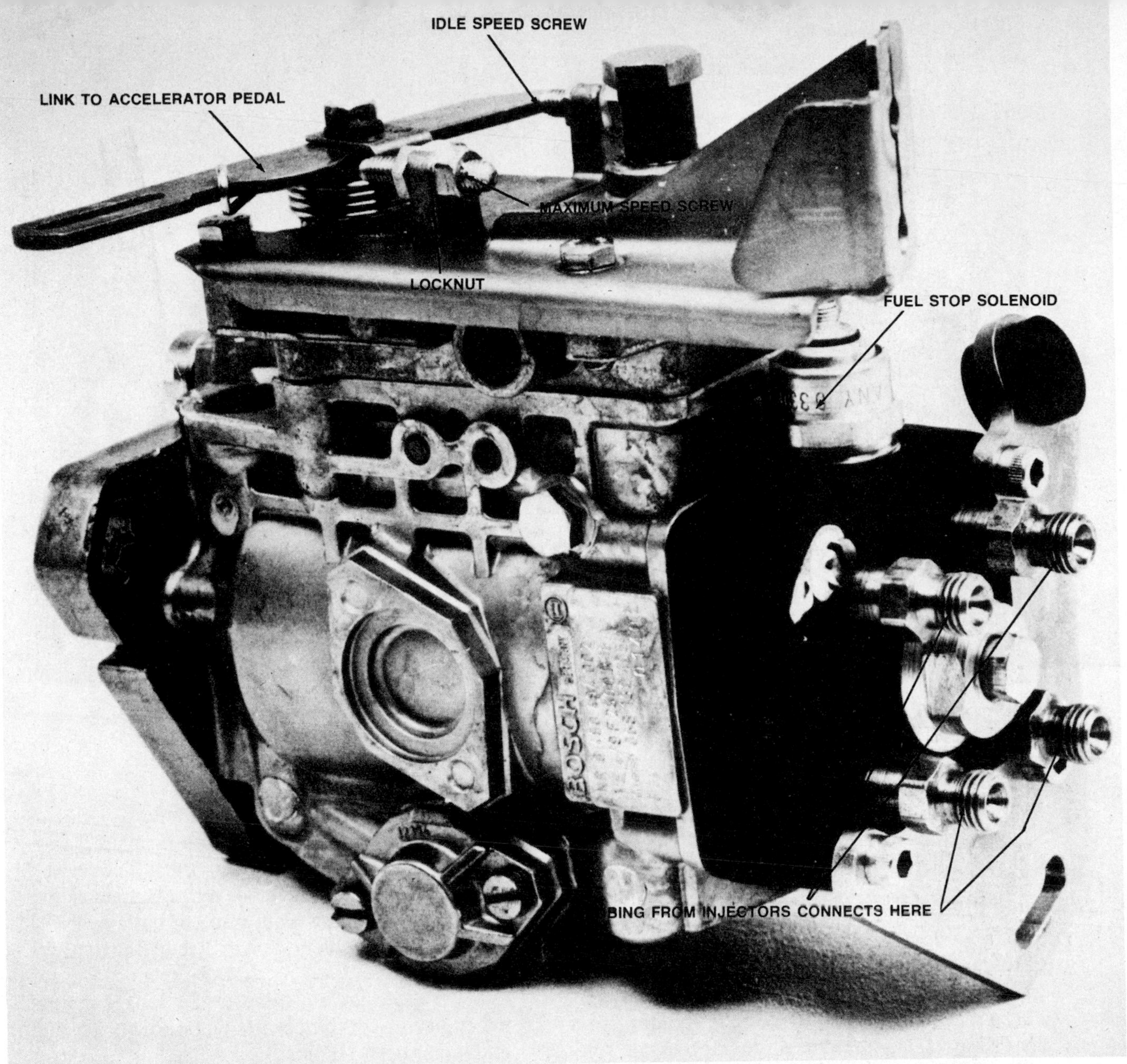

Fig. 6–8. This is the VW diesel fuel injection pump, with key parts identified. Do not confuse the locknut and adjusting screw for idle speed with the locknut and screw on other side of pump that is for setting maximum engine speed (necessary only after professional service of the pump). You can tell the idle speed screw by the fact that it touches the accelerator pedal linkage lever when the engine is idling.

(2) Buy a photoelectric tachometer. Just make a chalk mark on the crankshaft pulley, aim the tachometer at it, and the chalk mark will trigger the tachometer.

(3) Buy an electromagnetic tachometer. The GM diesel has a holder next to the crankshaft pulley, designed for a special probe used to measure ignition timing on the gasoline version of the same engine. The electromagnetic tachometer (see Fig. 2–53) also comes with a probe (and also measures ignition timing on gasoline engines with the tester). Just shove the probe into the holder and it will read engine speed on the diesel. Although this tachometer is not made for the VW diesel, it also can be used on that engine. The VW engine also is based on a gasoline powerplant, and there is an ignition timing probe hole (at

Fig. 6–9. The VW vibration sensor produces an engine speed signal that works with conventional tachometer.

the flywheel end of the engine). You will have to make a bracket to hold the probe in position.

Fuel Stop Solenoid

When you turn the key to shut off a gasoline engine you are cutting off electricity to the ignition system. A diesel, however, doesn't need electricity once it is running, so a special part must be used to stop it—a solenoid (electromagnetic) valve that blocks the fuel flow when you turn off the key (see Fig. 6–8).

If that solenoid does not energize when you turn the key to on, the engine won't start. If the solenoid is shorted, the engine won't turn off.

Before you go into any detailed engine inspection for a starting problem, have someone turn the key to on while you feel the solenoid (Fig. 6–9), the only part with an electrical connector on the fuel injection pump. If you can't feel or hear a click, and the engine won't start, the odds are the solenoid is defective. It can be unthreaded and replaced without disturbing the injection pump.

If the engine won't shut off, a short circuit in the solenoid wiring is surely the problem. As a temporary measure, disconnect the wire from the solenoid to stop the engine, and disconnect the battery ground cable(s) to prevent battery drain. Reconnect the solenoid wire, then the battery cable when you have to start the engine. Do not try to operate the car this way for longer than is necessary to get the car into a garage for tracing and repair of the short.

Note: Mercedes-Benz diesels have a vacuum-operated shutoff valve. If it fails, however, there is an underhood lever (with the label "Stop") that can be pushed to stop the engine.

Engine Testing

If a diesel starts to misfire, there are three common causes: defective fuel injector, low compression or water in the fuel.

To test for a defective injector, disconnect the fuel line at the injector, insert a plug, then reconnect the line. Start the engine and if the engine speed is no different, the plugged injector is the defective one. An injector is either threaded into the pre-chamber (remove with a deep socket) or force-fitted and held with a clamp (remove clamp, then take out injector with a special puller). Note: This test procedure will detect only an injector that has failed completely. One that is not spraying the correct amount of fuel must be checked on a special tester.

The injector is carefully sealed to prevent loss of compression, and when you take out an injector any seals in the hole or on the injector must be replaced. If you are removing an injector for other engine service, and will be reinstalling it, you also must be particularly careful not to damage the tip.

Whenever you disconnect a fuel line, plug it immediately to keep out dirt.

Compression Testing

It takes a special compression gauge to check a diesel engine, because the higher compression pressures developed are beyond the range of the gasoline engine gauge (500 p.s.i. vs. 250 p.s.i. for the typical gasoline gauge). Further, there is no spark plug hole, so the gauge requires special adapters to thread either into the glow plug holes (GM procedure) or into the injector holes (VW method). If you remove the VW injectors, remember to replace the seal, which can be withdrawn from the injector hole with a magnet (see Figs. 6–10, 6–11).

If a diesel engine compression reading is low, the engine will be hard starting if it starts at all, and perform very poorly.

Water in the Fuel

The problem of water in diesel fuel varies from area to area and service station to service station. All diesels are designed to pass through small quantities of water, but substantial amounts can cause both misfire and rusting of fuel injection system components.

Different cars have different provisions for the water problem. GM diesels are equipped with a water sensor in the fuel tank, wired to a dashboard light. Although this system is not on most other cars, it can be retrofitted with a kit available from many parts "jobbers" (parts stores

Fig. 6–10. Using compression gauge with adaptor that threads into fuel injector hole (VW). On most other passenger car diesels, the adaptor threads into the glow plug hole.

Fig. 6–11. After the compression test on VW diesel is complete, remove the seal with magnet as shown.

Fig. 6–12. If the dashboard warning light goes on signalling water in the diesel fuel, the tank must be siphoned. Locate the two fuel lines from the tank unit, disconnect the fuel hose from the narrower diameter one, shown in the photo, and attach the siphon to that line.

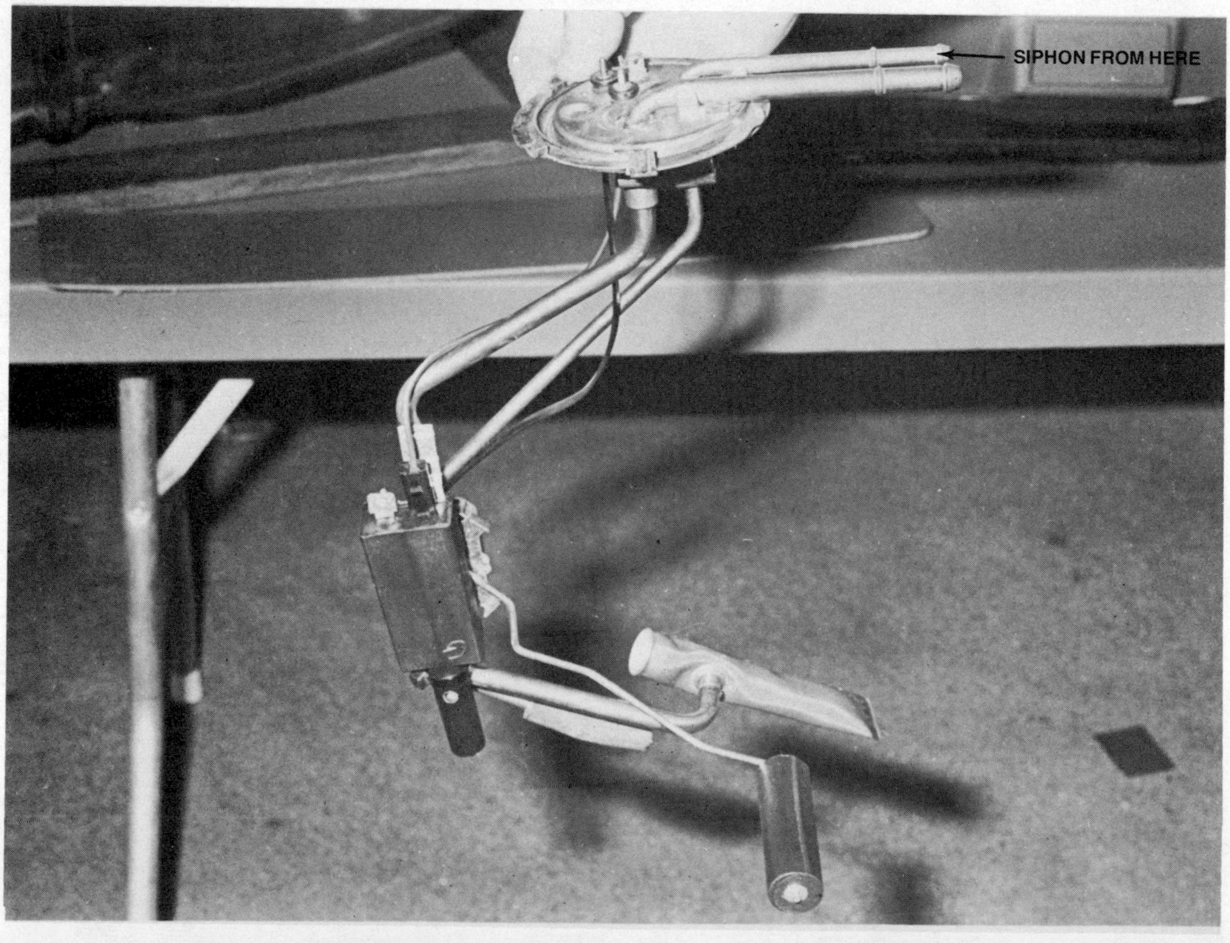

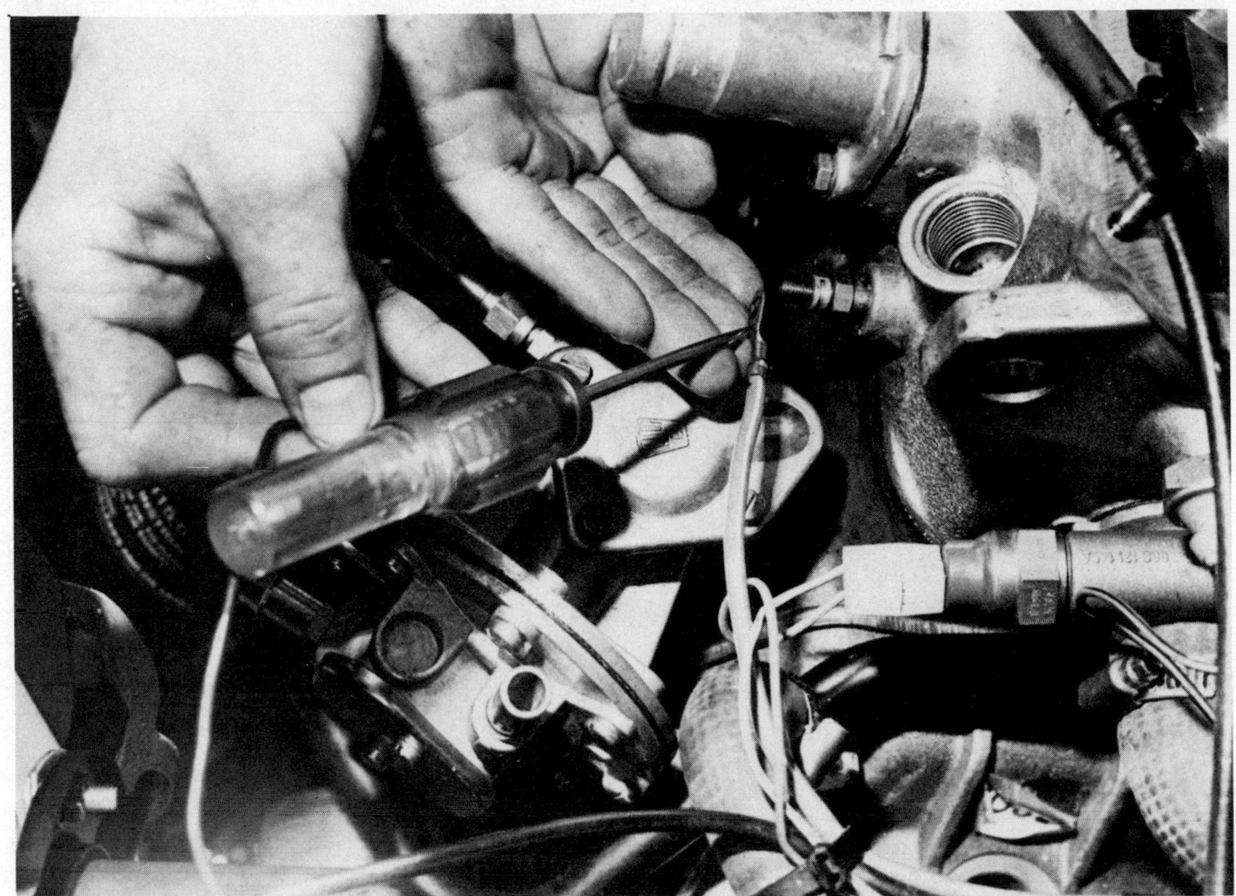

Fig. 6–13. Probing glow plugs' master connector disconnected for check with test lamp. Helper must hold key in test position for this test. Test lamp should light, even if for instant, unless engine is hot.

that service professional as well as weekend mechanics).

If you know you have water in the fuel, there's no question. Take the GM car into a service station, put it on a lift and have the water siphoned out of the tank. (The basic procedure should be a familiar one to stations that sell diesel fuel. See Fig. 6–12.) Volkswagens have a water drain cock in the base of the fuel filter. Just open the cock and allow the filter to drain until you see no water. Inasmuch as VWs do not have dashboard warning lights, unless you have had one retrofitted, you should periodically open the drain cock in the filter to check.

Checking Glow Plug Circuit

If a diesel fails to start only in cold weather, check to see if current is getting to the glow plugs. Disconnect the wire to a glow plug (see Fig. 6–4) or to the master connector if used (see Fig. 6–13). Connect a test lamp to the glow plug wire terminal or master connector wire and to ground. Have a helper hold the key in the glow position and the test lamp should light, even if briefly. Note: On most General Motors diesels, a fast-start glow plug system is used, and the current will flow to them for a maximum of six seconds (and that at 0° F.). On others, the time ranges from a few seconds at warm temperatures to a maximum of about 60 seconds.

If the test lamp doesn't come on at all, the problem is a defective connection in the circuit up to the glow plugs. You probably will give the job of tracing the wiring problem to a professional, but at least you'll know what basically is causing the starting problem.

7 THE COOLING SYSTEM

How It Works

THE BURNING OF GASOLINE (or diesel fuel) in the cylinders produces a tremendous amount of heat, and it is the job of the cooling system to get rid of the excess. On most cars a so-called water cooling system is used (see Fig. 7–1). A mixture of water and anti-freeze—called coolant—is circulated throughout the engine by a pump, called the water pump, and then pushed through a hose (called the upper radiator hose) into the radiator, a finned grouping of tubes that dissipates the heat into the surrounding air.

The coolant at the bottom of the radiator is pulled up into the water pump (through the lower radiator hose), circulated through the engine to pick up heat, and returned to begin the cycle again.

A fan is used to improve the airflow through the radiator when the car is moving slowly or standing still. On most cars the fan is bolted to the front of the water pump, just in back of the radiator, and both pump and fan are turned by a drive belt. On other cars an electric motor, triggered by a temperature sensor, operates a fan. This is commonly used if the engine is mounted sideways, such as on some front-wheel-drive cars, where it would be difficult to relocate the belt-driven water pump (see Fig. 7–2).

Heater

Not all the heat is dissipated into the atmosphere. During the winter some of the hot coolant also flows through a small radiator called the heater core. The heater core is located in ductwork in the passenger compartment, along with a small electric fan, called the fan blower. The fan blows air through the core, heating the air, which then flows into the passenger compartment to warm the occupants.

Thermostat

The temperature of the coolant is extremely important. If it rises slowly, the engine warms up slowly and with any real driving experience you know that a cold engine doesn't perform as well as a warm one. The reason is that the engine parts expand with heat and reach optimum size when warm. A warm engine is important for other reasons:

- *When the engine is cold,* unburned fuel sneaks past the pistons into the crankcase, diluting the engine oil. When the engine warms up fully, most of these unburned fuel vapors are removed from the oil by the PCV system (see *Chapter 4*). Unless the engine warms up quickly, a car used in short-trip driving may operate with fuel-diluted oil, which is harmful.
- *When the coolant is cold,* the heater performance is poor in winter. If the engine runs too hot, that's also harmful, for overheating can cause the engine to stall and may cause engine damage.

The thermostat has the job of both speeding engine warmup and helping to maintain coolant temperature. It's a little temperature-sensitive valve placed in the top of the engine just beneath the upper radiator hose. It is closed when the engine is cold so that coolant flow is limited to the engine, where the coolant gets hot rather quickly (see Fig. 7–3).

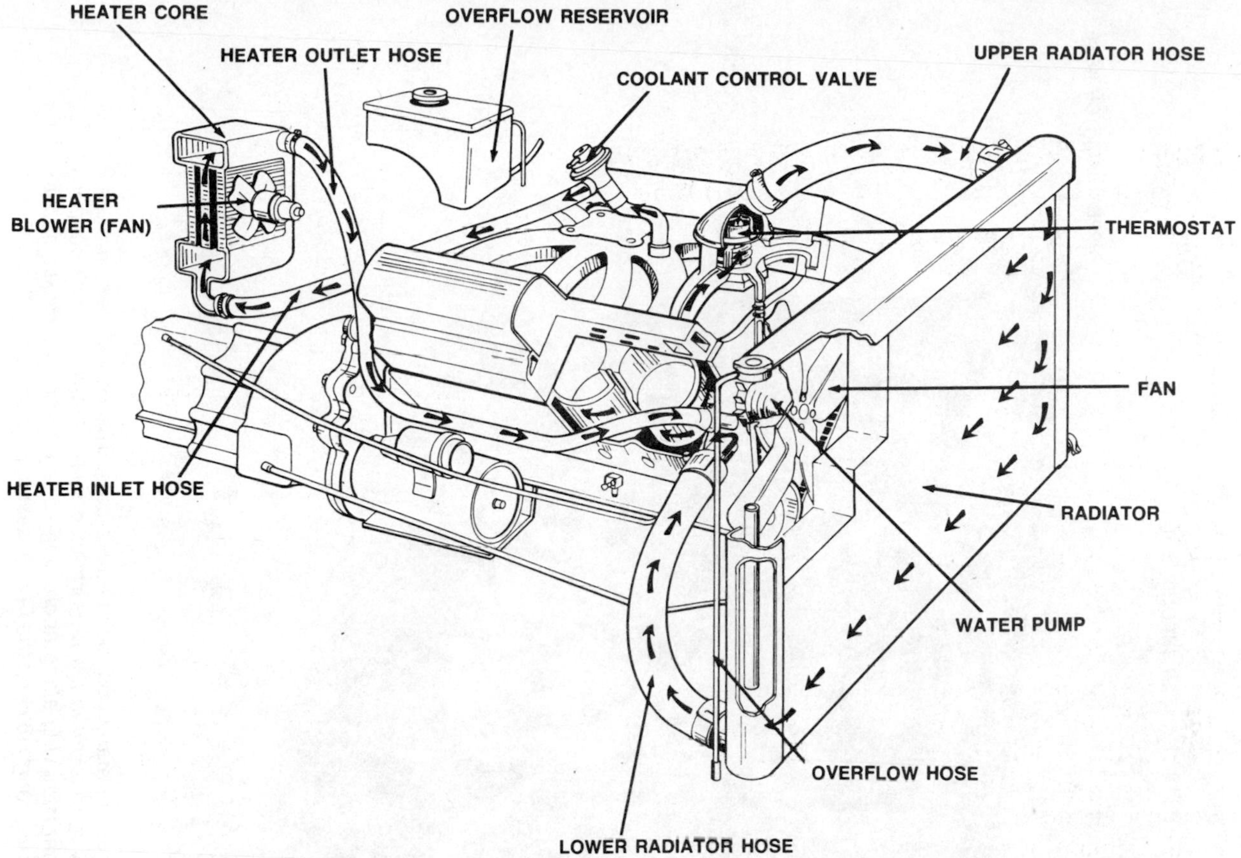

Fig. 7–1. The typical water-type cooling system. Mixture of water and anti-freeze is circulated by water pump through engine to pick up heat, then through radiator to dissipate it. The heater core, a miniature radiator, comes into the picture to dissipate heat into the passenger compartment when desired (coolant control valve permits flow when actuated). On many cars the overflow reservoir is used to catch hot coolant that would otherwise be lost.

When the engine reaches operating temperature, usually 180–205° F., the thermostat opens (see Fig. 7–4) and allows the coolant to flow through the upper hose into the radiator (see Figs. 7–5, 7–6).

Pressure Cap

Of course, the thermostat can open and close as necessary to maintain temperature in cold weather, but in the summer the heat buildup may be so great that even a wide-open thermostat may not prevent the coolant temperature from exceeding 205° F. In fact, on some cars it is not uncommon for temperatures to approach or exceed 250° F. in hot weather in heavy traffic. This is particularly true if the air conditioner is on, for the air conditioner rejects the heat it absorbs through a condenser, which is located in front of the radiator on most cars. The double dose of heat (from AC and radiator) reduces the effectiveness of both.

Therefore, the engine must be able to run at very high temperatures without boiling of the coolant. The pressure cap is an important contributor to this capability.

The cap contains a spring-loaded valve that limits hot coolant expansion, so pressure is built up in the system. Coolant under pressure has a higher boiling point, about 2.6° F. for each pound of pressure. The typical cap's valve will hold up to 16 p.s.i. before the coolant pressure forces it open. Boiling point of the coolant is raised some 40°F., from 212° to 252°F.

A further improvement in boiling point is provided by anti-freeze, which is now required

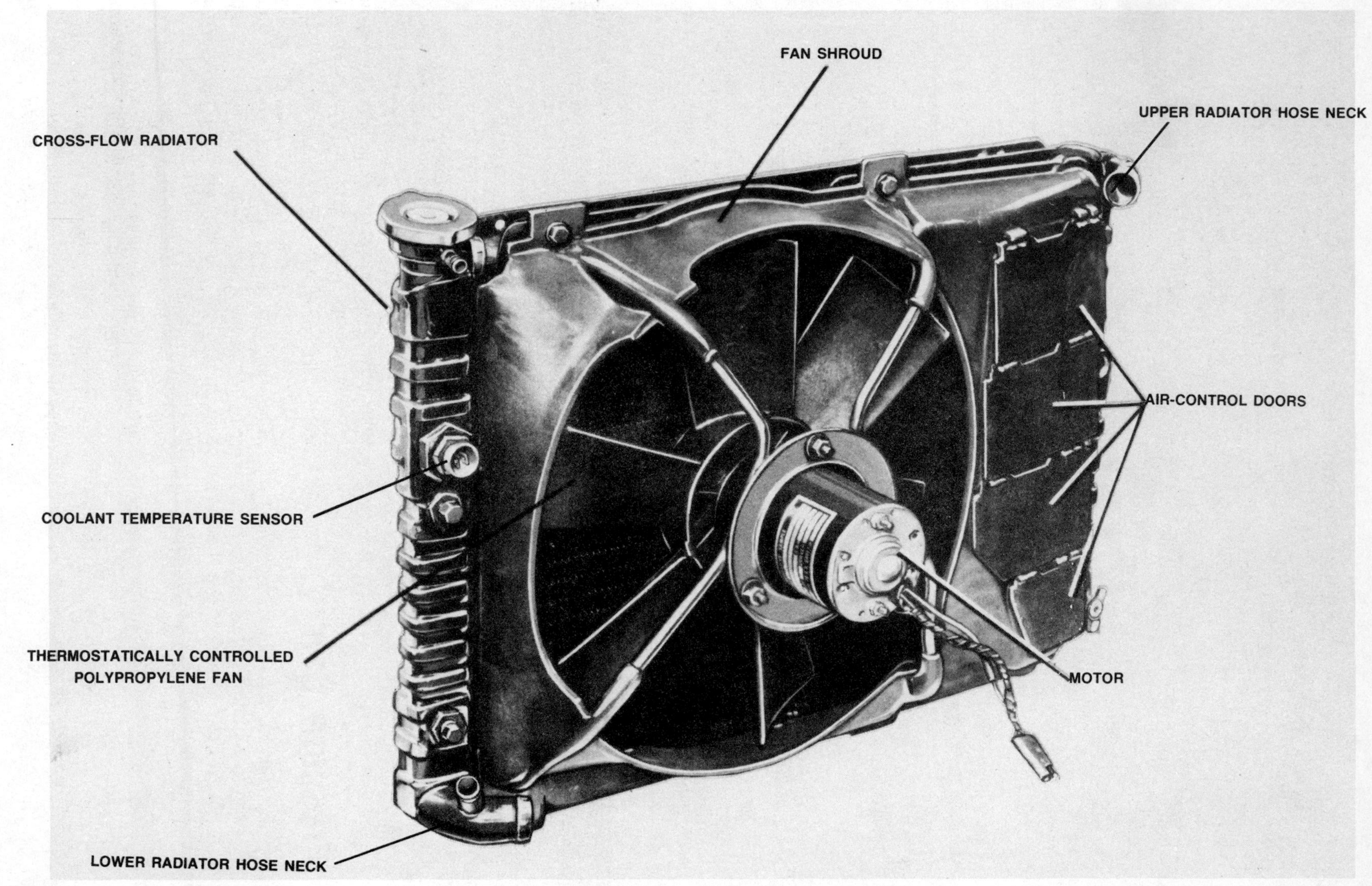

Fig. 7–2. Electric motor drives fan when engine is mounted sideways, as on many front-wheel-drive cars. The coolant temperature sensor in the radiator triggers the motor. The fan shroud directs air flow for maximum fan performance when it is on. The air control doors, a unique feature on this particular design, swing open when the car is moving along at cruising speed, permitting maximum air flow through radiator. In this mode, the coolant temperature may remain low enough to keep the fan off, saving power.

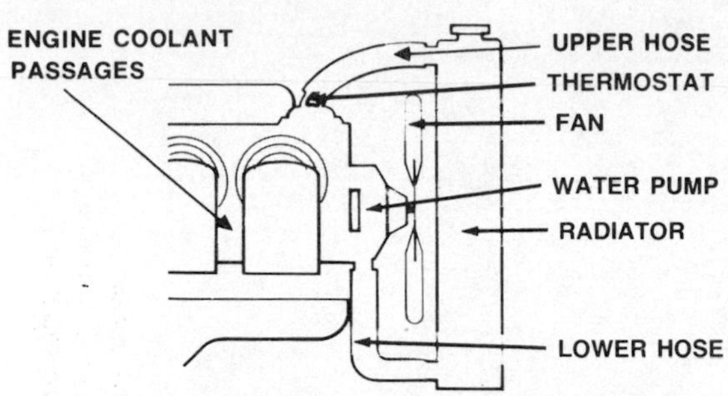

Fig. 7–3. With the thermostat closed, coolant flow is limited to the engine, so coolant gets hot quickly.

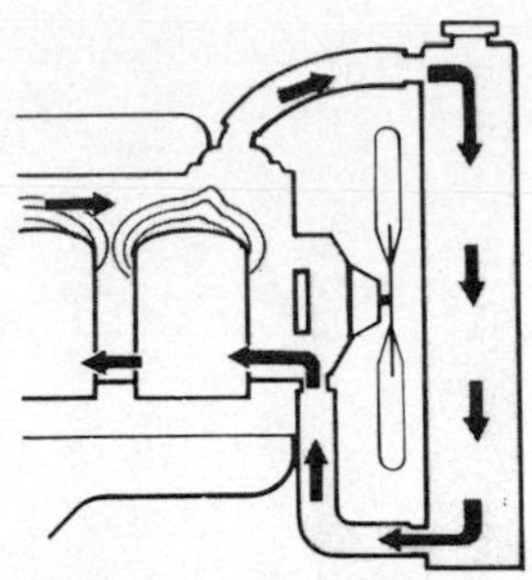

Fig. 7–5. With the thermostat open, coolant flows through upper radiator hose into radiator, then is drawn back into engine through lower radiator hose.

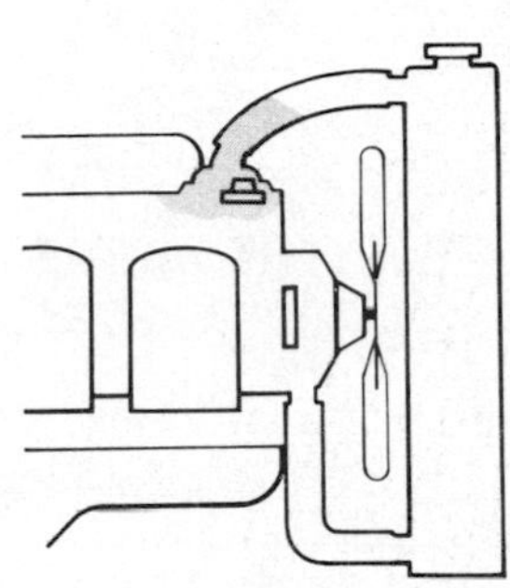

Fig. 7–4. When coolant reaches operating temperature, the thermostat opens.

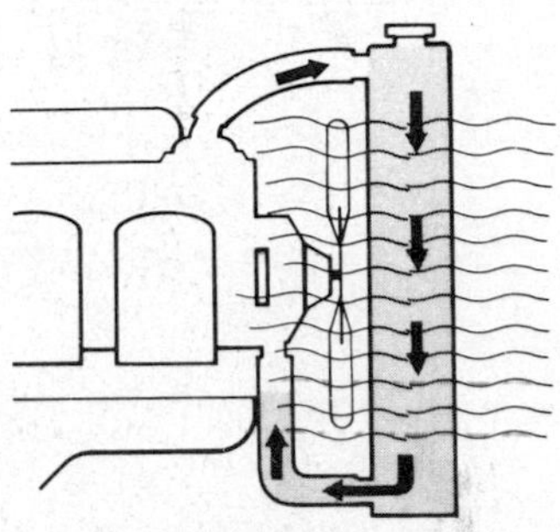

Fig. 7–6. Circulation of coolant through engine and radiator is shown. Spinning fan draws air through radiator, and coolant dissipates heat through radiator tubes and fins into the air.

as a year-round coolant. If you have a 50–50 mixture of water and anti-freeze, the boiling point is raised about an additional 13° F. With a 70–30 mixture of anti-freeze and water (the maximum recommended), the boiling point is raised from 212° to 238° F. Combine these increases with that provided by the pressure cap and the system may operate at 265° to 278° F. without boiling.

Overflow Reservoir

Under severe conditions, or if something is wrong in the system, the coolant may boil, push open the radiator cap valve, and escape. In most late-model cars, it does not flow to the ground, but into a catchpot called an overflow reservoir (see Figs. 7–1, 7–7).

When the engine and coolant cool off the coolant contracts, creating an empty space in the top of the radiator. This empty space is not filled with air, so it is a vacuum. With the conventional system (no overflow reservoir), a second valve in the radiator cap opens and admits air. With an overflow reservoir, air cannot normally enter. Instead, the vacuum draws in coolant from the reservoir and the radiator is automatically topped up once more.

Other Connections

On some cars the cooling system serves additional functions. By running hoses (usually joined into the heater hose connections), coolant can be used to warm up the carburetor's automatic choke, to warm the air/fuel mixture in the engine's intake manifold, and to cool exhaust gases before they are used for the exhaust gas recirculation system (see *Pollution Controls,* Chapter 1).

Cooling System Service

To minimize the chance of overheating in summer or freezeup and poor heater performance in cold weather, you should give your

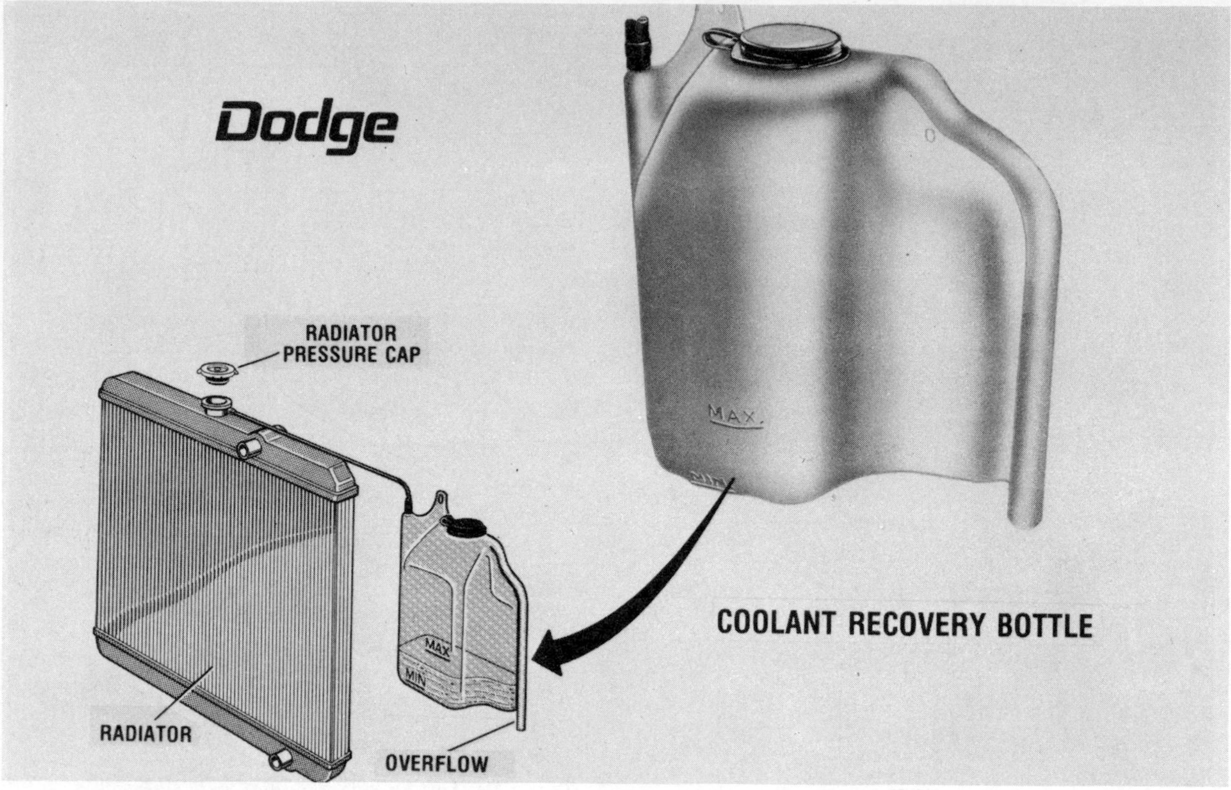

Fig. 7–7. Closeup look at the coolant overflow system. *When coolant gets hot it expands, pushing open the pressure valve in the radiator cap. Coolant flows into the recovery bottle (reservoir) and when the engine cools down, coolant contracts. Vacuum in cooling system then draws coolant from reservoir back into radiator to automatically top it up again.*

cooling system annual service. This consists of checking the radiator cap, hoses, hose connections, drive belt, radiator and fan, and changing the anti-freeze.

Changing the Anti-Freeze

Remove the radiator cap (press down and twist counterclockwise until you come to a stop, then press down again and twist some more). If the coolant is dirty, the system should be cleaned. Pour in a container of cooling system chemical flush and drive the car per the instructions given.

Next, flush out the system. There are many ways this can be done (see Figs. 7–8 to 7–10), but the simplest is with a flushing tee kit, available at a very low price from almost any auto supply store or discount house auto department. The tee, which is installed in the heater inlet hose, accepts a connection from a garden hose, and turns cooling system flushing into a five-minute job.

First, locate the heater inlet hose, which is the one connected to the top of the engine, not to the water pump (watch out, for on some engines the inlet hose connection is very close to the water pump). Just cut the hose at some convenient point, push the necks of the tee into each cut end, then secure with the hose clamps provided (see Figs. 7–11, 7–12).

Connect the garden hose, using the double-female adapter provided (see Fig. 7–13). Remove the radiator cap and press the water deflector into place, then aim it forward (see Fig. 7–14). Start the engine, turn the heater to high, then open the water bibcock. The water will flow through the tee into the heater inlet hose, through the heater, into the engine, through the lower radiator hose, and up the radiator and out (see Fig. 7–15). When the water runs clean for two minutes, shut the engine and the water bibcock.

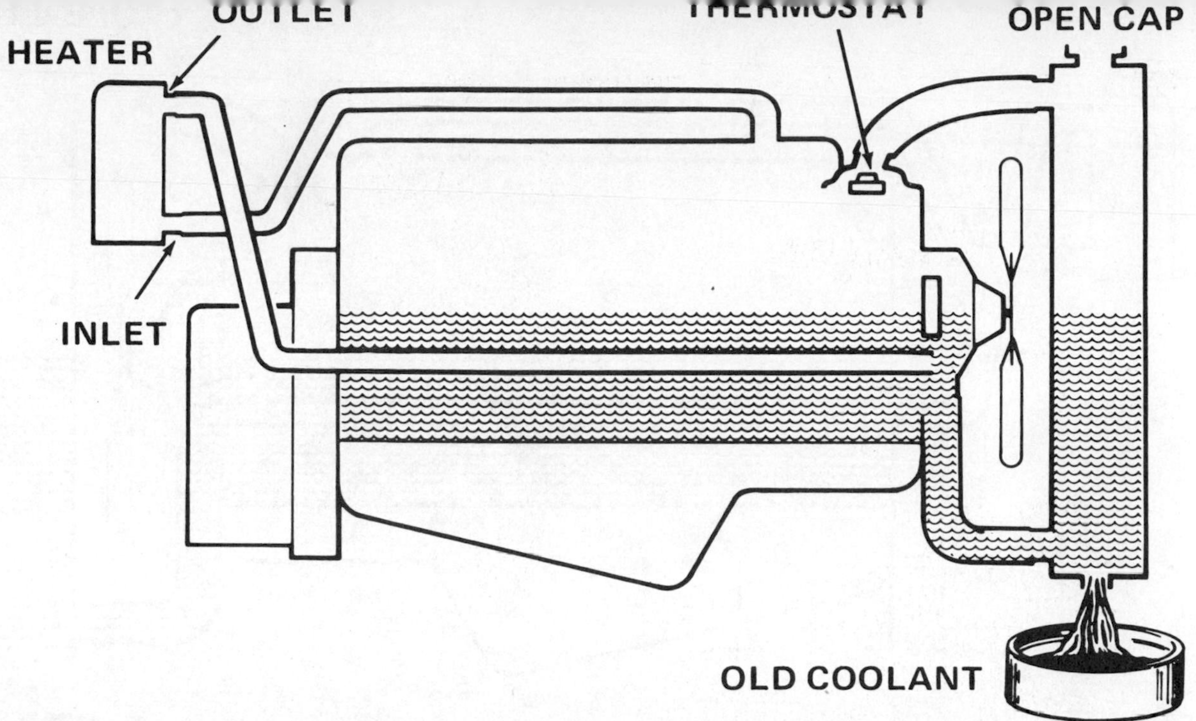

Fig. 7–8. Here's the first step in a system flushing procedure that can be performed without special tools. Just remove radiator cap and open petcock (see Fig. 7–16) or remove drain plug at bottom of radiator. Collect draining coolant in a bucket and look at it. If it's very dirty, this flushing procedure should be discontinued, for it is too mild. Instead use the flushing tee method described in this chapter. If coolant is not extremely dirty, continue with this method.

Fig. 7–9. Close petcock or refit drain plug and fill the system with water through the radiator fill neck. Reinstall the radiator cap (making sure it's on all the way). Start engine, turn heater control to maximum, and drive the car for 7–10 miles. Stop the engine.

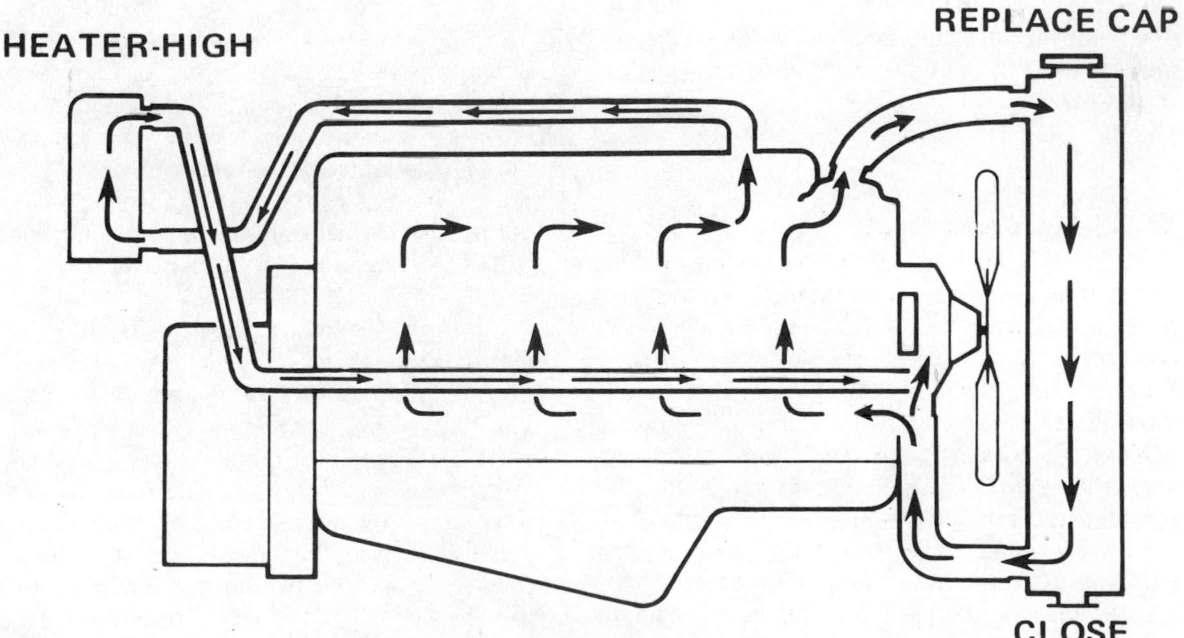

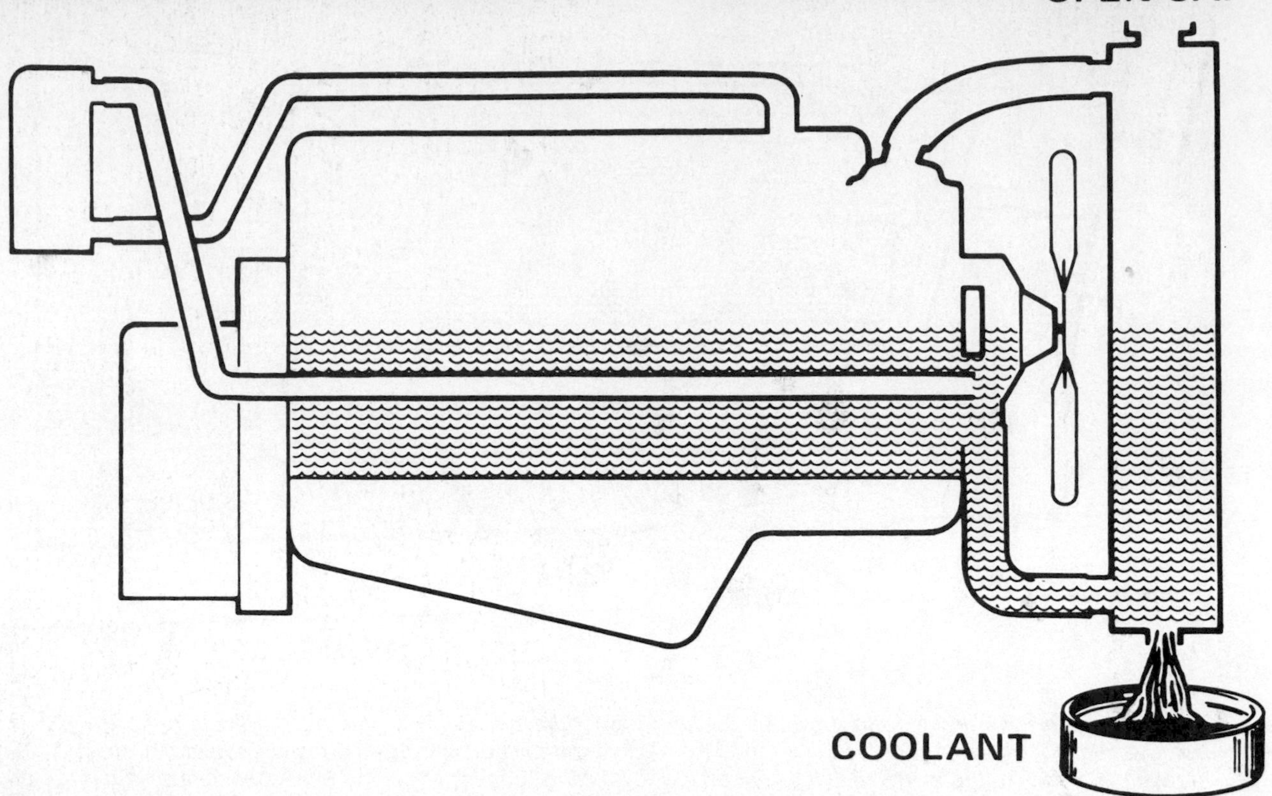

Fig. 7–10. Remove the radiator cap, open the petcock (or remove drain plug) and allow the radiator to drain once more. Then repeat the step shown in Fig. 7–9. If you do this four times, 94 percent of the old coolant will be removed; five times and 97 percent of the old coolant will be removed.

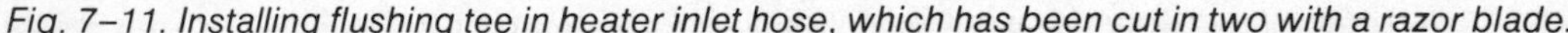

Fig. 7–11. Installing flushing tee in heater inlet hose, which has been cut in two with a razor blade.

Fig. 7–12. Securing the flushing tee with worm-drive hose clamps. Note the tee cap, which is threaded on when the system is not being flushed.

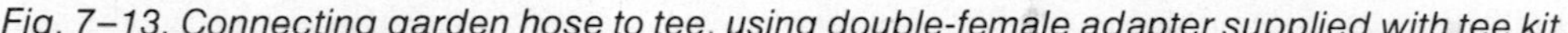

Fig. 7–13. Connecting garden hose to tee, using double-female adapter supplied with tee kit.

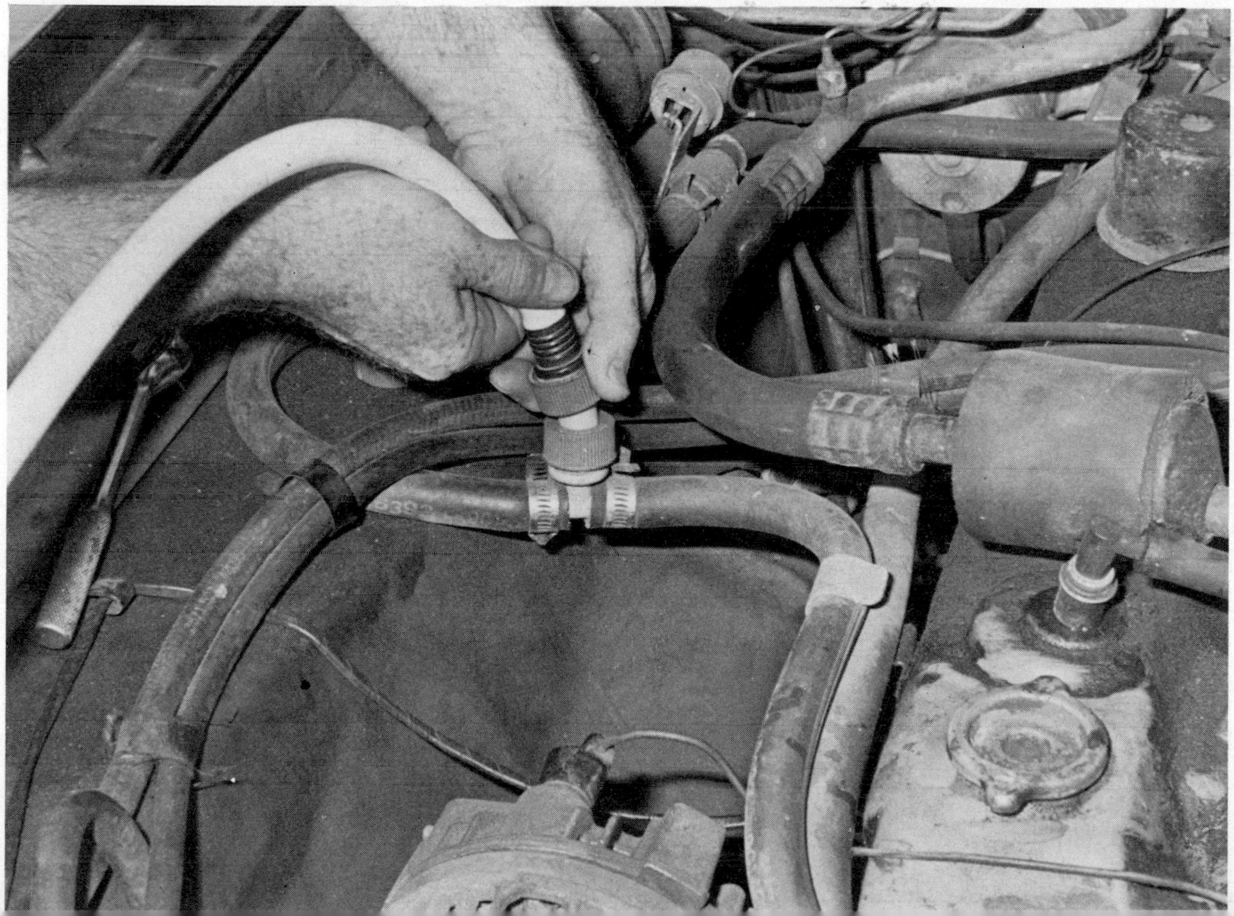

Fig. 7–14. Water deflector pressed into radiator cap neck, aimed forward.

Fig. 7–15. Illustration shows how water from hose flows through tee and heater hose, into engine and heater core, then up radiator and out the fill neck through the deflector.

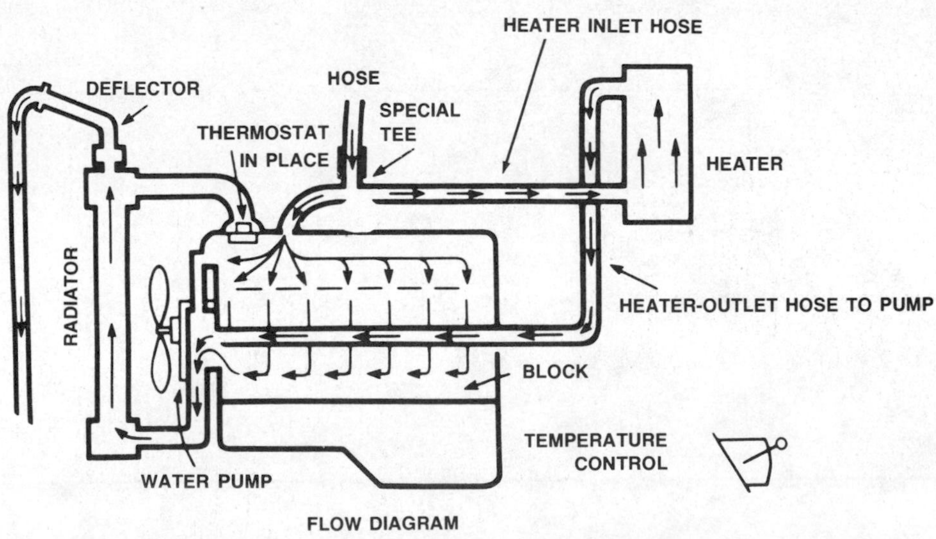

Anti-Freeze Requirements

Check manufacturer's specifications, usually in your owner's manual, for cooling system capacity. Fifty to 70 percent of that number should be anti-freeze. Example: Capacity is 16 quarts; minimum is eight quarts of anti-freeze, maximum is 11 quarts.

Adding Anti-Freeze

After flushing the system is filled with water. The typical radiator has a drain cock (see Fig. 7–16) or drain plug at the bottom, so begin by draining the radiator. If there is no plug or cock, you must disconnect the lower radiator hose.

On the typical car, only 25–40 percent of the coolant will drain out from the radiator. Since a 50–70 percent concentration of anti-freeze is desirable, just filling the radiator normally is not enough. There are several possible solutions.

• *Drain plugs*. Most engines have drain plugs threaded into the block. If you can find and remove these plugs, the block will drain. Unfortunately, they usually are rust-frozen in place.

• *Running the engine*. Disconnect the heater outlet hose (the one to the water pump) and plug the hose neck on the water pump with a cork or rubber plug. Run the engine (heater on high). Water will flow from the hose and you will be able to pour in additional anti-freeze to fill the system.

• *Flushing tee*. If you have a flushing tee, just leave it uncapped with the engine running and the heater on high, and some water will flow out. Note: With either of the previous two methods, some anti-freeze may mix with the water and flow out too, but the amount lost should be small. As soon as you have all the anti-freeze in, stop the engine and cap the tee. Or remove the plug and reconnect the heater hose immediately.

• *Overflow reservoir*. If you have an overflow reservoir, empty it, and you should be able to pour in enough anti-freeze to do the job. In time, the anti-freeze in the reservoir will mix with the water in the engine.

Fig. 7–16. Photo shows drain petcock at bottom of a typical radiator. Turn counterclockwise (grasping by ears with pliers, if necessary) to open.

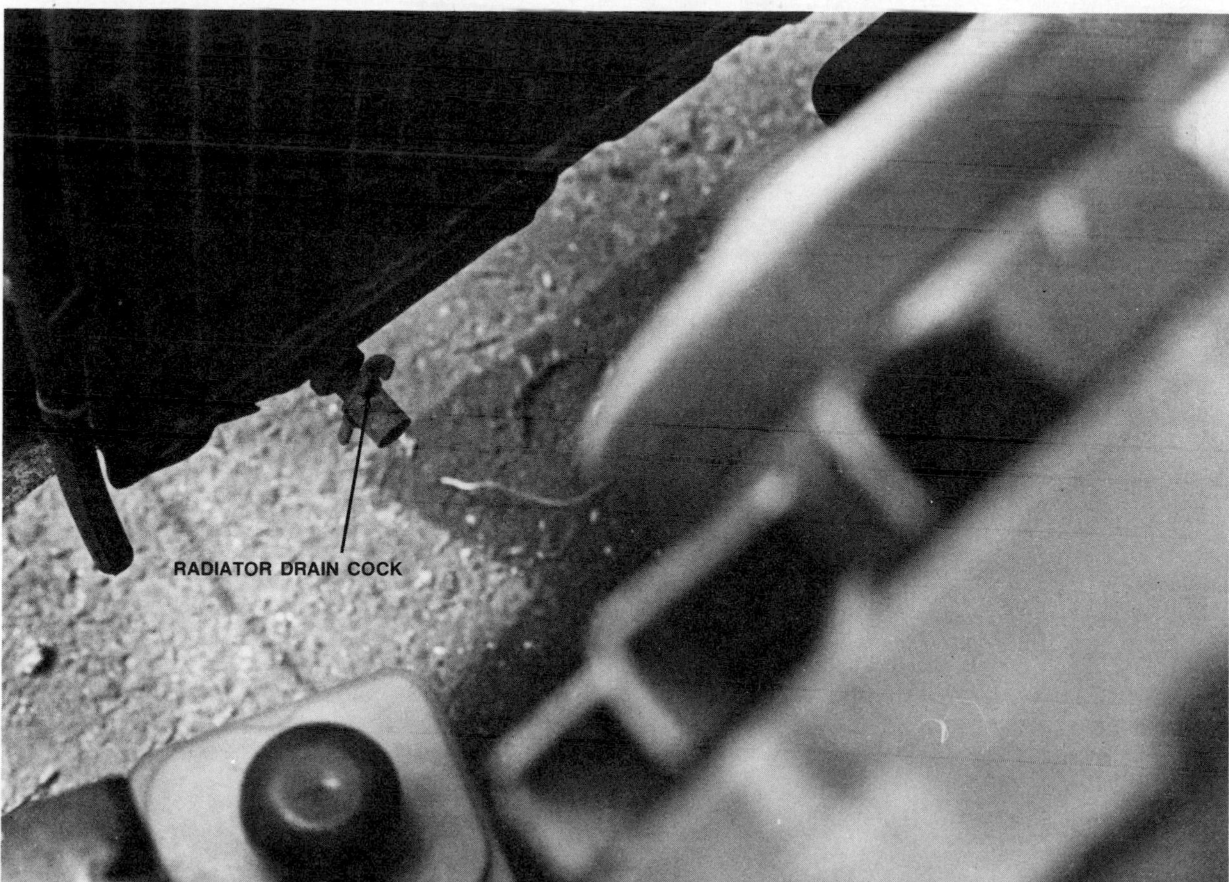

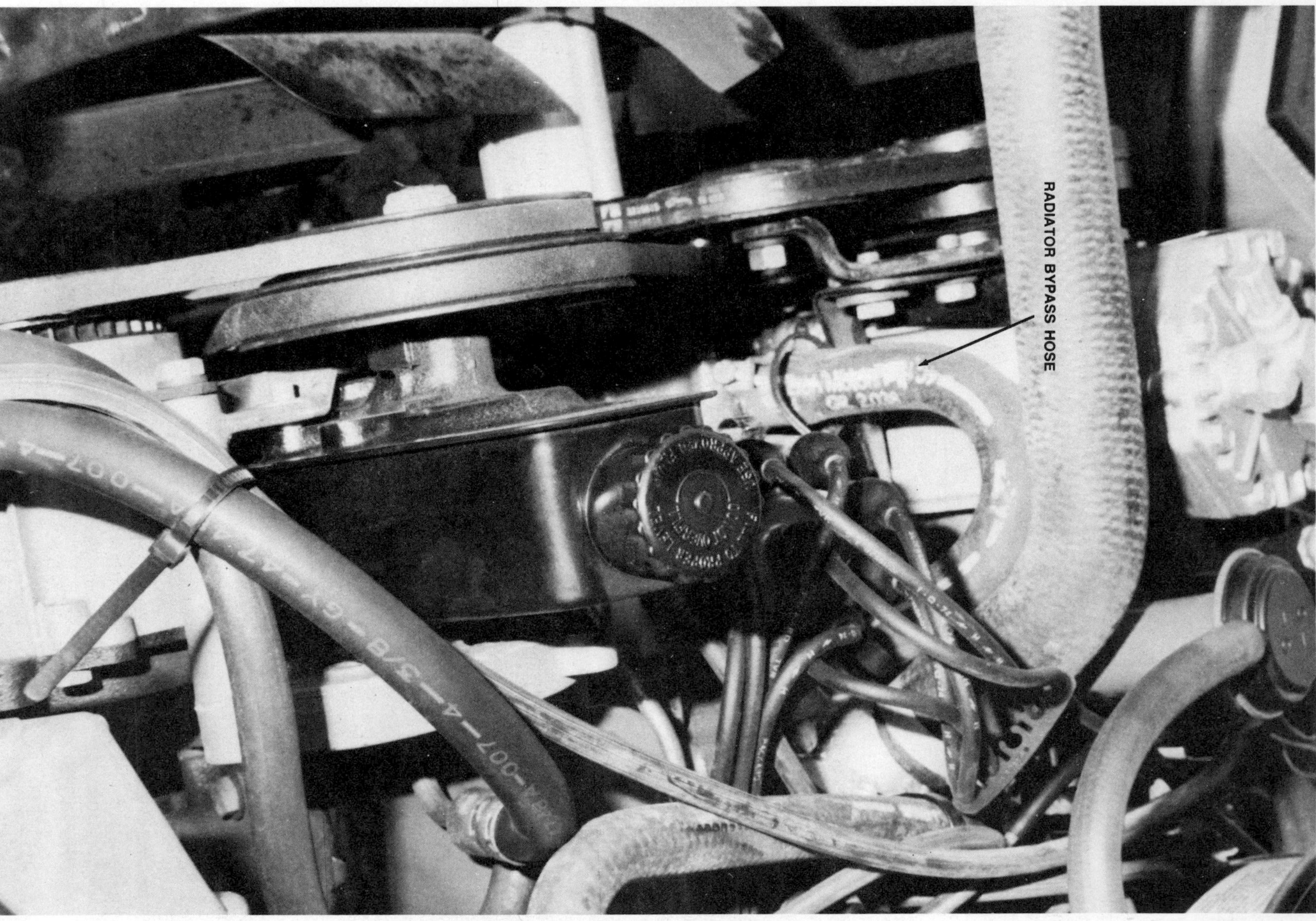

Fig. 7–17. Radiator bypass hose on engine. This little hose is subject to high engine-compartment heat, so is prone to failure.

Hoses and Connections

Feel each hose. If it's very soft, unusually hard, or if it is oil-soaked or cracked by heat, it should be replaced. One hose that should be given a *very* careful inspection is the radiator bypass hose, a little one used on many cars. It permits coolant to circulate through the engine when the thermostat is closed and, because of its location, is very prone to heat failure (see Fig. 7–17).

Replacing a Hose

To replace a radiator hose, begin by removing the hose clamp (see Figs. 7–18, 7–19). If it's corroded and won't loosen, cut it apart with tin snips. Caution: Don't try to pry it apart with a screwdriver against the radiator hose neck, for you might damage the neck and face an expensive repair. If necessary, cut the hose off at the neck, make longitudinal slits around the circumference of the hose remainder (on both sides of the clamp), and pull the cut sections out (they normally will tear through under the clamp). Soon you'll have clearance to remove the clamp.

Note: If you have spring clamps, obtain hose clamp pliers to compress them.

Wire-brush rust and dirt from the hose neck (see Fig. 7–20) and then fit the new hose into place. Install worm-drive hose clamps (which can be installed with the hose already in place). These tighten with a screwdriver and will be easy to remove, even years later. Center the clamp on the hose section that is on the neck.

Radiator Bypass Hose

Replacing a radiator bypass hose (see Fig. 7–21) is a similar operation, except that the hose may be physically difficult to install because of tight quarters. To reduce future replacements, obtain a premium hose made of silicone rubber. It should last the life of the car.

Heater Hoses

On most cars the heater hoses are readily accessible at both ends. If they are not accessible at the heater end (because the heater core is hidden in under-dash ductwork), a simple solution is to feel the entire length of each hose. You usually will find that the hose is in good condition

Fig. 7–18. Loosening retaining screw on band-type hose clamp.

Fig. 7–19. Once hose clamp, regardless of type, is loosened and out of the way, gently pry between hose and neck with screwdriver to free up hose.

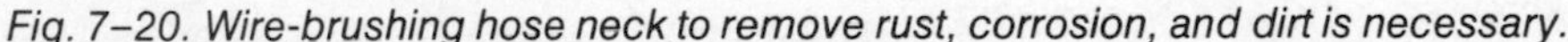

Fig. 7–20. Wire-brushing hose neck to remove rust, corrosion, and dirt is necessary.

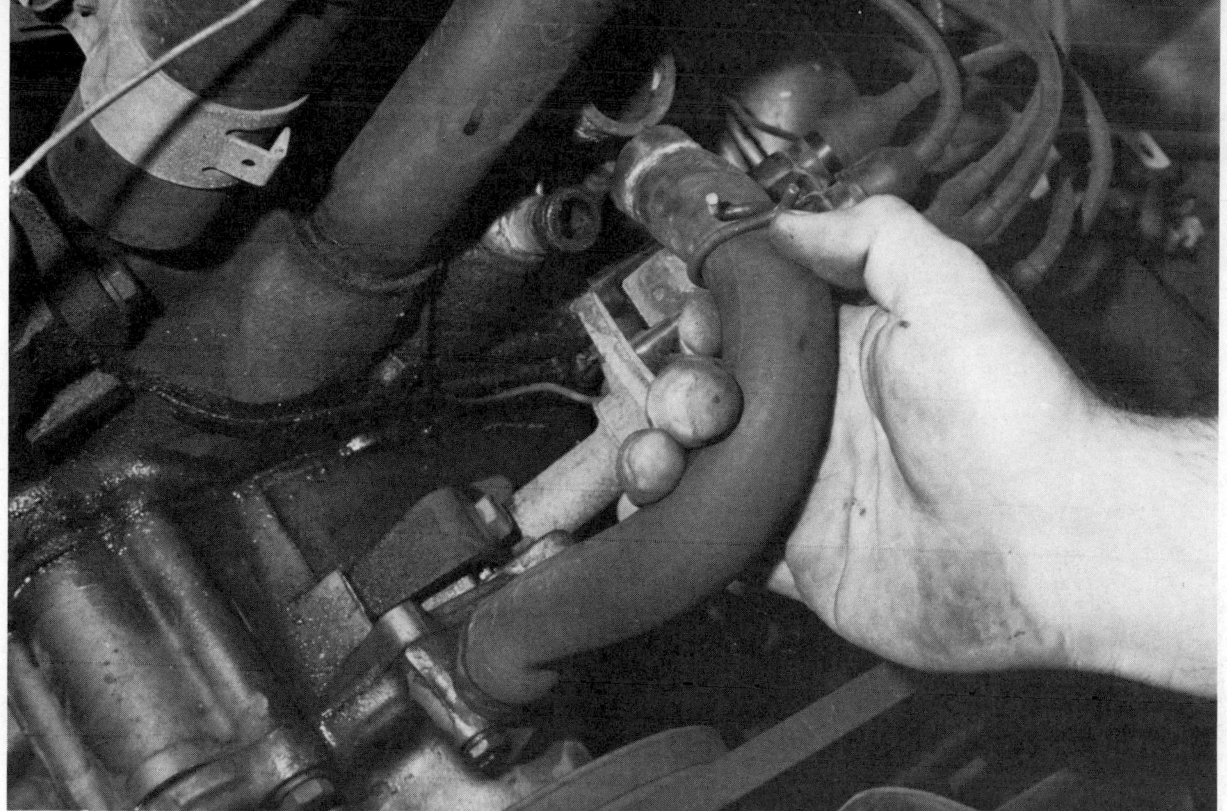

Fig. 7–21. Removing a bypass hose. Note that this particular hose was held by a spring clamp.

until it is close to the engine. Just cut the damaged section off and use a flushing tee as a connector between the good, old section and a brand-new section to the engine.

Overheating

Any engine will overheat in severe enough conditions, such as heavy traffic operation or operating a heavily-loaded car in mountainous or hilly areas, particularly with the AC on. Caution: When you overheat, let the engine cool before removing the radiator cap. Top up the radiator with coolant as required, then check the problem when you're home. If the overheating occurs under milder conditions, and the system has recently been flushed and filled with enough anti-freeze, check it as follows:

- *Leaks*. If there are any obvious leaks, they must be corrected. Look hard. You might find something as simple as a loose hose connection.
- *Radiator*. Is the front of the radiator obstructed by leaves, bugs, dirt, or other debris? Clean it off with a soap or detergent-water solution and a soft brush if necessary. On AC equipped cars, check the front of the AC condenser, the radiator-like part in front of the radiator.

Could the coolant level have been low? Even a healthy cooling system loses some coolant, and if you neglect it the level will drop. If the system has an overflow reservoir, it should automatically top up when it cools. If it doesn't, the problem is a defective gasket in the radiator cap that allows air to leak in instead of coolant being sucked in (see Fig. 7–22). Replace the cap.

Could the anti-freeze protection level have dropped severely? If you have been topping up with water for some time, the dilution of the anti-freeze lowers the concentration and the boiling point. Obtain an inexpensive hydrometer and check the freeze protection (see Fig. 7–23). If the freeze protection is very low (above 10° F.), the boilover protection also is cut so that in moderately severe operation overheating can result.

Perhaps the radiator is plugged. Here's a test procedure for downflow radiators (those with vertical coolant flow tubes. You can tell them by the fact the radiator cap and upper radiator hose are on the same side of the radiator). Start with a cool engine, radiator topped up. Connect a

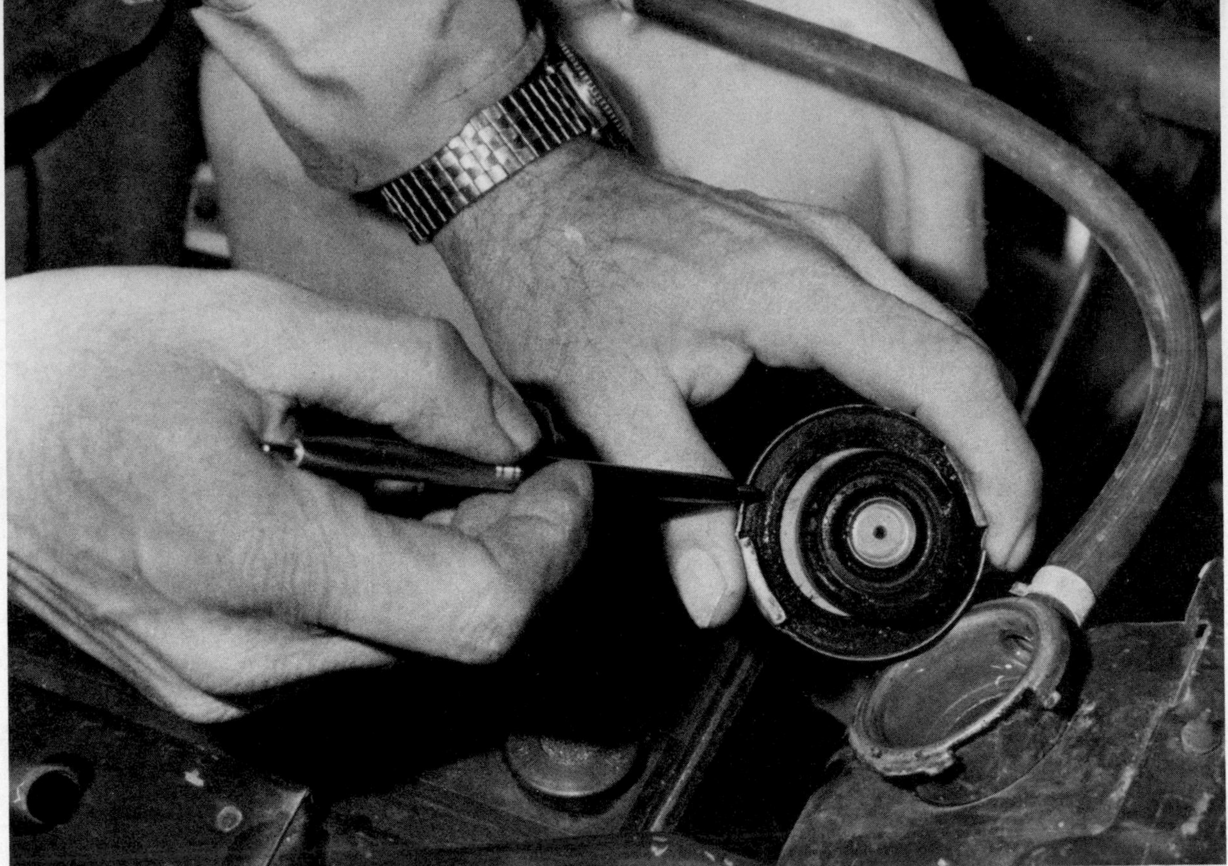

Fig. 7–22. Pen points to outer gasket of radiator cap on car with overflow reservoir. If outer gasket is deteriorated, air may be drawn in, instead of coolant from the reservoir, when the engine cools down.

tachometer (see *Chapter 2*). Run the engine at fast idle for a few minutes. Squeeze the upper radiator hose and, when you feel some pressure buildup (indicating the thermostat is open and coolant is starting to flow), remove the radiator cap and rev the engine to 3000 r.p.m. If coolant overflows the radiator fill neck, the radiator is apparently plugged. If the car has a crossflow radiator (horizontal tubes and radiator cap on the same side as the lower hose), make this test: Remove the radiator cap and gradually rev the engine to 3000 r.p.m. If the coolant level drops noticeably, there is a strong chance (combined with an overheating problem) that the radiator is plugged.

Removing the radiator is an obvious job on most cars. Disconnect the radiator hoses—the automatic transmission cooler lines on cars so equipped, the shroud if used. On transverse engine cars, also remove the electric fan assembly. Take out the two nuts or bolts that hold the radiator at the top and lift it out. Before you simply buy a replacement radiator, have the old one flow-tested, an inexpensive job that any radiator shop can perform. Unless the radiator is very badly plugged, the shop may be able to clean it with heavy-duty compounds that can not be used on the car.

• *Thermostat*. If the thermostat is stuck closed, the engine normally will overheat very quickly, perhaps in the first mile or so of driving on a hot day. Remove the thermostat as follows:

(1) Drain the radiator into a pan (if the coolant is clean and suitable for re-use).

(2) Remove the bolts or nuts holding the thermostat housing, the part to which the upper radiator hose connects on the engine (see Fig. 7–24). Note: On some VW–Audi models, the thermostat is accessible after removing a lower radiator hose housing from the bottom of the water pump.

(3) Lift up the thermostat housing and remove the thermostat. Note which way it faces so you can install correctly (see Figs. 7–25, 7–26).

(4) Suspend the thermostat in a pot of water on a stove, holding it with a piece of wire. When the water starts to boil the thermostat should be open. If you're not sure what part is moving on

Fig. 7–23. Checking anti-freeze protection level with hydrometer. Small sample is drawn into tool with syringe top. Marked float indicates protection level at the surface of the coolant.

the thermostat, douse it with cold water and see if anything moves. Whatever moves should also move, the opposite way, when the thermostat is in the hot water. If nothing happens, the thermostat is stuck and should be replaced.

(5) With a putty knife, scrape off gasket residue from the engine and/or thermostat housing mating surface. Put the new gasket in place, then install the thermostat (new or old as required), and refit the thermostat housing and the bolts or nuts. Tighten the nuts or bolts gradually in a criss-cross fashion.

• *Radiator cap.* If the cap fails to hold pres-

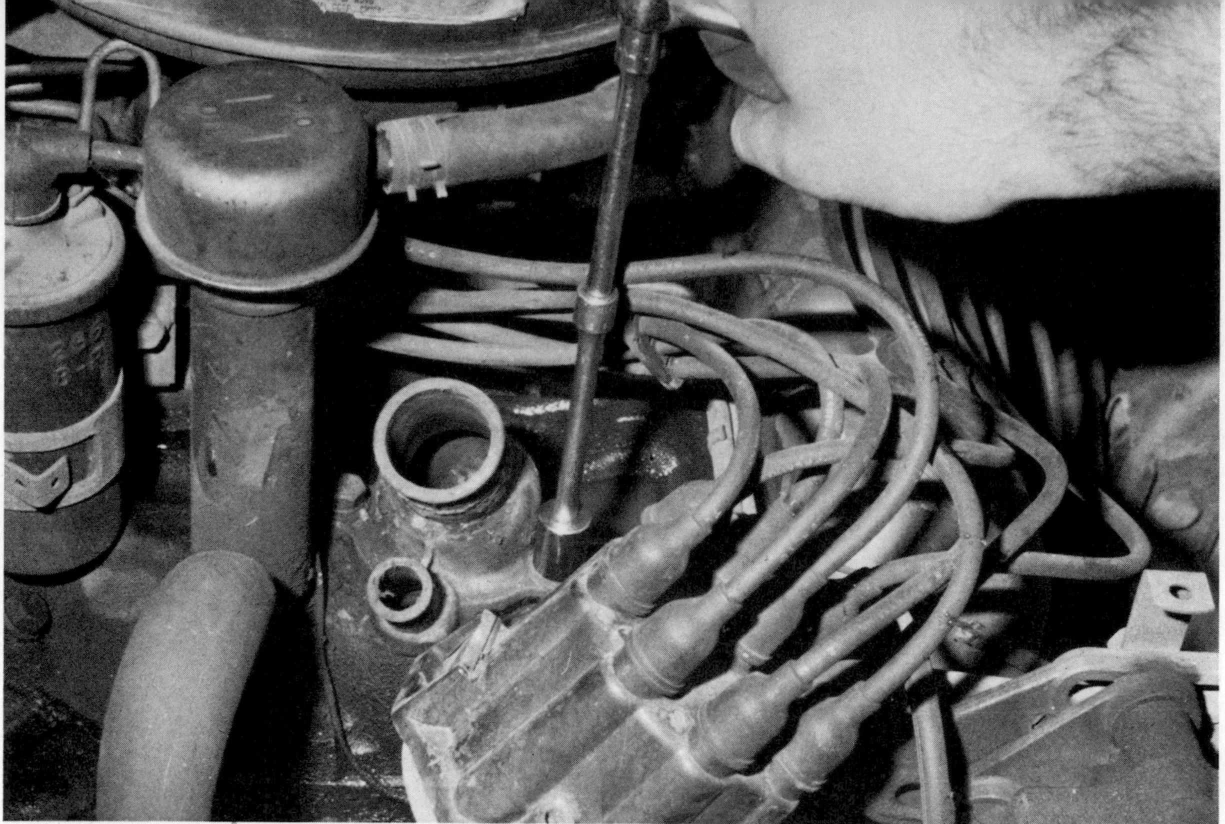

Fig. 7–24. Removing the bolts that hold thermostat housing to engine. Although upper radiator hose has been removed, this normally is not necessary unless hose is being replaced, too.

Fig. 7–25. Thermostat housing is lifted off and there's the thermostat underneath. Just lift it up and out after noting its position for correct installation of the existing stat or a replacement. Note: On some engines the thermostat is in the housing itself.

Fig. 7–26. This thermostat has a circular tab projecting from the bottom to seal off the water passage in the center. When replacing this thermostat, be sure the new one also has the tab to close off a passage to the coolant-heated intake manifold. If the passage is not closed off in very hot weather, the fuel mixture in the intake manifold will be overheated and poor engine performance will result.

Fig. 7–27. Cleaning radiator fill neck. Any dirt, rust, etc. on the sealing surface inside (about an inch or so from the top of the neck) can prevent the radiator cap valve from seating properly.

Fig. 7–28. Testing radiator cap with pressure tester. If the cap fails to hold specified pressure, it must be replaced.

Fig. 7–29. The finned circular disc in front of fan is the fan clutch.

sure, coolant will flow out. With an overflow reservoir it won't be lost, but the coolant level of a hot engine could drop so low that the engine still will overheat.

The cap may not hold pressure for one of two reasons:

(1) There is dirt or rust in the radiator neck, on the surface against which the cap's rubber seal seats. Clean the neck with a rag (see Fig. 7–27).

(2) The cap's valve is malfunctioning. Have a shop pressure-test it for you (see Fig. 7–28). A defective cap is one of the most common causes of coolant loss and overheating.

• *Drive belt*. A loose drive belt causes poor water pump and fan operation. Press down on the belt midway between water pump/fan and alternator pulleys, and if it deflects more than ¾ inch it should be adjusted as explained later in this chapter.

• *Fan*. Watch the fan spin as you view the engine from the side. If it seems to wobble, it or the pulley on which it is mounted is apparently defective. On cars with clutch fans (see Fig. 7–29) try to spin the fan by hand. If it spins freely (five or more revolutions with a single, forceful flip) it should be replaced. You may, after you have some experience, be able to replace it yourself.

• *Other causes*. There are many other causes of overheating, such as internal engine leakage and defective water pump. If the problem is not traced to something covered in this chapter it probably is a job for a professional.

Drive Belt Service

The typical engine may have as many as four drive belts wrapped around the pulleys of different accessories at the front of the engine. Belts drive the alternator, water pump/fan, power steering pump, anti-pollution air pump, and air conditioning compressor.

The problem is obvious if a belt snaps. In many cases, however, the belt is deteriorated from use, poor adjustment, or oil and road film splashed onto it.

To check a belt, look for oil soaking (if you find it, replace the belt). Twist the belt over on its side and look for glazing or cracks in the sidewall. If you find any, replace the belt. It's a lot easier to replace a belt at home than on the road. If a belt is an inner one (meaning other belts must come off first), home is the only place to be.

Removing a Belt

To remove a belt you must release tension on it, and there are many ways a belt may be tensioned.

The simplest and most common is that one of the accessories around which it wraps has a bolt passing through an elongated slot in a bracket (see Fig. 7–30). Just slacken this bolt, and one other mounting bolt, push the accessory in toward the engine to remove tension, and then take off the belt.

The belt may be tensioned by a stud adjuster (see Fig. 7–31). Slacken the locknuts on the accessory, turning the nut on the stud adjuster until the belt is loose enough for removal.

On VW Beetles, the generator pulley is actually a two-piece design. Place a screwdriver in the slot in the pulley to brace it and remove the pulley nut with a wrench. Pull off the pulley's outer half and the belt can be easily removed (see Figs. 7–32, 7–33).

There may be an idler pulley. This is a pulley that does nothing but guide the belt, and is adjustable to control tension. The idler pulley is commonly held by a locknut, and when this is loosened you can pivot the pulley to adjust tension. Often there is a hex or partial hex surface on the pulley stud, so you can put a wrench on it to move the pulley to remove tension and be able to take off the belt. The wrench also is used as a tensioning lever for tightening the new belt.

With the one-belt spring-loaded idler pulley there is just one special ribbed belt. Use a pry bar on the spring lever to release tension (see Fig. 7–34).

Belt Selection

When you buy a new belt, be very careful to get the right one. If your belt has cogs (see Fig. 7–35) or ribs, the replacement also must have them. Also, the width of the belt must duplicate the original. If you get a belt that's too wide, it will not fit completely into the V-groove of the pulley. Inasmuch as the V-belt transfers power from its sides to the sides of the pulley, this contact is extremely important. If you get a belt that's too narrow, it will fit all the way down into

Fig. 7–30. The most common drive-belt adjustment is as shown at this alternator. You slacken two bolts or nuts, one of which passes through the elongated slot, then apply and hold tension with pry bar as shown while tightening bolt also as shown. Complete job by tightening the accessory's other mounting bolt.

Fig. 7–31. To adjust belt at this power steering pump, slacken two locknuts, then turn stud adjuster nut as shown. Complete job by tightening the locknuts. On many newer cars (Chrysler front-drives' alternators, Ford V-6 single belt idler pulley, 1982 Cadillac), a similar adjuster called the jackscrew is used. Instead of turning a stud nut, you turn screw hex-head.

Fig. 7–32. VW Beetle uses the split pulley arrangement. Install screwdriver in pulley slot to hold pulley, then use wrench to remove pulley nut.

PULLEY OUTER HALF

SPARE WASHERS

Fig. 7–33. After removing nut, take off washer as shown, then pull off the outer half of the pulley. Tension is adjusted by installing or removing washers between the pulley halves. Spare washers are stored as shown.

Fig. 7–34. Single belt wraps around all accessories on this Ford engine. Insert pry bar into spring tang and yank to release tension on belt. Note that most pulleys are grooved to mate with ribs on one side of belt. Some Olds-built GM diesels also have this self-tensioning spring.

INSERT PRY BAR HERE

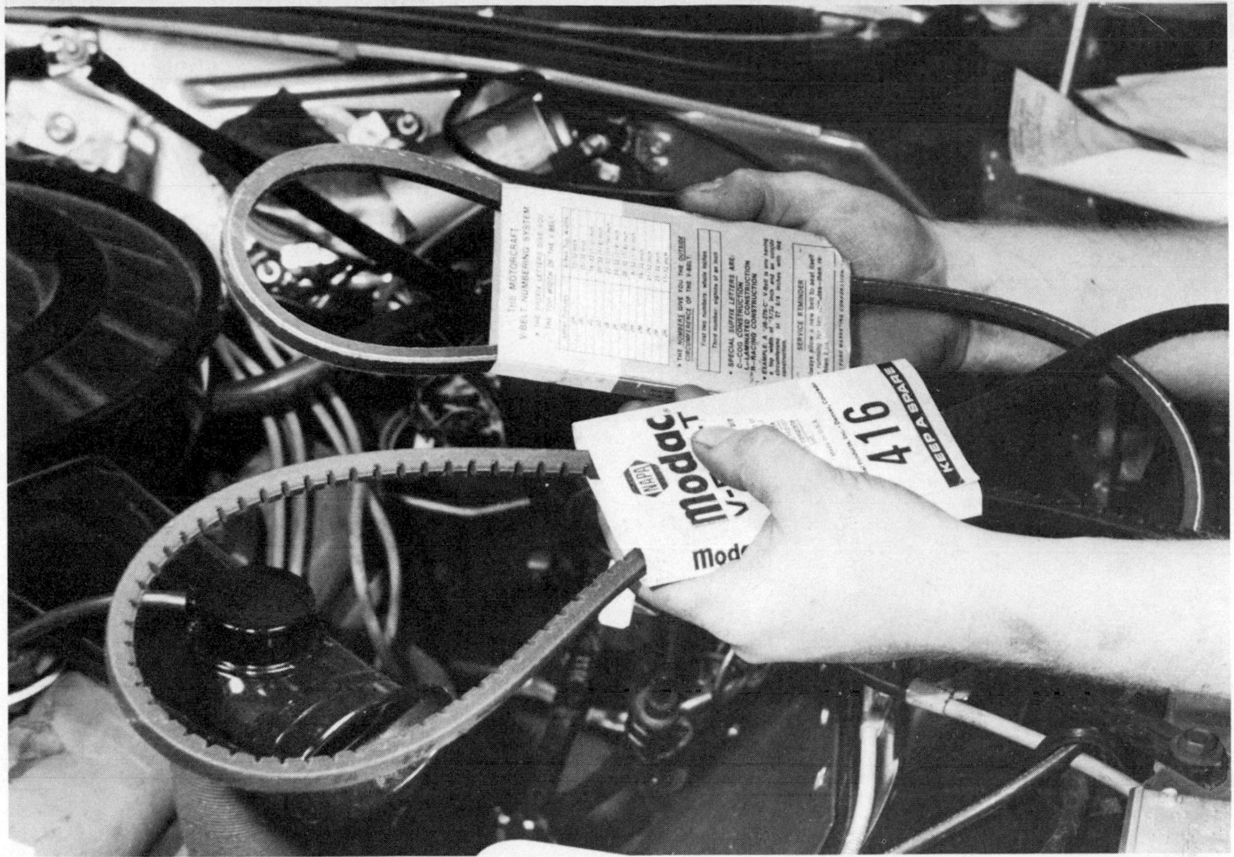

Fig. 7–35. Some belts have cogs. The cogs help the belt operate at sharp angles, so if the original is cogged, so should the replacement.

the pulley, make poor side-to-side contact, and the bottom of the belt may be damaged by contact with the V-groove.

In most cases, you will be asking the counterman at a parts store for the new belt. Because of performance problems that could result from the incorrect belt, double-check his selection for width and type. If your belt is snapped, you can't check overall circumference, but if it isn't, you can make an approximate comparison. The new belt should be no larger and preferably slightly smaller in circumference (the old belt may be larger because it has stretched in use). If you can't get the right belt, don't settle for one that the counterman "thinks" may do the job. Shop around and get what you need.

Emergency Belt

If you have an emergency situation, a special belt kit is available as a temporary replacement. You just wrap it around the pulleys, cut it to size, then attach the cut ends with a special crimp-together fitting.

Installing the Belt (Except VW Beetle or Ford One-Belt System)

Make sure the belt is neatly fitted into all the appropriate pulley grooves. On many engines it is all too easy to put a belt into a wrong pulley, particularly on the front of the crankshaft, which may have several pulley grooves.

Draw the belt tight with your hand at the adjustable pulley; then, if possible, slip the belt onto the adjustable pulley by hand. Can't do it? Try slipping a large screwdriver under the belt and prying it over the pulley edge into the groove. If the belt is twisted, you can align it without dislodging it from the groove. Once you have some general automotive experience you can crank the engine in very short bursts with the screwdriver under the belt and braced on the pulley edge. The belt will literally jump onto the pulley if you do it right. Caution: This technique takes "touch." The ignition system must be disconnected (see *Chapter 2*) to prevent the engine from starting, and the cranking bursts must be very brief.

Fig. 7–36. Belt tension gauge hooks onto belt to measure tension. The reading is compared with factory specifications.

Fig. 7–37. Square hole in power steering bracket accepts square drive from ratchet for belt tensioning.

SQUARE HOLE

Tensioning the Belt (Except VW Beetle or Ford One-Belt System)

Once the belt is on all pulleys it must be adjusted for correct tension. The most accurate way to measure tension is with a special gauge that hooks on the belt as shown (see Fig. 7–36). The gauge may be used both while setting tension or for post-adjustment checks.

With the elongated slot adjustment, exert force on the accessory to pivot it outward to apply tension, then tighten the nut of bolt to secure the adjustment. Finally, tighten the other accessory mounting bolt. Important notes: How you exert force is extremely important. If the part is stamped steel or aluminum with fins, use of a pry bar may cause damage. On most power steering pumps, for example, there is a special pry bracket with a square hole (see Fig. 7–37) so you can insert a ratchet with a long handle (or a pipe over the shank to extend the working length). Just insert the square drive end of the ratchet into the hole and gently pull on the wrench handle to tension the belt.

If the ratchet will not fit into the working area, check with an auto supply store to see if you can get a special thin bar with a square drive for belt adjustment. Or see if you can reach the square

Fig. 7–38. The hex-head on back side of power steering pump bracket accepts a socket. A wrench (in this case a torque wrench) with a long extension permits the socket to reach the hex-head. For this particular adjustment, the car manufacturer provides torque specifications, hence the use of the torque wrench, in addition to belt tension specifications.

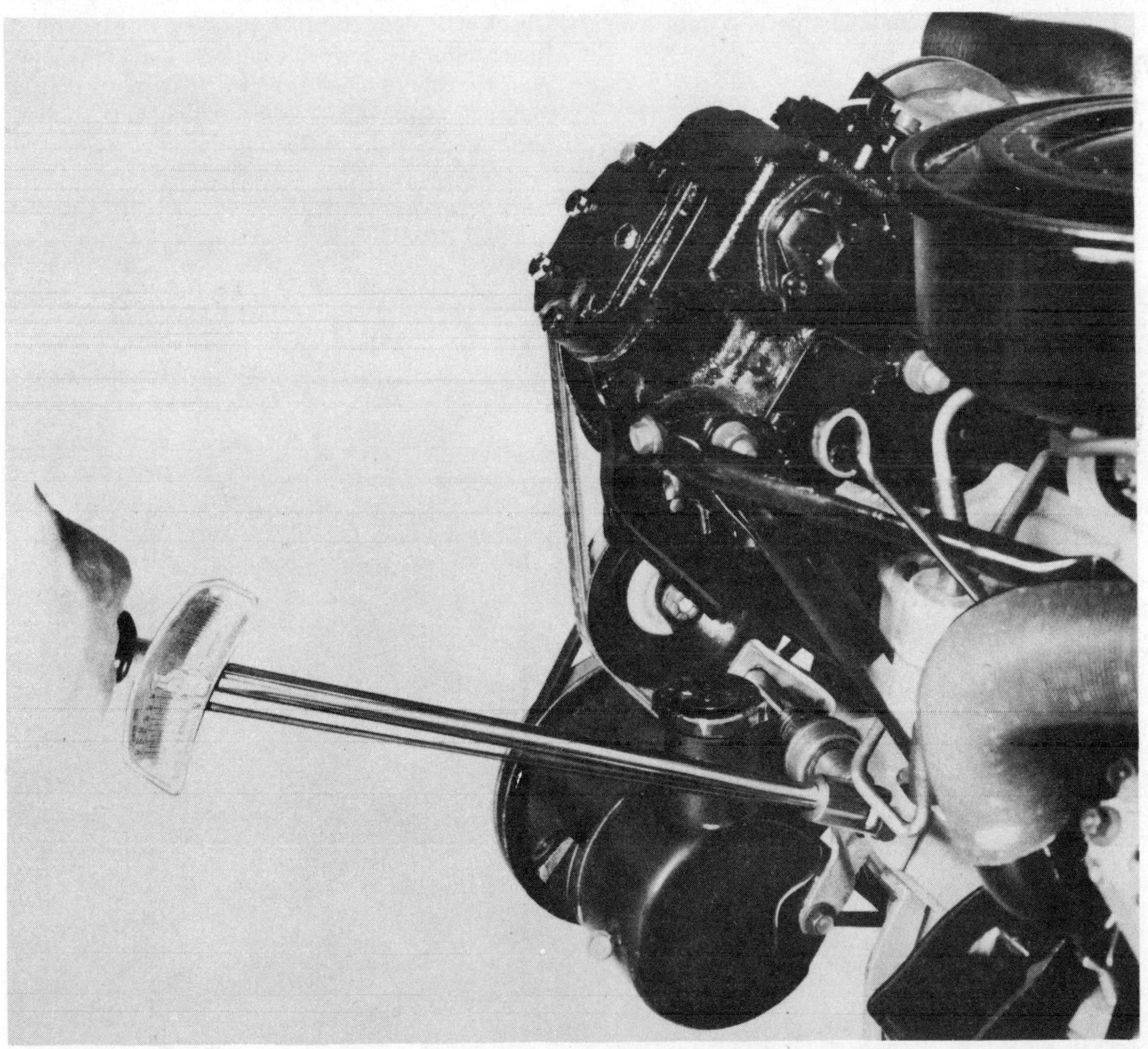

hole (or a hex-head as in Fig. 7–38) from the rear, using a long extension rod with the ratchet.

On some cars there is a hole in the accessory's mounting bracket (see Fig. 7–39) for a pry bar. By inserting it in the hole provided you are sure of prying against a safe location.

If there is no provision, and you are concerned, you can obtain a special strap wrench that locks safely around the accessory, so you can pull on the wrench shank to apply tension.

Stud Adjuster

If a stud adjuster is provided, such as on some power steering pumps, the adjustment is very simple. Just loosen the retaining nuts or bolts and turn the stud nut until tension is correct, as shown in Fig. 7–31. A similar adjuster is used on Chrysler front-drive cars' alternators (see *Chapter 14*) and on Ford V-6 and 1982 Cadillacs. Loosen a lockbolt on the underside of an adjusting bracket, and turn an adjusting bolt until tension is correct.

Idler Pulley

If there is an idler pulley adjustment there is some method provided for moving the pulley, usually with a wrench on the pulley's stud shaft.

VW Beetle Belt Adjustment

With the VW Beetle, just slip the belt around the crankshaft pulley (and the AC compressor if installed), holding it up to the pulley half remaining on the generator. Adjusting belt tension is a cut-and-try operation, and it is done simultaneously with the reinstallation of the outer pulley half. You must select a number of washers (extras are stored on the generator armature shaft, against the outside face of the outer pulley half, held in place by the pulley nut).

A practical approach is to use the number of washers you found between the pulley halves as a starting point. Assemble the pulley, store the extra washers on the shaft, then thread on the nut. Put the screwdriver in place to brace the pulley and then tighten the nut with a wrench. Check belt deflection by pressing down with a finger midway between generator and crankshaft pulley (see Fig. 7–40), and if it is more than a half-inch, remove a washer or two, reassemble, and check belt deflection again. Removing

Fig. 7–39. The finger points to hole in alternator bracket. Insert pry bar in this hole and it will be positioned correctly for tensioning belt.

Fig. 7–40. On the VW Beetle, check belt tension by pressing down with your finger midway between pulleys.

Fig. 7–41. A pry bar in the spring tang is used to release tension on the belt, permitting it to be disengaged from the idler pulley. Note the serpentine route of the belt with its smooth top side against the idler pulley, and also against the water pump pulley. Both these pulleys, therefore, are not grooved.

washers from between the pulley halves reduces the distance between them and causes the belt to ride higher in the groove, increasing tension. If tension is too great (little or no deflection), or if you can't assemble and tighten the pulley halves, insert washers between them to reduce tension.

Retightening

All drive belts stretch in use. Most of the belt stretch occurs in the first 15–20 minutes of operation, so after a brief test drive be sure to recheck tension and readjust if necessary.

Ford One-Belt System

With the Ford one-belt system, the spring-loaded idler pulley automatically sets and maintains correct tension. Once you have the belt wrapped around all other pulleys, use the pry bar in the spring tang to move the idler to the no-tension position (see Fig. 7–41). Slip the belt around the idler pulley and then release tension on the pry bar. The idler pulley will spring back. Note: Take care to insure that on those pulleys that accept the ribbed side of the belt, that it fits properly into the grooves on the pulley itself (see Fig. 7–42).

8 EXHAUST SYSTEM SERVICE

EXHAUST SYSTEM COMPONENTS are expensive, but if you inspect the system periodically (see Fig. 8–1) and make small repairs when they are first required, you can reduce expensive replacements to a minimum.

Jack up the car and support it with safety stands at front and rear; if you only have a pair of ramps, raise one end at a time. Inspect the muffler and the piping very carefully, looking for:

(1) deep dents (more than one-quarter of the way in, such as one-half-inch into a two-inch diameter pipe) that could cause restrictions preventing the engine from developing normal power, particularly at cruising speeds;

(2) holes or cracks anywhere in the pipes or muffler (see Fig. 8–2). Feel the top of piping and muffler for holes or cracks you might not be able to see.

Next, look at the clamps and hangers. If any are broken, rusted away or, in the case of hangers, the rubber support is cracked or broken, replacement should be made. If you don't, the exhaust system may come apart or fail and you'll face a major expense.

Exhaust System Service

Dents

If you find a dent in a catalytic converter, perhaps only the bottom cover need be replaced. Although this is a job for the car dealer (special equipment is required), it is far less costly than a new converter. Note: The bottom cover is not replaceable on all cars; check with your dealer.

If there are dents in piping or muffler, you can check to determine if they are causing a severe restriction by using a vacuum gauge. Remove a vacuum hose at the base of the carburetor or intake manifold (see Fig. 8–3), plug the hose end with a pencil, and connect the gauge hose to the carburetor neck. Run the engine at idle, and then have a helper hit the gas pedal. If the vacuum gauge reading drops to zero, and rises to a number that is less than the normal 16–22 inches, the exhaust is restricted. Note: For this test on late-model cars, disconnect the vacuum hose from the EGR (exhaust gas recirculation) valve and plug the hose end. Operation of the valve could give inaccurate results.

Note: This vacuum gauge test is a useful one for checking for exhaust restrictions even if there are no dents, but there is a loss of power at cruising speeds. Two other possible causes of exhaust restrictions are: plugged muffler or collapsed inner wall of a laminated (two-wall) exhaust pipe. This design is used on many cars to eliminate exhaust ringing noises. However, heat and cold can cause the inner wall to collapse, while the outer wall remains intact.

To determine where a restriction is located, you must disconnect the exhaust at different points and recheck engine vacuum. Once you make a disconnection that results in normal vacuum readings, you know the part you disconnected is plugged. Begin at the front of the muffler, then move forward, toward the front of the car.

Holes

If a hole in a muffler is no more than about two inches in diameter, it can normally be patched. The same is true for a hole up to a half inch, in a two-inch diameter pipe, provided that the opening is not at or very close to a clamp joint, or at a sharp bend in the pipe.

Patches are not approved by motor vehicle inspection laws in all states, so check before you buy. The two popular types of patches are the

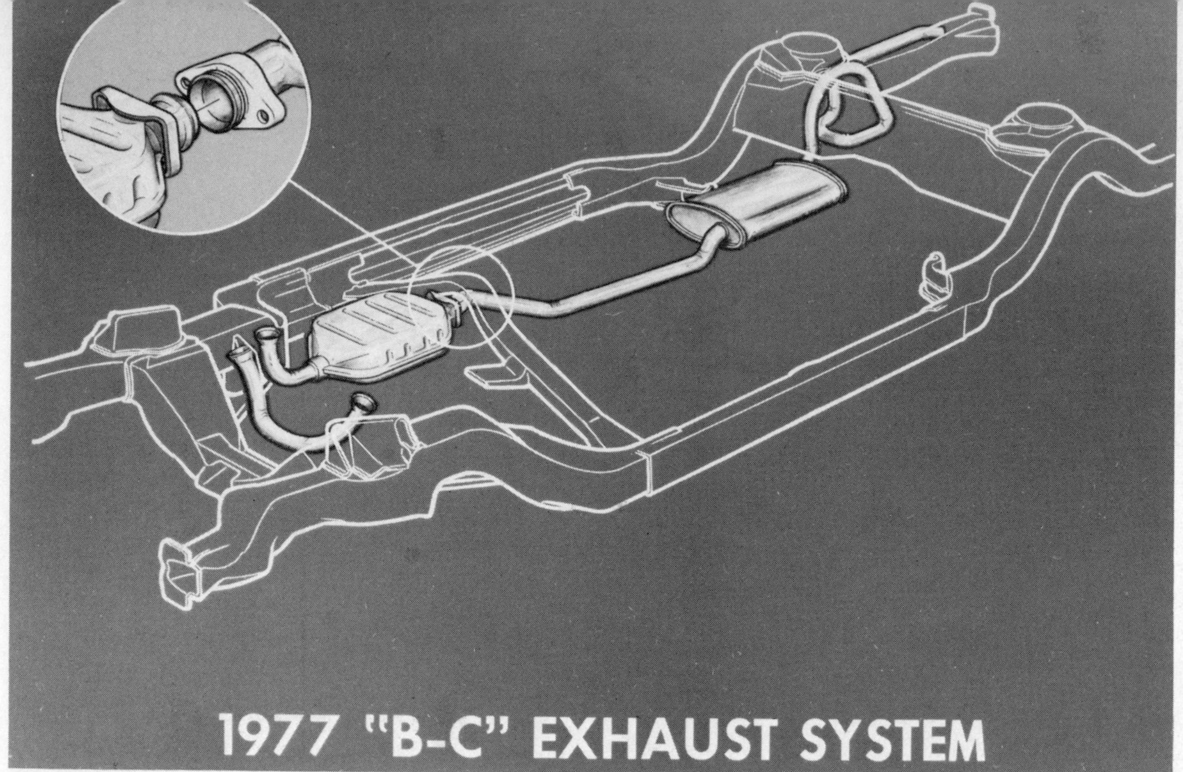

Fig. 8–1. This is a typical exhaust system with a catalytic converter, which is the cannister at the left. Note the detail illustration of the converter joint with the exhaust pipe designed for a tight seal. The cannister at the rear is the muffler.

asbestos bandage and the fiberglass bandage treated with high-temperature resin. The latter type has greater resistance to heat.

With either type, you place a piece of foil (provided in the kit) against the hole as a flame guard, then wrap the bandage around it, and secure the bandage with wire (see Fig. 8–4). After a day of driving, the patch will cure and the wire can be removed.

Pinholes

If an entire section is dotted with pinholes, the bandage patch is the best approach. When you have only a few isolated pinholes, however, you can repair them with a patching putty. Just sand away the rust around the hole and smear on a coat of the putty, working it into the hole. It will heat-cure as you drive.

Dents and Major Breaks

When a section of pipe is deeply dented, badly rusted through, or open right next to a clamp joint, you can't patch it. If the defective pipe section is straight or has a curve you can match up, a repair pipe section is the way to go (see Fig. 8–5). These repair sections come in many sizes and lengths, so check with a parts store to see if you can get what you need for a repair. If the bad section is only a couple of inches long, you may get by with just an exhaust pipe connector, a very inexpensive part that joins two pieces of pipe.

Installing a repair section begins with cutting out the damaged part of the old pipe. You may be able to do this with a hacksaw, or if clearances are limited, you may have to rent an exhaust pipe cutter (see Fig. 8–6). If you don't get a clean cut, rent a pipe expander to flare out the pipe end for a good connection (see Fig. 8–7). Just insert the expander into the end and turn the forcing screw with a wrench. Do not overexpand, or the repair pipe section won't fit over it. Sand any rust off the end of the pipe ends (see Fig. 8–8), and file the edges if necessary to remove burrs.

You'll need two clamps the right size for the pipe. Choose premium clamps if you can get them, for they will seal better. If you can't, buy a container of exhaust pipe sealer (a special putty) and apply a thin coat to each cut pipe end before you install the repair pipe and the clamps.

Fit the repair section into place and install a clamp at each end. Aim the clamp so the U-bolt ends face downward, for easy tightening.

Fig. 8–2. Check the entire exhaust system for breaks and dents. Note the break in the pipe just forward of the muffler in this illustration.

Fig. 8–3. A vacuum gauge test can tell you if the exhaust system is restricted.

Fig. 8–4. Foil patch covers hole in exhaust, then bandage is wrapped over foil and wired into place. Bandage cures from exhaust heat and wire can be removed.

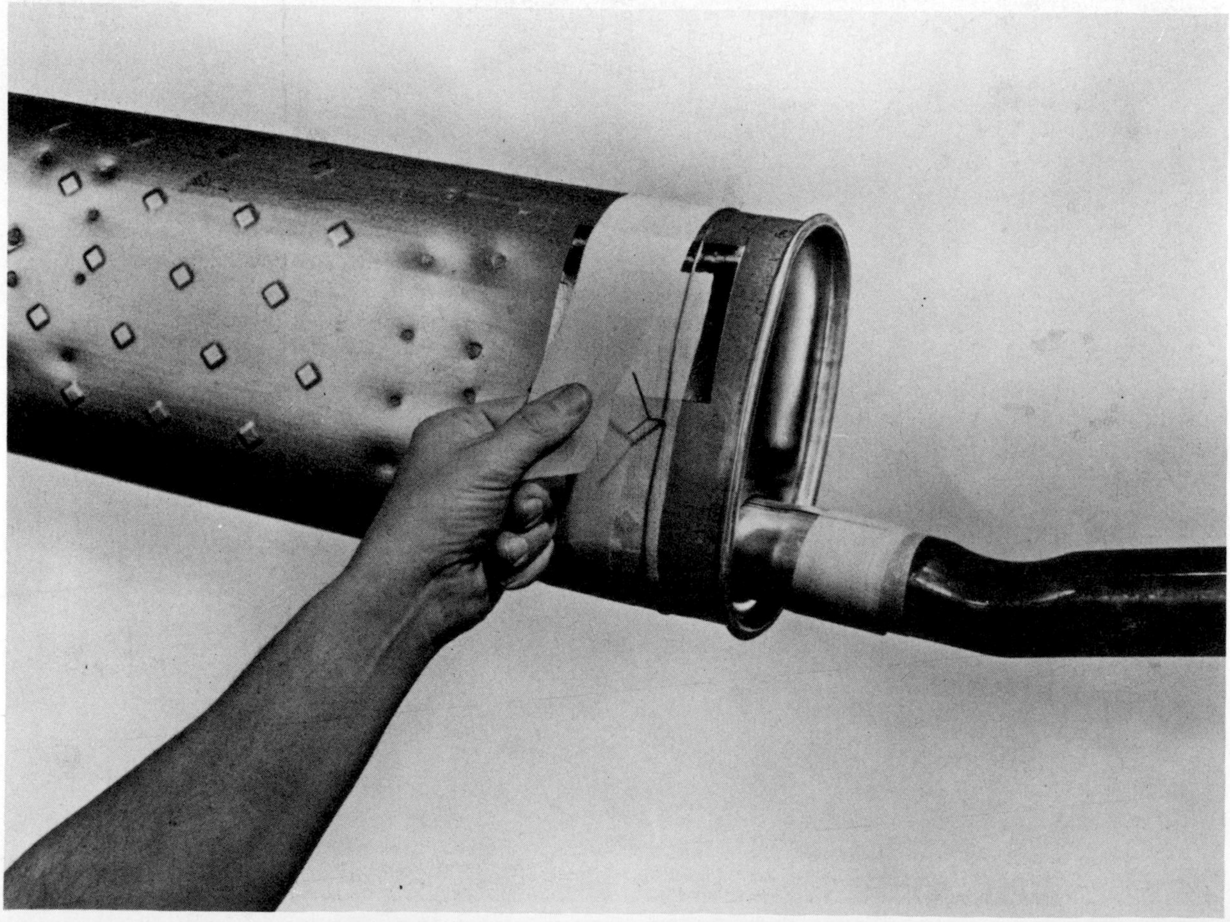

Fig. 8–5. If piping is good except in one area, pipe repair of the section is the most economical choice. This long pipe is a typical section. Short pipes are connectors that can be used for repair if dent or break is very small, and to connect pipes of the same or different sizes.

Fig. 8–6. Exhaust pipe cutter makes a clean cut and works well in tight quarters.

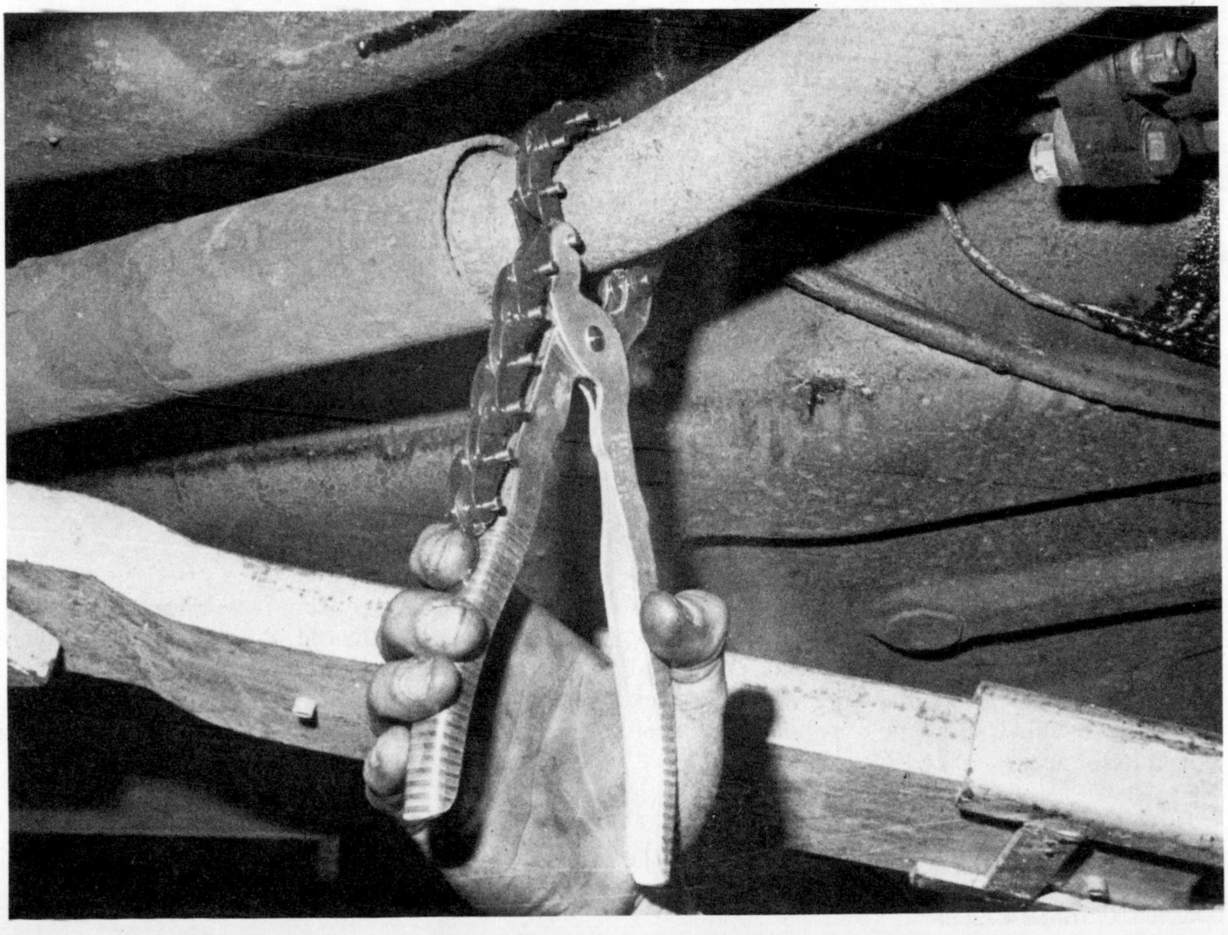

Fig. 8–7. Pipe expander flares out a pipe end or muffler neck if necessary for a good connection.

New Pipe

If you can't get a repair section, buy a complete new pipe (the cost still will be lower than if you took the car into a repair shop). In most cases, the pipe from the engine will be in good condition for some distance from the engine, so you don't really have to replace the whole thing, just the section that includes the deteriorated part.

Begin by cutting the old pipe at a point where it's good and straight. Next, remove the deteriorated portion from the muffler. Then remove the clamp holding the pipe to the muffler. Spray a rusty clamp with penetrating oil, then loosen with a wrench and remove. If the clamp refuses to budge, cut it apart with a hacksaw and discard.

Separate the muffler from the pipe. This can be easier said than done. If the pipe won't come out of the muffler neck, cut it off nearby and clamp vise-type locking pliers to a protrusion. Whack the pliers with a hammer (see Fig. 8–9). If the pipe is really stuck, obtain a muffler chisel and drive it in between the pipe and the muffler neck (see Fig. 8–10). Use a pipe expander to reshape the muffler neck.

Once you have the damaged section of old pipe removed, line up the new pipe at the muffler and measure carefully. The object is to cut the new pipe so that when you install the new section in the muffler the other end of the new section will line up (to within less than an inch) of the old, still good section. Note: Store the unused new section in case the old ever does fail.

Bridge the two sections with an exhaust connector (see Figs. 8–11, 8–12), a short pipe that fits over both ends of the exhaust pipe, new and old. Clean and expand the old pipe end if necessary. Clamp the connector at both ends and the repair is complete. If you're using standard clamps, apply a film of exhaust sealer putty to each cut pipe end before positioning the connector.

Welded System

If the exhaust system is welded to the muffler you cannot cut the old pipe off at the muffler. If the pipe is good in the area of the muffler, cut as far forward of it as possible. If it isn't, the muffler must also be replaced as explained later in this chapter.

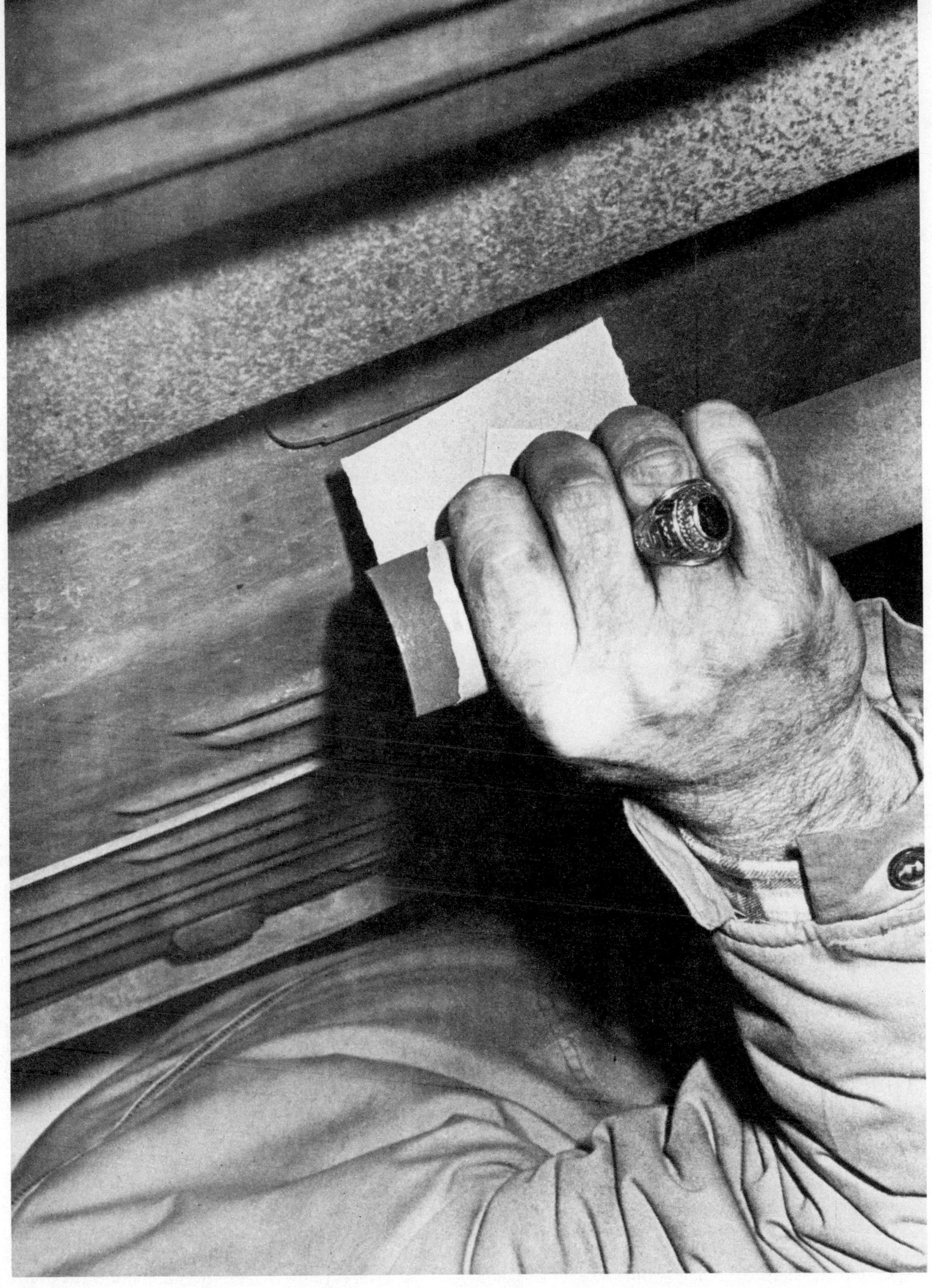

Fig. 8–8. Sand rust off old pipe end as shown before inserting into a new pipe, connector, repair section, or muffler.

Fig. 8–9. Using vise-type locking pliers and hammer to remove stuck piece of pipe from a muffler.

Laminated Pipe

If your problem is a collapsed inner wall of a laminated or two-wall pipe, replace with a single wall pipe (the exhaust ringing will go away as the exhaust soot coats the inside of the pipe) or obtain a laminated pipe with vents in the outer wall. It should be far more resistant to the problem.

Replacing Hangers

The mounting of exhaust hangers should be obvious when you're looking underneath. If any are rusted out, or have rotted or cracked rubber, you realize that immediate replacement could save expensive parts.

New universal pipe hangers come in several varieties, most of them cheap and short-lived (see Fig. 8–13): If you look hard, however, you can get high-quality hangers that can work at a substantial horizontal angle (up to 45°) and can swivel to any horizontal angle (see Fig. 8–14). This makes installation easy when, as is common, parts do not align perfectly.

To remove a defective hanger, begin by spraying the upper end (where the bolt holds it to the body) with penetrating oil, then remove. If penetrating oil also will free up the lower end where it clamps to the exhaust system, that's a plus. If that doesn't work, cut it apart with a hacksaw.

Note: If you have an import with a rubber "doughnut"–shaped hanger, replace only with a quality brand, which will be made of high–temperature rubber. If necessary for good alignment, bend the doughnut–retaining tangs on the body, not on the exhaust system.

Some late-model American cars (primarily front-drives) have hangers similar to the import car doughnut. They have rubber sections that are push-fits on rods that are built into hanger brackets (see Fig. 8–15). To remove, just pull them off, or pry them off with a screwdriver.

Replacing Muffler and Tailpipe

Replacing the muffler and tailpipe is the most common exhaust system job, and it's well within the ability of a weekend mechanic. If you're using jack and stands, jack up the car and place the safety stands under reinforced (box) sections

Fig. 8–10. Using muffler chisel to cut stuck piece of pipe in muffler. After cutting, the piece may be removable with ordinary pliers or with vise pliers and hammer.

Fig. 8–11. Lining up old pipe and repair section. Note exhaust connector on old pipe.

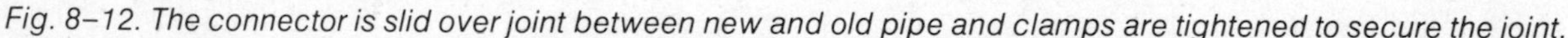

Fig. 8–12. The connector is slid over joint between new and old pipe and clamps are tightened to secure the joint.

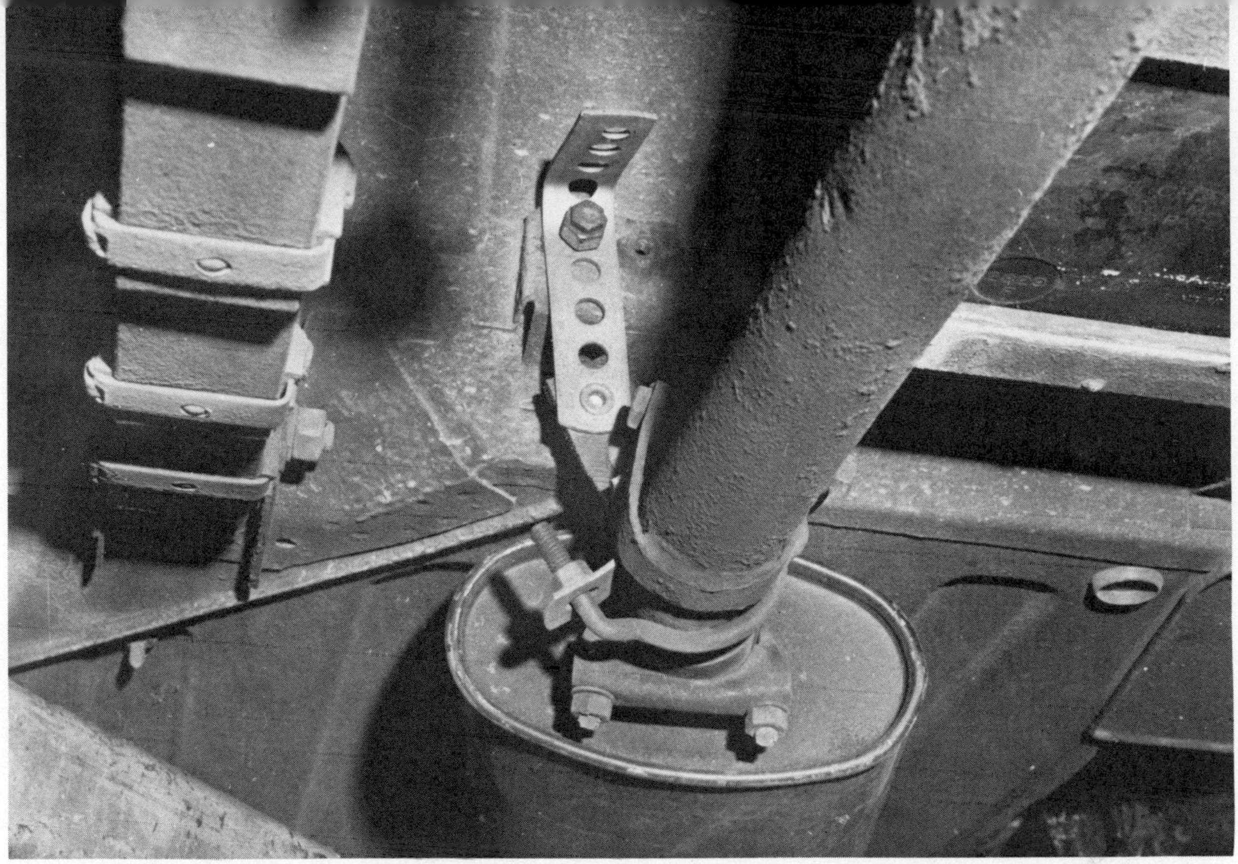

Fig. 8–13. Exhaust hanger shown is inexpensive, but its life is limited if it must work at anything but a very slight angle.

Fig. 8–14. This premium exhaust hanger has an end that can be turned 360 degrees and heavy-duty straps in rubber section, so it will hold even at severe angles.

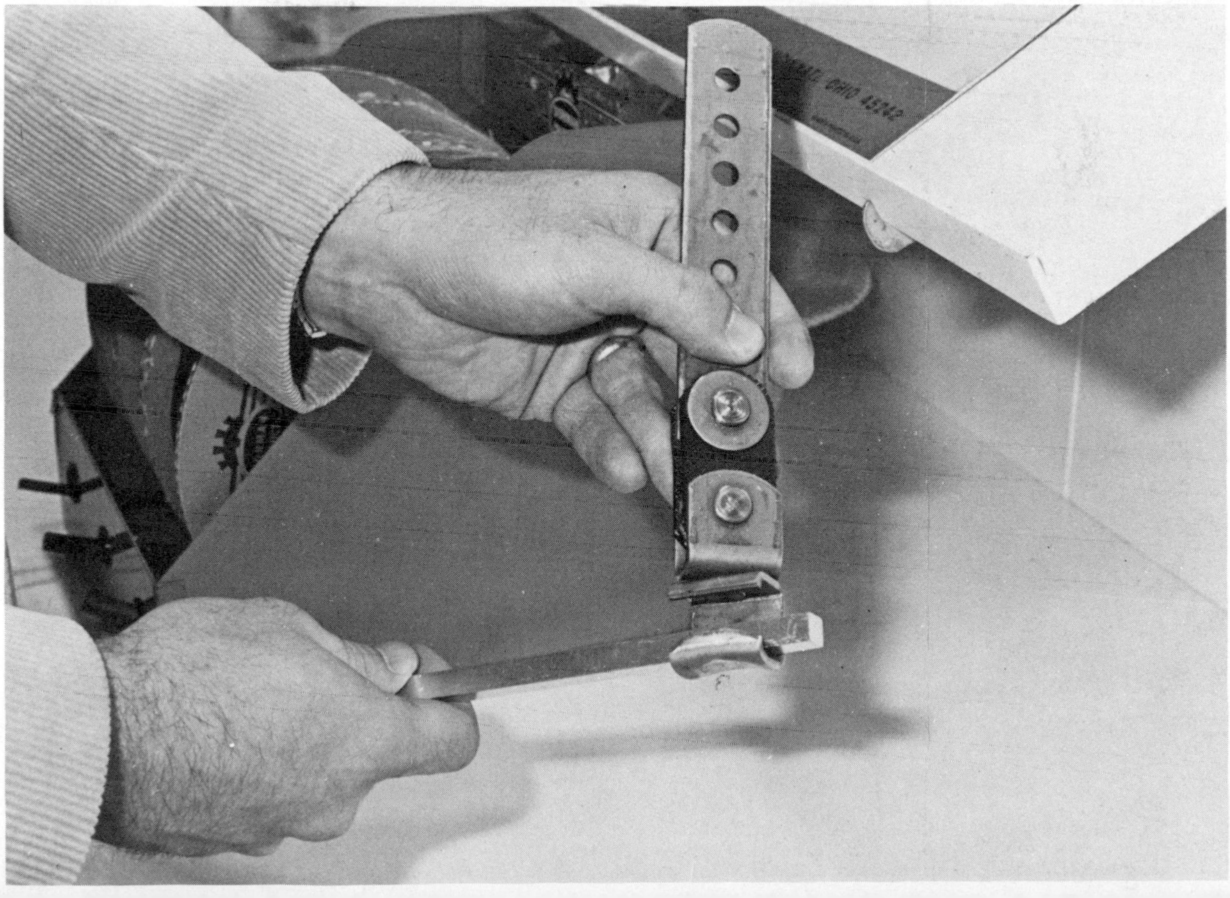

of the chassis, rather than under the usual locations on the suspension or rear axle. This will permit the suspension to drop and there will be more working clearance for you.

Begin by removing the clamp that holds the muffler to the exhaust pipe. Next, spray the joint with penetrating solvent, then knock the muffler off the exhaust pipe with a hammer. Disconnect the rear hanger from the tailpipe and, if possible, remove muffler and tailpipe together. If underbody clearances make this impossible, you have two choices: salvage or disposal. If the tailpipe is salvageable, spray the joint with penetrating oil and try to knock the muffler forward, off the rear joint. If the tailpipe is to be replaced, just cut it with a hacksaw at any convenient point.

If the system is welded at the exhaust pipe-muffler joint, cut the exhaust pipe where it tapers down to single thickness, a few inches from the muffler (that's where the muffler neck ends, but because of the welding, the neck is obviously not a separate part). Many replacement mufflers have a slightly longer inlet neck to compensate for the fact that part of the pipe was cut. If the one you obtain does not, also buy an exhaust connector that is sized at one end to fit into the muffler and at the other end to fit over the exhaust pipe. Every joint between pipes must be clamped, so if you use a connector you'll need two clamps.

The best procedure for installing the new parts depends on the shape of the tailpipe. If it's long and curved and you couldn't pull out the old parts together, install the tailpipe first, just letting it hang over the rear axle. Next, fit the muffler to the exhaust pipe. If necessary, prepare the exhaust pipe by sanding the cut end, flaring if required, and coating with a film of exhaust sealer. Install but do not tighten the clamps. Then install the tailpipe on the rear muffler neck, and fit the clamps but do not tighten.

Install the hanger(s) and connect them to the exhaust system. If necessary to reduce the working angle of the hanger, rotate the muffler and/or tailpipe. It's important to have some clearance between the exhaust system parts and the body, suspension, brake lines, etc., so try to duplicate the original positions as closely as possible. When you think it looks safe, tighten everything only enough to hold it in place. Lower the car to the ground, get underneath, and recheck to see if everything is clear. If not, turn muffler and tailpipe as required to get the clearances. Only then should you give everything a final tightening.

Fig. 8–15. Some American cars have exhaust hangers somewhat similar to the European rubber doughnuts. The one shown has a rubber segment that fits on tubular necks that extend from the upper and lower metal sections of the hanger. Just pull or pry off the rubber segment when disconnecting the exhaust system.

9 YOUR CAR'S ELECTRICAL SYSTEM

THE ELECTRICAL SYSTEM of the automobile is direct current (DC), compared with the alternating current (AC) used in the house. The automobile does have an AC generator (alternator), but the current, in order to be used, is changed to DC by electronic components called diodes. The battery charger which plugs into household current also is an AC device, but it too has electronic components that convert the current to DC to charge the battery.

The DC circuit is basically very simple. A wire is connected to a source of current at one end and run to the current "consumer" (bulb, motor, radio, etc.) at the other. The source of all current is the battery, and you undoubtedly have noticed that it has two posts or terminals, one marked positive, one marked negative. The positive post is the source, but as a practical matter current can be picked up anywhere in the wiring that is in some way directly connected to the battery. Example: The battery positive post is connected to the starter, and from the starter there is often a wire that goes into a multiple connector. Any of the wires that emerge from that connector are also "hot," that is, they can carry current because they are in some form of metal-to-metal contact with a wire from the battery. In many cases, "hot" wires go into fuse boxes, or into switches which when closed will complete circuits and make additional wires capable of carrying current. Example: When you turn on the ignition key you complete a circuit to the ignition system, and when you turn it still further, you complete a circuit to the starter solenoid and the engine cranks.

In order to flow, current must have a complete path from the source and back to the source. When there is such a complete path, it is called a circuit (See Fig. 9–1).

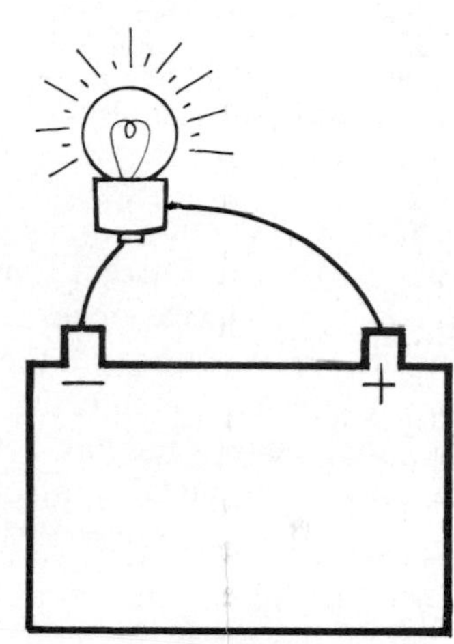

Fig. 9–1. This simple battery circuit runs from positive terminal through wire to bulb terminal, through bulb filament to second terminal, through wire to battery negative terminal.

Electrical Circuits

In the automobile we use the metal car body, chassis, and engine as if it were one big wire, in a system called the electrical ground. The car battery's second post or terminal is connected by cable to either the body, engine, or both (if not both, there is a separate cable elsewhere to connect body and engine).

Everywhere else, the circuit back to the battery is completed by grounding to the car body or engine. This is commonly done in either of two ways:

(1) A wire is connected to the body or engine.

(2) The part itself is made of metal, and when it is held to the body or engine by mounting screws the mounting screws provide the ground, through the metal part (see Fig. 9–2). Inside the metal

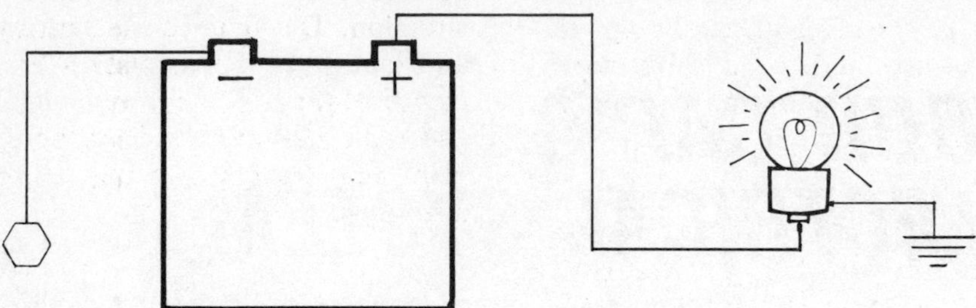

Fig. 9–2. This battery circuit uses a ground. The negative terminal of the battery is connected by wire to a metal part of the engine or body such as a bolt. The positive terminal has a wire that runs to the bulb that fits into a socket. The other terminal of the bulb touches the metal socket that is attached to the car body, and so is in metal-to-metal contact with it, completing the circuit.

part there is a physical connection, either by wire or by a metal tab, from the current consumer. As an example, a light bulb has a socket that is made in part of metal. When you install the bulb, the metal retainer for the glass is in contact with the metal part of the socket, and the socket itself somehow is in metal-to-metal contact with the car body.

The Switch

The switch (see Fig. 9–3) is nothing more than a mechanical break in a piece of wire. When we want to, we connect a metal tab to both ends of the break, and the wire is now "complete." The closing of a switch may join ends of a hot wire or a ground wire. When you press the horn button, for example, you are completing a ground circuit.

What Goes Wrong

Two things can happen to a circuit: it can break, forming an open circuit, or it can "short out."

• *Open circuit.* A switch is a form of open circuit. When you operate the switch, you close, or complete, the circuit. That's normal. When a wire breaks or a wiring terminal becomes so severely corroded that current won't pass through, an open circuit also is created, but this one can't be closed and the part to which the wiring is connected won't work. If the wire goes into a harness or is otherwise inaccessible, the actual break in the wire may be difficult if not practically impossible to find. Finding the break is not important, for once the existence of the break

Fig. 9–3. This battery circuit uses a ground and a switch. When the switch is open, the wiring from battery positive terminal to the bulb is open, so the circuit is incomplete and the bulb does not light. When the switch is closed, the circuit is complete and the bulb lights.

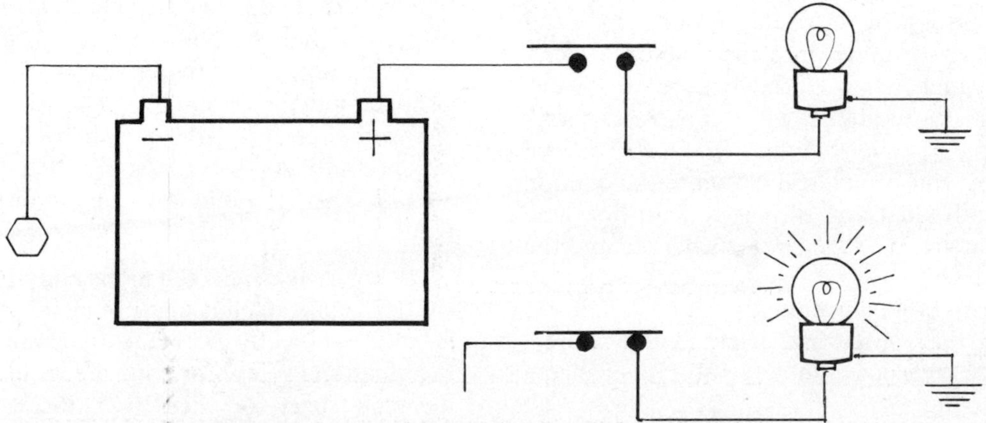

becomes obvious, a mechanic runs a new wire (outside the harness) to correct the problem. A corroded wiring terminal may be wire-brushed clean to correct a problem, or it may be cut off and a new terminal attached. Auto parts stores sell replacement terminals that can be attached to a wire with a pair of pliers.

• *Short circuit*. Electricity takes the path of least resistance, and the only way to force the electricity to go into a part and do some work (such as light a bulb or operate a motor) is to hold it in the wire until it reaches the part. Holding in the electricity is the job of that plastic covering, the insulation. If the insulation is cut or otherwise damaged, the electricity may be able to find an easier path through the insulation break to an electrical ground, such as a metal part of the car body or engine, for example, and although the circuit to ground is complete, it ends short of the part the electricity is supposed to operate (see Fig. 9–4).

Because there is little resistance to the flow of electricity through a break in insulation, and because the electricity may flow even when all switches are turned off, the electricity flow will be very substantial and the battery will often quickly run down.

In some cases you may lift the hood and see smouldering insulation at the break. You may be able to cure the problem in such an obvious situation. Disconnect the battery ground cable and tape over the insulation break with plastic electrical tape. If the wire itself is damaged, cut away the damaged section and splice in a replacement, using crimp-together connectors sold at auto parts stores.

How to Buy A Battery

The battery (see Fig. 9–5) is the storage unit of the car's electrical system. It accepts electrical energy from the charging system (the alternator) and converts it into the chemical energy it is designed to store. When you turn the key to crank the engine or turn on the radio or other electrical accessory, it converts that chemical energy back into electricity and transmits it through cables to the part that needs it.

If you cut apart a battery, you'd find a series of plates (made of lead and a stiffening alloy, such as antimony, calcium, and, in some cases, strontium) bathed in an acid solution. Actually, there are groups of plates, each group called a cell and providing two volts (the measurement of electri-

Fig. 9–4. This is a short circuit. If the wiring insulation breaks and the bare wire touches against the car body, electricity can get to ground without going through the bulb. Although the bulb does not light, electricity is consumed.

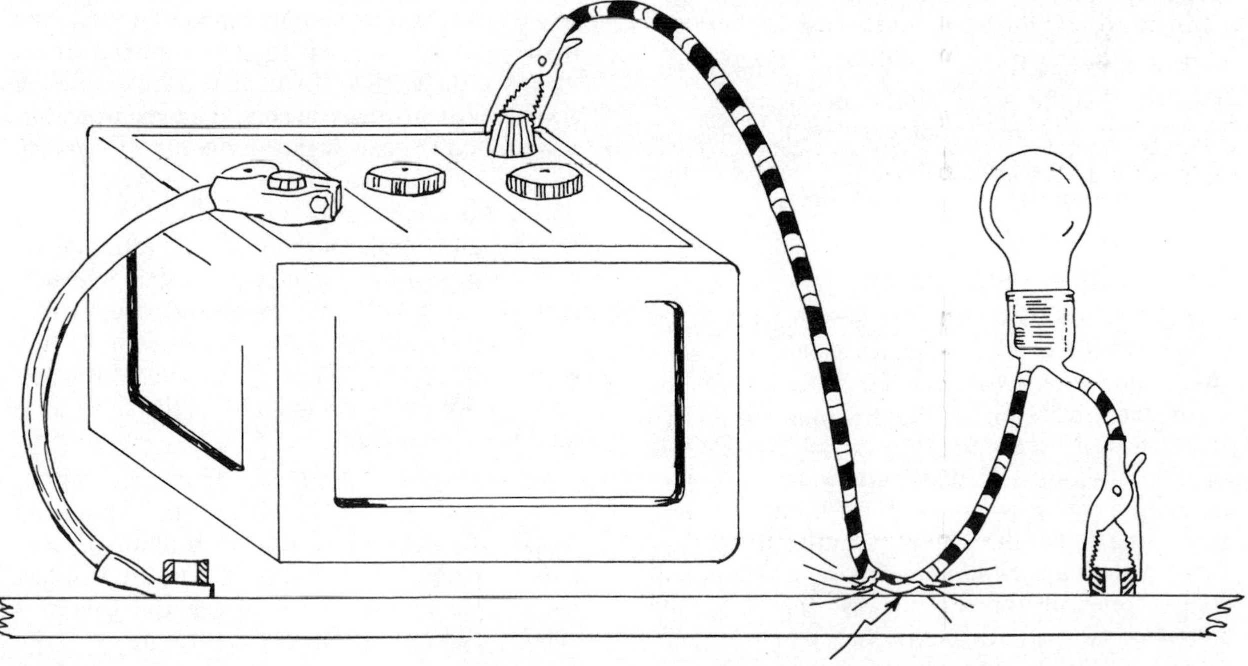

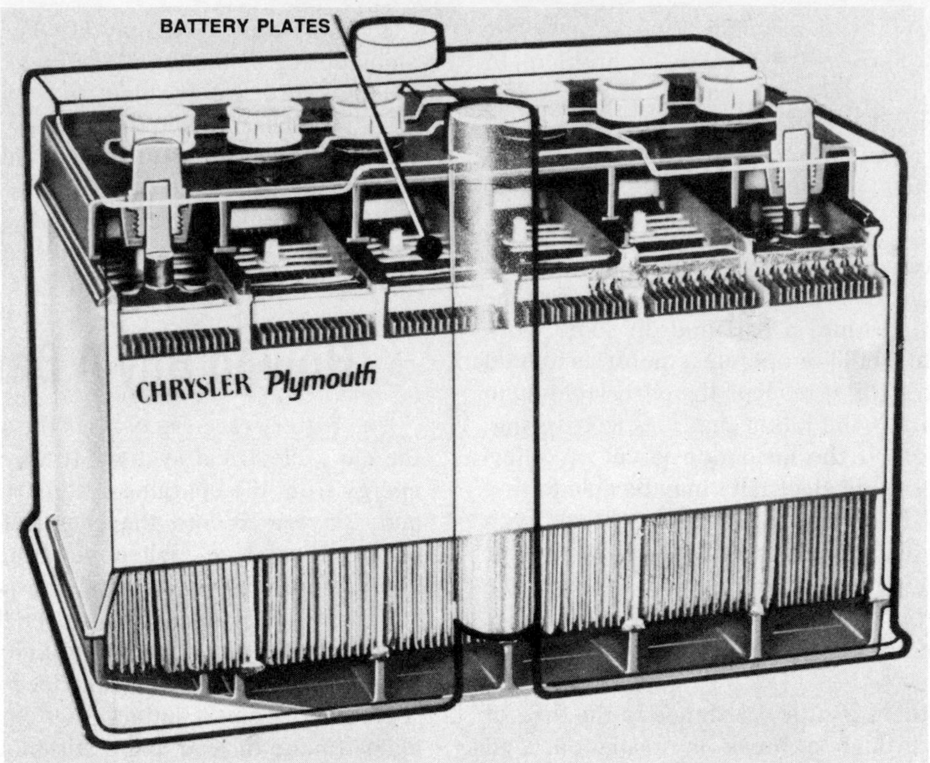

Fig. 9–5. Cutaway view of a battery showing the plates.

cal pressure). Put three cells together and you have a six-volt battery; put six together and you have a 12–volt battery. All batteries of the same voltage are not the same. One significant difference is in the amount of amps—electric current—they can supply.

The internal details of battery construction might be helpful if you could cut apart a battery before you bought it, but you can't, so instead you should concern yourself only with the battery ratings and any external features.

Rating Battery Performance

• *Amp-hour capacity*. The amp-hour rating of a battery tells you how long it can survive a slow discharge, such as if you left the car with the dashboard lights on. A battery that can last 20 hours with a 2.5 amp discharge is given a rating of 50-amp-hours. If that battery supplied five amps, the rating would be 100-amp-hours. Generally, the higher the amp-hour rating, the greater is the power available under more severe conditions. Generally is not always; therefore, the following two ratings are probably more significant to you.

• *Zero cold rating*. The number of amps the battery can deliver for 30 seconds during cranking at 0° F. without going dead. A rule of thumb is that the battery should have a cold rating in number of amps equal to the number of cubic inches of the engine. If you have a 240-cubic-inch six-cylinder engine, you should get a battery with a 240-amp zero cold rating if you have to start the car in cold winter weather.

Note: If you live in an unusually cold area, you may be able to ask for the −20 cold rating, which is the number of amps the battery can deliver for cranking at −20° F. Although this specification is not readily available everywhere, it may be in areas with severe winters. The rule of thumb—one amp for every cubic inch—is the same as for the zero cold rating.

Many engines—both on American and imported cars—are rated in liters rather than cubic inches. To extend the rule of thumb to metric measurement, just multiply the engine displacement in liters by 60. A 1.8-liter engine, therefore, needs a battery with a 108 amp rating.

As a practical matter, the smallest batteries (in cranking amps) available are generally well over 200, so you can hardly get a battery too small for an economy car from a standpoint of this rating.

• *Reserve Capacity.* This tells you how long a fully-charged battery can provide fairly substantial amount of current—25 amps—if you have a sudden charging system failure, such as a snapped drive belt or alternator or regulator malfunction. The amount of current is an estimate of what it would take to run the ignition system, headlamps, wipers, and heater, and gives you an idea if you can limp the car home when the charging system warning light goes on or the ammeter suddenly shows discharge. An hour rating is an acceptable figure.

Battery Maintenance

You can buy a conventional battery, a low-maintenance type or a so-called no-maintenance battery.

The low-maintenance type is the kind you see advertised as "normally never needs water during the warranty period." This is a battery designed for low water usage, having a large water capacity, so it just might make it without water. This type does have some sort of cell cap so you can add water if necessary, and it is a good idea to check the water level at least once a year.

The no-maintenance type has a smooth top and no provision for water additions. Though not guaranteed for life, it is guaranteed for some given amount of time. Some of the no-maintenance batteries have a charge indicator built into the top. Note: These batteries are not really sealed (there are hard-to-see vents built into the sides), so don't tip them over during any removal for underhood service, or you may lose some water you can't put back in.

The standard battery is just what you have known all these years. It needs water additions at least once every few months. If you have a smaller car, particularly with a four-cylinder engine, it is probably all the battery you will need for starting, which is your primary consideration. Its lower price makes it a good investment, even if it carries only a modest warranty, such as 24 months.

Side-Terminal vs. Top-Post

If your car comes with a side-terminal battery, instead of the conventional top posts, you may feel you have to get the same thing as a replacement. You do not! If your best buy is one with top posts, that's what you should get, for adapters are readily available at auto parts stores and are very inexpensive. The top post design has one important advantage you also should consider: It's a lot easier to use for jump-starting and for connecting a charger or test equipment.

Battery Warranties

There are many types of battery warranties, and you should read them carefully and understand them before you buy.

Lifetime. This means that the seller will replace the battery if it fails, unless that failure is caused by an electrical problem in your car, for as long as you own that car. The words YOU OWN and THAT CAR are important. You can't transfer the warranty when you sell the car and you can't transfer the battery to a new car. The price for such a battery with such a warranty is measurably higher, so unless you have every intention of keeping your car for many years, it really doesn't pay. The battery is a top-line design, but there's nothing lifetime about its construction. The seller is betting that most people will sell the car before the battery dies, and the extra charge covers the cost of servicing those who don't.

Simple pro-rata. A pro-rata warranty means that if a battery with a 36-month warranty fails at 18 months, you get a new one at half the current price. There also is a free replacement clause in the deal, usually limited to the first 30 days or three months, or sometimes even six months.

Initial free replacement, then pro-rata. The free replacement is for a longer period, often up to 18 months; then it's pro-rata. This is a good bet if you're planning to keep the car for about a year and a half, for it costs very little more than simple pro-rata. You should understand that once you're past the 18 months free replacement, you're instantly into pro-rata that includes that period. For example, you have a 48-month battery warranty with 18 months free replacement. The battery fails at 19 months. You must pay 19/48ths of the cost of a replacement, not for just one month of use.

When looking for a new battery, consider what you need to start the car, reserve capacity (particularly if you drive a lot on roads without service facilities), price, and warranty. Maintenance characteristics are not significant, because all battery cables loosen from road shock and periodic attention is required, even if water additions are reduced. Look for the best overall value, but don't buy more or less than you really need. Just because a huge capacity battery with a long warranty is on sale for $20 off the regular price doesn't mean it's worth buying. It may still be $10 more than you really need for the little four-cylinder car you plan to dump in a year.

Used Batteries

If you are planning to sell your car very shortly, you can take a small gamble and save big money by buying a used battery from a wrecking yard. A good battery from a late-model wreck should last for a good long time, and it will probably cost as little as one-fourth the price of a new one. Reconditioned batteries are available in some areas of the country, but quality of battery reconditioning is very uneven, so you are taking a chance unless the local rebuilder has a good reputation.

Battery Service

Batteries require care, particularly for winter starting. A battery at 0° F. develops only forty percent of the power it would (in the same state of charge) as at 80° F.

If your battery is a conventional type, periodically check the water level and add water as required. It is not necessary to use distilled water unless the tap water in your area is unusually hard.

Exterior cleaning. Clean the top of a conventional battery with a solution of baking soda and water (one tablespoon baking soda to an eight-ounce glass of water) whenever it is dirty.

If the battery hold-down bracket is corroded away, or if it has loosened, the battery may vibrate as the car moves down the road. This vibration could crack it and allow fluid to leak out. The battery then would be ruined. Wire-brush corrosion off the bracket, coat it with corrosion-resistant paint, then tighten securely. If the corrosion is severe, replace the hold-down.

State of charge. The state of charge of the battery is extremely important in cold weather. A substantially discharged battery could freeze internally and crack apart. If the battery is a no-maintenance type without fill caps, but has an external indicator, you can make an easy visual inspection. If it is the conventional type, you can check its condition with a hydrometer or an electronic tester.

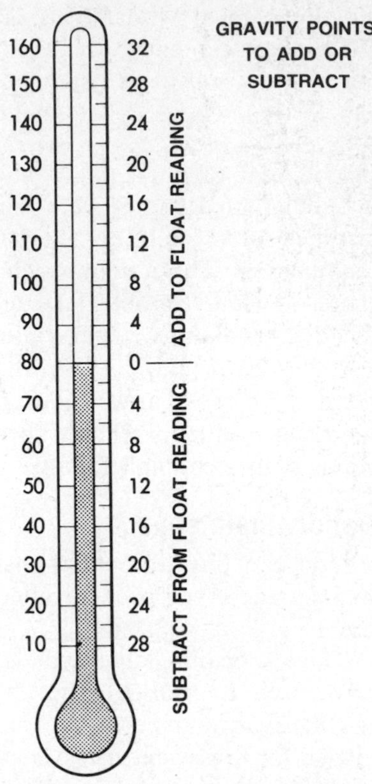

Fig. 9–6. Temperature correction chart for battery hydrometer readings. If you have a reading of 1.220 and the battery temperature is 80°F., you make no change. If it's 100° you add 8 to raise the number to 1.228.

The hydrometer is a glass device with a float built in. Just remove a cell cap, draw out a small sample with the syringe top, and read the float at the surface line of the fluid. If it is below 1.220 (see temperature correction chart, Fig. 9–6), the battery needs a recharge. If there is a difference between high and low readings of more than .025, the low cell is defective and the battery must be replaced.

The sealed-top battery may have a built-in hydrometer (see Fig. 9–7). This design is often misinterpreted. Here is what the appearance of the plastic circle in the top of the battery means:

• *Green*. A green dot or any green appearance means the battery can be tested, such as with a voltmeter. It does not mean the battery is good. The hydrometer goes into just one cell, and if another cell is bad, the battery won't work prop-

Fig. 9–7. This sealed-top battery has a temperature-compensated hydrometer built through the top into one cell.

erly even though the green dot is there.

• *Dark; green dot not visible.* If there's a cranking problem, it may be caused by a defect in the electrical system, including the charging system, as well as the battery. Check these too.

• *Yellow or clear.* This has often been interpreted to mean the battery should be replaced. Actually, it means that the battery may not be in great shape, but may well be able to continue in service. If, however, there is a cranking problem that can't be obviously traced to bad cables or a defective starter, replace the battery.

If these sound a bit fuzzy, you must understand that the sealed-top battery can only be checked accurately with equipment that load-tests it, and this is equipment the weekend mechanic will not find to be cost-effective. If you have a battery with removeable caps, however, you can test it with a low-cost electronic tester.

With the electronic tester, you simply insert the probes into adjacent cell holes and the tester should light. If it does not, the cell is low. If all cells are low, try a recharge. If only one is low, the cell is apparently defective.

Recharge. To recharge a battery you need a charger that plugs into household current. A small charger (maximum 10 amps) will do the job, given enough time, and inasmuch as slow recharging is best for the battery, even very small chargers are not necessarily a bad investment. A 24- to 48-hour recharge with a four-amp charger, for example, will do just fine.

If after a slow recharge all cells do not read in the normal range, replace the battery. Note: All recharging should be done with the caps off, cells topped up with water, and adequate ventilation.

Cleaning cable terminals. One of the best things you can do for a battery is clean and tighten the cable terminals and the battery posts. Connections that are clean and tight will permit easier current flow and make cold weather starting much more feasible.

To remove the cables on top-post batteries slacken the nut that secures the cable terminal to the battery and twist it back and forth with an upward motion. If the terminal is stuck, tap it back and forth with a hammer to free it up. Still having trouble? Don't try prying or yanking, or you could ruin the battery post and the battery. Buy a terminal puller, an inexpensive tool that makes cable terminal removal safe and easy (see Fig. 9–8).

With a wire brush, clean corrosion from the inside and outside of the cable terminal and the entire surface of the post (see Figs. 9–9, 9–10). Refit the cable terminal and tighten securely. Spray the connection with a rust preventive lubricant, such as Auto-Pak 4-Way, and the job is done.

If the cable terminal is distorted and will not tighten properly, buy a new terminal, cut off the old with a hacksaw, and install the replacement (see Fig. 9–11) on the cable (attachment is an obvious procedure). This is much easier than replacing an entire cable, particularly the cable that goes to the starter motor.

Although *side-terminal batteries* are less susceptible to corrosion, the smaller contact surface between the cable and the battery terminals means that even a small amount of corrosion, perhaps not even visible to the naked eye, can prevent a car from starting. Remove the retaining screw, separate the cable and battery terminals, wire brush the contact surfaces carefully (see Fig. 9–12), and then refit and tighten. Although this type of terminal has a very long life, if necessary it also can be cut off the cable and a replacement installed.

Alternators and Generators

Charging System Service

The alternator or AC generator is the heart of the charging system. Driven by a belt from the engine crankshaft, it converts the mechanical energy of rotation into AC current. The automobile uses direct current (DC), so the AC is passed through electronic check valves—called diodes—which convert the AC to DC. The DC current then flows through cables to operate the electrical system and recharge the battery.

To control the output of the alternator, a device called a voltage regulator is used. On most cars the regulator is an electronic device, in many cases built into the back of the alternator. On older cars it is a mechanical device with contact points that close to permit current flow and open when the battery is charged and no current is needed. In actual practice, the points vibrate open and closed so fast that you couldn't even see that it was happening. With an electronic regulator there are no moving parts, and transistors electronically open and close the circuit from the alternator.

Two things can happen to the charging system: It can produce too much current or too little. If it produces too much, the battery will overcharge and water will "gas" out of the battery. If it produces too little, the battery will run down, and in the extreme case (no current at all) the car won't start or will stop running once the battery is dead.

• *Overcharge*. If your battery runs dry very often, overcharge is the likely cause. The easiest way to check this is with an electrical evaluator, a tester sold at auto supply stores (see Figs. 9–13, 9–14). You use it right at the battery or alternator and simple lights (actually voltage sensitive light-emitting diodes—LEDs) tell you whether the system is charging normally, overcharging, or undercharging. Overcharge is always caused by a defective regulator and the cure is to replace it. If your car does not have a separate regulator, you often must replace the alternator. Both cases and their solutions are explained later.

• *Undercharge or no charge*. The more common problem is that the alternator does not keep the battery fully charged or does not provide any charge at all.

First, check the drive belt. A snapped belt will result in no charge. A slipping belt will result in undercharge. See Chapter 7 for details on belt service.

Next, if the car has a separate voltage regulator, the problem may be caused by a defective electrical ground. The ground is provided by the regulator mounting screws, which thread into the body. As a simple test, connect a jumper wire (a piece of wire with an alligator clip at each end) from one of the mounting screws to another suitable grounding point, such as a metal screw on the engine. If the charging system now operates properly, remove both mounting screws, wire-brush the threads, reinstall, and tighten securely. If brushing the threads doesn't work, make up a grounding wire from a piece of wire and a pair of crimp-on terminals (see Fig. 9–15).

Alternator Output Test

If the problem is not in the regulator ground or drive belt, it is either in the alternator or regulator. To make a suitable test, you will need an ammeter. Proceed as follows:

(1) Disconnect the battery ground cable, the one that is bolted to the car body and/or engine block (see Fig. 9–16).

(2) Disconnect the wire from the armature terminal (also called battery terminal) on the back of the alternator (the correct terminal is the one with the thicker wire, and may have an A, ARM, or B or BAT embossed next to it). Connect the wire from the ammeter positive terminal to the alternator terminal, and the wire from the ammeter negative terminal to the wire you disconnected. If you have a combination amp-voltmeter, the ammeter section likely has just a

Fig. 9–8. Using terminal puller on top post battery.

Fig. 9–9. Using a wire brush to clean inside of battery cable terminal. Also clean exterior of terminal.

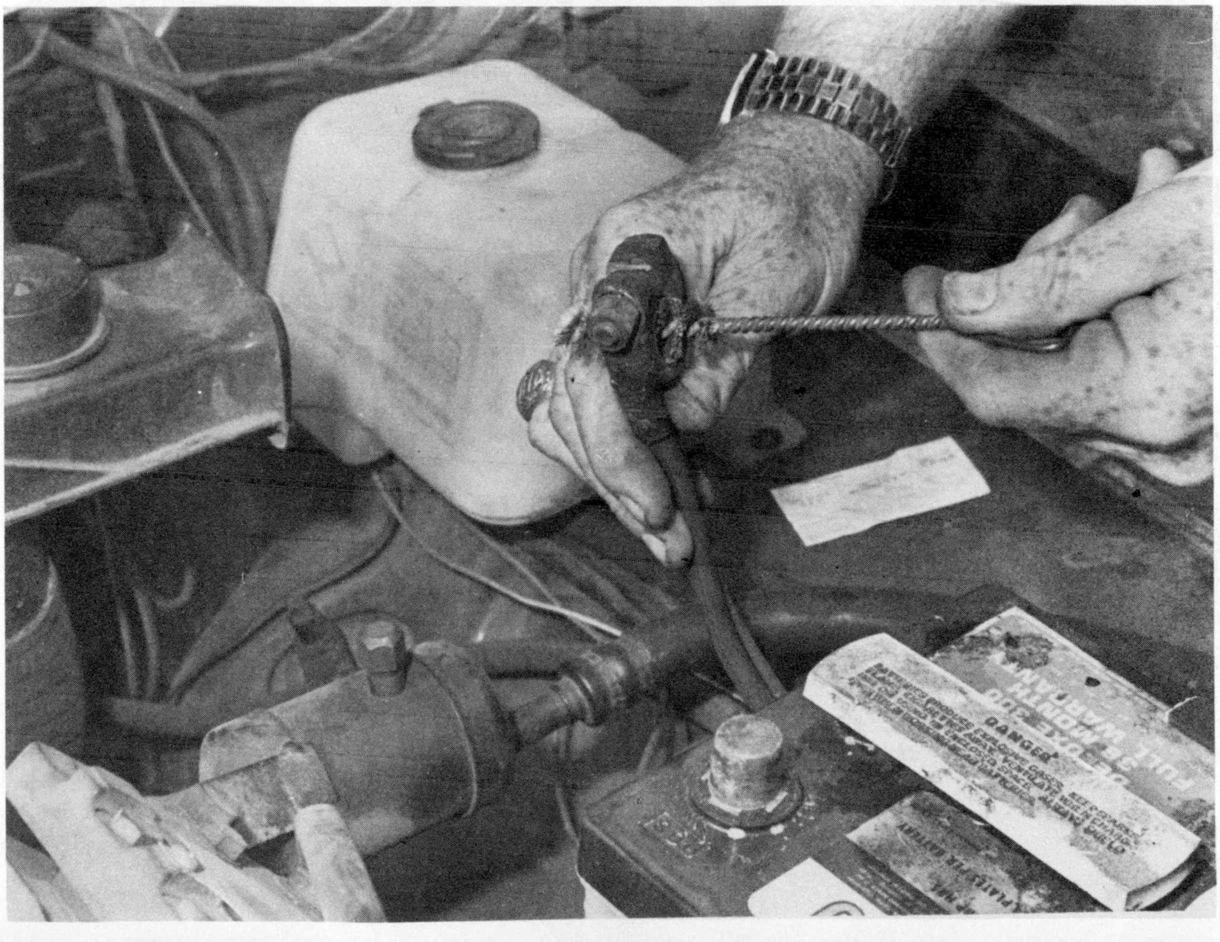

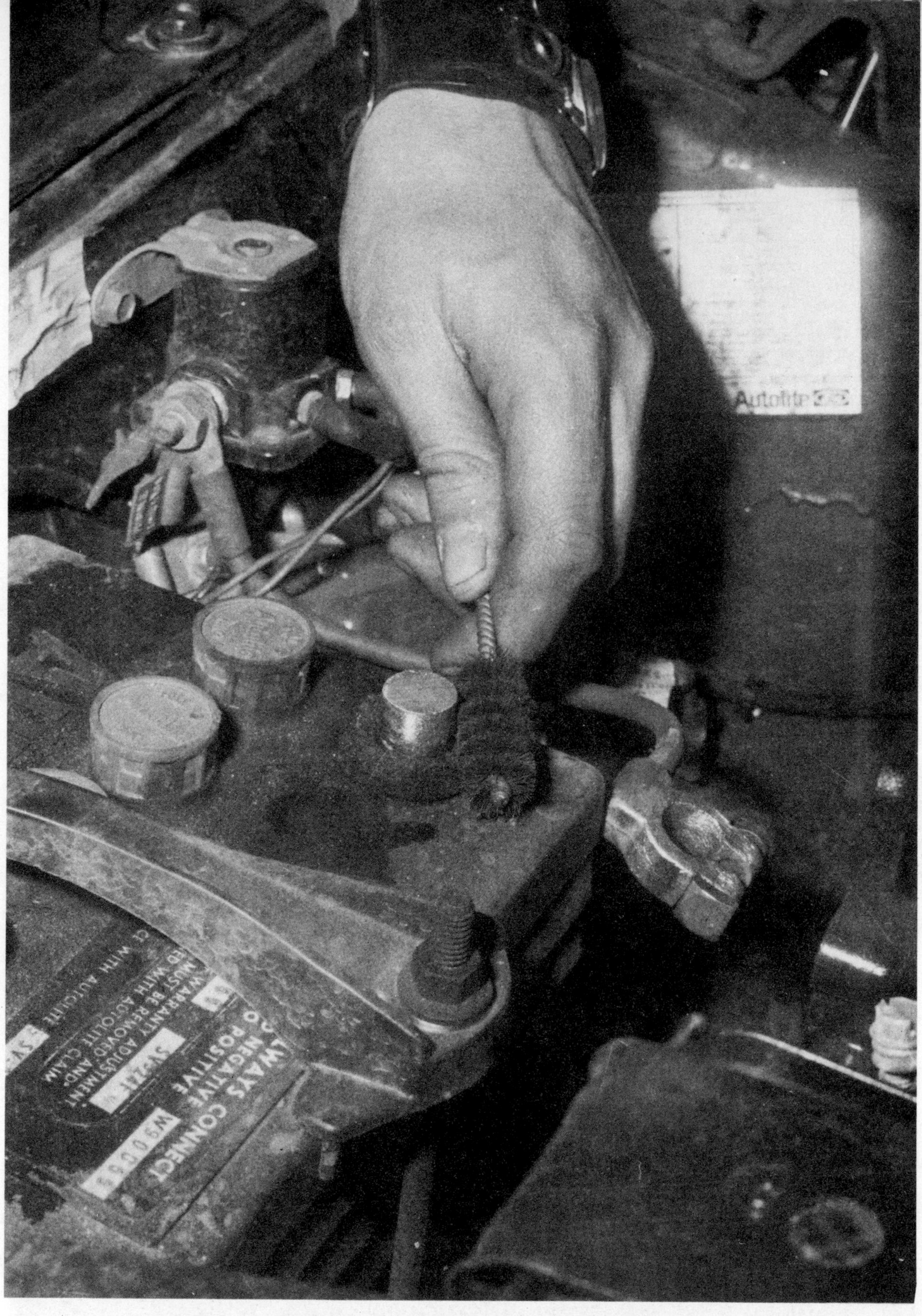

Fig. 9–10. Using a wire brush to clean the battery post.

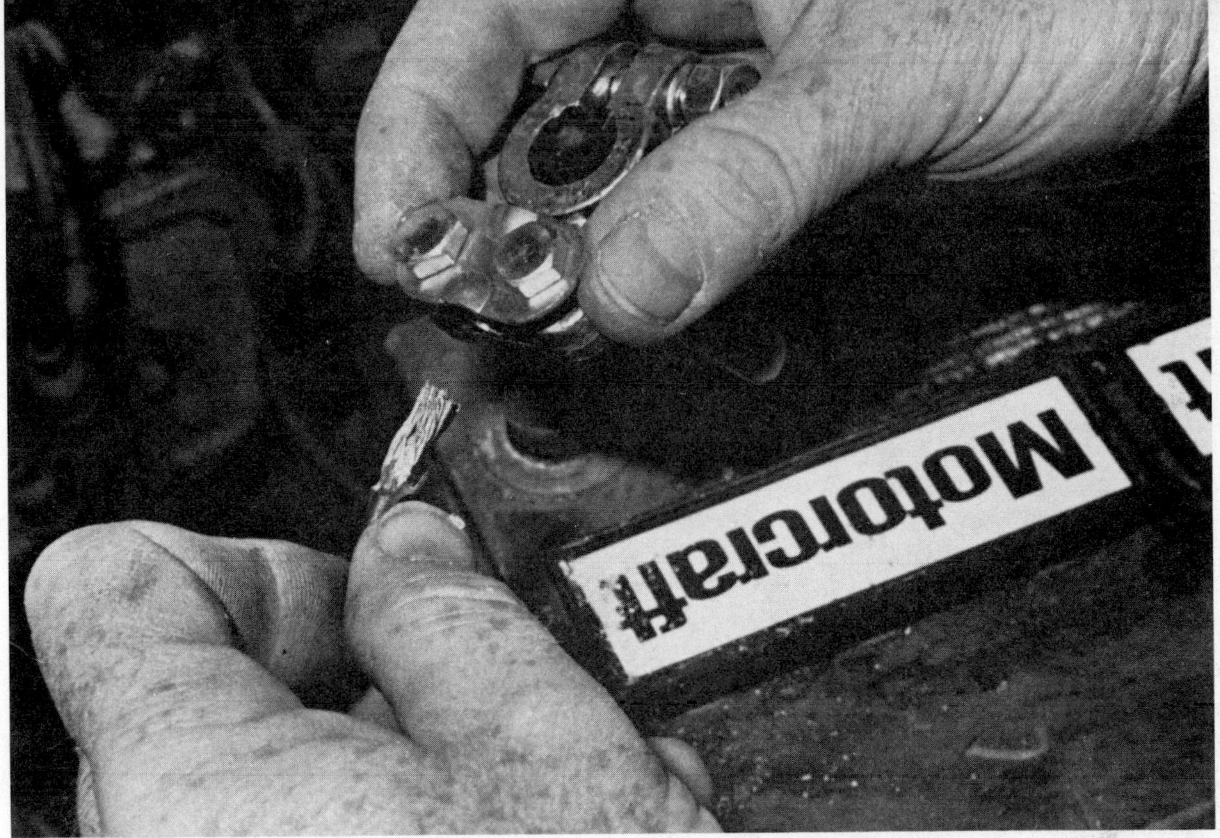

Fig. 9–11. If a battery cable terminal is corroded, cut it off, bare the end of the wire, and fit a replacement terminal. The bared wire goes between the plate (part with two screws) and the base of the terminal.

Fig. 9–12. Clean the disc contact surfaces on cable terminal and battery with a wire brush.

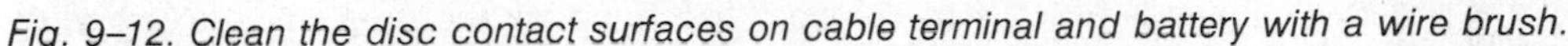

Fig. 9–13. Here, a pencil-type charging system tester is used at the battery.

Fig. 9–14. This tester is easily connected at the alternator. For most tests, no disconnections of car wiring at the alternator are necessary.

single lead with a rectangular tab—called a shunt—at the end. The shunt tab is shaped so that one end goes on the alternator armature terminal and the other attaches to the wire you disconnected from that terminal. Make these connections, then move the meter switch to the AMPS position.

(3) Disconnect the thin wire from the field terminal (marked F or FLD on the alternator) and connect a jumper wire from the armature terminal to the field terminal as shown in Fig. 9–17. On cars whose alternators have built-in regulators, ground out the regulator with a screwdriver or special tool into the hole in the back of the alternator (see Fig. 9–18). Reconnect the battery ground cable. Turn on all accessories.

(4) Start the engine, rev it for an instant and the ammeter reading should reach or exceed the rated output, which is a minimum of 35 amps on smaller cars. Shut the engine down immediately and turn off accessories.

If the alternator output is not satisfactory, replace the alternator. If output is normal, replace the voltage regulator (on those with an external type).

Replacing the Alternator

Replacing an alternator is a simple procedure. Begin by disconnecting the battery ground cable and any wiring at the back of the alternator.

The typical alternator is held by two bolts, one of which passes through an adjusting bracket. Loosen both bolts (see Figs. 9–19, 9–20), then pivot the alternator in toward the engine and disconnect the drive belt from the pulley (see Fig. 9–21). Remove both bolts and lift out the alternator (see Fig. 9–22).

Obtain a factory rebuilt alternator (using your old one as an exchange unit) from a parts store that serves professional mechanics. You could

Fig. 9–15. Installing regulator grounding wire. Although this step is rarely necessary, it may solve a problem if the regulator mounting bolts are not providing a good electrical ground.

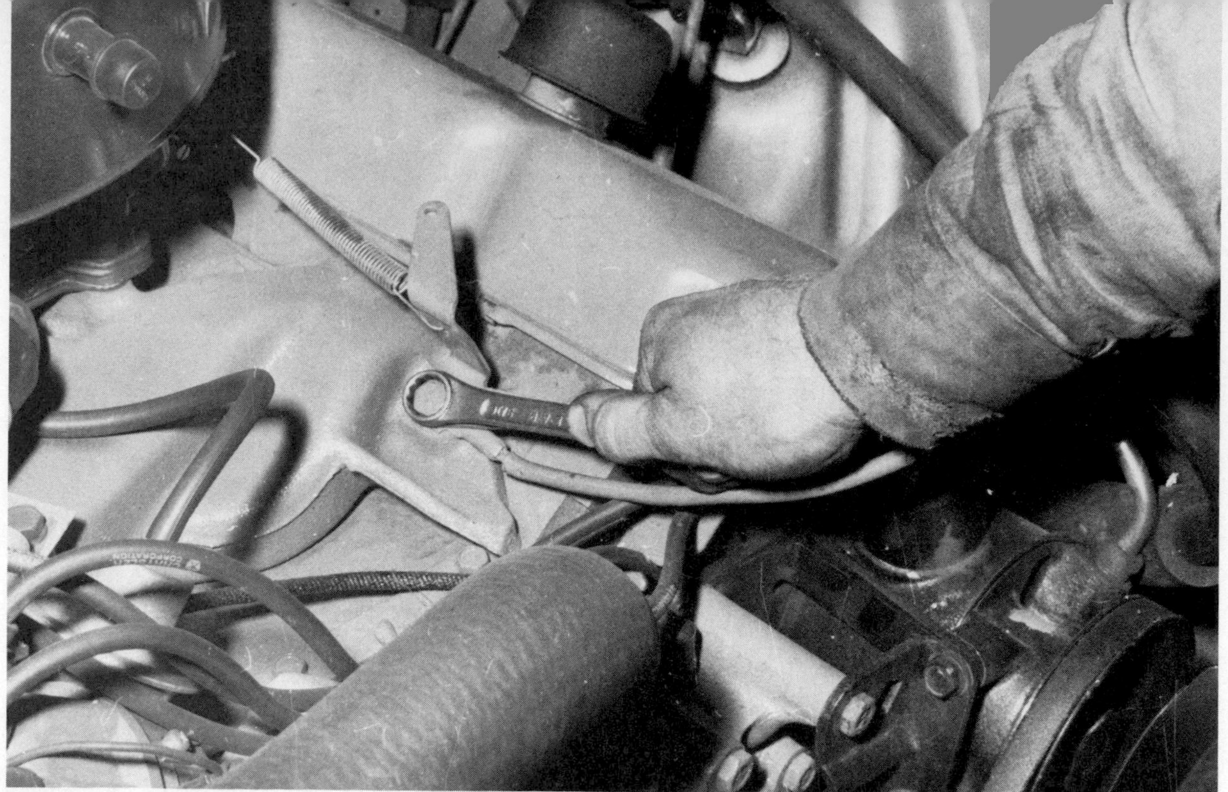

Fig. 9–16. Disconnecting battery ground cable from engine. Note: Most ground cables are either also grounded to the body, or are split into two wires, one that grounds at the engine, the other at the body.

Fig. 9–17. Field and armature wires are disconnected from alternator and jumper wire is connected to the alternator's armature and field terminals. The test instrument used is a combination ammeter-voltmeter that has only a single wire lead with a "shunt tab" for ammeter instead of positive and negative wire leads for an ammeter-only instrument.

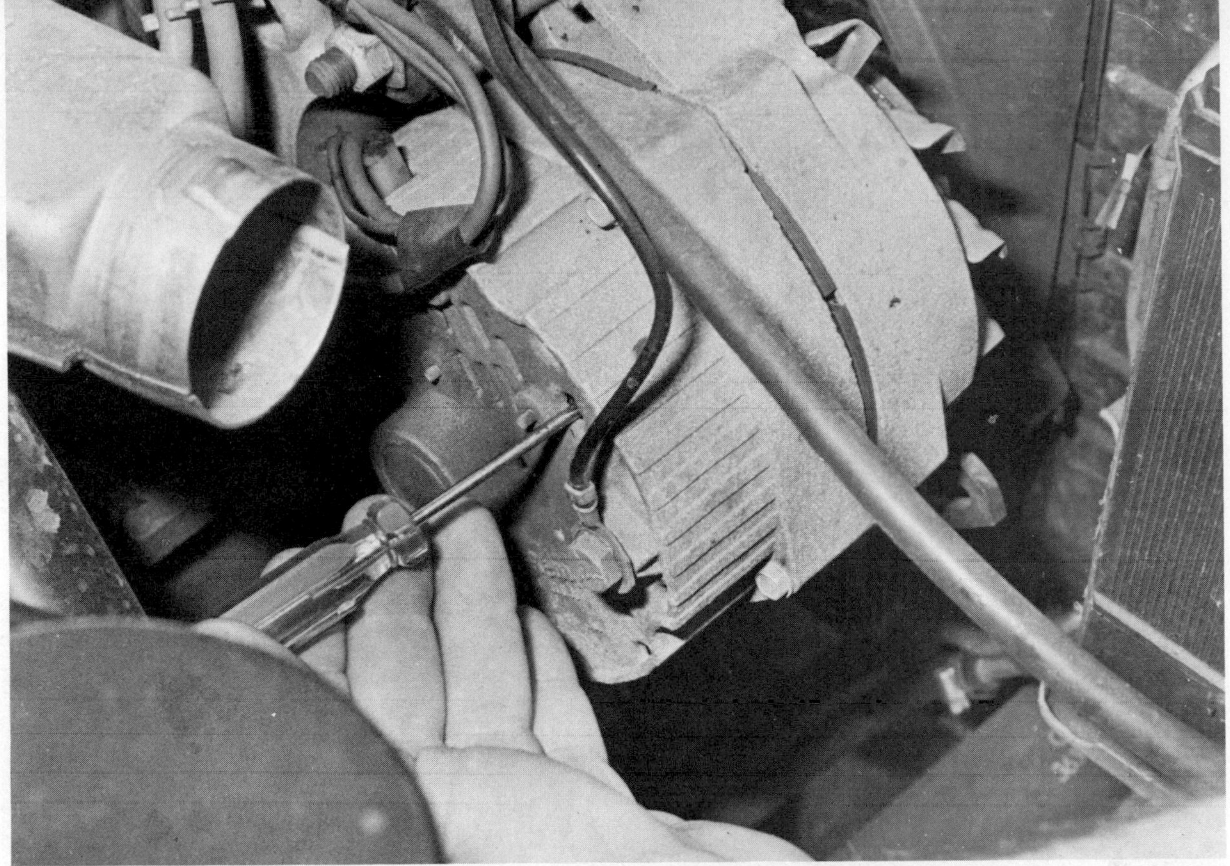

Fig. 9–18. Inserting screwdriver into hole in back of Delco (GM) alternator with built-in regulator. When the screwdriver tip comes into contact with tab, lean the shank of the screwdriver against the alternator body. Regulator is now eliminated from charging circuit, so if ammeter output now is normal, your problem is the regulator.

get a new alternator, but the price will be at least three times higher, with a long wait for delivery.

Install the rebuilt alternator with both mounting bolts, but do not tighten. Engage the drive belt (note: make sure it is properly installed on all pulleys, not just the alternator one), then adjust drive belt tension as explained in Chapter 7.

Connect wiring at the back of the alternator, and refit the battery ground cable.

Replacing the Regulator

If the alternator passes the output test, your problem is a defective regulator. To replace, simply disconnect the battery ground cable and the electrical connector at the regulator. Remove the regulator mounting screws and install a new regulator and the screws. Reattach the regulator electrical connector and then reconnect the battery ground cable.

On alternators with built-in regulators, either obtain a rebuilt or take the alternator to an auto electrical repair shop for disassembly and installation of a new regulator.

Starter

When you turn the key to crank the engine, what you are actually doing is completing a circuit from the battery to an electric motor and an electromagnetic switch on top of the motor.

When this switch (called the solenoid) receives current, it closes, pushing a fork or lever. The fork or lever moves a little gear—the pinion—on the end of the motor shaft into engagement with another gear—the ring—on the engine's flywheel (see Figs. 9–23 to 9–25).

The solenoid serves a second purpose. When it closes it completes a thick wire circuit (thick wire carries more current) from the battery to the motor portion of the starter.

As the motor's shaft turns, the little pinion gear turns the big flywheel ring gear, and that's

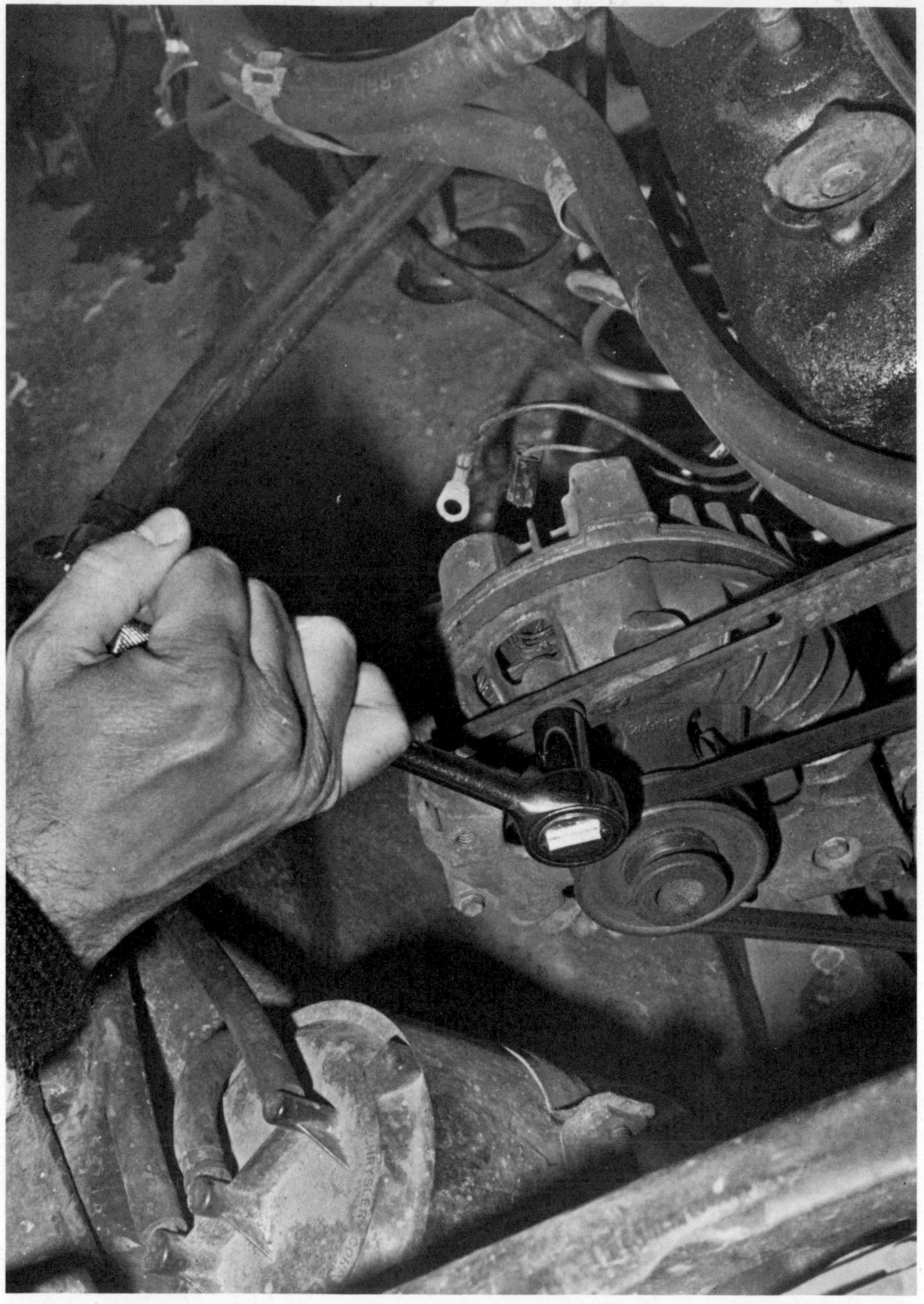

Fig. 9–19. Loosening alternator mounting bolt that passes through slot in adjusting bracket. Remove this bolt.

Fig. 9–20. Loosening second alternator mounting bolt. Remove this bolt, too.

Fig. 9–21. With both mounting bolts out and wiring disconnected from back of alternator, slip belt off the pulley.

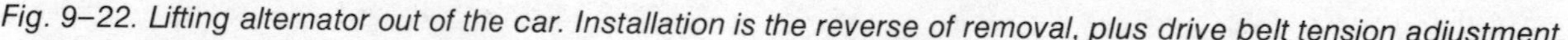

Fig. 9–22. Lifting alternator out of the car. Installation is the reverse of removal, plus drive belt tension adjustment.

what starts the crankshaft and all other parts of the engine into motion.

When the engine finally runs on its own, the flywheel spins so comparatively fast (even at idle speed) that its gear tries to drive the little gear on the starter. If this were allowed to happen, the starter would quickly be damaged.

To prevent this, a two-section device called an overrunning clutch is used. When power is supplied by the starter, the front half of the overrunning clutch locks onto the rear half and the power of the starting motor is transferred to the little gear. When the little gear is running faster than the starter (because the flywheel is spinning it), the front half of the overrunning clutch automatically disengages from the rear half. All this action takes place in an instant, because once the engine starts the normal driver releases the key. Immediately, the starter solenoid disengages and the little pinion gear retracts from the ring gear on the flywheel.

You can visually identify this starter by the solenoid on top (see Fig. 9–26) and by the two-wire connections, a thin wire to trigger the solenoid and a thick one to carry current for the starter motor.

• *Other starters.* In another popular design, the overrunning clutch is used, but instead of a solenoid another type of electromagnet is used to move the fork that pushes the little pinion gear into mesh with the flywheel ring gear.

This type has only a single electrical connection to the starter. With this design, widely used on Ford and American Motors products, there is a device called a relay, mounted close to the battery. When you turn the key you actually complete a circuit to the relay, which is another form of electromagnetic switch. The relay closes and completes a circuit from the battery to the starter motor (see Fig. 9–27).

Ford and AMC also use a solenoid starter on some models, but with the relay. In this case, the relay closes and completes a circuit from the battery to a starter motor terminal on the solenoid. A metal link connects from this terminal to the solenoid terminal to simultaneously activate the solenoid.

Most Chrysler Corp. cars also use a starter relay, although it differs from the Ford–AMC type (as explained later).

Starter Problems

The only function of the starter is to crank the engine. If the engine cranks reasonably well (as a driver, the normal sound should be familiar to you), any failure to start is not the fault of the starter.

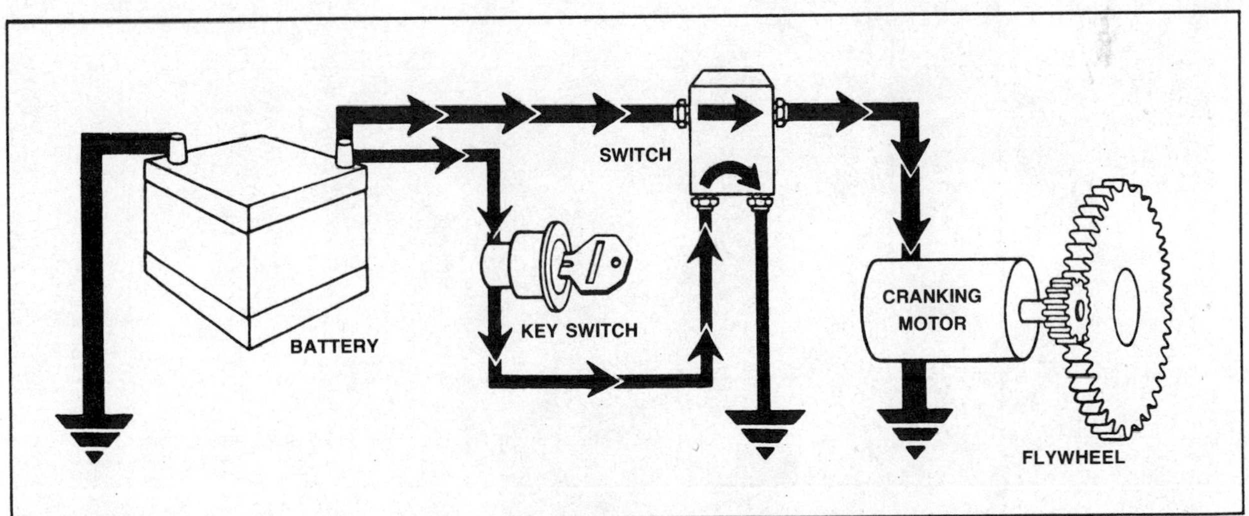

Fig. 9–23. This is a schematic of the starting circuit. When key is turned, it closes a switch that completes a circuit from the battery to a second switch (actually called a solenoid or relay). When the solenoid or relay is energized, it closes and completes a second circuit (capable of carrying more current) from battery to solenoid or relay to the cranking (starter) motor. The starter spins a little gear that turns the large gear on the engine flywheel.

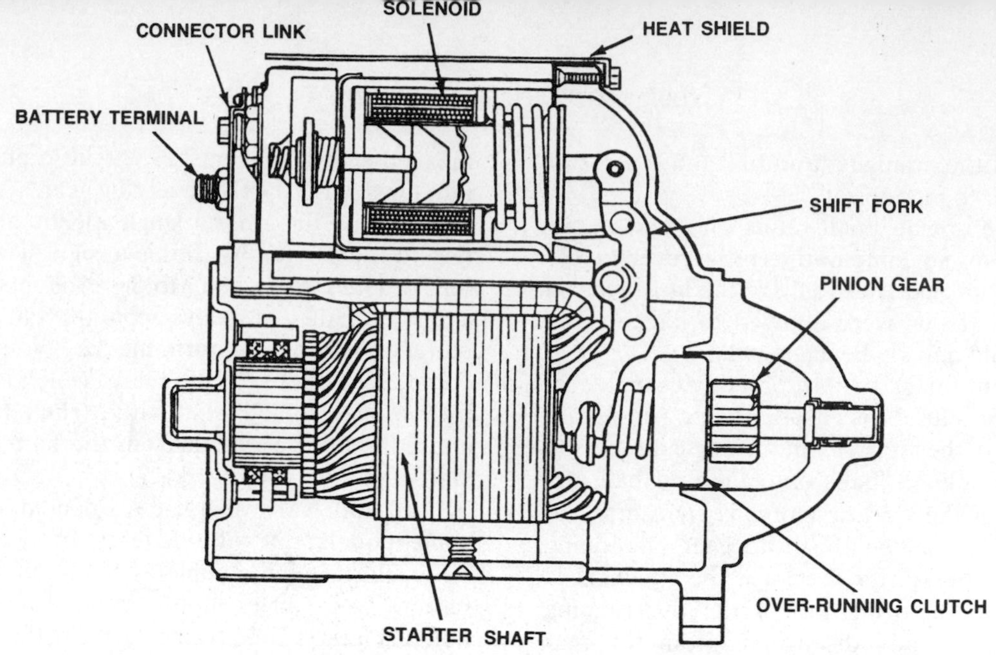

Fig. 9–24. Cutaway view of a solenoid-type starter motor. When the solenoid is actuated, it not only supplies current to the starter motor but also moves a shift fork that pushes the pinion gear forward into mesh with the engine flywheel gear.

Fig. 9–25. A rear view of the engine with transmission removed shows flywheel and starter motor.

FLYWHEEL RING GEAR

STARTER MOTOR

Fig. 9–26. External view of starter motor with solenoid atop it.

The following may or may not be starter problems:

• *Engine cranks very slowly.* This could be the starter, a weak battery, corroded connections at the battery, poor cable connections at the starter or the battery ground on the engine, very thick engine oil, or a defective engine.

• *Starter spins but the engine does not crank.* If you hear a fast whirring noise from the starter, but not the familiar sound of the engine cranking, the starter is either not in engagement with the ring gear or the overrunning clutch is not holding. Replace the starter.

• *Engine does not crank at all.* This could be a starter problem, but might also be caused by a dead battery, poor cable connections, or a defective neutral safety or clutch switch. On all cars with automatic transmission, a neutral safety switch is installed to prevent cranking the engine in anything but neutral or park positions. If this switch is maladjusted or fails, the starting circuit is not complete. Many manual transmission cars have a switch on the clutch linkage so the engine will crank only if the clutch pedal is fully depressed. If this switch is maladjusted or fails, the effect is the same as a neutral safety switch problem on an automatic.

• *Erratic and/or noisy cranking.* If you get unusual noises and erratic cranking, the problem might be in the starter or a ring gear that is loose on the flywheel (so that it turns but the engine doesn't crank or cranks erratically).

Troubleshooting Techniques

• *Slow or no cranking.* If the battery and cable connections are good, check the starter current draw with an induction ammeter, an inexpensive little tester that clips onto the battery cable to the starter (see Fig. 9–28). If the current draw is more than 150–300 amps in warm weather, up to 375 in cold weather, it is excessive (exception is the diesel engine; check manufacturer's specifications, for they may be much higher).

• *High current draw.* High current draw may be caused by very thick engine oil. Do you have a sludged up engine from poor oil change practice? Do you have a couple of cans of engine oil thickening additive in the crankcase? Or could the engine itself be the problem?

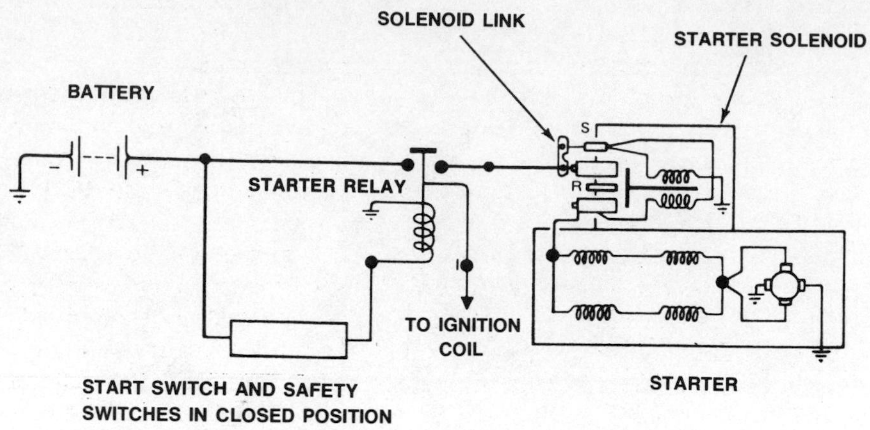

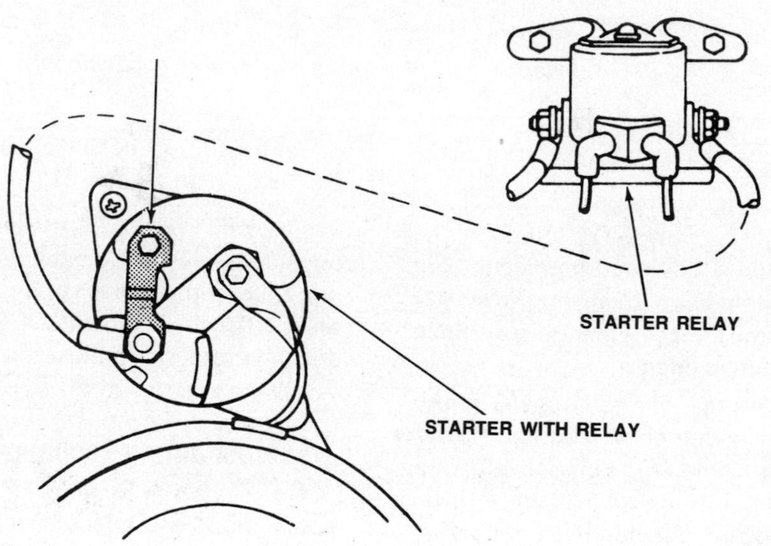

Fig. 9–27. Schematic shows starting circuit with a relay system and a solenoid starter, too. Drawing shows the metal link from the solenoid's starter motor terminal to a second terminal, which is the current source to close the solenoid. When the solenoid closes, current can flow through the cable terminal to operate the starter motor.

Remove the starter (as explained later) to check it out. With the starter removed, test it with a good battery and jumper cables. Connect one booster-type jumper cable from the starter frame to the battery ground post, and another from the battery starter terminal to the thick wire terminal on the starter or solenoid. Have an assistant help you hold the starter in place, and connect a thin jumper wire from the thick starter terminal to the solenoid S terminal. The starter should energize and you should see the pinion gear move forward and spin enthusiastically. If it hangs up or if the starter turns sluggishly or not at all, replace the starter.

If the starter spins well outside the car, but not when installed on the engine, the problem is caused by either: (1) excessively thick engine oil; (2) defective battery cable(s); (3) if a battery other than the one in the car was used for the test, perhaps the battery is the cause.

Test the battery state of charge as described earlier in this chapter. Inspect the cables for corroded terminals, damaged insulation (on the cable to the starter), frayed or broken wires (on an uninsulated ground cable) and loose connections.

• *Normal current draw*. If the current draw is normal, remove the starter and look at the pinion and ring gear teeth (you will have to move the flywheel with a screwdriver on the ring gear to see all the teeth). If you see damaged teeth, or if the flywheel ring gear is loose, the transmission must be removed, the flywheel taken off, and a new ring gear pressed on. This is a job for a professional.

• *No current draw*. If there is no current draw, the starting circuit is not complete. Check the connections from the battery, and if they are complete make this basic test on all cars with solenoids on the starter: run a jumper wire from the battery to the solenoid's S terminal. If the engine cranks, you know that the solenoid, starter, and battery are okay. The problem is in the neutral safety switch (on automatic transmissions), the clutch switch (on some manual transmissions), the key switch, or possibly the relay (on cars with a relay).

If the engine does not have a solenoid on top of the starter, such as Ford and AMC cars with the relay, a similar test can be made.

• *Starter relay (Ford, Chrysler, AMC)*. If your car is one with a starter relay, it makes a good starting point simply because it's normally quite accessible in the engine compartment. Tests at

Fig. 9–28. Using an induction ammeter that clips onto the battery starter cable, you'll find the amperage draw was 300 in cool weather, within specifications.

the relay also may pinpoint problems in the key switch and clutch or neutral safety switch.

With the Ford-type relay (see Fig. 9–29), disconnect the thin wire and connect a jumper wire from the battery starter terminal to the thin wire terminal of the relay. If the engine now cranks, the problem is in the key switch or the neutral safety/clutch switch or their wiring. Pinpointing the problem any further requires you to locate the neutral safety or clutch switch, remove the wiring connector, and attach a jumper wire across the connector's two terminals. If the engine now cranks with the key, the neutral safety/clutch switch is the problem. If it doesn't, the key switch is at fault.

Note: Ford products with steering column levers do not have a neutral safety switch (a lock in the column prevents turning the key except when the lever is in neutral or park). Only console shifts have the switch. Clutch and neutral safety switch locations will be discussed later in this section.

If the engine doesn't crank with the jumper wire from the battery to the relay thin-wire terminal, the relay itself is apparently defective. Double-check by connecting a booster cable from one relay thick-wire terminal to the other (these are the battery cable terminals). If the engine now cranks, you have confirmed that the relay is the problem.

If there are two thin-wire terminals, the S terminal is the one you want.

Some AMC cars use the Ford-type relay, and if there is a light blue wire with white tracer, it goes to the neutral safety or clutch switch. To bypass that switch, ground the terminal to which this wire connects, and if the engine now cranks, the neutral safety or clutch switch is the problem. With the other test procedures, you can isolate a key switch or relay problem. Note: On some Ford-type relays, the second terminal wire carries current to the ignition coil. Do not confuse this with the AMC blue/white wire to the neutral safety or clutch switch.

On Chrysler Corp. cars, the starter relay is usually on the driver's side of the firewall (the sheet metal rear of the engine compartment). Run a jumper wire from the battery starter terminal to this relay's I terminal, and if the engine cranks the key switch is the problem. Connect a

Fig. 9–29. Ford-type starter relay.

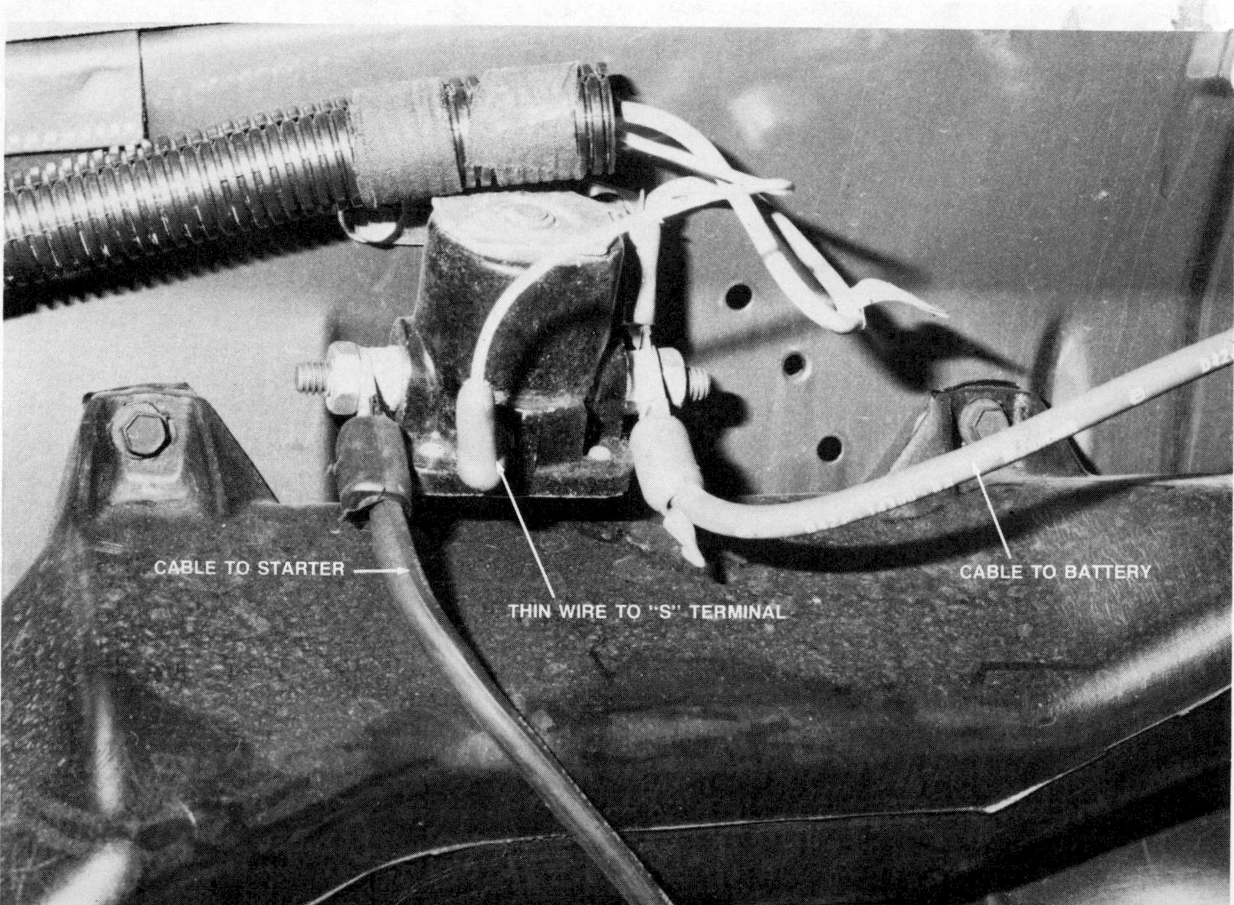

jumper wire from the relay G terminal to ground, and if the engine now cranks the neutral safety or clutch switch is at fault. If the engine still doesn't crank, run a jumper wire from the battery starter terminal to the starter solenoid's S terminal, and if the engine now cranks the relay is defective.

• *Neutral safety/clutch switch locations.* The clutch switch is usually under the dashboard and is the only switch you will find that bears against the clutch pedal linkage. If it has two wires, disconnect them and join them together with a jumper wire to bypass the switch. If it has only one, connect it to an electrical ground with a jumper.

The neutral safety switch on console floor shifts is typically under the shift lever cover, which may be held by just four to six screws. On steering column cars, if it is used, it is somewhere on the column near the bottom, under the dash or under the hood. It may be distinguished by the fact that it moves when you move the shift lever.

Note: On many cars you can tell if it's bad without much trouble, for it commonly also operates the backup lights and, on some cars, the warning light that goes on if you put the car into Drive without your belts fastened. Should these indicators fail simultaneously, the neutral safety switch is obviously the problem.

If not, look for the switch on the column on GM and older Ford products. It is threaded into the transmission on Chrysler and AMC cars, but can be tested at the easy-to-reach starter relay on those models.

Caution: For safety, only crank the engine when the transmission is in Neutral or Park, even though it would crank in the other positions with the switch bypassed. Have a mechanic replace the switch as quickly as possible.

• *Key switch.* If you bypass the neutral safety switch and can't crank the engine after the test procedures outlined earlier, the key switch is suspect. Replacing it usually is a job for a professional, however. Note: On many cars the switch is not in the key assembly, but at the base of the steering column, operated by a rod from the key lock assembly. On this type, widely used on late models, you can replace it. Consult a service manual for your car to find out how.

• *Replacing a starter.* In many cases, the solution comes down to replacing the starter. Whether or not you can replace a starter depends entirely on its physical location. On some cars, starter removal is blocked by other components which must be taken off first. On others, you just remove two or three bolts (see Fig. 9–30) and lift the starter out. You will have to look at the starter and its location and decide if you can do the job.

On all but a few cars, the car must be jacked up and the starter removed from underneath.

Always begin with the battery ground cable disconnected. Next, disconnect wiring at the starter, then remove the bolts and take out the starter for testing and/or replacement. You will undoubtedly select a rebuilt starter because of its much lower cost and easier availability. Have it bench-tested before installation.

Horn Service

Horns are prone to any one of three problems; they won't blow, they won't stop, or, when they blow, the sound is weak. Correcting these faults is usually something a weekend mechanic can handle.

You will be able to do a job on the horn circuit much more easily if you know it before something goes wrong. Learn it now, beginning with the basic operation (see Fig. 9–31).

The current circuit is from the fuse box to a type of electromagnetic switch called a relay. The circuit cannot be completed to the horn itself until the relay is electrically grounded, which is done by the horn switch in the car. When you press the horn button you complete the circuit by providing an electrical ground for the relay.

The current then flows through the relay to the horn, a spring-loaded diaphragm device operated by a pair of electrical contact points that are part of another electromagnet.

The contact points are normally closed, and when current flows through them an electromagnet is formed that pulls them apart, breaking the circuit. The diaphragm spring then pulls them back together and the electromagnet is again formed. This happens many times per second, and each time the diaphragm flexes it makes a sound. A cone-shaped metal section of the horn

Fig. 9–30. Remove mounting bolts and disconnect wires to starter to remove it. Battery ground cable should be disconnected first.

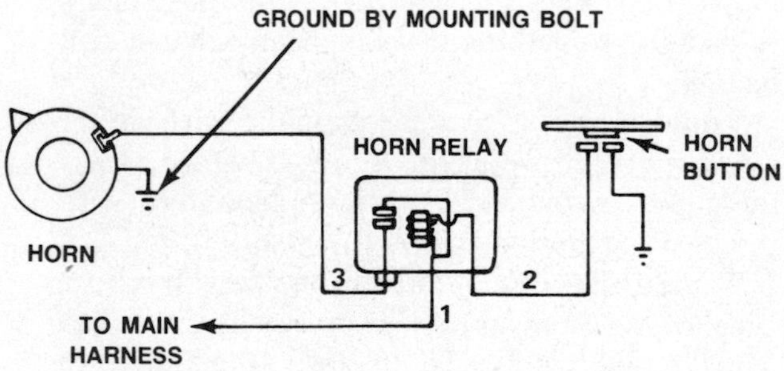

Fig. 9–31. Typical horn circuit with relay. When you press the horn button, you complete an electrical ground for the horn relay. Current then flows through the relay to the horn, which is electrically grounded by its mounting bolt.

amplifies the sounds, which occur so fast that it seems like a single, continuous sound.

Component Locations

To check out a horn circuit you should know where the components are located. You can find your horn (if you do this when it works) by having someone press the button while you trace the noise. Or you can just look under the hood at the front of the engine compartment.

The horn relay is usually, but not always, plugged into the fuse box or is taped to the wiring harness somewhere near the box. If you are foresighted enough to check it out before it goes bad, just disconnect the wire from the horn and have someone press the horn button. You'll hear a click, which is the operation of the relay. Just trace it to a little metal can—the relay.

The horn button can be built into the steering wheel or a multipurpose stalk switch on the steering column.

Troubleshooting Horn Problems

• *Horn won't stop blowing*. This is a maddening problem because you've got all that noise. First, disconnect the battery ground cable so you can collect your thoughts. Next, disconnect the horn and reconnect the battery cable, so although the horn won't work you can do some testing.

First, press the horn button, and if you don't hear the relay click the problem is a stuck relay or a defective horn button.

You can determine which it is by having a helper disconnect the battery ground cable, then reconnect it as you listen at the relay. If you hear the relay click when the cable is touched to the battery post, the problem is a stuck horn button. If you don't, the problem is a stuck relay. If you wish, you can double-check as follows:

Remove the relay and test it alone at the battery with jumper wires (you'll need two). Connect the relay terminal marked B or 1 to the battery starter terminal, the one marked S or 3 to an electrical ground. You should hear the relay click. If you don't, replace it.

If the relay tests good, the problem is in the horn button. On the rim-blow type, the cure is to replace the steering wheel, but that's too expensive. Just get to the wire from the relay's S or 3 terminal and clip it off near the relay. Splice on a new piece of wire and run it along the steering column to a replacement horn switch, which clamps onto the top of the steering column jacket. Run a second wire from the auxiliary horn switch to a suitable electrical ground, such as a metal screw into a metal part under the dashboard.

Other horn buttons may be replaced less expensively, or you could just remove the button section from the wheel and look for a bent tang making contact and causing the problem. That's something you might fix with a screwdriver. If you can't figure out how to fix the switch, you can, as with the rim blow type, just run a new switch. Note: When splicing in a new piece of wire, use a crimp-together electrical connector, available at auto supply stores, and cover the connection with plastic electrical tape.

• *Horn won't blow*. This could be caused by a defective horn switch, bad relay, or defective horn. To check the horn itself, just disconnect the wire and run a jumper wire from the horn terminal to the battery starter terminal. If the horn now blows, it is good.

Next, with the horn still disconnected, push the horn button and listen for a click. If you don't hear one, remove the relay for testing at the battery, and if it doesn't click there, replace the relay.

Finally, if the relay is good, you're back to a horn button problem.

• *Horn sounds weak*. This problem is caused by the horn. You could just unbolt the horn and install a replacement, but you might also be able to make an adjustment. Look at the horn and you'll see a locknut and adjusting screw, or a self-locking adjusting screw (see Fig. 9–32). With everything connected, turn the screw counterclockwise until the horn just stops blowing, then clockwise until it has a clear sound. Don't make adjustments with the horn blowing, but blow between adjustments to check.

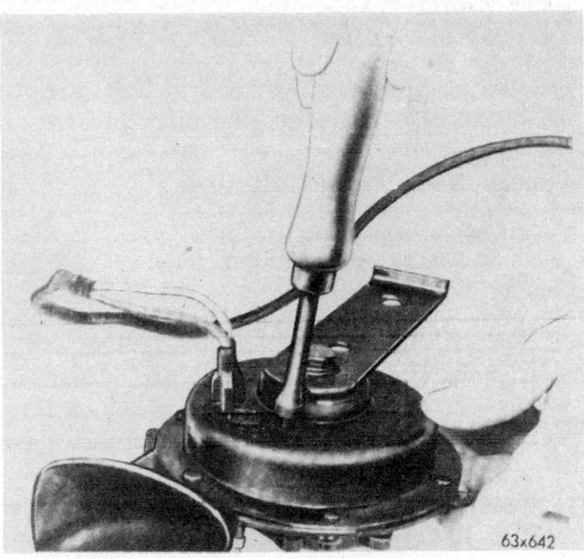

Fig. 9–32. Typical horn with adjusting screw. This type has hex-head adjuster; others have locknut and screw slot adjuster.

• *Intermittent operation*. Like intermittent problems of any kind, an intermittent horn problem can be impossible to trace when the horn is working. You must test when it is not.

• *Wiring problems*. The test procedures all are based on component problems, rather than wiring difficulties, for wiring problems with a horn circuit are comparatively less common. If you find that the horn and relay are good, but the horn doesn't blow, it is possible that the problem is not in the horn button, but in the wire from the relay to the button, either an internal break in the wire (causing no-blow) or bared insulation short-circuiting against the steering column (causing continuous blow). Installing the accessory horn button that clamps to the steering column will also handle this problem, just as if the horn button were the cause. You might try snaking a new wire up to the horn button and through the steering column jacket, but this may not be a simple job. Occasionally, a weak-sounding horn problem may be traced to a poor wiring connection at the horn terminal.

Automotive Lamps

Your car has many lamps: for parking, directional signals, stop indicators, dashboard and other interior convenience, and of course the headlamps. Replacing them is work you can do at a substantial savings, for the price of bulbs is much lower at the discount house than from the garage or service station.

Headlamps

You can replace headlamps without disturbing the aim. First, make sure the problem is a defective headlamp. If a single headlamp works on high or low beam (but not the other), the problem is likely in the headlamp itself. If both headlamps work on one beam but not the other, the problem is likely in the wiring of the headlamp switch or the dimmer switch. On four-lamp systems, a simultaneous failure of two headlamps (high or low beam) is statistically unlikely; if they don't work, suspect a switch or wiring problem.

• *Replacement procedure*. The exact method of getting to the headlamp varies from car to car but usually it's obvious. In most cases there is a molding or chrome ring held by screws or force-fit into place (see Fig. 9–33). Once you remove it, look carefully at how the headlamp is held: You'll see a metal ring held by three screws; those and only those screws should be touched (see Figs. 9–34, 9–35). There are two other screws, spring–loaded, which do not go through the ring, that control headlamp aim.

If the metal retaining ring screws are tight, don't force them. Spray them with penetrating solvent until they can be worked loose. Remove the ring and then disconnect the headlamp from the electrical connector in back (see Fig. 9–35). Install the new headlamp, making sure you have the correct side up (or headlamp aim will be completely distorted). Most headlamps have a number somewhere on the lens, to make positioning easy.

• *Headlamp aim*. Headlamp aim can be set accurately only with headlamp aimers, and the cost of even the least expensive portable type is probably more than you want to spend. If you have a private house with a driveway, however, you can maintain the aim of your headlamps (immediately after they have been set professionally) with this simple procedure:

Place your car at the foot of the driveway, about 25 feet from the garage door. Turn on the headlamps and mark the center of the beams on the garage door. If your driveway slopes too sharply, park your car just outside the garage door and use the inside wall of your garage instead. Mark the location of your car exactly, either by paint marks, an embedded stake, etc., so you can put your car in the same position at any time.

If your headlamp beam ever is below your marks on the garage door or wall, turn the top adjusting screw clockwise (to raise the beam); above the marks, turn it counterclockwise (to lower it). If the beam is to one side or the other, turn the side adjusting screw to bring it back into alignment.

Exterior Lamps

Most exterior lamps are easily accessible, either from the back (just untwist the bulb or socket and remove) or after removing the front cover lens held by a couple of screws (see Figs. 9–37, 9–38).

To remove the typical exterior bulb push it in slightly, then twist, and it will release and spring up. If there isn't enough of the glass protruding to get a good grip, buy an inexpensive bulb removal kit, a set of sleeves that fit over the glass and usually prevent damage (see Fig. 9–39). If you do break the glass, use needle-nose pliers to grasp the broken filament in the base, push down, and twist to release (see Fig. 9–40).

If the filament of the old bulb appears to be good, test the bulb before buying a replacement. Just set the base on one battery terminal, then run

Fig. 9–33. Removing headlamp molding after taking out retaining screws.

Fig. 9–34. Removing the screws that hold the headlamp retaining ring. Do not confuse retaining screws with headlamp adjusting screws or you will disturb the headlamp aim.

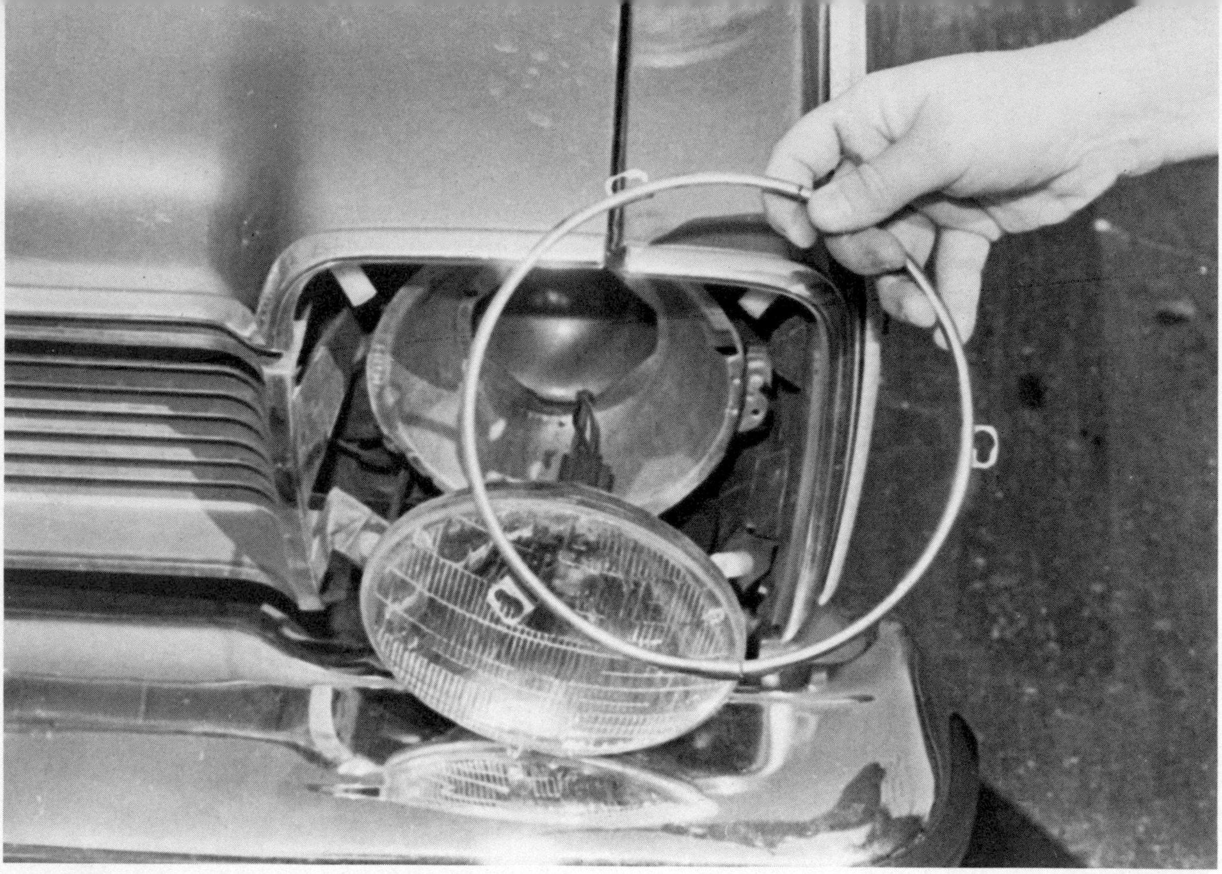

Fig. 9–35. Retaining screws removed, ring can be withdrawn as shown.

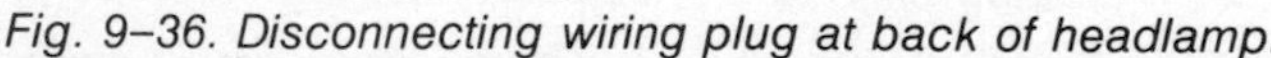

Fig. 9–36. Disconnecting wiring plug at back of headlamp.

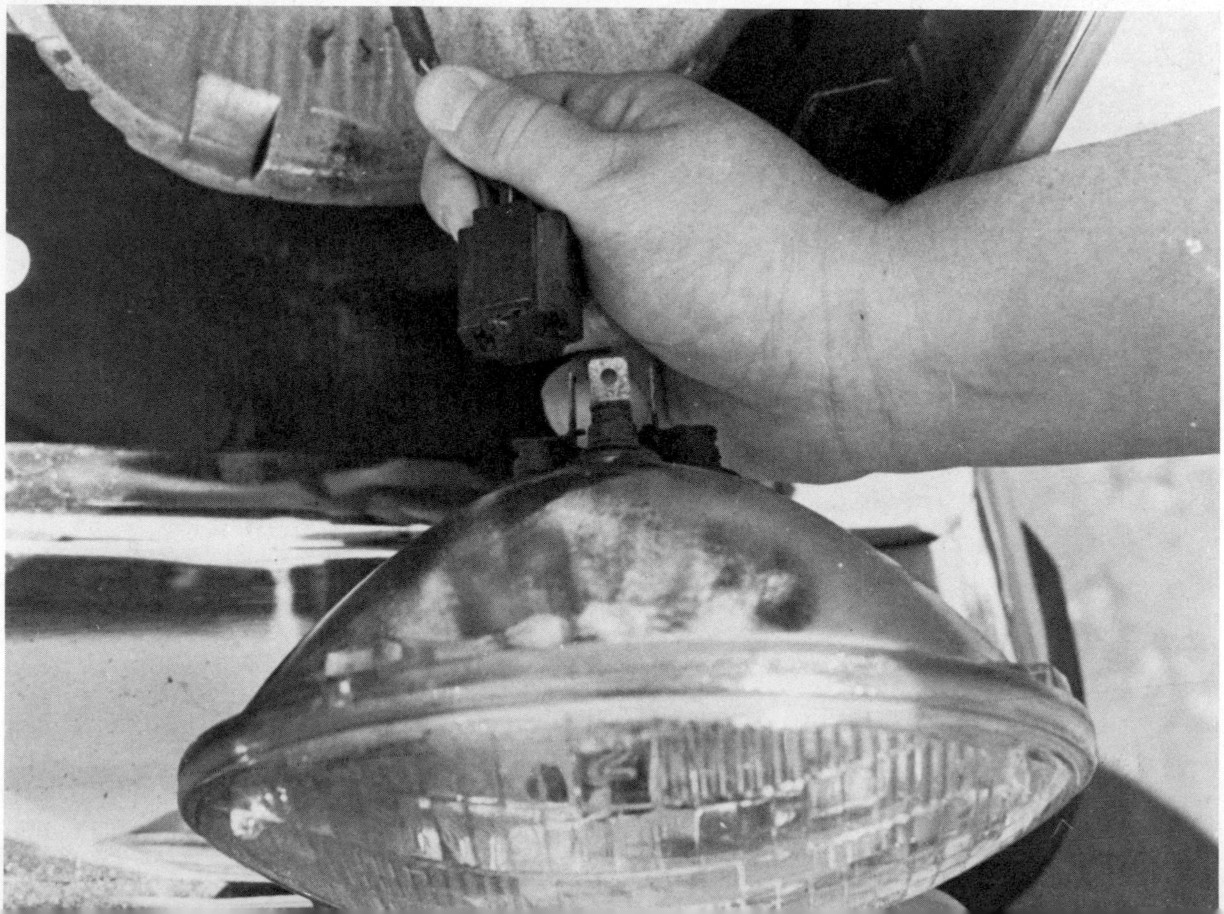

Fig. 9–37. Gaining access to exterior bulb by removing front lens held by screws.

Fig. 9–38. Gaining access to exterior bulb by prying out entire socket-lens assembly, which is held by spring clips in rectangular holes.

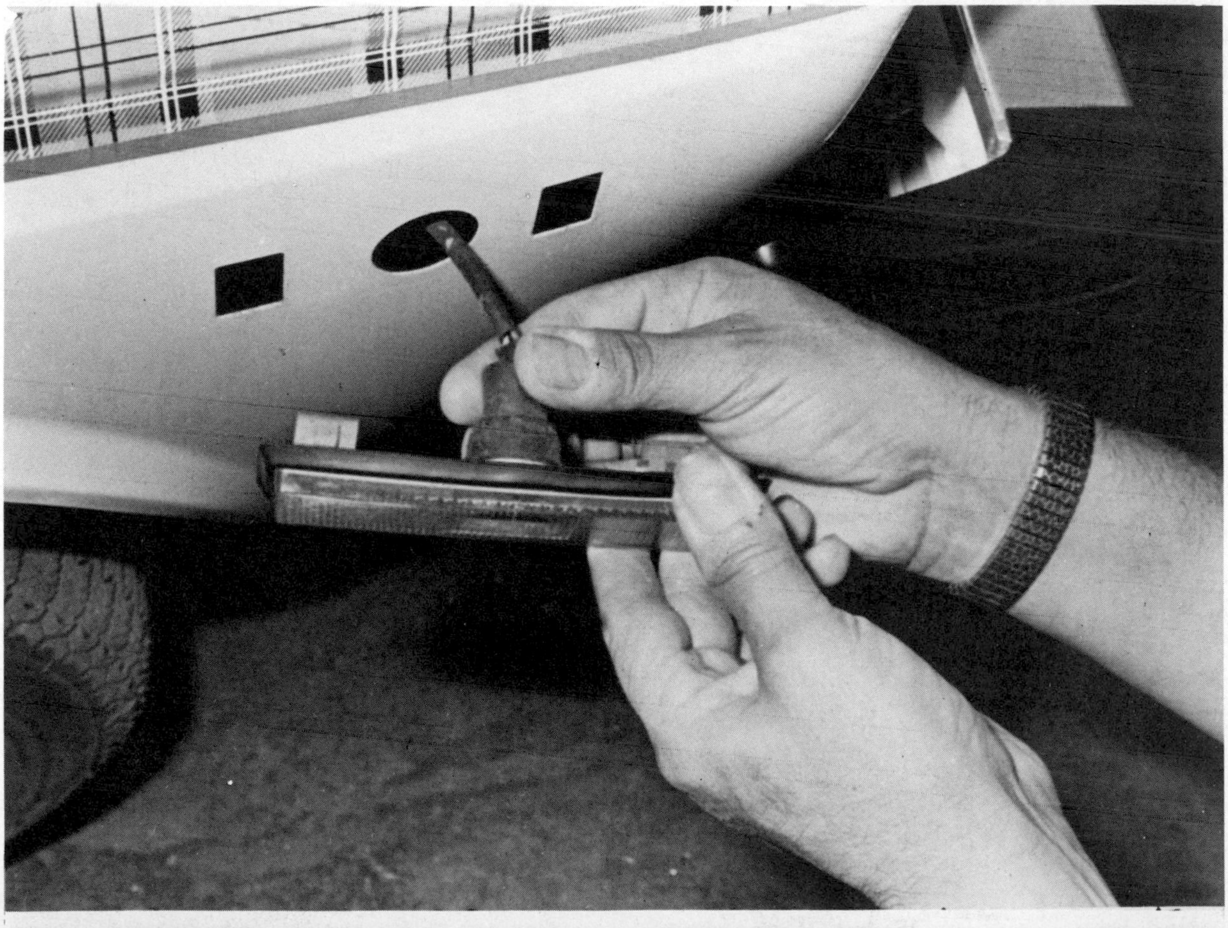

Fig. 9–39. Using plastic holder to remove bulb.

Fig. 9–40. Using needle-nose pliers (jaws wrapped with masking tape to improve grip) to remove base of broken bulb.

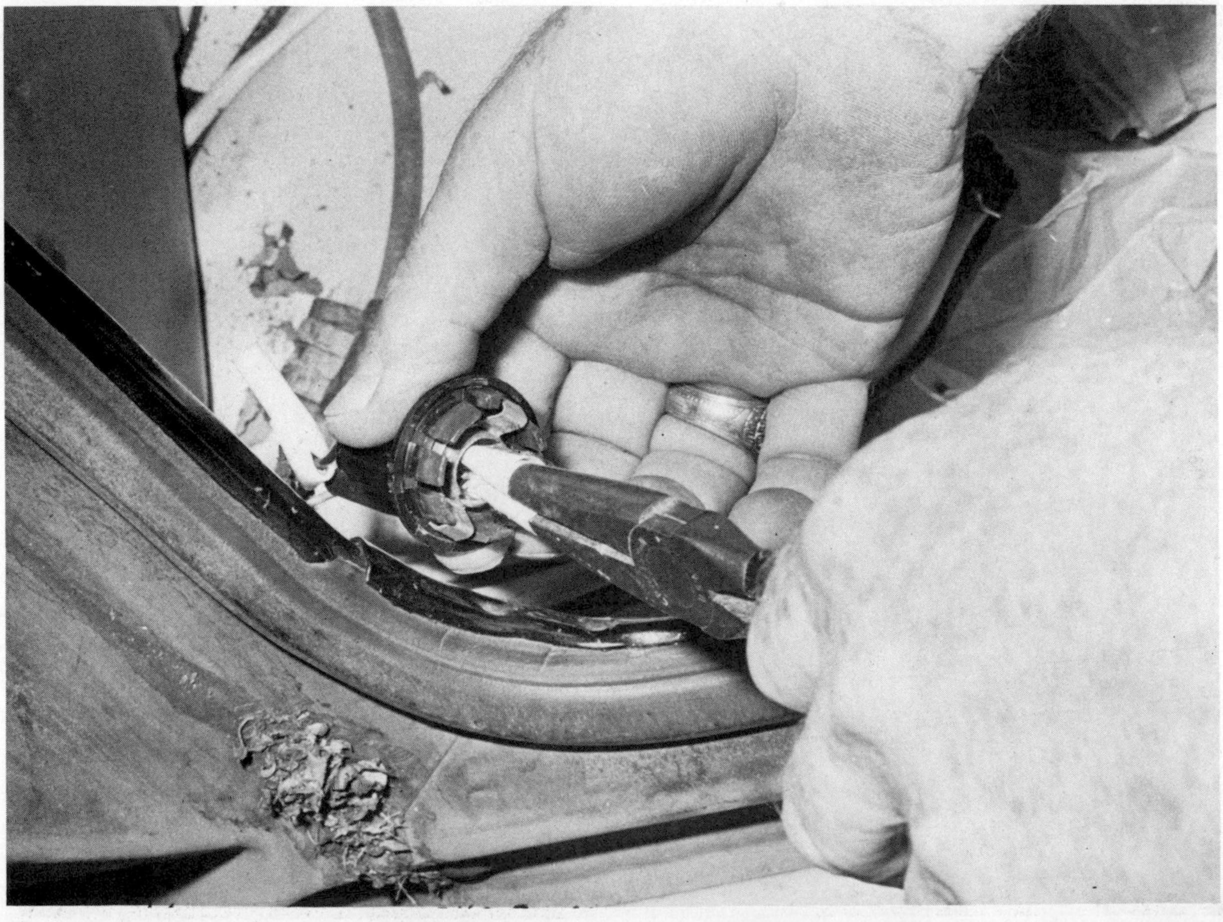

a jumper wire from the other battery terminal to the metal side of the bulb. If the bulb lights, it is good. With two–filament bulbs (those used for headlamps or both stop and tail lighting), both filaments should light (see Fig. 9–41).

• *Socket service*. If the inside of the socket is physically good, that is, no broken spring contacts, just wire-brush it (see Fig. 9–42) and spray with penetrating oil to clear out any corrosion. See broken contacts? Replace the socket.

If the bulb still doesn't work in the socket, a wiring problem apparently exists, and that is a job for a professional mechanic.

In almost all cases, however, the problem is in the bulb or the socket.

• *Installing a bulb*. The exterior "bayonet" bulb has protruding tangs on the side that engage slots in the socket. On many bulbs, the tangs and slots are at different heights and you must aim the bulb into the socket correctly to engage them when you push down and twist to seat the bulb.

Interior Bulbs

The dome light and the dashboard bulbs are the ones common to all cars, and in most cases they are easy to install if you know how.

• *Dome light*. Removing the lens is the only tricky part of dome light service. The lens is held somewhat differently on every car, but if you don't see anything obvious, try pushing in on the lens and twisting it, first one way, then the other, and it may release and drop out, giving you access to the bulb. If the lens is not flexible and there are no screws, it must be pried out (see Fig. 9–43).

• *Dashboard lights*. The dash lights may be reachable from behind the gauges. See if you can fit your hand up in there and remove the bulb. Some twist out and others pull out; all you can do is feel your way around and try both techniques.

If access from the back is difficult, the odds are one of the following:

(1) The gauge cluster comes out after removal of a few screws and perhaps unplugging the

Fig. 9–41. Checking out headlamp at car battery. Note that one electrical prong of headlamp rests on battery post while a jumper wire connects from another prong to other battery post. Other car bulbs can be checked in a similar manner. However, you should understand that the other car bulbs are wired to the yellow metal side, so the jumper wire should be touched against this side while the base of the bulb is on the battery post.

Fig. 9–42. Wire-brushing corrosion from bulb socket.

speedometer cable (see Fig. 9–44).

(2) The dashboard is front-service. You remove cover bezels held by easily accessible screws and you then can remove any gauge or switch from the front (see Fig. 9–45).

(3) The bulbs are front-service, just remove the cover bezels.

(4) The bulbs are accessible after removing a speaker grille from the top of the dash.

(5) The lights are fiber-optic, that is, special fibers that carry light from an accessible location where the bulb is to an inaccessible one. One bulb may light many runs of optical fibers. Check with your dealer service department on this one, if none of the other options pan out.

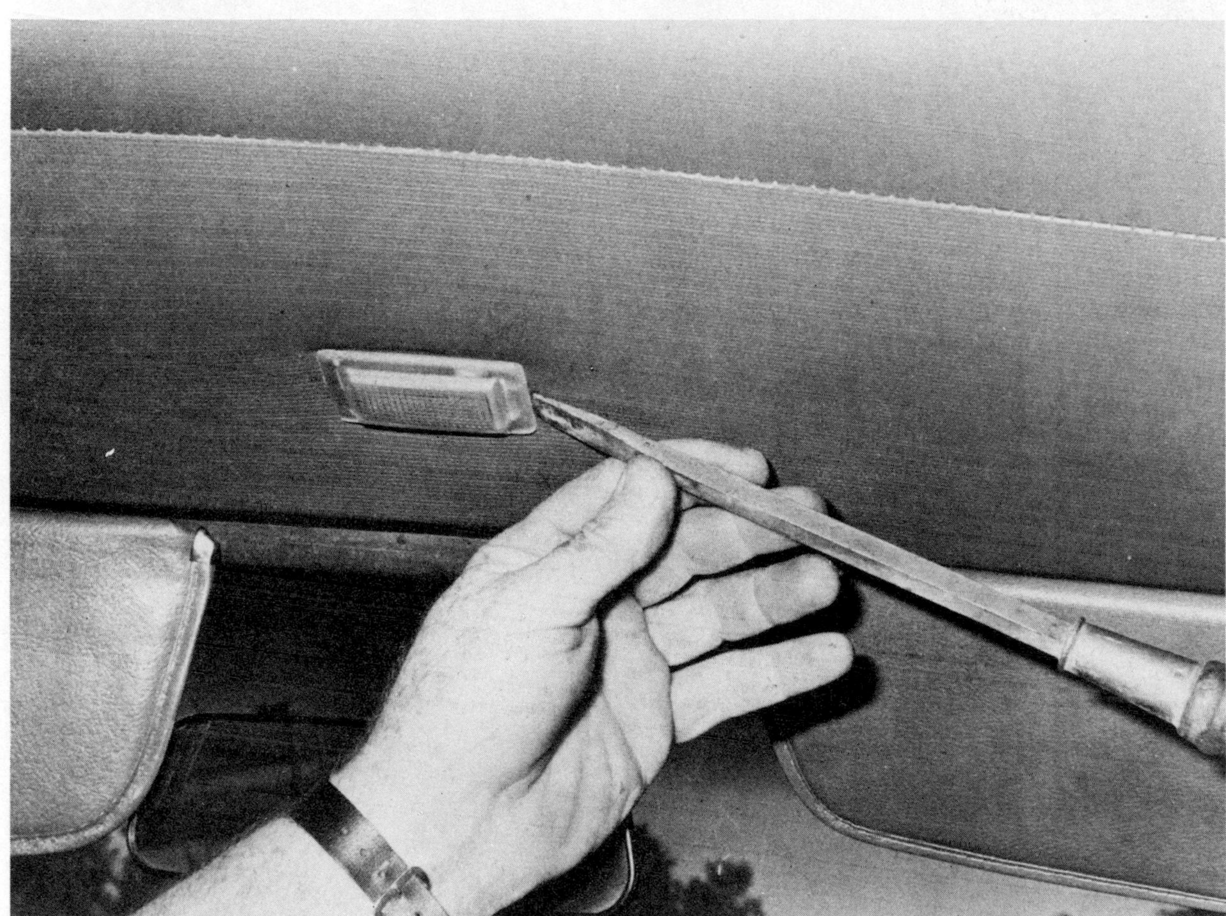

Fig. 9–43. Dome light lens isn't flexible and there are no retaining screws, so it must come out by prying with screwdriver as shown. Be careful not to damage headliner.

Fig. 9–44. This gauge cluster comes out after removing four external screws, pulling forward, then reaching in back to disconnect electrical connector and speedometer cable. On some cars you may have to disconnect cable before you can pull cluster forward. On others the cable housing may be caged and you can pull the cluster forward without a cable disconnection.

Fig. 9–45. This is a front-service dashboard. Cover bezels and moldings are unscrewed and removed, and everything—including bulbs—is accessible.

10 AIR CONDITIONING SERVICE AND REPAIR

ALTHOUGH MANY AIR CONDITIONING REPAIRS require special equipment and training, most aspects of routine maintenance—including simple recharging with refrigerant gas—is work that the weekend mechanic can do. All you need are some inexpensive tools and a basic understanding of the system and some of its components.

How It Works

There are three basic components in the automotive air conditioning system: an evaporator, a compressor, and a condenser (see Fig. 10–1). They are connected by hoses and form a closed system (not open to the air) that is filled with a gas called Refrigerant 12 (Freon is its most popular trade name). Refrigerant 12 has the unique characteristic of boiling at −21.7° F. at normal atmospheric pressure.

The compressor is a pump that circulates the refrigerant gas through the system.

As a liquid, Freon flows into the evaporator, a part that resembles a tiny radiator, located in the air conditioning ductwork (in the passenger compartment or at the rear of the engine compartment). In the duct it absorbs heat from the surrounding air and boils, turning into a gas once more. This is much like a pot of water on the stove. It absorbs heat from the stove and at 212° degrees F. turns to a gas (steam). The change from liquid to gas is the way many fluids absorb heat.

Inasmuch as cold is the absence of heat, the air in the ductwork becomes cold. A fan called the blower then blows this cold air into the passenger compartment.

The heated Freon (now a gas) is drawn from the evaporator by the compressor, which compresses it and pushes it into the condenser, that finned radiator-like part in front of the cooling system radiator. Here the air flow through the condenser cools the refrigerant gas and turns it into a liquid. It then flows as a liquid to the evaporator to continue the cooling operation.

The air conditioning system has many control valves, switches, and other parts, but for routine service the only ones that you need to understand are the expansion valve and the receiver-dryer (or a cousin called the accumulator). A system will have one or the other.

The expansion valve is a device that meters the flow of refrigerant into the evaporator. Actually, it is the dividing line between liquid Freon from the condenser and the gas in the evaporator. Some cars do not have a valve, just a part with a calibrated passage. This is called an expansion tube.

The receiver-dryer is a cannister that holds extra refrigerant liquid to provide extra supply on demand and compensate for minor leaks. It is located in the high-pressure side of the system, adjacent to the condenser on most cars. On some General Motors cars, it is mounted in the rear of the engine compartment and also contains some system control valves, including the expansion valve, and is called a VIR (valves-in-receiver) (see Fig. 10–2).

The receiver-dryer holds a bag of dessicant to absorb moisture in the system; hence the term dryer. On many cars the receiver-dryer has a tiny window (see Fig. 10–3) so you can actually see into the sealed refrigeration system. How this is used is explained later in this chapter.

On many newer cars, those with an expansion tube instead of the expansion valve, a part called

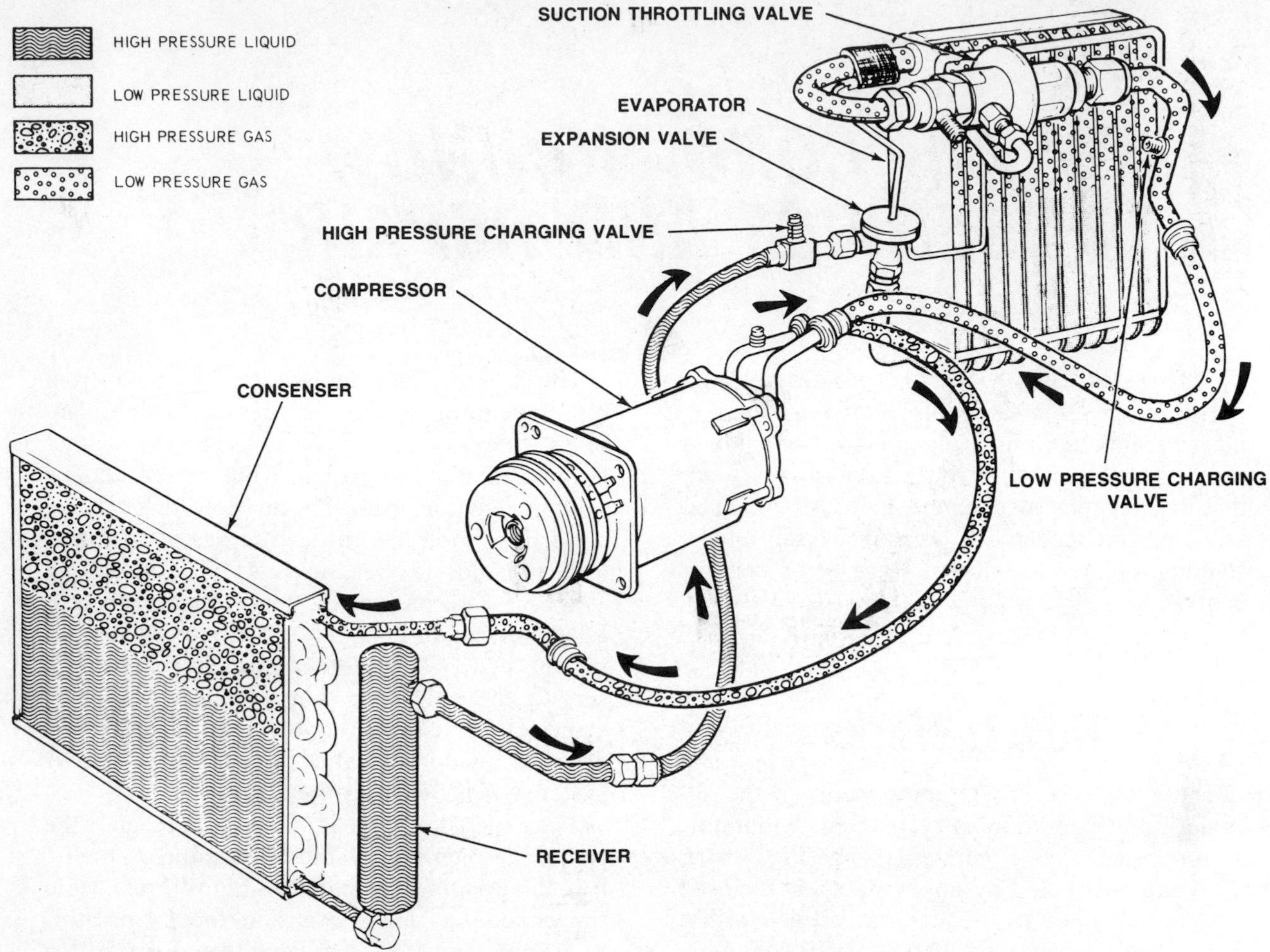

Fig. 10–1. This is a schematic of the refrigeration system of an automobile air conditioner. The condenser is in front of the radiator and the evaporator is in the ductwork, either under the dashboard in the passenger compartment or under the hood at the rear of the engine compartment. Heat is taken from the air at the evaporator, then rejected to the atmosphere through the condenser. Compressor circulates refrigerant gas, called Freon 12, between the evaporator and condenser.

an accumulator is used instead of the receiver-dryer. The expansion tube system is more likely to contain some liquid in the Freon, and this liquid could damage the compressor. The accumulator, located on the low-pressure side of the system, separates the liquid from the gas to prevent this. It also removes moisture from the system.

AC Sections

The air conditioning system is divided into two sections: the high-pressure side and the low-pressure side. The compressor separates the sections, with the compressor inlet (where the gas is drawn from the evaporator) being the low-pressure side, and the compressor outlet (where the gas is compressed and pushed into the condenser) being the high-pressure side.

The evaporator and the line connecting it to the compressor inlet, therefore, constitute the low pressure section. The line from the compressor outlet, through the condenser, and up to the evaporator is the high-pressure side.

Checking the System

If your air conditioning system is not performing properly, all you want to know is if the problem is one you can correct, or if it's some-

Fig. 10–2. This is a receiver-dryer that includes air conditioning control valves. It is called a VIR for valves-in-receiver.

Fig. 10–3. The little window the finger points to is a sight glass. It enables you to see if refrigerant gas is circulating in the system. Most newer cars do not have it, however.

thing that requires professional service.

First, check the front of the condenser. If it's plugged with road film, bugs, leaves, or other debris, it can't cool the refrigerant gas. Clean it with a pail of household detergent and water, and a soft brush.

Next, check the compressor drive belt. If it's loose or so badly deteriorated that it's slipping on the pulley, performance also will be poor. Adjust the belt if necessary (see *Chapter* 7).

Finally, sit in the car, run the engine at fast idle, and turn on the system. Within five minutes you should feel cold air. If it's a very humid day, the air won't be as cold as on a dry day, but that's normal. If the AC performance only seems poor because the blower won't go up to high speed, that's an electrical problem (check for a blown fuse before taking the job to a shop). If the AC works normally for 10–15 minutes, then starts blowing warm air, the problem is moisture in the system. The sealed system must be discharged and moisture vacuumed out of the system using a special pump. This is a job for a professional shop. Note: Most GM cars built before 1977 have a fuse called a thermal limiter under the hood in wiring near the compressor. This fuse is supposed to blow only when refrigerant has been lost, but may blow when the system is otherwise good. A new thermal limiter is cheap, so if the system fails completely buy a new one and see if it cures the problem. With the new one in your hand you should have no trouble locating the fuse holder and installing it.

If the AC performance is just very weak, however, the likely cause is loss of refrigerant gas. No seal is perfect, and within a few years most systems need some additional refrigerant.

Begin by checking the system pressure at the test valves with the gauge.

Test Valves

The pressures developed in the high- and low-pressure sides of the system are important indicators of what is happening inside. To permit measurement of these pressures, all auto air-conditioning systems have valves to which pressure gauges can be connected. There normally is one for the high-pressure side and one for the low, although many Chrysler Corp. cars have a second valve on the low-pressure side.

Professional mechanics use somewhat elaborate gauge sets to measure pressure, but you can find out all you need to know with an inexpensive

Fig. 10–4. This inexpensive pressure gauge, much like a tire gauge, permits you to check pressures inside the air conditioning system.

tire-type air conditioning gauge (see Fig. 10–4).

Finding the Test Valves

When looking for them, start at the compressor. With one exception, they resemble tire valves (including the thread-on cap). On many cars they are on the compressor itself or in the connecting lines very close to it.

• *Chrysler Corp. cars.* On most cars the high-pressure test valve is on a cylindrical can (called a muffler) in a line very close to the compressor (see Fig. 10–5) and the low-pressure test valve is on the compressor. If you see two valves on the compressor, the one you want is NOT on a flat plate held by eight bolts. It is the one next to the tubing line that goes to the condenser or receiver-dryer.

• *General Motors cars.* On cars with a receiver-dryer that includes the expansion valve (and another valve), you'll find the low-pressure valve at the top and the high-pressure one at mid-point on the cylinder (see Fig. 10–6).

On cars with an accumulator (see Fig. 10–7), the low-pressure valve is on the accumulator; the high-pressure one is on a line that is physically nearby.

On older models with standard air conditioning, both valves are on the compressor. On older models with automatic temperature control systems, the high-pressure valve is at the expansion valve (just follow the line from the receiver-dryer until you come to it). The low-pressure valve is at a part called the POA valve (just follow the low-pressure line—the thicker hose—from the compressor until you come to it).

• *Ford products.* This car line can present the most difficult problems for the novice, for there are three possibilities:

(1) The high-pressure valve is close to the compressor, in the line that goes to the condenser; the low-pressure is in the line close to a part called the suction throttling valve (just follow the low-pressure line—the thicker hose—from the compressor until you come to it). Caution: There is a valve on the suction throttling valve that will resemble the one for testing. It is used for testing, but it's not the one you want (see Fig. 10–1).

(2) Both test valves are in the lines very close to the compressor. The low-pressure one is in the thicker hose.

(3) There are two valves on the compressor (see Fig. 10–8); however, they are manually controlled, rather than the tire-type. You cannot use the tire-type gauge on these valves, but

Fig. 10–5. Fingers point to pressure test points on a Chrysler air conditioning system. At left is the low-pressure point into which refrigerant may be added. At right is the high-pressure point with its cap removed and placed just to the left.

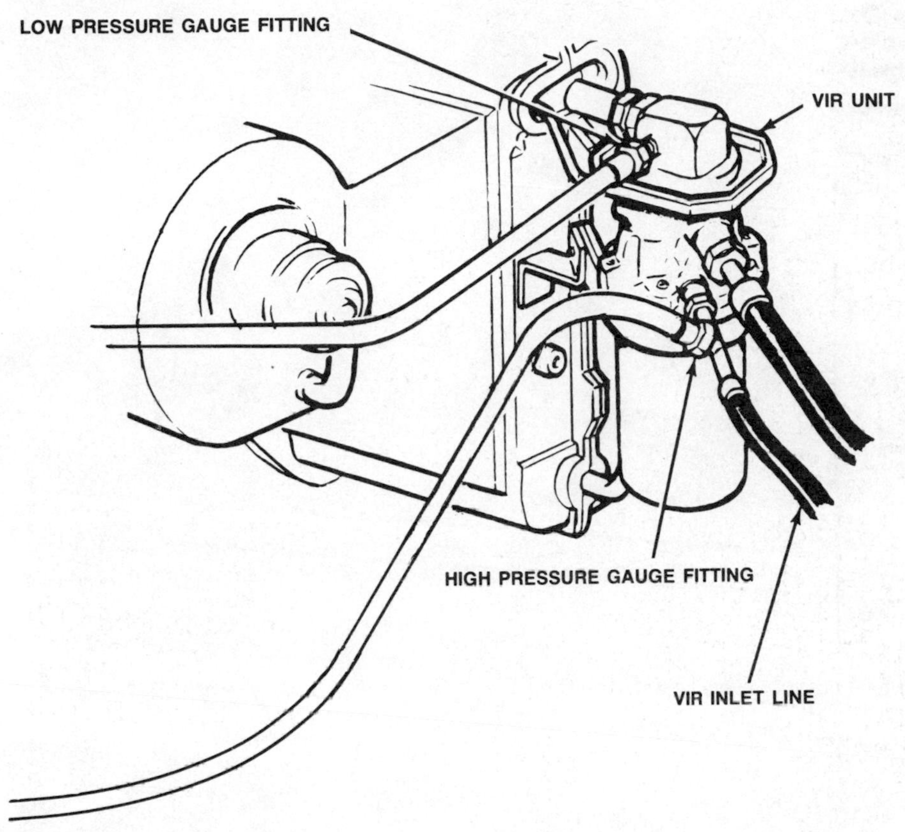

Fig. 10–6. This illustration shows location of pressure test points on General Motors VIR system.

instead must have a professional gauge set. Further, the ports for adding refrigerant gas are not made for use with the kits packaged for use by weekend mechanics. Leave the service on this design to a professional.

Determining Need

To determine if your system requires additional refrigerant gas, check the sight glass on systems so equipped. Run the engine and turn on the AC. You should see slight bubbling in the glass. After five minutes the bubbling should stop. If it does not, additional refrigerant gas is necessary.

If you see no bubbling at the start or as the unit continues to operate, and there is absolutely no cooling, the system probably has lost all its gas charge from leakage. The leak that allows the refrigerant to be lost relatively quickly also is enough to admit moisture, so the best procedure is to get the system vacuum-pumped and recharged by a professional.

If your car does not have a sight glass, locate the high-pressure test valve, run the engine and AC, remove the cap, and take a pressure reading with the tire-type gauge. Pressure readings are

Fig. 10–7. Fingers point to pressure test points on General Motors system with accumulator (the cylindrical can at left). Low-pressure point is on accumulator; high-pressure point is on narrow-diameter air conditioning hose.

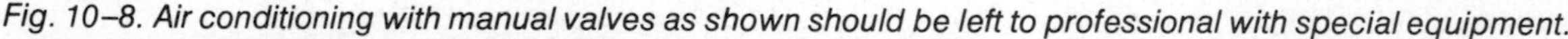

Fig. 10–8. Air conditioning with manual valves as shown should be left to professional with special equipment.

dependent on many factors, including humidity, and it is possible to get a normal high-pressure reading even if something is wrong. However, see if you get a reading of more than 120 p.s.i., and preferably closer to 185–210 p.s.i.

Next, remove the cap on the low-pressure valve and take a reading there. Normal pressure is 15–30 p.s.i. If both readings are very low, and there is effectively no real cooling from the system, inspect the expansion valve exterior. If it's frosted at the inlet, see a professional. If not, adding Freon should do the job if there are no leaks of significance in the system.

Checking for Leaks

Before you add refrigerant gas, you can check for leaks in the system in any of these ways:

• *Brush a soap-water solution around each hose connection.* If you see bubbling, you've found a leak. Tighten the connection with wrenches and repeat the test. If you stop the leak, you can proceed to add refrigerant to the system. If you can't eliminate a leak by normal tightening, the hose must be replaced and the system completely vacuum-pumped and recharged. A new receiver-dryer also may be necessary. See a professional.

• *Use a leak detector.* Two inexpensive types use butane or propane (see Fig. 10–9). You just light the detector and run a search tube around each connection in the system. Leaks of refrigerant gas will cause the butane or propane to change color. Details on color changes are included with each tester. These leak detectors are

Fig. 10–9. Using butane leak detector at air conditioning hose connection. If butane flame changes color, leak is indicated. Propane leak detector works on similar principle.

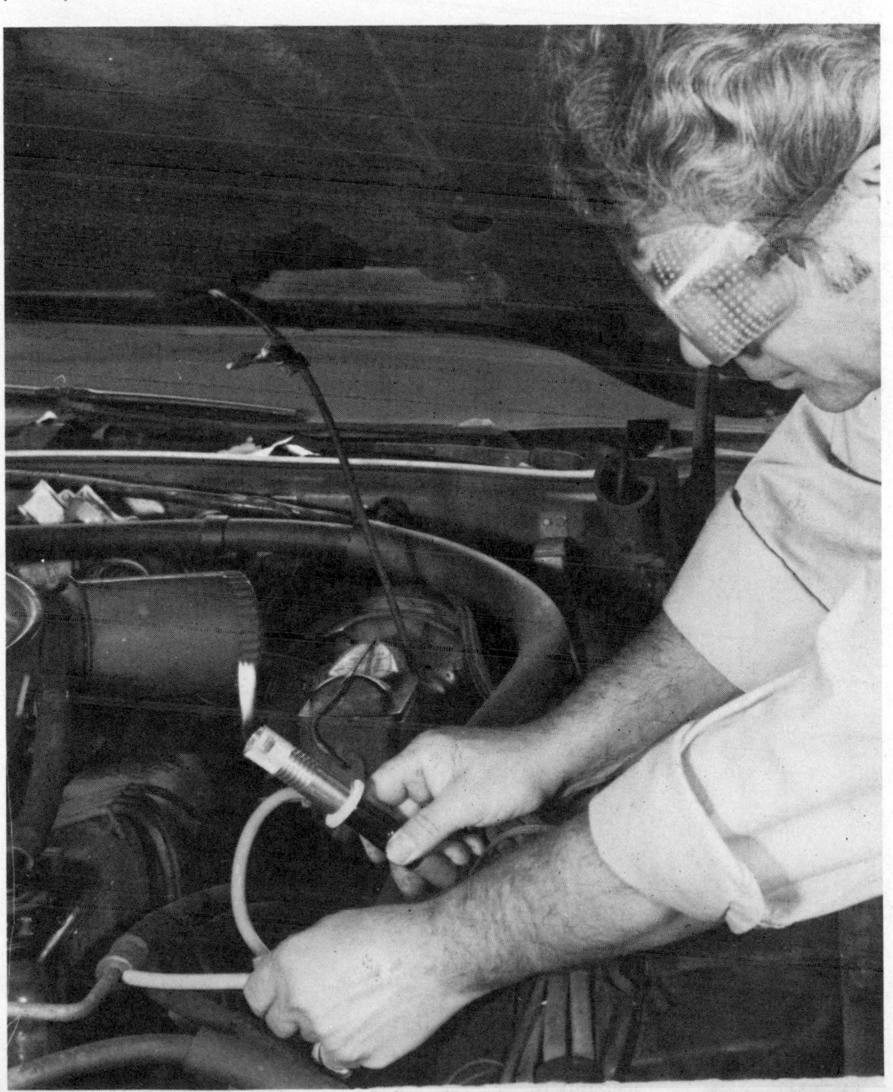

far more accurate than soap solution, and will pick up leaks that the soapy water won't.

Adding Refrigerant Gas

Refrigerant gas comes in 15-oz. cans. You should always add refrigerant into the low-pressure side of the system using a charging kit, which includes a special tap-valve assembly for the can and a hose.

Close the valve on the refrigerant can tap and install the tap onto the can, following the kit manufacturer's instructions. Connect the hose to the can tap valve.

Thread the other end of the hose onto the tire-type valve as finger tight as you can make it.

Wearing gloves (or using a pair of pliers to keep your hands away), loosen the hose connection at the refrigerant can tap valve to allow some refrigerant gas from the system to purge air from the hose. As soon as the gas pours out (a second or so), tighten the hose connection at the tap valve.

Run the engine at fast idle (2000 r.p.m.) and turn on the AC. Open the tap valve on the can and the refrigerant gas will flow into the system (see Fig. 10–10). If you keep your hands wrapped around the can to warm it, the flow of gas will be faster. When you feel the can is empty, close the valve and disconnect the hose from the low-pressure valve. Note: You will feel the can get very cold as it empties (and the weight also will drop). If the can is very slow to empty (more than 10 minutes), heat it in a pot of water to 120° F. (no higher, for it can explode at 125–130° F.).

Two cans of refrigerant gas normally will restore performance when the problem is a simple loss. Do not, however, attempt refrigerant gas additions as some sort of cure-all when the problem is not clear-cut. A low refrigerant problem usually is obvious, and that's when refrigerant addition will give you results.

Fig. 10–10. Installing refrigerant into the low-pressure side of the air conditioning system.

11 YOUR CAR'S BRAKE SYSTEM

THE BRAKING SYSTEM on a modern car is actually three systems in one: a hydraulic system that transfers movement of the brake pedal to each wheel; a friction system at each wheel that is actuated by the hydraulics to stop the car; and a mechanical linkage from the parking brake pedal or handle to the friction system.

How the Brakes Work

Hydraulic System

When you step on the brake pedal you push a rod that goes into the master cylinder, a fluid-filled (hydraulic) cylinder with pistons. The movement of the rod pushes on the pistons, which push on the fluid.

This movement is transferred through tubing that runs from the master cylinder to each wheel (see Figs. 11–1, 11–2). At the wheel the hydraulic pressure goes to one of these:

- *A wheel cylinder* (see Fig. 11–3). It's a cylinder with a piston at each end, positioned by a spring between them. Hydraulic pressure pushes the pistons outward, to act on the friction mechanism.
- *A caliper* (see Fig. 11–4). It's like a C-clamp, but instead of a forcing screw, there's a piston in one side. Hydraulic pressure pushes the piston out to apply the friction mechanism.

The fluid used in a braking system is a special formulation that will not boil at the temperatures built up from the heat of friction braking. The hydraulic system should not leak, but over a period of time minor loss of fluid may occur. A reservoir at the top of the master cylinder permits easy addition of fluid (see Fig. 11–5). On a new car you should use a top quality fluid, one with a boiling point of at least 450° F.

Friction System

The friction system is what actually brings each wheel to a stop, and there are two types in popular use:

- *Drum type*. In this design the wheel is bolted to a drum, which is shaped somewhat like a jar lid. As the wheel spins, so does the drum. The friction material is called brake lining and it is attached to a half-moon-shaped part called a shoe. The wheel cylinder is used in this system, and when its pistons are pushed outward they force the shoes against the inside surface of the drum (see Fig. 11–6) bringing the drum (and the wheel) to a stop. When you release the pedal, coil springs retract the shoes (see Fig. 11–3).

All drum brake systems must have a provision to adjust the position of the shoes to compensate for wear of the linings. On older cars (and a very few current models), manual adjustment is required. At the top or bottom of the brake, between the two shoes, is a threaded adjuster called a star wheel (see Fig. 11–3). Turning it one way unthreads it to move the shoes closer to the drum. Turning it the other way threads it in, moving the shoes further from the drum.

On most current cars, the drum brake shoes adjust to compensate for wear automatically, by means of a lever that turns the star wheel (see Fig. 11–3) to move the shoes closer to the drum.

- *Disc type*. The clamp-like caliper is positioned over the edge of a part shaped like a phonograph record, called the disc or rotor (see Fig. 11–4). The wheel is bolted to the disc and spins with it. The friction material, as in the drum system, is called brake lining, and it is attached to a flat steel pad, called the pad or shoe. Although most brake linings are made with asbestos, there is a trend to linings that are made with

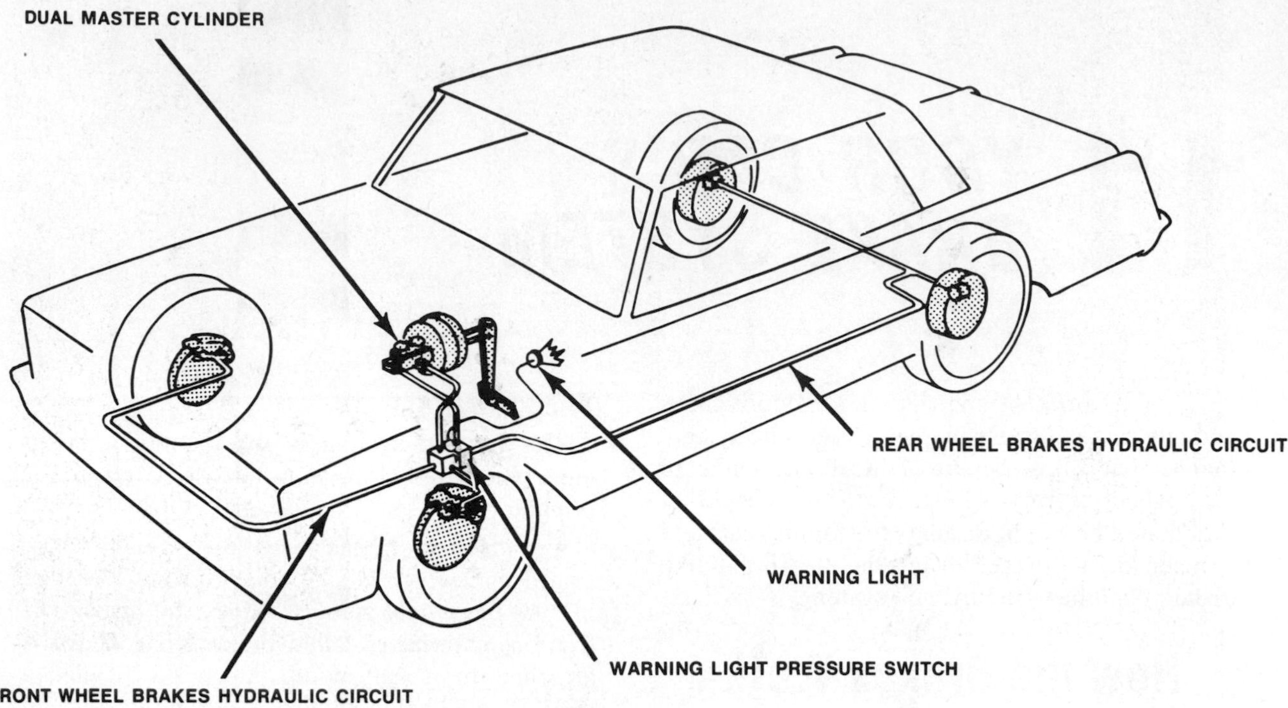

Fig. 11–1. The hydraulic portion of the braking system. When you step on the pedal, fluid pressure is transferred from the master cylinder through tubing to the front and rear wheel brakes. If there is a failure in either fronts or rears, a warning light pressure switch is activated, but braking still is available from the unaffected pair of brakes.

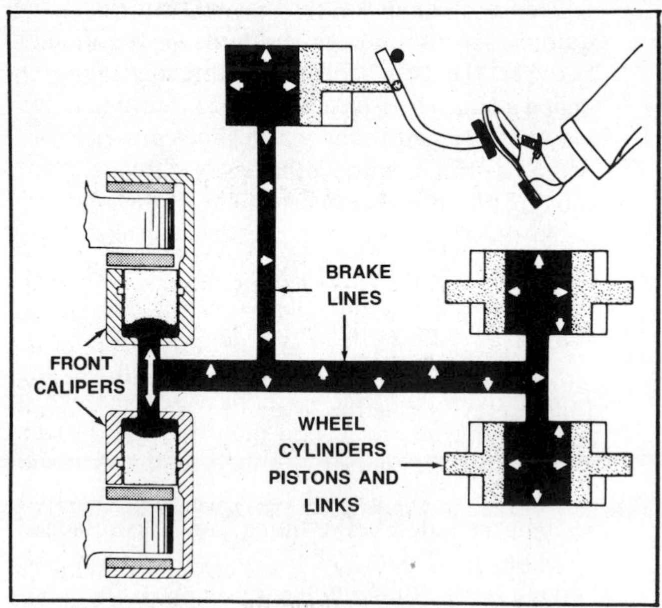

Fig. 11–2. Closeup look at a basic hydraulic system. A rod connected to the brake pedal pushes on a piston in the master cylinder. Pressure is transferred through the brake lines to calipers at the front wheel disc brakes and wheel cylinders at the rear wheel drum brakes. The typical caliper is a single-piston clamp that applies friction material parts to the sides of a disc. The typical wheel cylinder has a piston in each end that pushes out against friction material parts called brake shoes.

Fig. 11–3. Here's what a drum brake looks like when the drum is removed. The pistons inside the wheel cylinder push out (against retracting springs) to force the shoes with their lining against the drum. To compensate for wear of the friction material a star wheel adjuster is used at the bottom. When the car is braked in reverse, the star wheel is operated automatically by a mechanism that includes an automatic adjusting lever and cable.

metallic powder and fibers, called "semi-metallics." The reasons for the switch are that asbestos particles are considered a health hazard, and the semi-metallic linings last longer and are more fade-resistant. However, they are more expensive. There are two shoes at each wheel, one in each side of the caliper (on each side of the disc). One shoe rests against the caliper piston; the other is directly opposite and rests against a flange on the caliper.

When hydraulic pressure pushes out the caliper piston, it forces one shoe against the side of the disc. Simultaneously, the caliper slides (along rods or guides) and draws the shoe on the opposite side against that side of the disc. This provides the two-sided clamping action to stop the disc, and therefore the wheel.

Parking Brake

The parking brake operates by a mechanical cable and linkage setup that applies the rear brakes. If the car has self-adjusting drum brakes (or disc brakes, which are inherently self-adjusting) to compensate for brake lining wear, the parking brake theoretically needs no attention.

However, as a practical matter the cables stretch, and soon you find yourself at the limit of parking brake application, yet the car won't hold. In this case the cable must be shortened, as explained in this chapter.

Disc Brake Service

The single piston sliding caliper front disc brakes (see Fig. 11–4) have been used on virtually all American cars since 1976, and were used on many imports even earlier. Inasmuch as the front brakes do at least 60 percent of the braking, they wear out faster than the rears. It is not uncommon for rear brakes to last more than twice as long as the fronts.

A weekend mechanic can service disc brakes. This isn't a job you should attempt as your first effort, but once you have general experience with tools and have done the routine under-the-

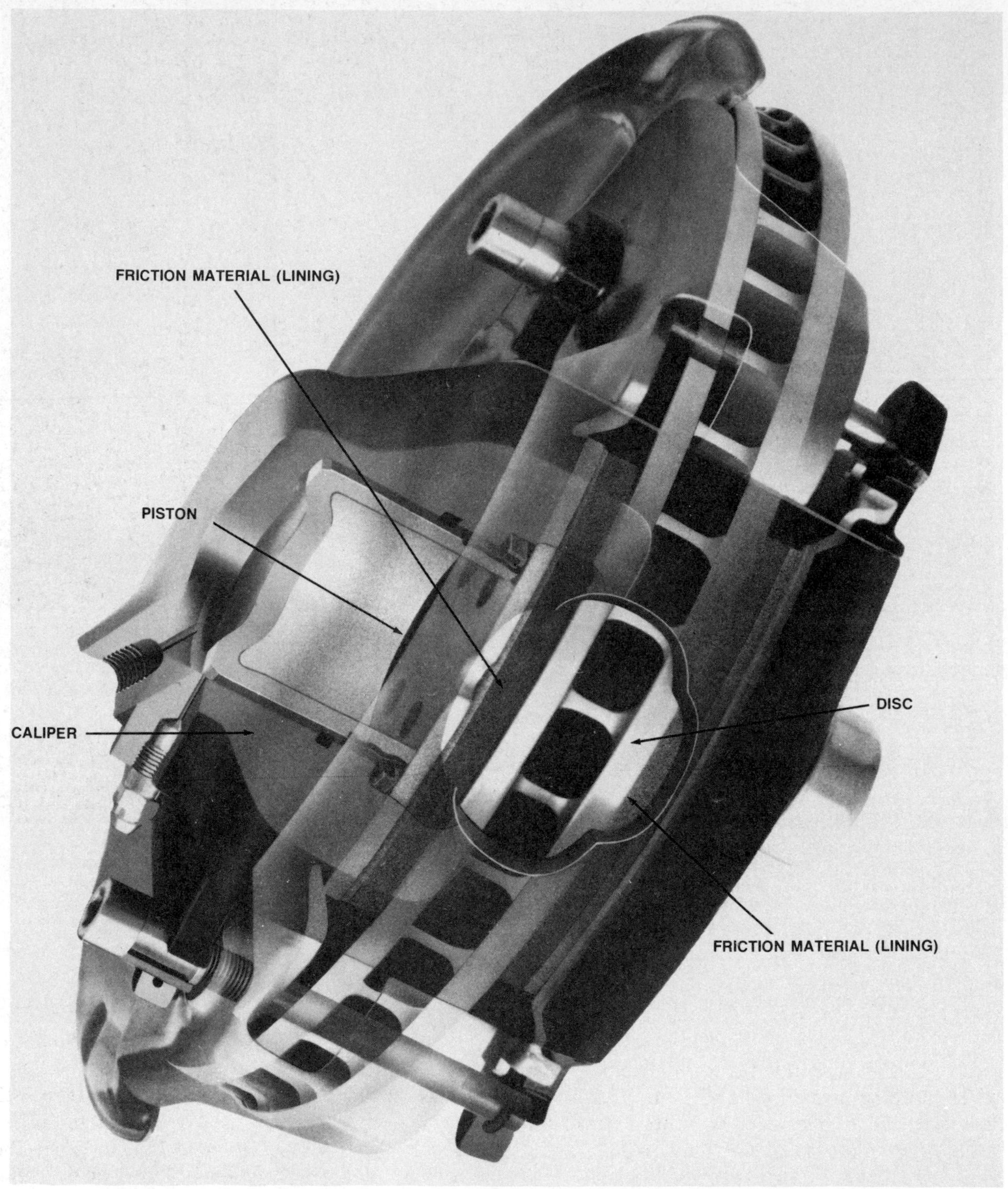

Fig. 11–4. This is a phantom view of a single piston disc brake. A friction material shoe is on each side of the disc, covered by the caliper. When the brakes are applied, hydraulic pressure pushes the caliper piston out to force the shoes against the sides of the disc. Because the disc and wheel are bolted together, the wheel comes to a stop when the disc does.

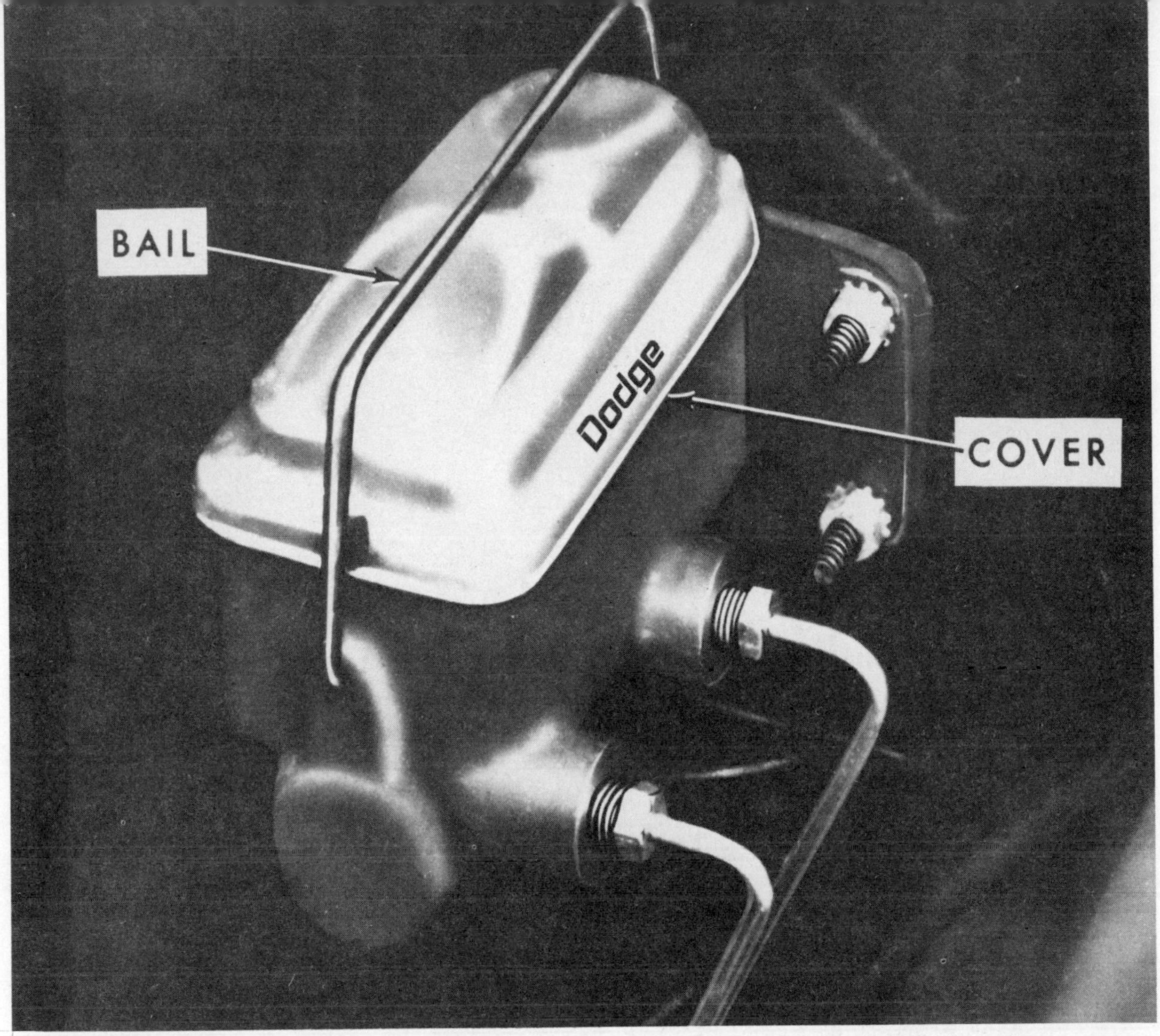

Fig. 11–5. Here's a typical master cylinder. When the bail wire is pried away to the side, the cover can be lifted off so the amount of fluid in the master cylinder reservoir can be checked. Note at the right the two brake lines coming out of the side. One goes to the front brakes, the other to the rears. The master cylinder itself is also separated into two sections, so no matter where a partial failure occurs the car retains some braking.

hood services there is no reason why you should be afraid. You must be careful, of course, for we're talking about a key safety item, and if you aren't confident, don't attempt it. There is, however, nothing mysterious about the brakes, and many weekend mechanics do routine brake service. Start with simple shoe replacement on a car that has enjoyed normal brake performance, and if you're careful all will go well.

One thing is in your favor: If you aren't sure about how things go together, just look at the wheel on the other side of the car. It will have a completely assembled model.

Brake Inspection

On some cars there is an inspection hole in the side of the caliper, through which you can see one or both brake shoes. If you see that the lining is worn to 1/16 inch or less, the shoes should be replaced.

Begin by loosening the wheel nuts, then jack up the car (see *Chapter 5*), support on safety stands, and remove the wheels. You'll see the disc and the caliper, which is the hydraulic clamp that forces the shoes against the sides of the disc.

Inspect the disc on both sides. Spin it, and if it wobbles (in-and-out movement called "runout"), check wheel bearings (see *Chapter 12*), and if they aren't the problem the disc itself is defective. Inspect the disc for deep scores (deep enough to completely stop a fingernail) as you move your finger along the surface. If you find

Fig. 11–6. This is a typical drum removed from a rear wheel. Notice the smooth inside surface against which the shoes are pushed by the wheel cylinder. Both rear wheel and drum are bolted to the rear axle shaft, so when the drum is stopped by the brakes, so is the rear wheel.

such scores on the otherwise smooth surface that comes in contact with the brake shoes, the disc must be resurfaced. If the fingernail can glide in and out of the score, it can be ignored. Discs are much more tolerant of scores than drums.

You also should check the disc for minimum thickness, variations in thickness, and "runout." Inexpensive plastic gauges (see Fig. 11–7) may be available for measuring thickness, or you can use an outside micrometer available for rent at some auto supply stores and rental centers.

Measure the thickness of the disc (about an inch in from the edge) at four to six equally-spaced points around the circumference. The readings should meet the manufacturer's specification for minimum thickness (stamped somewhere on the disc) and they should not differ from each other by more than .0007 inch. If the disc fails the minimum thickness measurement, it must be replaced. If it passes that check, but the difference between the highest and lowest readings on the micrometer is more than .0007 inch, the disc must be resurfaced.

Measuring disc runout calls for a dial indicator (see Fig. 11–8), an inexpensive tool, but still one that you may choose to rent. It also can be used for checking wheel bearings and engine valve guides, so if you're planning to do a lot of weekend mechanic work consider buying one.

With the dial indicator attached to a fixed point, such as the steering knuckle, position the dial head plunger so it is just touching the brake shoe contact surface of the disc, as illustrated. Turn the dial until it reads zero, then turn the disc. If the dial needle moves more than .005 inch, the disc runout is excessive. Tighten the wheel bearing nut (see *Chapter 12*) on all cars with adjustable wheel bearings and repeat the test. If runout still is excessive the disc must be resurfaced, if possible, or replaced. Removing the disc is discussed later in this chapter.

Preparing to Remove the Caliper

You can gain access to the brake shoes on the modern sliding caliper only after removing the caliper from its mounting over the disc.

Fig. 11–7. Inexpensive plastic gauge may be used to measure disc for thickness. If you can't obtain an inexpensive gauge such as this, you can rent a micrometer for very little from a rental center. The cost of renting a micrometer, dial indicator and other tools for the brake job could be under $10. The minimum acceptable thickness of a disc is stamped somewhere on the part.

Fig. 11–8. Checking a disc for runout (wobble) can be done with a dial indicator. The type shown has a bracket that clamps to the suspension and a link arrangement, so it can be positioned as shown. The little rod that projects from the indicator is brought into contact with side of disc, then the indicator is zeroed and the disc is turned by hand. Runout is shown by changes in readings on dial indicator.

If there is any rust on the outside edges of the disc, however, the brake shoes will catch and make removal impossible. Therefore, you must provide clearance by pushing the caliper piston back into its bore. This will allow the brake shoe that bears against it to also come back, and when you remove the caliper there should be lots of room to complete the job.

First, check the brake master cylinder reservoir level (see Fig. 11–9). If it's topped up, siphon out some fluid from the larger (front) reservoir, or if you're not sure, both reservoirs. When you push back the piston, fluid will be forced up into the reservoir, and if it's full already it will overflow.

Next, attach a large C-clamp so that it braces against the disc or a protruding part of the brake shoe on one side and against the caliper on the other (see Fig. 11–10). Close the clamp and the caliper will be squeezed against the disc, forcing the piston back into its bore. When you can't turn down the clamp any more, the piston should be all the way in. If you can't turn down the clamp at all, the piston is apparently stuck in the bore. Removing it is probably best left to the professional, but if you want to try, the procedure is described in the following section.

Taking Off the Caliper

Most calipers are held to a bracket on the steering knuckle assembly by the rods along which they slide. The rods are secured by some form of fastener at one end, such as a clip, nut, cotter pin, etc., or the rod threads into the caliper bracket on the steering knuckle. Typical fastening arrangements are shown in Figs. 11–11 to 11–14. Remove the fastener or unthread the rod. Next, pull the rod itself if necessary, using pliers on the end with the head. Lift the caliper up (see Fig. 11–15) and out and secure it to the suspension with coat hanger wire. Do not let the caliper hang free or the brake hose will be strained.

Some cars have guides—called ways—machined into the caliper mounting bracket (see Fig. 11–16) instead of rods. On this arrangement, remove the bolt-on part of the lower way (see Figs. 11–17, 11–18), then lift the caliper up and off, and hang it to the suspension. Occasionally,

Fig. 11–9. Prying bail wire off the top of the master cylinder cover. Cover then is lifted for fluid level inspection. If reservoir is full, siphon off some fluid from each of the two chambers so they won't flood over when disc brake caliper pistons are pushed back.

Fig. 11–10. Using C-clamp on caliper to push back the piston. This step will permit easy removal of the caliper from its position over the disc.

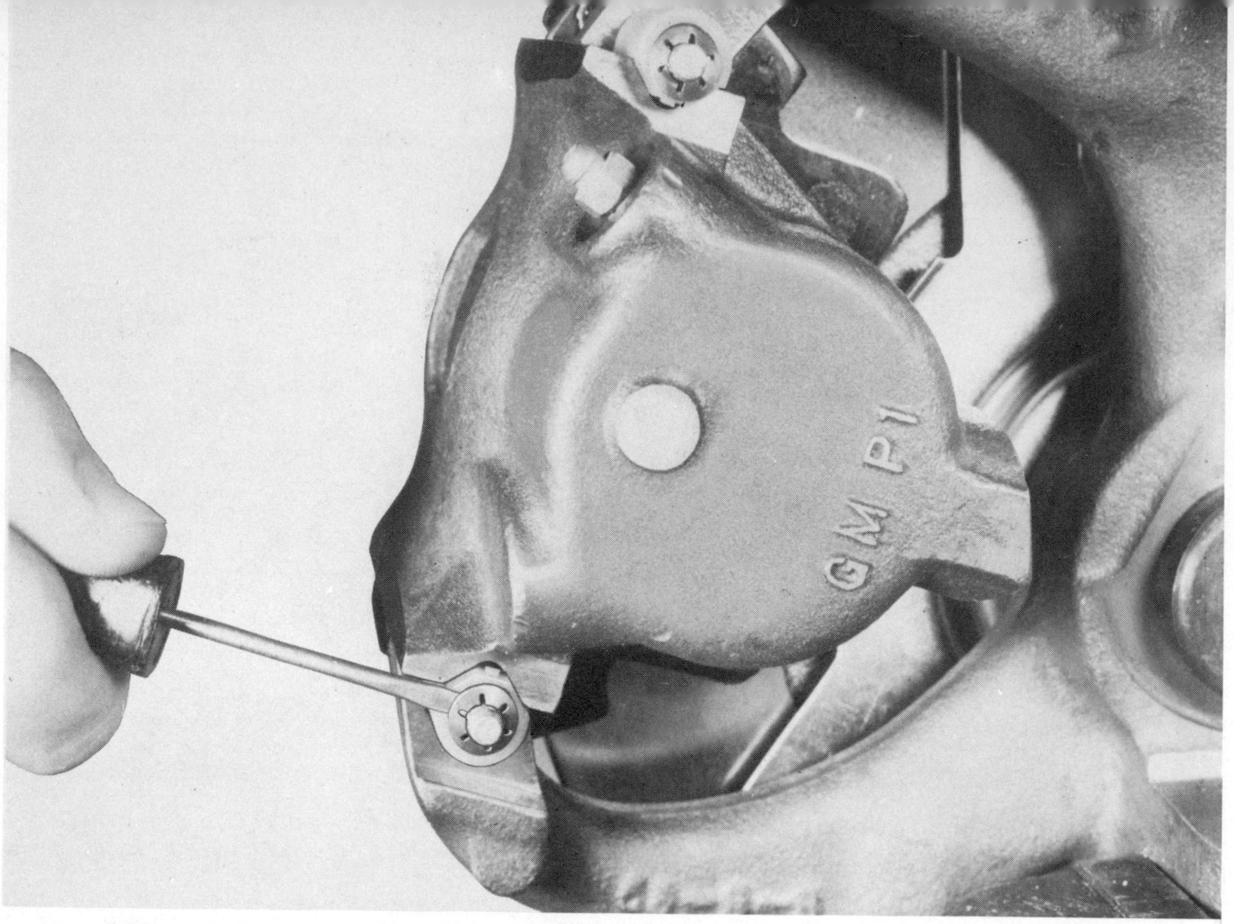

Fig. 11–11. If you see that the caliper rods have these "speed nut" type washers on the ends, pry them off with a screwdriver. Install new speed nuts when reassembling.

Fig. 11–12. After speed nut is off, pull out rod with pliers as shown.

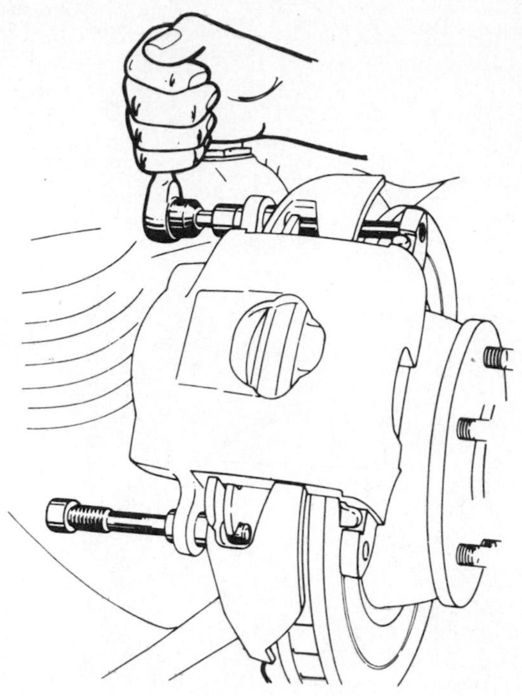

Fig. 11–13. If rod is threaded in, remove with Allen bit, extension and ratchet, as shown.

Fig. 11–14. Perhaps caliper is held by large cotter pin. Straighten the dual ends, and then pull out from the single "head" end with pliers as shown.

Fig. 11–15. Lifting caliper and brake shoes off the disc.

the mounting of the brake hose is such that you must disconnect it from the caliper to get the caliper off. If this is done, plug the end of the hose to minimize fluid loss and entry of air. You will have to bleed air from the brakes at the conclusion of the job.

If you can't force the caliper piston back in, you must first check the edge of the disc circumference for rust, and if there's a rust ridge it must be sanded off. Then remove the caliper from its bracket on the steering knuckle, and pull it away from the disc. If you're lucky, it won't be a hard pull. If you still can't get it off, you must remove the disc and caliper together.

Disc Removal

Once the caliper is off, the front disc can be removed for service, if necessary. On the conventional rear-drive car, remove the dust cup, wheel bearing nut, and outer wheel bearing, then pull the disc off, as explained in detail in *Chapter 12*. On front-wheel-drive cars or any cars with sealed front wheel bearings, the disc simply unbolts from the hub.

If you are struggling with a caliper stuck on the disc, follow the procedure for disc removal. Disconnect the brake hose at the caliper (and plug the hose end), then take the disc and caliper to an automotive machine shop that has the equipment necessary to separate them. The machine shop also may be able to remove the stuck piston and rebuild the caliper, or at least obtain a rebuilt caliper for you.

When you reinstall a disc you must adjust the wheel bearings, unless they are the no-service type. If you disconnected the brake hose don't forget to reconnect and bleed the brakes, as explained later in this chapter.

Caliper Inspection

Inspect the caliper before proceeding further. If there is fluid seepage or if the dust boot is cut or unseated, odds are that some dirt has gotten in. Have a machine shop disassemble it and replace the piston seal and dust boot for you.

If the caliper passes inspection, clean out the inside (keeping clear of the piston and dust boot) with a wire brush. If the piston was not pushed back all the way before removal, complete the job now.

On calipers with machined ways, inspect the ways for rust. If they're rusty, remove the rubber O-rings around them (if used), clean with a wire

Fig. 11–16. This is a machined-way caliper. Guide rails (ways) are machined into the edges of the caliper and the mounting bracket. Bottom of this type is held by bolt. On other types, both top and bottom rail sections unbolt from the caliper bracket.

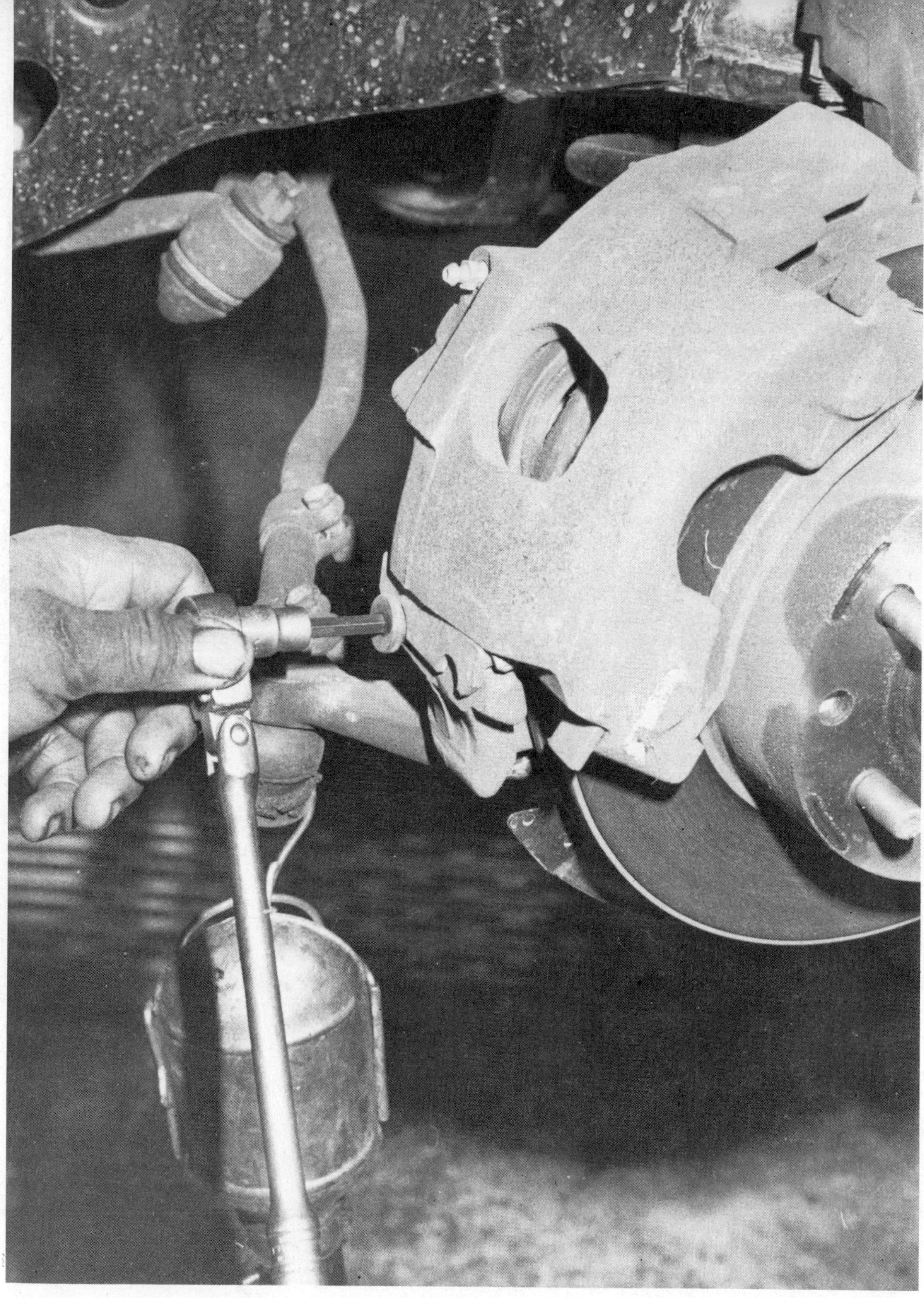

Fig. 11–17. Unbolting bottom way from caliper.

Fig. 11–18. Driving out the unbolted way with hammer and punch.

brush, then refit the O-rings. Smear a film of machined–ways lubricant on the ways. This new lube is available in squeeze tubes. If you can't get it locally, use an aerosol silicone lubricant.

Installing Brake Shoes

In most cases, the disc can remain in place and the caliper will come off with no problem, permitting you to proceed directly to brake shoe replacement. Observe how the old ones fit. For example, one shoe (usually the one that bears against the piston) may have a curved spring that scratches against the disc when the lining is worn down, to alert the driver. If the replacement shoes have such a spring, be sure to install the shoe in the same location (such as against the piston).

The shoe that bears against the piston may be held to the piston by a spring clip. Notice how it fits, so you can correctly install it later (see Fig. 11–19).

On most machined way calipers, the inner shoe (that bears against the piston) does not come off with the caliper. Instead, it remains on the caliper bracket. With the caliper off, it can easily be removed (see Fig. 11–20).

In some designs there is a tab on the back of the shoe that engages a slot near the outer circumference of the piston, to keep the shoe in proper alignment. Don't miss this one.

On many cars, particularly General Motors and Chrysler products, the outer shoe (the one that does not bear against the piston) has a projecting ear on the metal back. Clinch the ear of this new shoe to the caliper, using large pliers (see Fig. 11–21). Note: This can also be done with the caliper mounted in place, as a final step before installing the wheels. If the shoe ear is too thick for bending with vise pliers, bend it by tapping with a hammer, so it is a force fit, and holds the shoe in place (see Fig. 11–22). If the shoe won't fit on with moderate force, clamp it on using old shoes as supports and a C-clamp.

To prevent brake squeak, many discs have silencer pads that bear against the back of the shoe. If yours do not, obtain a tube of paste rubber silencer and apply it to the backs of the new shoes before installation, following the manufacturer's instructions.

Caliper Slide Rods' Bushings and Guides

If the caliper slides along rods through rubber bushings or plastic guides in the caliper (see Fig. 11–23), replace the bushings or guides in which they move. Before installing the rods spray the inner surfaces of the guide or bushings with an aerosol silicone lubricant.

On some cars with the rod arrangement you will think you need three hands to hold the caliper, position the plastic guides, and start the rods. It really can be done with only two, however.

Always use new retainers for the rods if they are not the thread-in type.

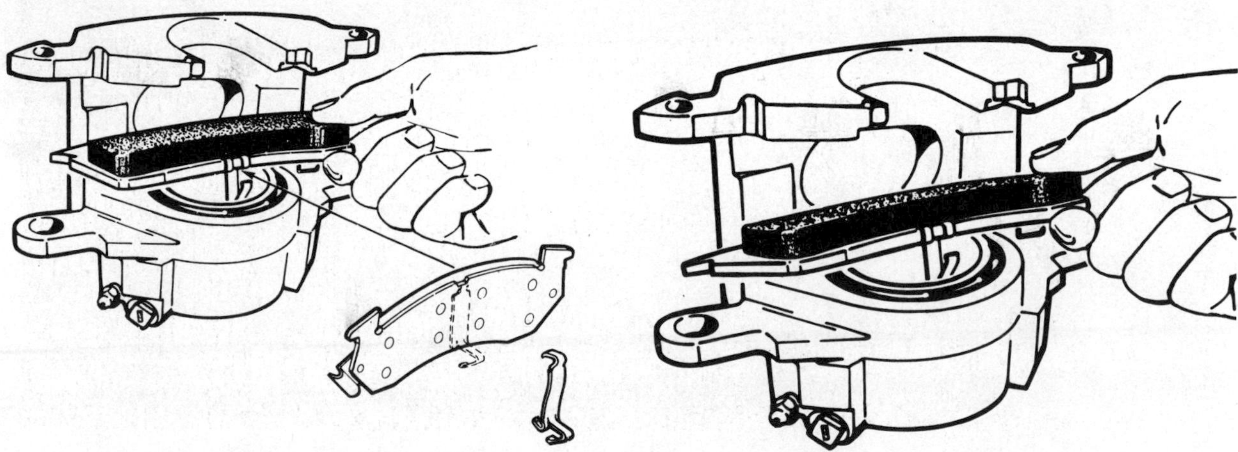

Fig. 11–19. When removing brake shoe that bears against piston, look for a spring as shown and note how it fits so you can correctly reinstall it.

Fig. 11–20. Removing inner shoe left in place when machined-ways caliper is out.

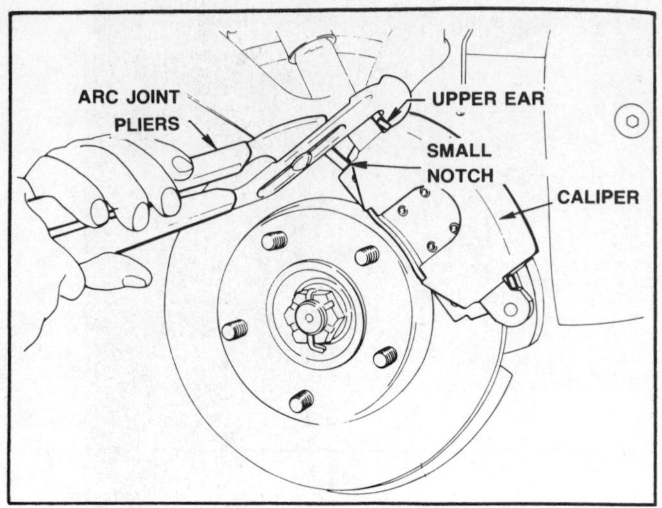

Fig. 11–21. Clinching ear of new outboard shoe with pliers, a procedure typically required on General Motor cars. Note: If this procedure allows shoe rattling on GM "J" cars, an alternate procedure is available. Install lug nuts and tighten. With screwdriver between outboard shoe flange and "hat" of rotor, pry against shoe. Have helper apply brakes. Position 8-oz. ball peen hammer against shoe flange tab and strike ball peen with 16-oz. brass hammer to bend tab to 45-degree angle.

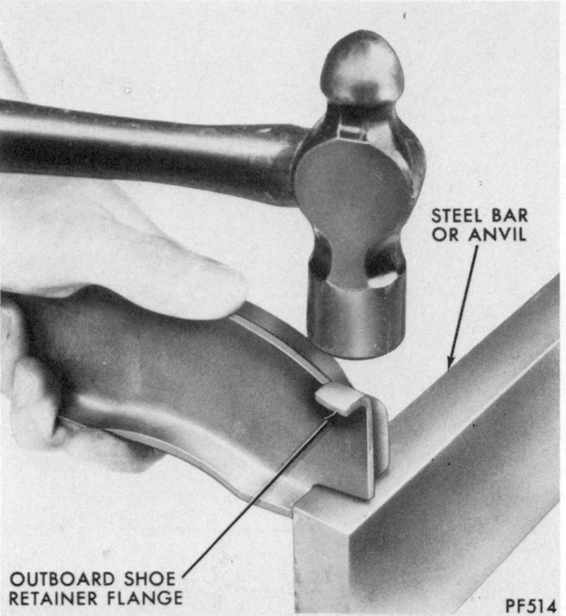

Fig. 11–22. Tapping ear of outboard shoe with a hammer with shoe braced on a steel bar, so that shoe is a force fit on the outboard side of the caliper. If the shoe ear is too tight to simply push into place, clamp it on, using old shoes between the clamp and caliper and shoe.

Fig. 11–23. Plastic guides or rubber bushings should be replaced when doing a brake job. Also, lubricate with silicone spray before installation of the rods.

Fig. 11–24. Note piece of cardboard in caliper. It is on the inboard side against the piston to protect the piston, on a machined-ways caliper, from damage by the inboard shoe, which is put into place first.

Machined Ways Caliper

The shoe that bears against the caliper piston on the machined ways arrangement fits onto the bracket and the caliper (with the other shoe in it) is installed afterward.

Put the brake shoe that bears against the piston onto the bracket, where it will hold in place. Then install the caliper (with the other shoe attached inside) on the bracket. To prevent the possibility of the shoe's metal back damaging the piston dust boot, insert a thin piece of cardboard in the caliper and hold it against the piston as you refit the caliper (see Fig. 11–24). Once the edge of the brake shoe has cleared the piston, pull out the piece of cardboard. Set the caliper in position on the bracket and then bolt on the removable part of the lower way. Note: Tap caliper with hammer if necessary to seat it for installation of the lower way (see Fig. 11–25).

Install the wheels and tighten the lugnuts. Note: On pre-1979 Olds Toronado and Cadillac Eldorado, the lugnuts must be very tight—130 lbs.-ft. With any front-wheel-drive car it's a good idea to check specifications and use a torque wrench to be sure.

Note: If the caliper was disconnected from the brake line, bleed the system of air as described later in this chapter.

Initial braking may be slightly erratic. Break in the brakes by driving the car at 30–40 mph and making moderate stops every half mile. After about ten stops under these conditions the brake action should be normal.

Fixed Caliper Disc Brakes

Some older cars and one late model, the Chevrolet Corvette, have a fixed caliper disc brake system (see Figs. 11–26, 11–27). In this arrangement, the caliper does not move. Rather, there is a piston on each side, or two pistons on each side, that push each brake shoe into the sides of the disc.

You do not have to remove the caliper to change the brake shoes on most of these systems. First, inspect and measure the disc. If it requires service or replacement, leave the job to a professional. If the disc is okay, take off the wheel, remove a retainer (held by a spring clip,

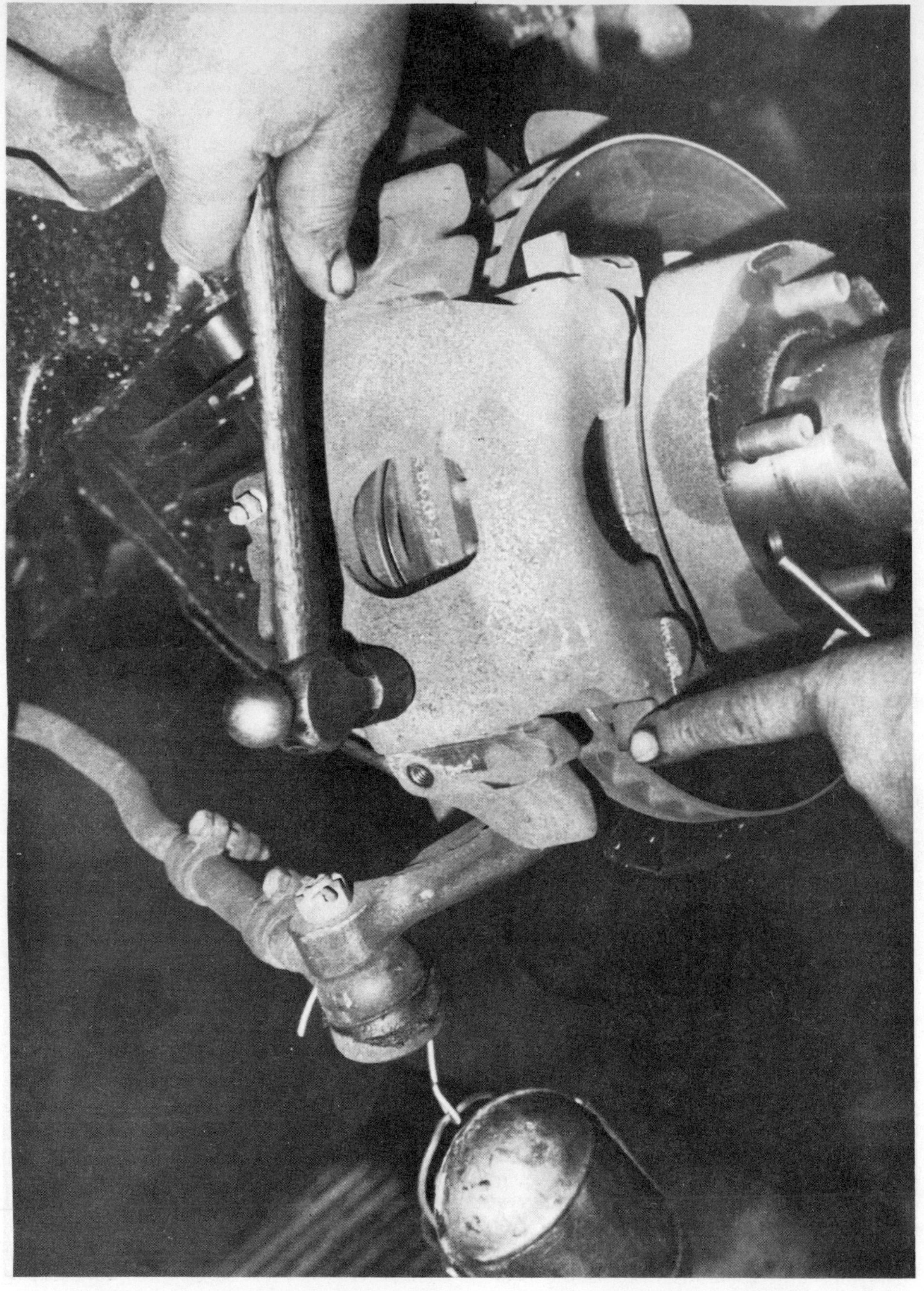

Fig. 11–25. Tap machined-ways caliper into place with hammer if necessary to seat it for proper installation of the removable way.

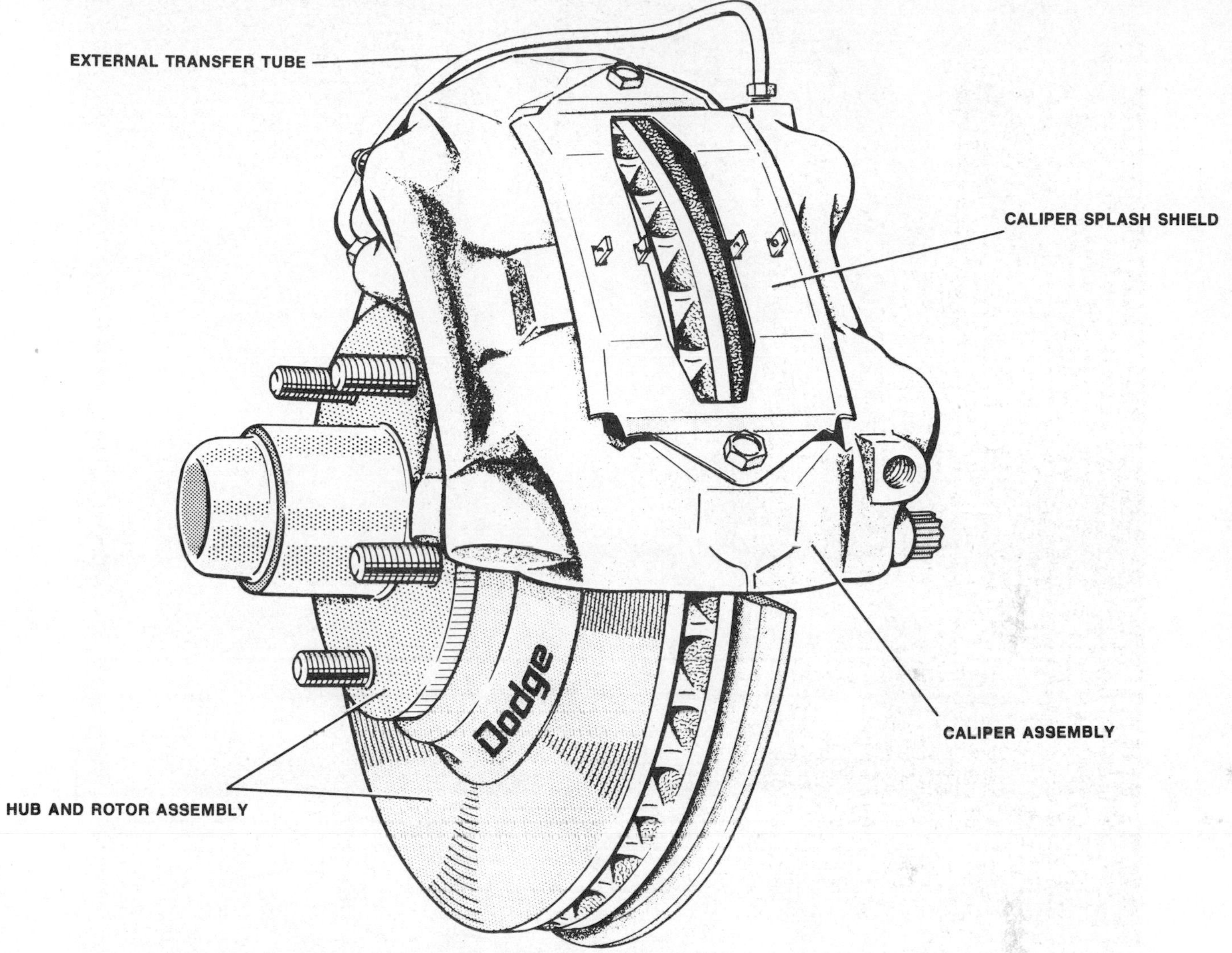

Fig. 11–26. This is a typical fixed caliper disc brake. When splash shield is removed, the brake shoes can be pulled from the side of the caliper. Therefore it is not necessary to disturb the caliper for simple replacement of brake shoes.

cotter pin, etc. from the caliper outer end, then remove the shoe retaining rods (see Fig. 11–28). Now you can pull out the shoes with vise-type locking pliers or a screwdriver (see Fig. 11–29). Grind off the remaining lining from one shoe and reinsert it temporarily, as a brace, and then insert a large screwdriver or two between the shoe's metal plate and the disc (grinding off the lining will give you the clearance to insert the screwdrivers). As with the sliding caliper system, if the master cylinder reservoir is full, siphon some fluid from the reservoir to prevent overflow when you push back the pistons. Pry with the screwdrivers to force the piston back into the caliper bore (see Fig. 11–30). Repeat the procedure for the other side. Now there's clearance to install the new shoes with their full-thickness lining.

Rear Disc Brakes

The rear discs of General Motors' sliding caliper system are serviced in a manner somewhat similar to the fronts. The only difference is the way in which the piston is pushed back.

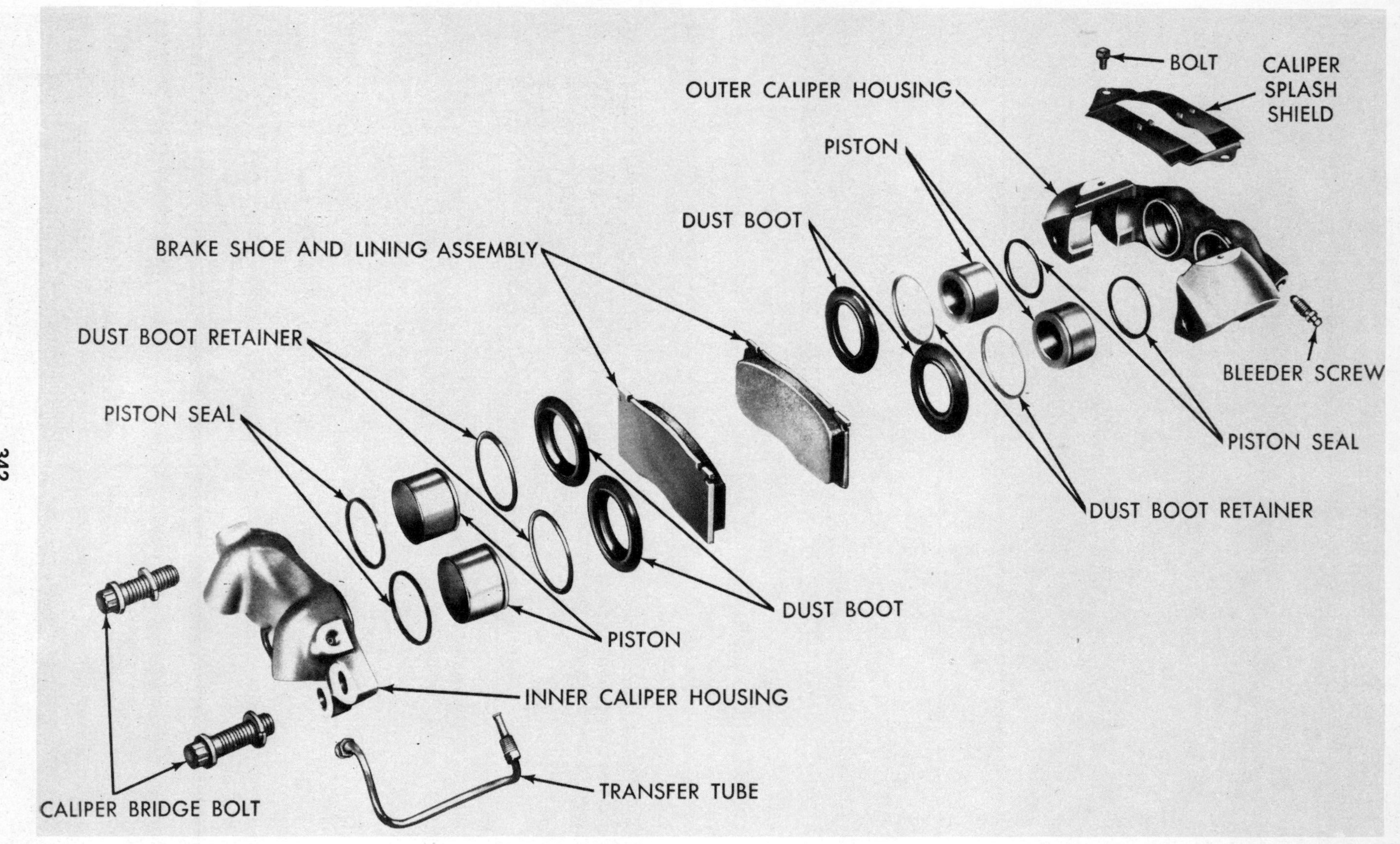

Fig. 11–27. Disassembled view of the typical four-piston fixed caliper. Some imported cars with fixed calipers have had only a single piston on each side.

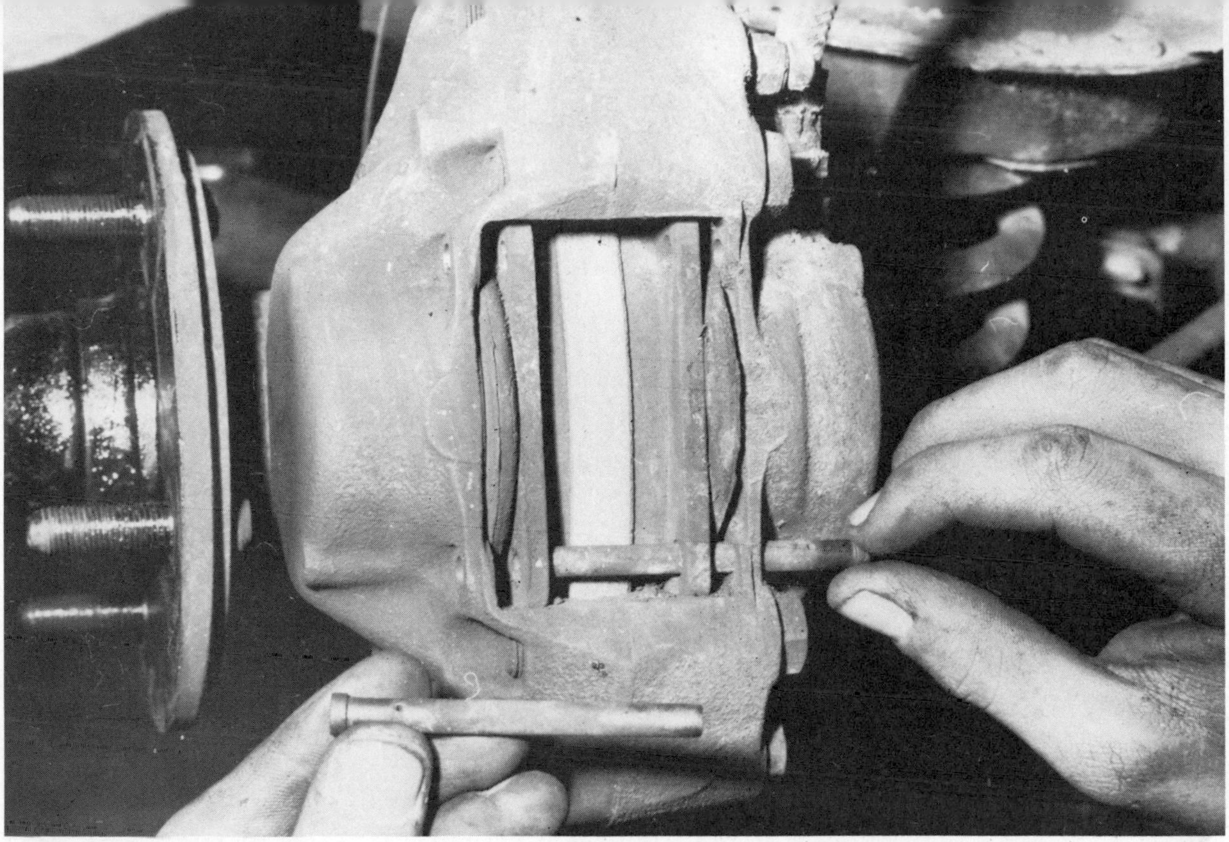

Fig. 11–28. The splash shield is off and now the retaining rods are being removed from a fixed caliper.

Fig. 11–29. On this particular fixed caliper there are holes in the brake shoe steel plates, so you can pry out the shoes with a screwdriver as shown. On other fixed calipers you may have to yank out the shoes with vise-type locking pliers.

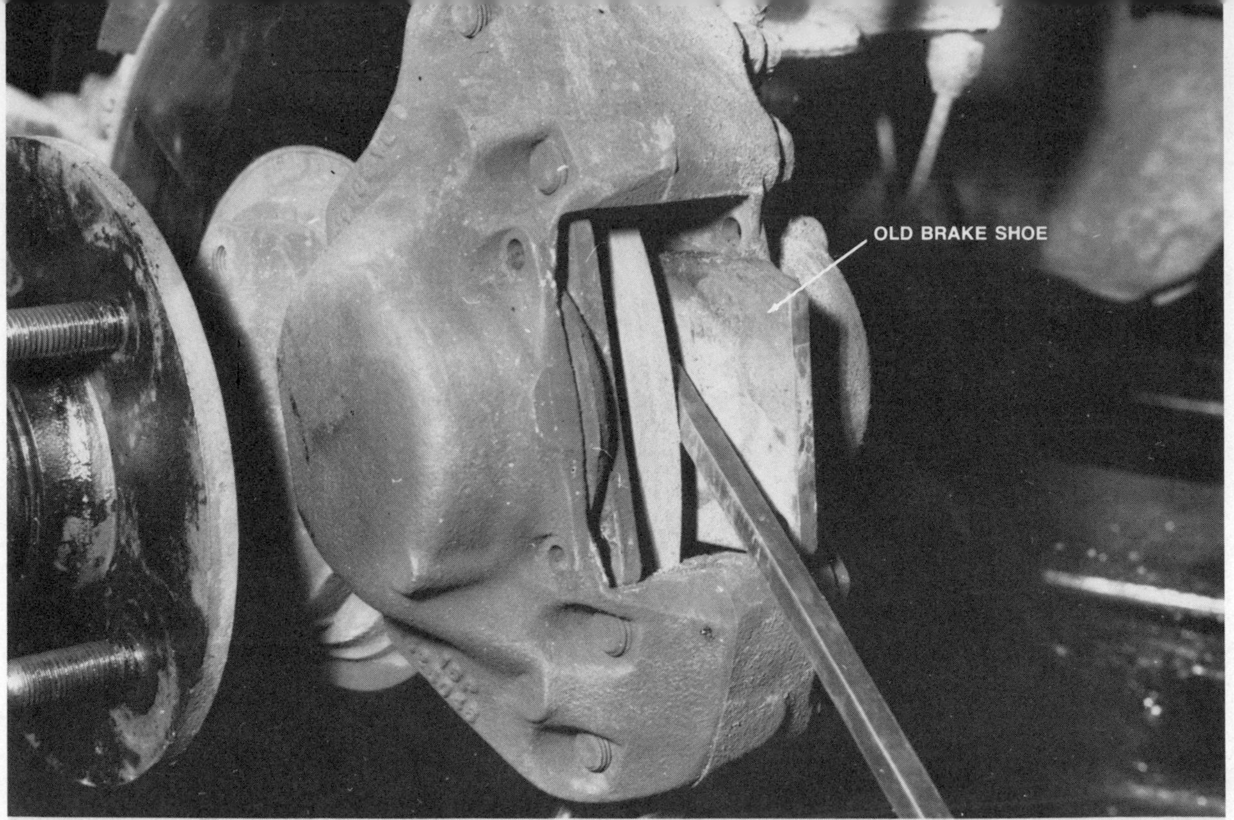

Fig. 11–30. Once the old shoes are out of the fixed caliper, grind off the lining remaining on one of them and reinsert as shown. With the lining ground off, there is clearance to insert a screwdriver between the shoe and the disc, as shown, to pry and so push back the pistons.

Fig. 11–31. Unthreading the hex nut on the back of the rear disc brake caliper of a General Motors car. The rear brake cable must first be disconnected. Although the caliper is shown removed from the car, this was done for illustrative purposes only.

First, remove the parking brake cable lever from the back of the caliper. Next, unthread the hexagonal nut on the back (see Fig. 11–31). Using a C-clamp, push back the caliper. The additional step is necessary because the piston assembly is combined with a mechanical adjuster for the parking brake function.

The rear shoe, which rests against the caliper piston, may be located by a spring clip. Before removing the shoe, note how the clip is installed so it can be refitted properly.

Ford products have a similar arrangement, but to turn back the piston you must use a special tool on the piston itself. This fact, and other requirements of Ford rear disc brake service, make the job best left to a professional.

Drum Brake Service

As with disc brakes, drum brake service should only be attempted after you have had

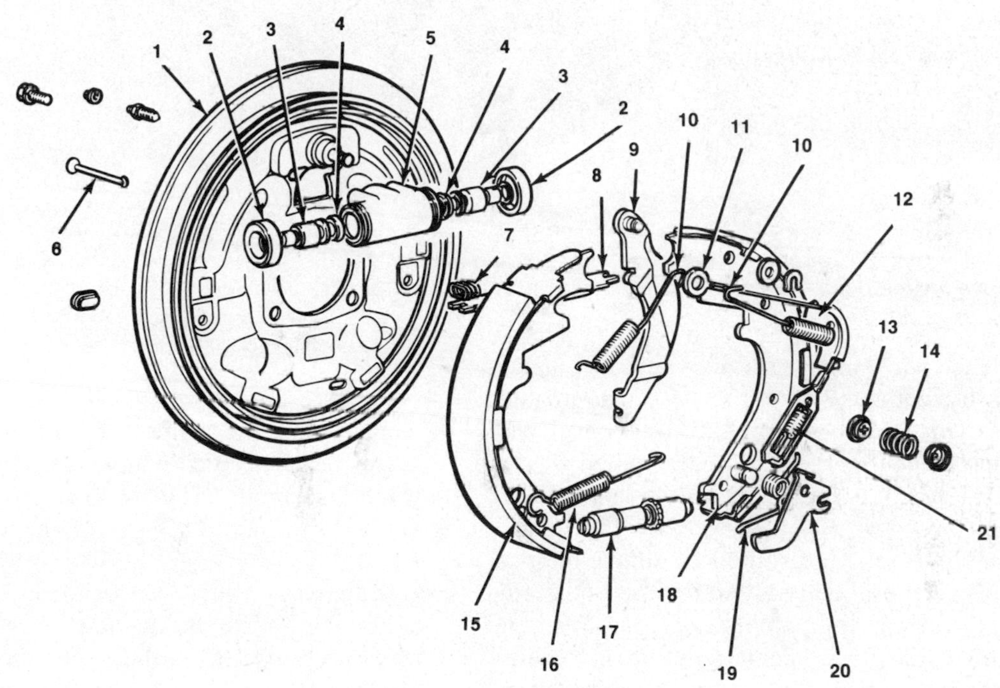

Fig. 11–32. Disassembled view of a rear drum brake. Parts are: (1) backing plate; (2) wheel cylinder dust cover; (3) wheel cylinder piston; (4) wheel cylinder piston cup seal; (5) wheel cylinder body; (6) brake shoe holddown pin; (7) anti-rattle spring; (8) parking brake strut; (9) parking brake lever; (10) brake shoe return spring; (11) brake shoe automatic adjusting cable; (12) brake shoe automatic adjusting cable guide; (13) brake shoe holddown pin spring cup; (14) brake shoe holddown pin spring; (15) primary brake shoe; (16) brake shoe automatic adjusting lever spring; (17) brake shoe star wheel adjuster; (18) secondary brake shoe; (19) spring; (20) brake shoe automatic adjusting lever; (21) over center travel assembly part of automatic adjuster.

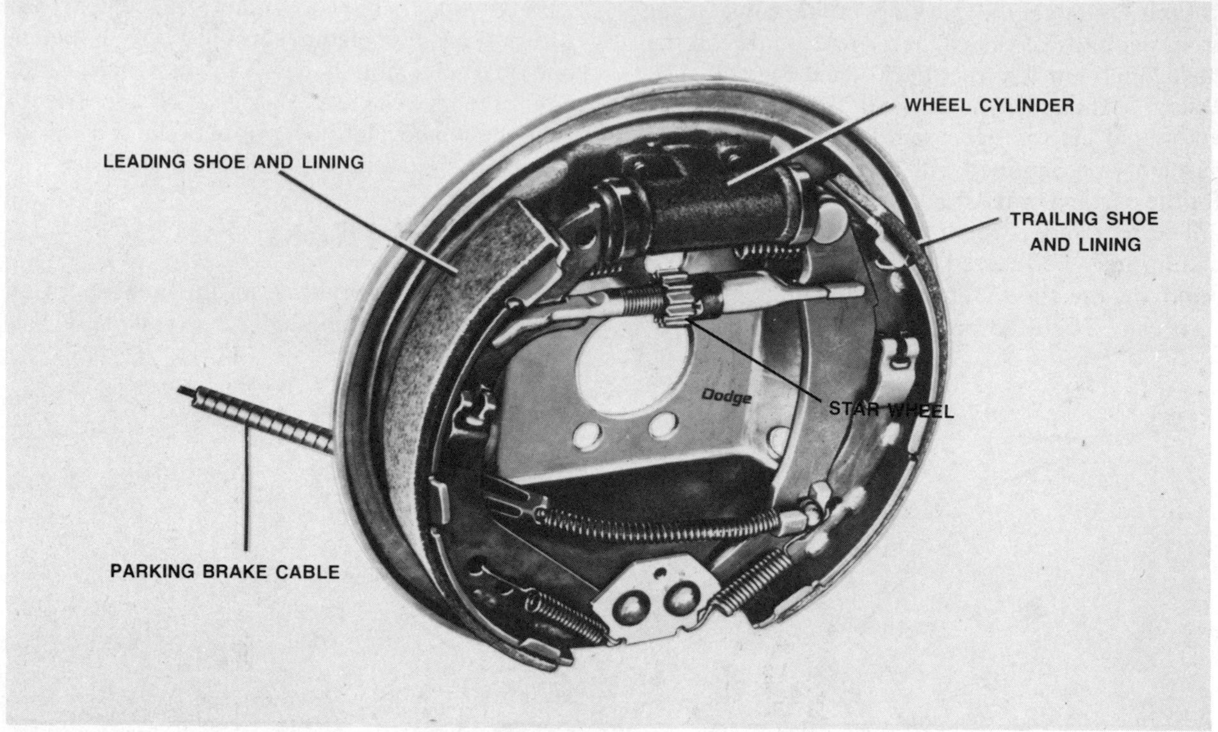

Fig. 11–33. This is a leading-trailing shoe type rear drum brake. Notice that the star wheel is not at the bottom, but just above the midpoint. This particular type does not have an automatic adjuster.

experience with simpler services and have a good familiarity with use of tools. Also as with disc brakes, you can disassemble at one wheel with the knowledge that there is an assembled version of what you are taking apart on the other side of the car.

Your first brake job should be simple replacement of worn brake shoes, and this is something that is within the ability of the weekend mechanic. The only special equipment requirement may be a tool that is used to remove and reinstall brake shoe springs.

The drum brakes on virtually all cars are self-adjusting, but there are some exceptions, as explained later in this chapter. Before beginning, study Figs. 11–3, 11–32, 11–33 carefully, so the parts will look familiar when you see them.

Front Drum Brakes

All late-model cars have front disc brakes, but if you have an older car, it may have front drum brakes instead of discs.

For access to the brake shoes, begin by removing the hub caps or wheel covers, then jack up the car (see *Chapter 5*) and support it on safety stands.

Remove the dust cap in the center of the wheel, then the wheel bearing nut and outer wheel bearing, and finally, the brake drum and wheel assembly. This service procedure is described in Chapter 12 under *Wheel Bearing Service*.

If the brake drum won't slide off, it apparently is binding on the brake shoes. You must, therefore, back off the brake shoes from the drum, working through an access slot in the backing plate (see Fig. 11–34), the steel plate to which the brake shoes are attached. Pry out the plastic or rubber plug in this slot, then look inside with a penlight and you'll see what Fig. 11–35 is all about. Insert a thin screwdriver to push away the lever, then another screwdriver, or an inexpensive tool called a brake spoon, into the slot to engage the slots between the teeth or the shoe adjuster (called a star wheel). In most cases you lift the handle of the spoon upward to back off the brakes. After a few operations with the spoon, the drum should be free.

You now have access to the front brake shoes (see Fig. 11–3). If the linings on the brake shoes

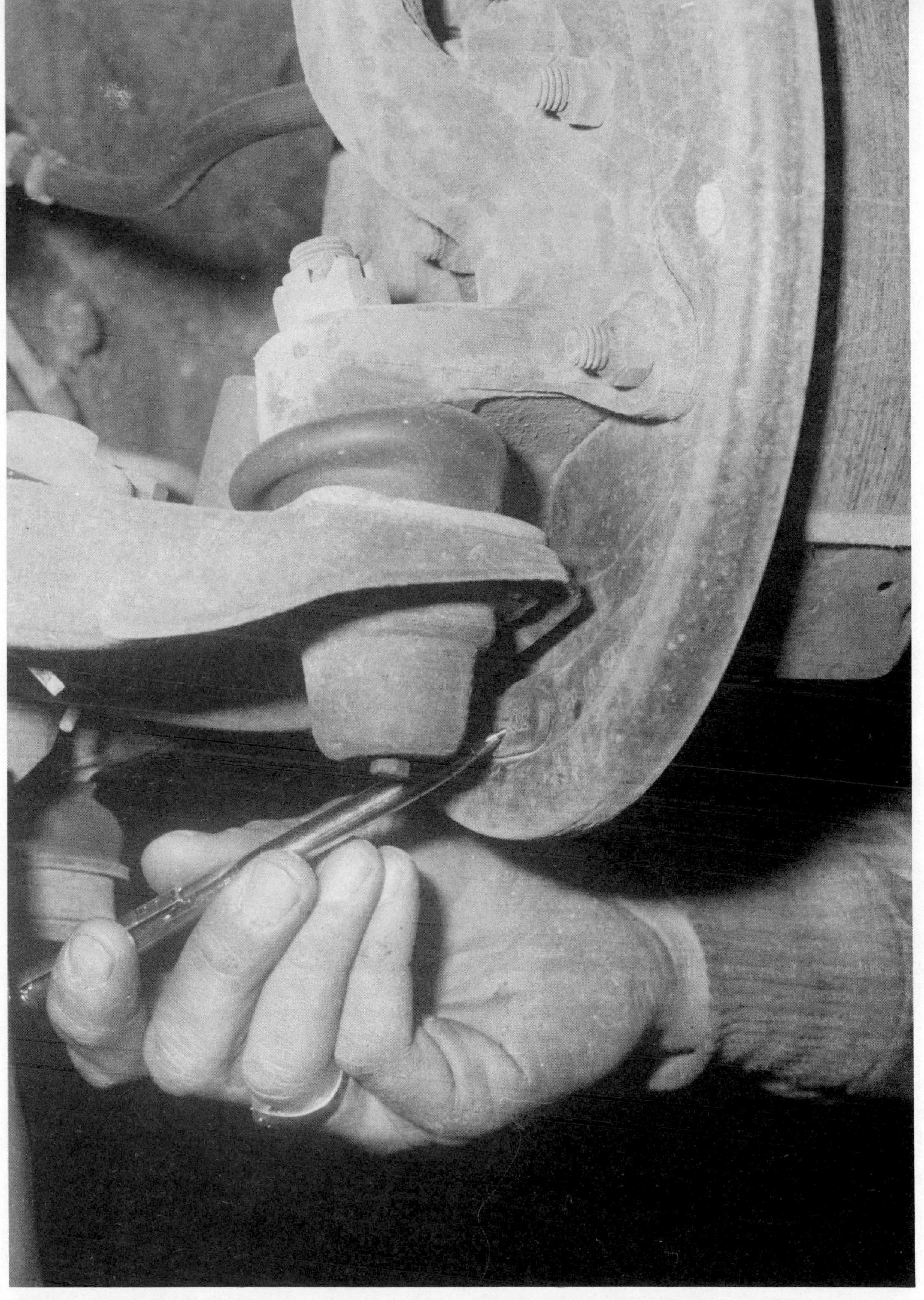

Fig. 11–34. Pen points to rubber plug in the inboard side of the brake backing plate. Pry out the plug for access to brake shoe star wheel adjuster if necessary to back off the brake shoes in order to remove the drum.

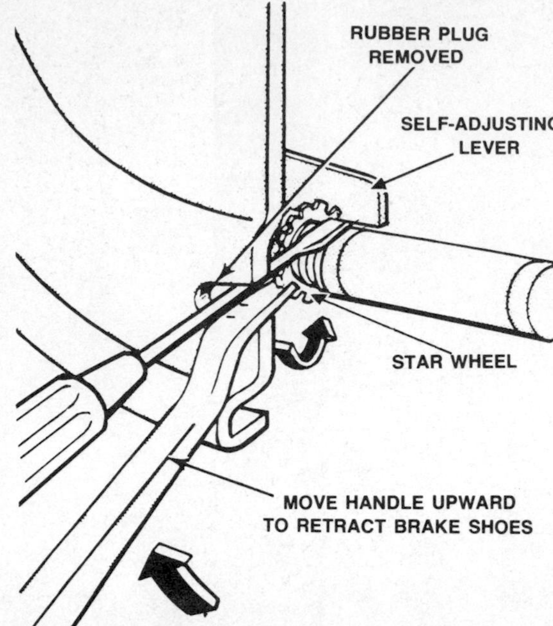

Fig. 11–35. Retracting brake shoes with drum on are shown in this illustration. A thin screwdriver is used to push away the automatic adjusting lever and brake spoon is used to turn star wheel. A screwdriver can be used instead of spoon.

are worn down to $^1/_{16}$ inch at their thinnest point (see Fig. 11–36), the shoes should be replaced. If the linings are held to the shoes by rivets, the $^1/_{16}$-inch thickness should be above the heads of the rivets.

The normal wear pattern of the linings is somewhat greater in the center. If you see an irregular wear pattern, such as tapered from inboard toward outboard side, or very thin at one or both ends, the shoes and/or drums are warped. You will automatically replace the shoes and you also should take the drum into an automotive machine shop for measurement and resurfacing if necessary. Note: If one drum needs resurfacing, have the one on the opposite side resurfaced too, and have the shop fit new shoes with oversize linings to match. The machine shop is typically an adjunct to an auto parts store, so this is normally no problem.

Also check the drum's brake shoe contact surface with your finger (see Fig. 11–37). If there are any scores deep enough to catch a fingernail,

Fig. 11–36. Measuring thickness of brake lining with ruler. The ruler shown is one that is included with a set of feeler gauges and has very small gradations.

Fig. 11–37. Feeling brake drum surface with finger. If there are scratches on which a fingernail hangs, the drum must be resurfaced.

take the drum in for resurfacing—and the opposite side drum too.

Look for any cracks in the drum, and if you find any that go through, replace the drum. Drums are not inexpensive, so consider a used one from a wrecking yard (see *Chapter 16*).

Replacing the Shoes

Begin by removing the brake shoe return springs (also called retracting springs as in Fig. 11–38). Then disengage the self-adjusting mechanism, the method for which depends on the type.

If it's a simple cable type, just remove the connecting spring from the lever to the primary shoe, then disconnect the cable.

If there is a cable with an overtravel assembly (designed to prevent overadjustment of the brakes under abnormal conditions) as shown in Fig. 11–32, begin by sliding the eye of the cable off the anchor at the top, then remove the cable guide from the shoe. Next, disconnect the cable from the opposite end of the adjuster lever and rotate the adjuster lever clockwise. You'll be able to disengage it from its retaining pin and a connecting spring.

With a lever type self-adjuster, just remove the two springs between the lever and the secondary shoe and the wire-like actuating link that connects to the anchor pin at the top and the lever at the bottom. You'll then be able to take out the lever itself.

Other self-adjusting lever mechanisms also are used, but no matter what the individual design the method of removal should be reasonably obvious. If there are springs connected to the lever, the usual procedure is to take them off, then disengage the lever and any other links.

Important caution: On many cars the brake shoe return springs for primary and secondary shoes are not the same, although visually very close. Be careful not to confuse them. In fact, to be sure you don't get mixed up, tape the springs to a piece of cardboard, in the same relative position as on the car. Then double-check yourself by referring to the wheel assembly on the opposite side, keeping in mind that a primary shoe spring on the driver's side is on your left as you face it, but on the passenger's side it's on

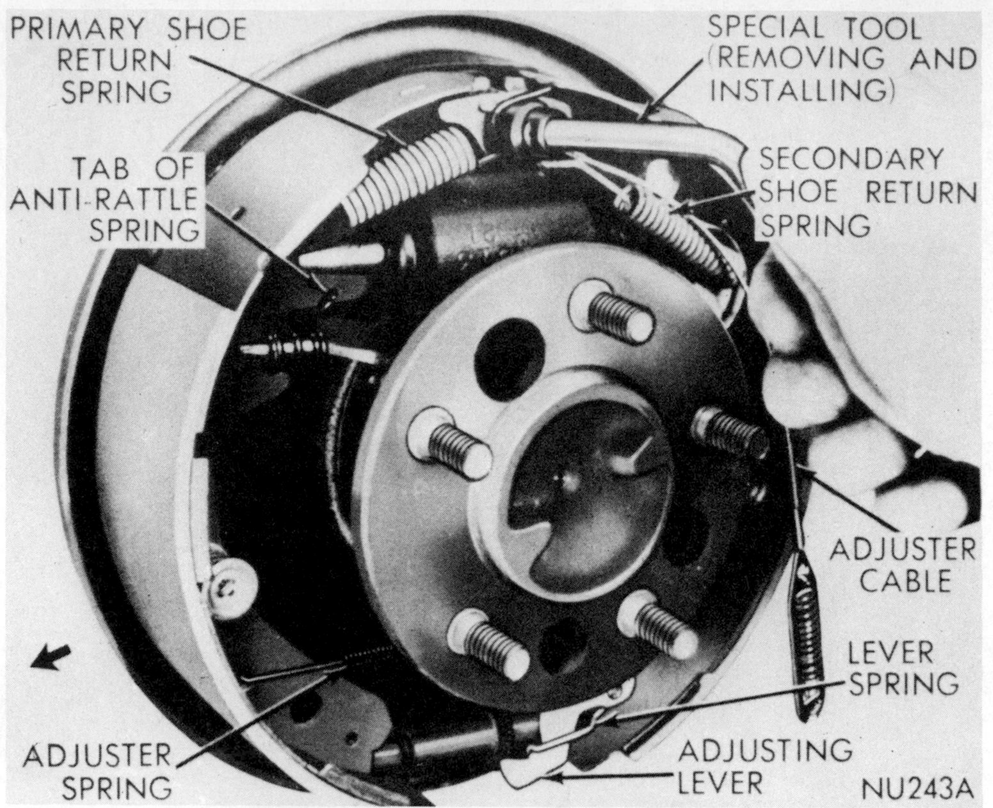

Fig. 11–38. Removing brake shoe return springs with special tool, an inexpensive item available in most auto parts stores. Other types of special tools, including a form of pliers, also are available for this job. The tool is shown at the anchor end of the return spring on a rear drum brake, but the principle is the same for front and rear drum brakes of this type. Note: This tool cannot be used on General Motors transverse-engine front-drive cars because there is no anchor pin against which the tool can react. Instead, there is a riveted-in-place shoe rest at the top. To remove and reinstall the springs without a struggle, use a special tool with a hook attached. The special tool, sold in auto parts stores, reacts against a rivet head, as the hook pulls on the spring (see illustration at bottom).

Fig. 11–39. Using the special tool to disengage the retainer from the shoe holddown pin.

Fig. 11–40. Spreading the brake shoes apart from the anchor and wheel cylinder to remove them from the wheel. Caution: If the innards of the wheel cylinder start to come out, stop immediately and wrap wire around them and the wheel cylinder body to prevent this. Or you can obtain special spring clips from an auto parts store to hold the wheel cylinder together.

your right, and vice versa for the secondary springs.

• *Hold down springs*. The brake shoes are held in the center by little spring-loaded hold-down assemblies. You can remove them with the special tool by just pushing in on the top retainer and turning 90 degrees, which will turn the retainer so that the head of a through pin will pass through the key slot in the retainer (see Fig. 11–39). Without this special tool, you can do the job by pressing down on the retainer with needle-nose pliers and using the plier jaws to grasp and turn the head of the retainer pin.

Finally, spread the shoes away from the wheel cylinder and lift them out, with the star wheel and connecting spring at the bottom still attached (see Fig. 11–40).

• *Star wheel*. The star wheel is the adjuster. As you will be able to see with the drum off, the movement of the self-adjusting lever rotates just the star wheel, unthreading it to push out the brake shoes closer to the drum to compensate for lining wear. When you reached in with two thin screwdrivers or an awl and screwdriver to back off the shoes, what you did was thread down the star wheel and the return springs pulled them away from the drum surface.

If any of the teeth on the star wheel are damaged, the star wheel must be replaced. If you try to re-use it, the self-adjusting lever may simply ride over damaged teeth and not turn the star wheel.

Even if the star wheel teeth are in good condition, unthread the assembly and spray the threads with penetrating lubricant. Then thread it together and unthread it several times to work in the lube (see Fig. 11–41).

• *Backing plate*. Clean with solvent and wipe off the outboard side of the brake backing plate. If there are any raised platforms on the backing plate, lube them with a thin film of silicone grease.

• *Wheel cylinder*. The wheel cylinder is the terminal of the hydraulic system. Inspect it by gently prying back the dust covers at each side with a thin screwdriver. If you see any evidence of brake fluid leakage, you must replace the wheel cylinder. If the dust boots are cut or heat-cracked, replace them. Refer to *Brake Hydraulic System,* later in this chapter. Note: A beginner who might not wish to tackle replacement of a defective wheel cylinder should make the inspection before removal of the shoes (see Fig. 11–42).

• *Reassembly with new shoes*. Before you leave for the parts store to exchange your old brake shoes for replacements, check the shoes for any dowel pins or guides for the self-adjusting mechanism. On many shoes they are removable parts (unbolt or knock out with a hammer and punch) that must be installed on the replacement shoes.

When you buy new brake shoes, ask the parts supply store for its premium line. It may cost three times the price of the cheap line, but you get exactly what you pay for. The better lining will not only last a lot longer, it will provide better stopping for its entire life.

You may be offered a choice of riveted vs. bonded linings. At one time mechanics always took the riveted because the bonding agents were of inconsistent quality and sometimes there was separation of lining and shoe. This is not a problem today. The bonded lining will last longer because it can be worn down almost to the shoe, whereas the riveted lining must be changed when it is close to the tops of the rivet heads.

If the brake springs have a bluish tinge (indicating they were overheated) or if the springs were not replaced at the last brake job, get a new set now. It's an inexpensive purchase that adds a measure of security to the job.

Reassembly of the brake parts is basically the reverse of disassembly. Work with clean hands and a clean backing plate to prevent grease from getting on the linings. If any grease does, lift up the glob with a putty knife and sand down the lining to remove any that got into the surface.

A simple starting point is to install the shoes with the hold-down assemblies, then insert the star wheel adjuster and connecting spring at the bottom. Add the self-adjusting mechanism and springs, and finally, the shoe return springs. Be very careful (wear safety glasses) when installing stiff springs such as the shoe returns. If one slips off the tool it might cause injury. Note: The star wheel adjuster should be threaded all the way in.

Refit the brake drum and wheel assembly, and the wheel bearing and adjusting nut, and adjust the bearings as explained in Chapter 12.

Caution: The self-adjusting drum brakes adjust only when the car is braked while moving in reverse. Do not back the car up until you are assured the road is clear, then brake gently the

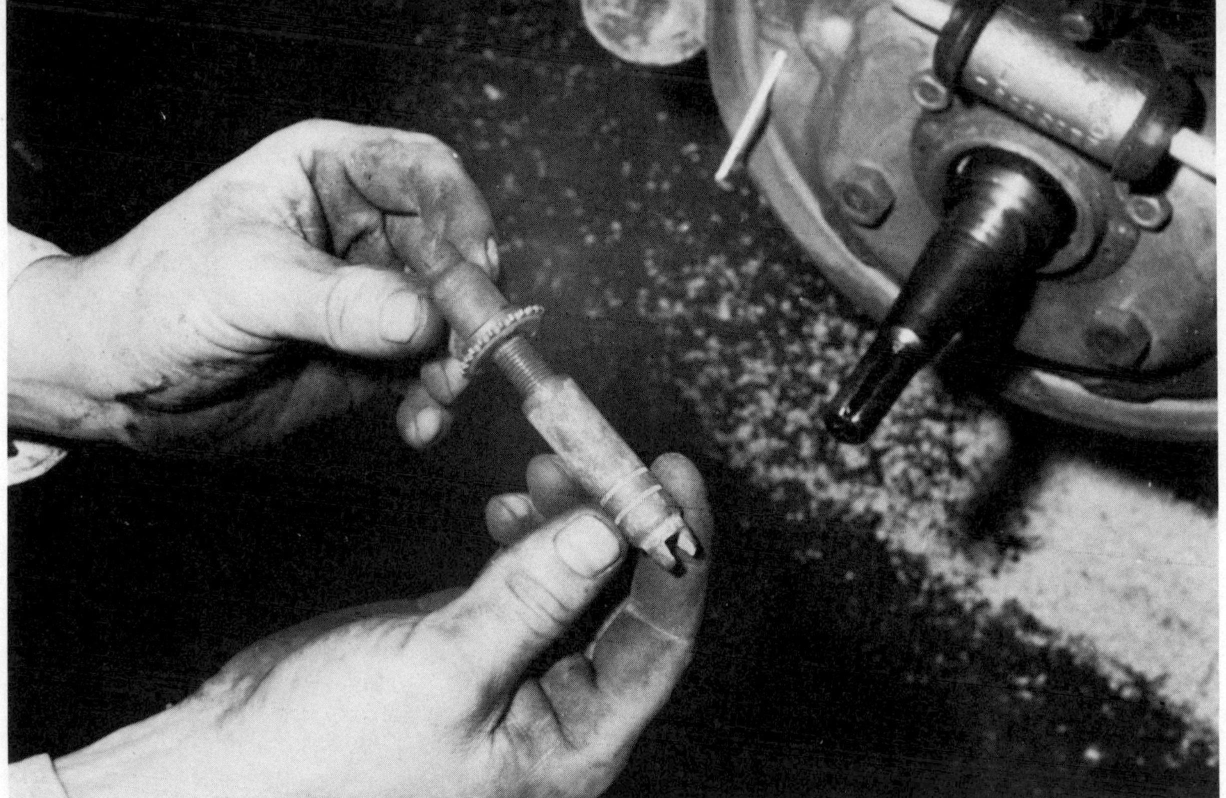

Fig. 11–41. Unthread the two sections of the star wheel adjuster and lubricate the threads with penetrating solvent, then thread them together and apart as often as necessary to work the solvent in. Stop when the parts thread absolutely freely.

first few times, taking it as easy as possible on the brakes for the first fifty miles or so to give the linings a chance to wear in.

Rear Drum Brakes—Rear Drive Cars

Although there are similarities in the service of front and rear drum brakes, there also are important differences.

Perhaps the most important is getting off the drum. Loosen the wheel lugnuts on the ground, then jack up the car (see *Chapter 5*), support it on safety stands, and remove the wheels.

Look at the studs and see if there are any tiny clips on them, to hold the drum. If there are, pry them off with a screwdriver and discard. Or if there is a screw holding the drum, remove it.

Try to pull the drum straight off. If it is stuck, turn the drum slowly and listen for a scraping noise. Have trouble turning the drum and hear a scraping noise? Well, there is some resistance to turning a rear drum because you're also turning a shaft and gears in the rear axle, but if there's real binding and a scraping noise the brake shoes are probably in contact with the drum. Back them off as described earlier for front drum brakes. If there is no slot in the backing plate for access, look for a lanced area in the drum or backing plate and knock it in with a hammer and punch to obtain an opening. Turn the drum until the opening in the drum is aligned with the star wheel, hook onto the self-adjusting lever with stiff wire, and turn the star wheel with a screwdriver or brake spoon. Note: This opening must be sealed; plugs are available at auto supply stores. Discard the lanced piece after removing the drum.

If there is no access slot or lanced area for a star wheel, look for what appear to be hex-head bolts on the brake backing plate, near the outer circumference (where the shoes are located). These hex-heads operate cam adjusters for the brake shoes. Turn them all the way until they stop, one at a time. Then spin the brake drum, and if it spins, the shoe is backed off. If it is locked, turn the hex-head all the way in the opposite direction, and the shoe will be backed off (see Figs. 11–43, 11–44).

If there's no scraping noise or unusual resistance when turning the drum, the problem is that it's stuck to the axle shaft hub. You have four possible approaches:

(1) Douse the joint between the hub and the big hole in the center of the drum with penetrating

Fig. 11–42. The wheel cylinder dust cover can be pried away gently, as shown, to check for fluid seepage. If there is none, the cylinder may be assumed to be in good condition. This can be done with the brake shoes off, or before you start the job (as illustrated), when everything is in place.

solvent. Also douse the joints between the lugnut studs and the drum. Allow the solvent to soak in and pry with big screwdrivers (180 degrees apart) between the brake drum and the backing plate. It may take several treatments over several hours to free up the drum (see Fig. 11–45).

(2) You can use a brake drum puller. In combination with penetrating solvent treatments this should pull even the most stubborn drum. The puller is an inexpensive item that will easily justify its existence, but if you don't want to buy you probably can rent one. The puller shown in Fig. 11–46 is typical and easy to use. The jaws grasp the edges of the drum, and you turn down the forcing screw with a T-handle (the forcing screw pushes against the axle shaft hub while the jaws pull out on the drum).

(3) On cast-iron drums you can apply heat from a propane torch around the areas where the binding exists (stud and hub holes in the drum). Heat causes the holes to expand and break the rust bond. Note: Do not use heat to break a corrosion bond on an aluminum drum. If in doubt, put a magnet to the drum's exterior surface. It will not be drawn to aluminum. Caution: Work in a well-ventilated area when using a torch at the rear, and aim the flame away from the gas tank. Gas fumes and flame are an explosive combination.

(4) On some imports, push in a brake hydraulic pressure regulator lever (see Fig. 11–47).

• *Shoe replacement*. Once you have the drum off, you can proceed to disassemble the brake as you would a front brake, with these exceptions:

The rear brake also serves as the parking brake. A set of cables and levers from the parking brake handle or pedal operates the rear brakes mechanically, to hold the car for parking. You therefore must disconnect the parking brake from the secondary brake shoe.

A typical arrangement is a clip (possibly with a spring washer) holding a lever to a brake shoe. The lever is hooked onto the parking brake cable (see Fig. 11–48). Disengaging the parking brake lever from the shoe is usually obvious, such as by prying out the spring clip with a screwdriver (see Fig. 11–49).

If the lever is held to the shoe by a pin, you may have to remove that pin from the old shoe and install it on the new one. Note: On some cars it's easier to disconnect the cable from the lever, remove the lever with the shoe, then separate.

On some rear brake systems the two shoes are the same. If you might re-use the brake shoes, you should mark one of them so you can reinstall in the same position.

On some rear drum brakes, a single return spring is used for both shoes. Limited clearances and the shape of the spring make the standard brake spring tools unusable. Instead, do the job with needle-nose pliers and be prepared for a bit of a struggle, particularly on reinstallation. It may be helpful to have an assistant. While you stretch the spring to align the hook with the eye in the shoe, he pushes on the hook with a screwdriver or punch to force it into the eye.

Manual Adjustment Brakes

Not all cars have self-adjusting drum brakes. If your owner's manual lists rear brakes for periodic adjustment, look for one of the following:

• *Access hole in backing plate and a star wheel adjuster*. In some cases it is at the top, just under the wheel cylinder. See Fig. 11–33.

• *Hex-head bolts for cam adjusters on the backing plate*. They resemble the arrangement for the cam-type automatic adjusters explained earlier, but only bear against the shoes, rather than have a pin fitted into them.

The adjustment procedure is simple: Turn the adjuster while spinning the wheel forward by hand (car jacked up and on safety stands). As soon as the wheel starts to drag noticeably, back off (turn the adjuster the opposite way) until the wheel just spins freely. With two adjusters (one for each shoe), set each one separately.

Rear Brakes—Front Wheel Drive Cars

The rear brakes on front-wheel-drive cars somewhat resemble the front drum brakes on older cars. You remove the wheel, but then it's a matter of removing the wheel bearing dust cap, wheel bearing nut and outer bearing, and pulling off the drum to gain access to the brake shoes. On some newer models, the wheel bearings are not adjustable, so the drum unbolts from the assembly that holds the wheel bearings.

Although most rear drum brakes on front-drive cars have star wheels accessible through a hole in the backing plate, there are exceptions. A significant one, introduced in 1980 on VW Rabbits, is a tapered-wedge arrangement (Fig. 11–50). To retract the shoes on this model, you must remove the wheel and turn the brake drum so that a lugbolt hole aligns with the bottom of the wedge. Then insert a screwdriver and pry up on the wedge, against the pressure of its spring. As you do, the stronger spring between the two shoes will compress, retracting them.

Note: Whenever you are working on drum brake designs with a single spring, you cannot use a brake spring tool to lever it back on. Instead you must grasp the spring close to where it begins to coil, using needle-nose pliers as close to the tips of the jaws as possible. Then pull the spring until the hook end aligns with the hole in the brake shoe, and push the hook end into the shoe hole with a screwdriver. You may find that the spring is so strong that you will want to hold the needle-nose pliers with two hands, in which case you will need a helper to push the hook end of the spring into the shoe hole.

Brake Hydraulic System

The hydraulic system transfers foot pressure on the brake pedal to operate the friction mechanism at each wheel to stop the car, as explained at the start of this chapter.

If your brake warning light comes on, it means there has been a failure at one or more wheels. The failure could be in either the hydraulic or friction systems.

To check the hydraulic portion, remove the master cylinder reservoir cover (see Fig. 11–9) and see if you have lost fluid. A small drop in the fluid level is normal on cars with disc brakes, but if the level is quite low, there is a leak in the system.

Inspect brake hoses and tubing connections from the master cylinder while a helper presses down on the pedal. If you can't find a leak, remove the brake drums and then the disc calipers, as explained in earlier segments of this chapter, to inspect for leakage.

A leaking wheel cylinder or caliper or a defective master cylinder should be replaced. You

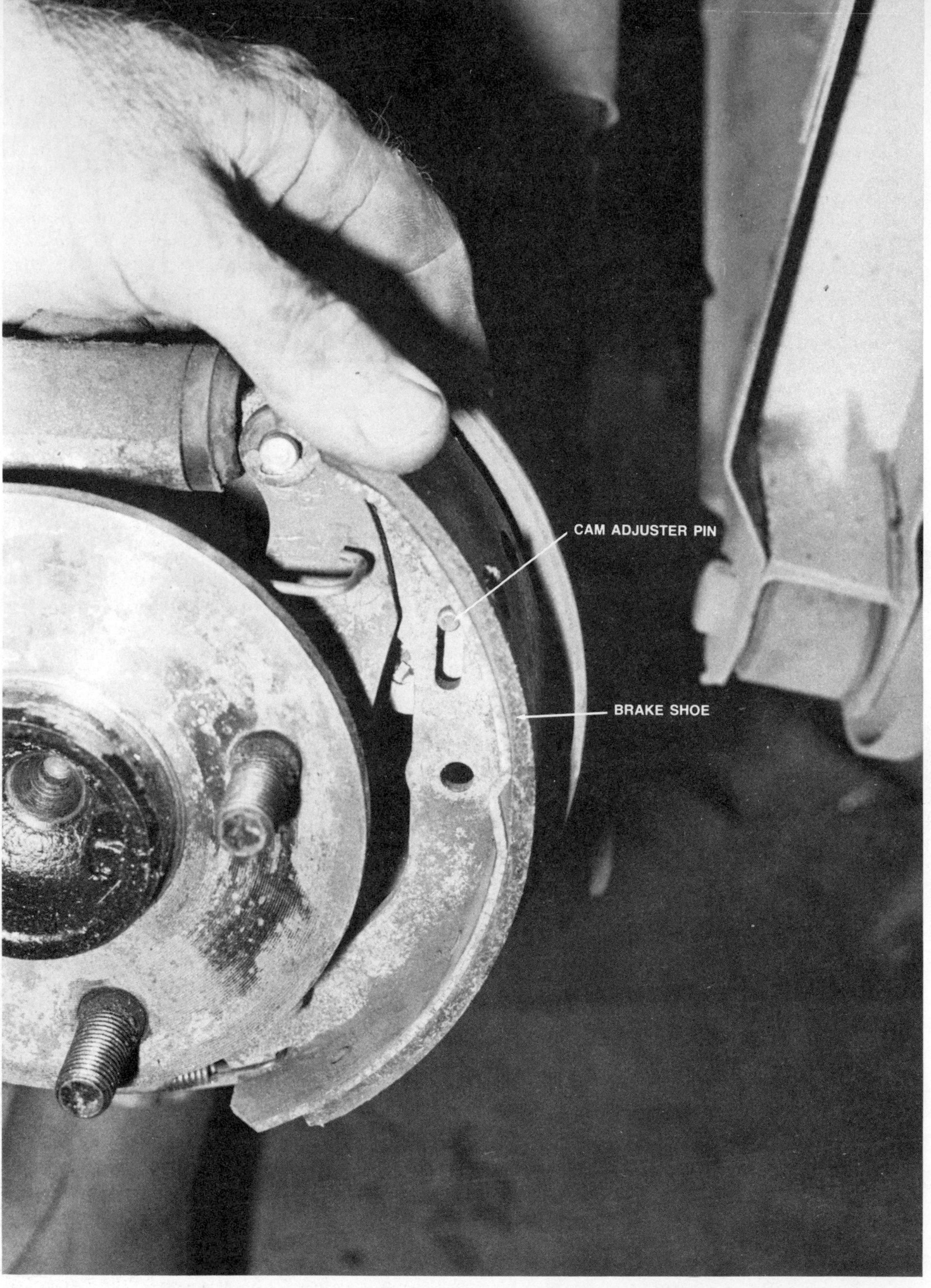

Fig. 11–43. Some self-adjusting drum brake systems do not use the star wheel. This one has a cam adjuster with a pin that operates in a slot in the brake shoe.

Fig. 11–44. With the brake shoes removed you can see the cam adjuster and pin. Turning a hex-head on the inboard side of brake backing plate with a wrench spreads or retracts the shoes.

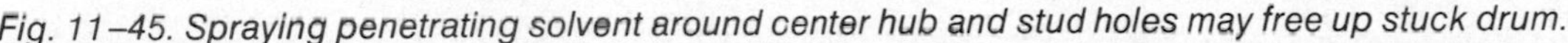

Fig. 11–45. Spraying penetrating solvent around center hub and stud holes may free up stuck drum.

Fig. 11–46. Special brake drum puller available in auto supply stores can be used to pull drums when solvent treatment doesn't work. Jaws grip against inboard edge of drum while forcing screw is turned down against axle shaft hub.

might be able to rebuild, but this is an operation requiring some experience and therefore not recommended for a beginning weekend mechanic. Inexpensive rebuilt components are readily available from auto parts stores.

Note: Whenever you do any work on the brake system that involves draining old fluid, discard that fluid, even if it is visually clean. Brake fluid attracts moisture in the air and this lowers the boiling point of the fluid, which reduces its effectiveness. As explained earlier, buy only a high-quality brake fluid, with a boiling point of 450 degrees F. or higher. Keep the fluid container tightly capped when not in use, to prevent the fluid inside from attracting moisture.

Replacing Calipers

Remove the caliper as explained earlier in this chapter, then disconnect the brake line at the caliper with a wrench. Plug the end of the brake line with a rubber eraser or something similar to minimize fluid loss and entry of air. Squirt brake fluid slowly into the replacement caliper to displace as much air as possible, then reconnect to the brake line. Reinstall the caliper and complete the job by bleeding the brakes, as explained later in this section.

Replacing Wheel Cylinders

Most wheel cylinders are held by bolts which are accessible either from the inboard or outboard side of the brake backing plate. First disconnect the brake line to the cylinder, working at the inboard side of the backing plate with a wrench. Plug the end of the line to minimize air entry and loss of fluid. Remove the drum, take out the bolts, and withdraw the wheel cylinder. If you see no bolts, look for a circular locking retainer on the inboard side. To remove it, insert awls or pins of ⅛-inch diameter or less into the access slots between the wheel cylinder and the retainer's lock tabs. Bend both tabs away with the tools, at the same time, until the tabs spring over the wheel cylinder and release it. See Fig. 11–51).

Installation of a wheel cylinder is basically the reverse of removal. You must complete the job by bleeding the brakes.

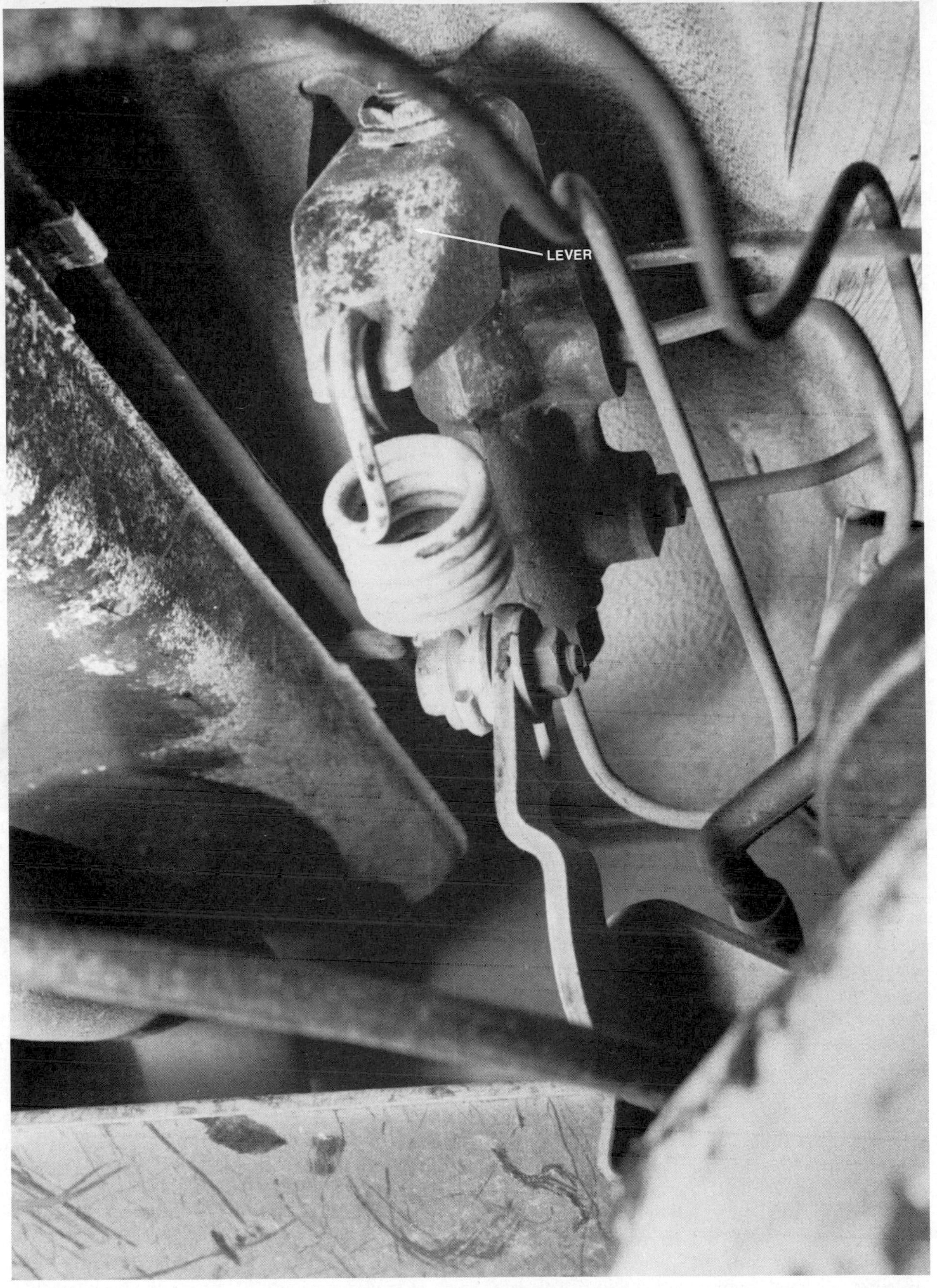

Fig. 11–47. If you have an import with this type of brake pressure regulator, push lever toward spring before attempting to remove the drum.

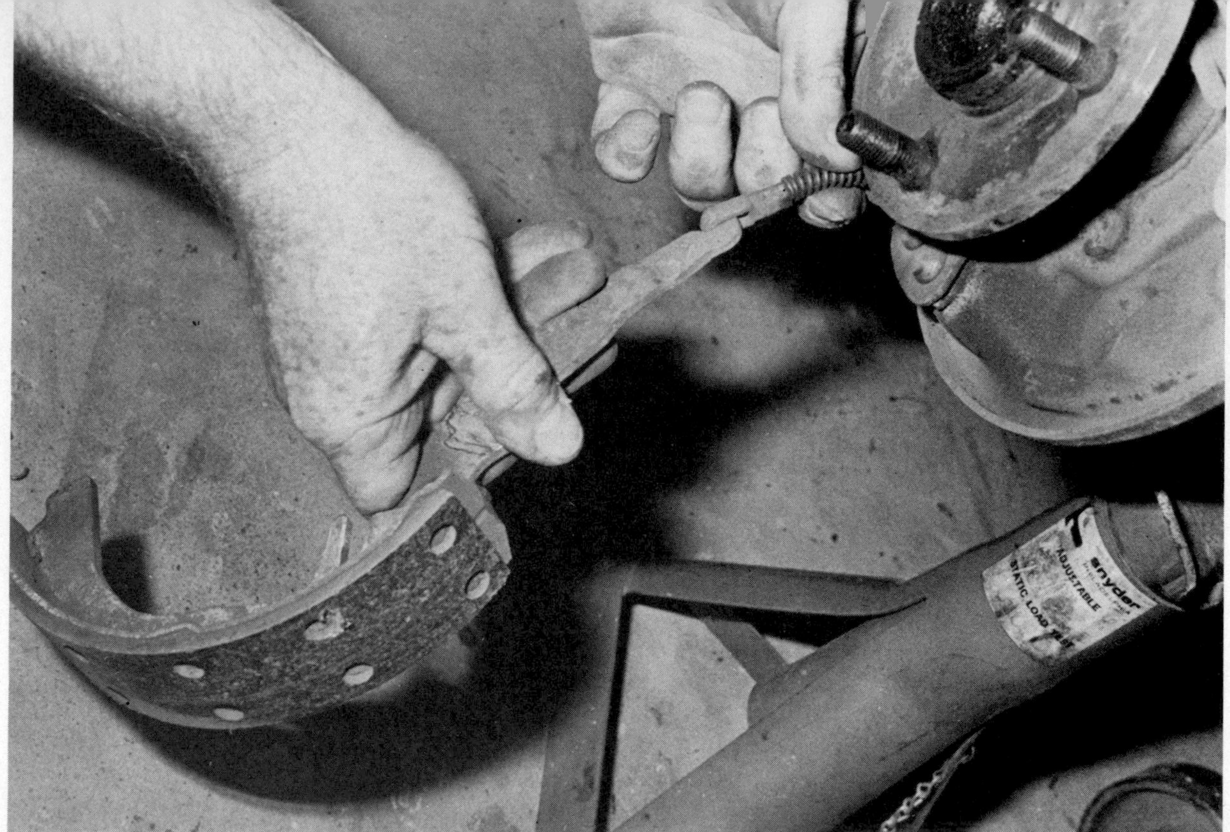

Fig. 11–48. Parking brake lever from brake shoe is hooked to parking brake cable. Just disengage this type with brake shoe removed.

Bleeding the Brakes

There are many methods of bleeding the brakes, and many special tools. Here are two ways particularly suitable for weekend mechanics.

• *Use of a Mityvac vacuum pump brake bleeding kit.* This kit (see Fig. 11–52) permits one-man bleeding. Just set up the kit as shown, making sure the jar lid is on tight. Attach a cone adapter to the open end of the long piece of tubing. Put a suitable box wrench on the brake bleeder valve of the caliper or wheel cylinder. A suitable wrench means one meant for brake bleeder valves, typically with an offset shank so it can easily reach the recessed bleeders used on many cars. There are several possible sizes, so measure the bleeder hex with an ordinary wrench before you buy. Note: Some rear disc calipers have two bleeder valves; do each one separately.

Open the valve fitting (turn wrench counterclockwise) ¾ turn and begin operating the trigger-type hand pump immediately. The vacuum you create by operating the pump will hold the cone tip in place on the bleeder valve. After you have drawn out an inch of fluid into the pump jar, close off the valve and top up the brake master cylinder reservoir with brake fluid.

Start with the wheel furthest from the master cylinder (usually right rear), then go the next furthest, etc. The usual sequence is right rear, left rear, right front, and left front. Common exceptions to this recommended sequence are General Motors front-drive compacts and subcompacts (right rear, left front, left rear, right front), AMC four-wheel-drives and Honda Accord (left front, right rear, right front and left rear), air-cooled Volkswagens (right front, left front, right rear, left rear), and Chevy Corvette and most Toyota Corollas (left rear, right rear, left front and right front). The left side is the driver's side of the car. If, however, you have only introduced some air into one wheel brake, such as when a caliper or wheel cylinder is replaced, you may start at that wheel. After bleeding out the specified amount of fluid, you normally will also have removed all the air in it too.

Check your work by hitting the brake pedal. If it's firm, all air is out. If necessary, repeat the process, until the pedal is firm.

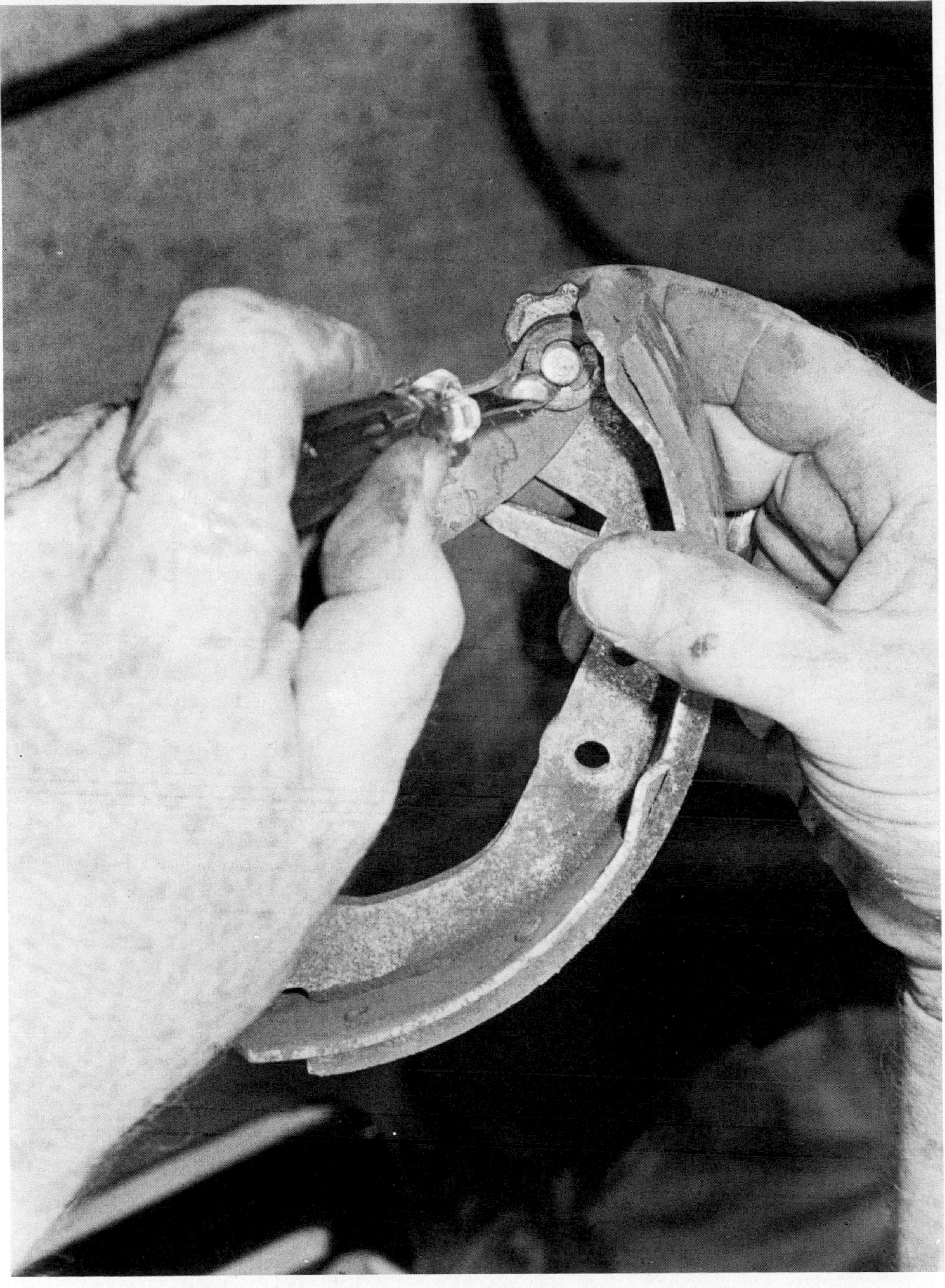

Fig. 11-49. To separate this old brake shoe from the parking brake lever, pry away spring clip with screwdriver as shown.

Fig. 11–50. As of 1980, the drum brakes (rear) on VWs are an unusual self-adjusting design. As the linings wear, the tapered wedge is pulled down by a spring to hold the shoes partly out, so much of the free space between the shoes and drum is taken up.

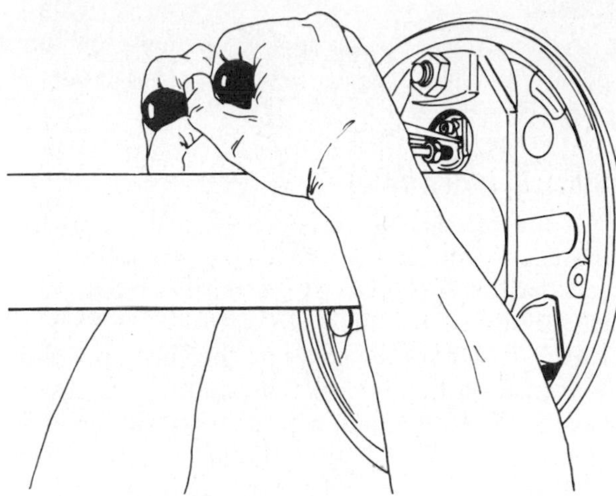

Fig. 11–51. Prying out retainer for wheel cylinder on some cars. Working from inboard side of brake backing plate, insert awls or pins ⅛-inch diameter or less into access slots between wheel cylinder and retainer locking tabs. Bend both tabs away simultaneously until the tabs spring over the wheel cylinder shoulder. When installing a new wheel cylinder, use a new retainer. Hold the wheel cylinder against the backing plate by inserting a block between the wheel cylinder and the axle flange. Install the new retainer over the wheel cylinder, lining up the tabs with the wheel cylinder tab grooves, then drive the retainer into place (snapped under the abutment) with a 1⅛-inch socket and a long extension (10 inches). If there is limited clearance and/or you do not have a suitable socket for installation, you can obtain an inexpensive tool from an auto parts store for this job. The caliper-like tool spreads the tangs of the clip for easy removal and installation. See below.

• *Manual bleeding*. This requires a helper and a special clip, for the stem of a part called the metering valve (on cars with front discs, rear drums only). Just follow the brake lines from the front wheels until they come to a junction block. Have a helper hit the brake pedal and you should see a dowel-like stem either eject or move into the junction block, which is really the valve assembly.

If the stem is ejected when the brake pedal is depressed, you must hold it out. If it goes in, you must keep it in. Inexpensive special clips for this purpose (see Fig. 11–53) are available from auto supply stores. Note: If you have your helper pump the pedal hard, the special clip need not be used.

You'll also need a hose that fits tightly over the bleeder valve nipple and a clear glass jar. Fill the jar half-way with brake fluid, force the hose end onto the bleeder valve nipple, and submerge the other hose end into the jar of fluid.

Have a helper pump the brake pedal, then hold steady pressure down on it. Open the brake bleeder valve with a wrench. The helper's foot will push the pedal to the floor. He must not lift his foot yet. As the pedal goes to the floor, fluid will be expelled through the hose into the jar, and with this fluid are trapped air bubbles.

Close the valve. Now the helper can lift his foot and pump the pedal once more, then hold it down. Open the valve again and watch the fluid in the jar. If there are air bubbles, you must close the valve and have the helper pump again, then hold down on the pedal. If there are no air bubbles, close the valve and have him test the pedal for firmness. You normally will save the pedal test for when you have bled all four wheels. Note: Top up the master cylinder reservoir after bleeding at a wheel.

Follow the sequence outlined for the Mityvac procedure. Remove the metering valve clip when you are done. If you encounter trouble getting air from a rear brake, this procedure may help on some cars, particularly Chrysler products: Crack a front bleeder screw open and apply the brakes. The warning light valve will shift (and the warning light will come on and stay on), but any entrapped air should now come out with bleeding at the rear wheels. Bleed the front brakes, then apply the brakes with moderate force, and the warning switch valve will recenter and the light will go out.

Note: Normally, it will take three or four pedal-pumping sequences at each wheel to get rid of all the trapped air.

Stuck Bleeder Valve

If the bleeder valve is frozen and cannot be loosened with the wrench, try a special tool called a Tap-A-Socket. It's a socket wrench, sold in various sizes and lengths to fit all bleeder valves. You put the socket on the valve, position a wrench on the hex section, and tap the anvil-like head with a hammer as you pull on the wrench (see Fig. 11–54). It will free up most stuck valves. If the tool doesn't work, and bleeding is absolutely necessary, you must remove the wheel cylinder or caliper and have the old bleeder valve drilled out and a replacement installed. Kits for this job are available at many auto supply stores, but installation really isn't a job for a weekend mechanic. The machine shop of an auto parts store, however, can handle it for you.

Brake Master Cylinder

The brake master cylinder is normally long-lived, but eventually the seals inside deteriorate and fail to hold hydraulic pressure. When this happens, the pedal sinks slowly to the floor under foot pressure, making the problem obvious.

The typical brake master cylinder is held by two to four nuts or bolts and is readily accessible at the firewall (rear sheet metal surface of the engine compartment) on the driver's side. Just undo the tubing connections (then plug the tubing ends to minimize air entry and loss of fluid). Undo the nuts or bolts and withdraw the master cylinder.

If your car has power brakes, make sure the replacement you buy is for power brakes.

Bleeding the Master Cylinder

The brake master cylinder must be bled before installation to remove most of the air inside. Attempting to bleed it by working at the wheels alone could take many hours and still not be successful.

You can bleed either off or on the car. To bleed on the car, install the master cylinder on the car and tighten the bolts but do not connect the brake lines. To bleed off the car, mount the master

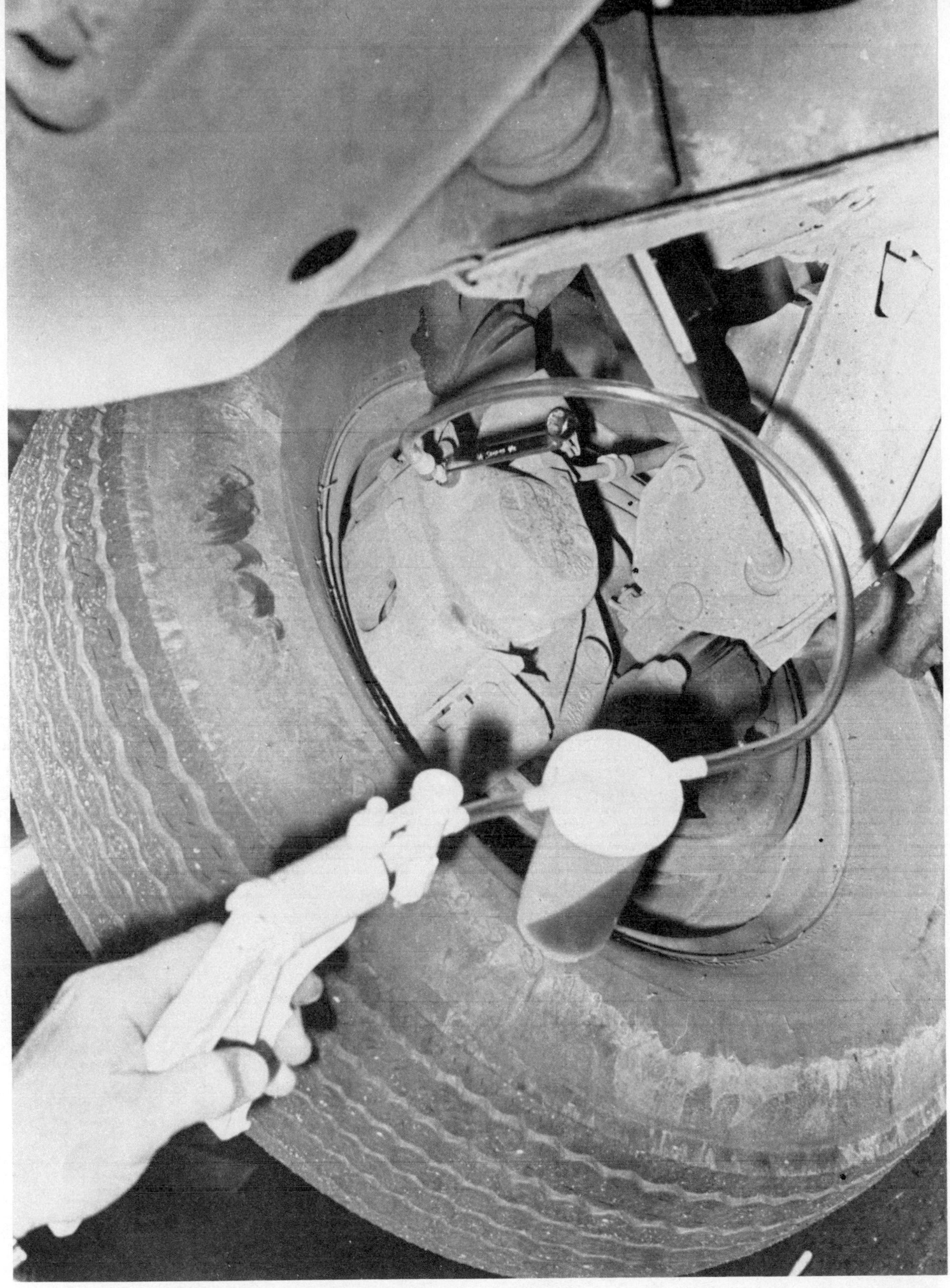

Fig. 11–52. Using a Mityvac pump designed for brake bleeding. After an inch or so of brake fluid has been sucked from the brake line into the reservoir with the pump, the line should be free of air. Lock the brake valve, using the wrench already in place as shown.

Fig. 11–53. Special clip on metering valve stem holds it in open position so brakes can be bled.

Fig. 11–54. Using a Tap-A-Socket to free up stuck brake bleeder valve. Socket is placed on valve and, as shown, the socket head is tapped with hammer while open-end or box wrench on the socket is pulled.

cylinder in a vise and obtain a wooden dowel to use as an out-of-car manually-operated pushrod.

Obtain a master cylinder bleeding kit, which consists of flexible tubing and fittings to connect to the tubing connections on the component (see Fig. 11–55). Aim the flexible tubing into the master cylinder reservoir, which should be filled to the half-way mark with clean brake fluid.

Operate the master cylinder by pushing in and releasing the dowel as shown, or have a helper pump the brake pedal. When air bubbles stop forming in the reservoirs, the master cylinder bleeding is complete. Connect the lines from the wheel brakes and check the pedal for firmness. Some final, minor bleeding at the wheels may be necessary.

If the car has disc brakes or is a 1973-76 Chrysler, Dodge, or Plymouth with drums, there is a minor complication. When you release pressure on the wooden dowel or the brake pedal, brake fluid will be drawn back into the tube and it will just move forward in the tube when you do apply pressure.

Therefore, pinch the hoses before you release pressure on the dowel or brake pedal to prevent this. Let go when you have pressure applied.

Note: You may be able to obtain brake master cylinders that are "self-bleeding." This type has special plastic outlet fittings installed (see Fig. 11–56) that eliminate the need to pinch the hoses. The special outlet fittings also may be obtained in a master cylinder bleeding tool kit, available in auto parts stores.

Parking Brakes

Because the parking brake is just a mechanical linkage to the rear brakes, it normally needs no special attention. It is common, however, for the parking brake cables to stretch, and so require adjustment to shorten them.

All parking brake cable arrangements have some provision, usually a type of turnbuckle arrangement, to shorten the operating length of the cables to compensate for stretch (see Figs. 11–57, 11–58).

The normal location for the adjustment is near the equalizer, a bracket-like part at which the cable from the passenger compartment lever and the cables to the rear brakes meet. This is under the car in most cases, but if you can't find it underneath and you have a console-type parking brake, lift the console and you'll probably find it there (see Fig. 11–59).

Usually there is a locknut and an adjusting nut. Loosen the locknut and turn the adjuster, which will pull on the rear cables. Give the adjusting nut a couple of turns, then tighten the locknut and try the parking brake lever. It should catch on about the fifth notch or after a few inches of move-

Fig. 11–55. Bleeding kit made up of tubes (plus fittings that thread into master cylinder) and a dowel. With master cylinder in vise and reservoirs full, pump dowel until there are no more air bubbles from tubing into reservoirs.

ment. If it locks up almost immediately, the adjustment is too tight. Back off a turn or two and recheck.

It should be reasonably obvious which way to turn the adjusting nut to shorten the effective length of the cables (normally clockwise), but if it isn't, a bit of trial-and-error will soon demonstrate the proper direction.

Power Brakes

The power brakes on most cars is a large vacuum diaphragm device that uses engine vacuum to reduce the pedal pressures (see Fig. 11–60). If the diaphragm unit fails, the pedal pressures required to stop the car increase dramatically, beyond those required for a car without power brakes. In some cases, failure also can allow engine vacuum to suck brake fluid from the master cylinder through the vacuum diaphragm into the intake manifold. Of course, a hard pedal also may be caused by defects in the friction and hydraulic systems, so check these first.

A power assist defect may be in the unit or in the vacuum supply. Disconnect the vacuum hose at the power unit (see Fig. 11–61) and feel the end for vacuum (and any tell-tale sign of brake fluid). If the vacuum is weak, check the hose for kinking. If the vacuum is good, disconnect the hose at the intake manifold and test vacuum at the source with a vacuum gauge. Normal manifold vacuum at idle is 16–22 inches. If noticeably lower, the engine requires service.

If you have normal vacuum to the power brake unit and a hard pedal, the problem may be in the vacuum valve or its retaining grommet (see Fig. 11–62). If you can get a replacement valve and grommet, try it before condemning the complete power unit. If the valve and grommet are not the problem or are not available, the only cure is to change the power brake unit. Note: Unless you are an experienced weekend mechanic who has done most other brake work, leave replacement to a professional.

The replacement procedure on most cars is normally straightforward. Disconnect the power brake pushrod from the brake pedal, then unbolt

Fig. 11–56. This master cylinder comes with special bleeding kit that does the job with the master cylinder on the car if desired. Special fittings and flexible hose combine to operate as check valves, simplifying the bleeding process.

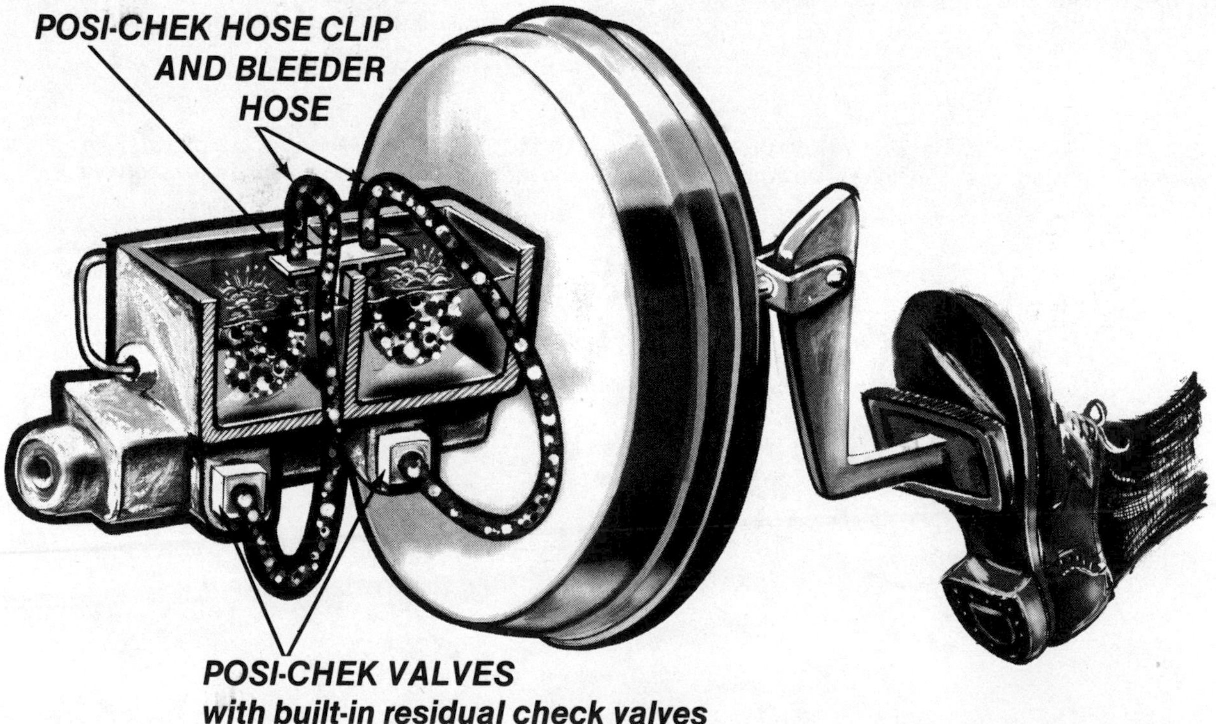

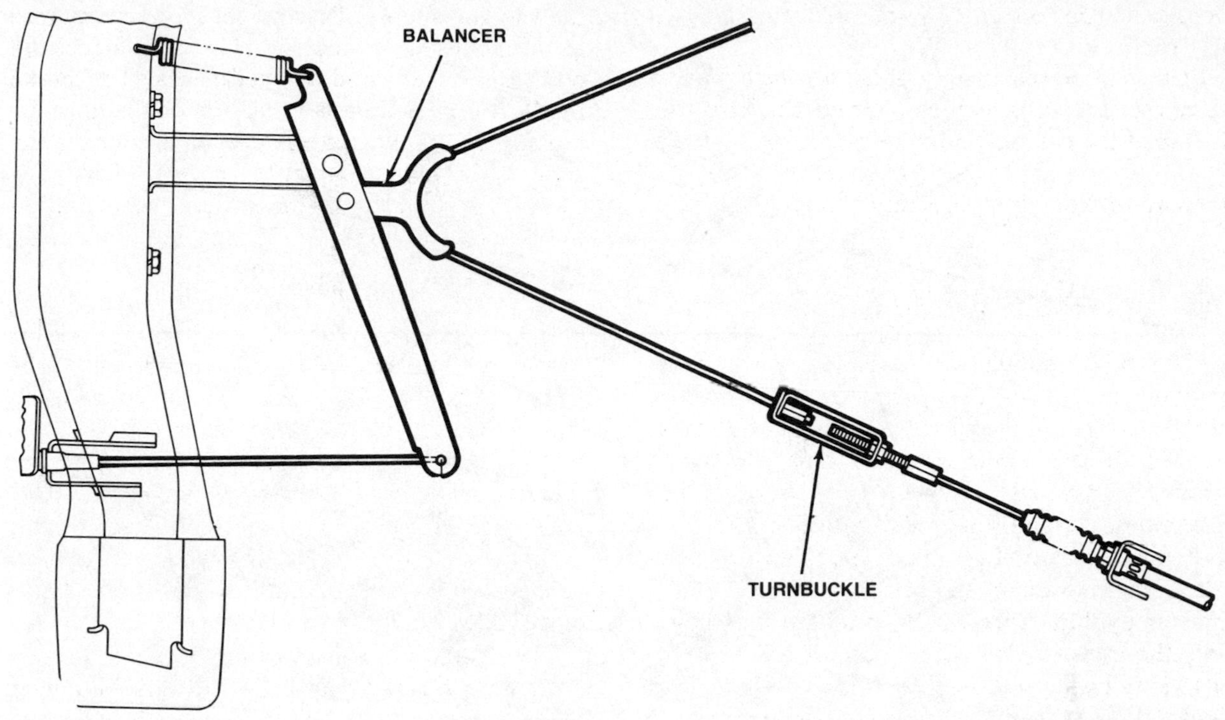

Fig. 11–57. This parking brake has a turnbuckle adjuster to compensate for cable stretch. Just turn the turnbuckle and cable slack is taken up.

Fig. 11–58. These are two typical brake cable arrangements in which turning the adjusting nut takes up cable slack.

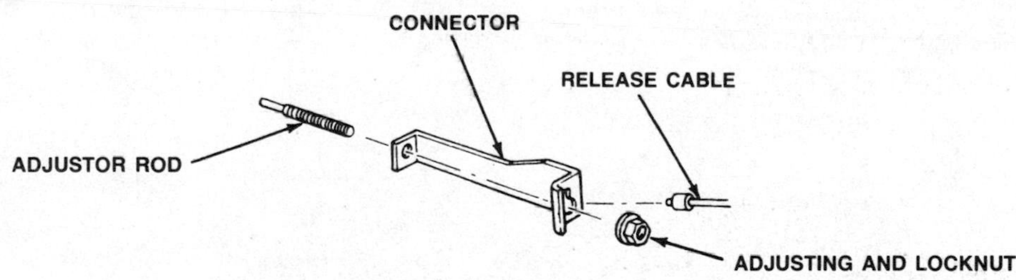

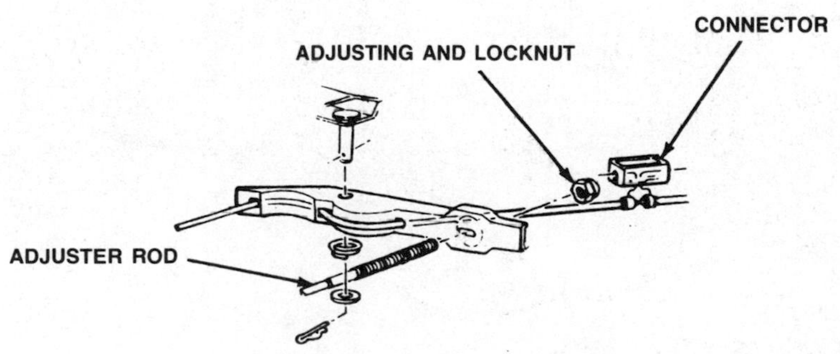

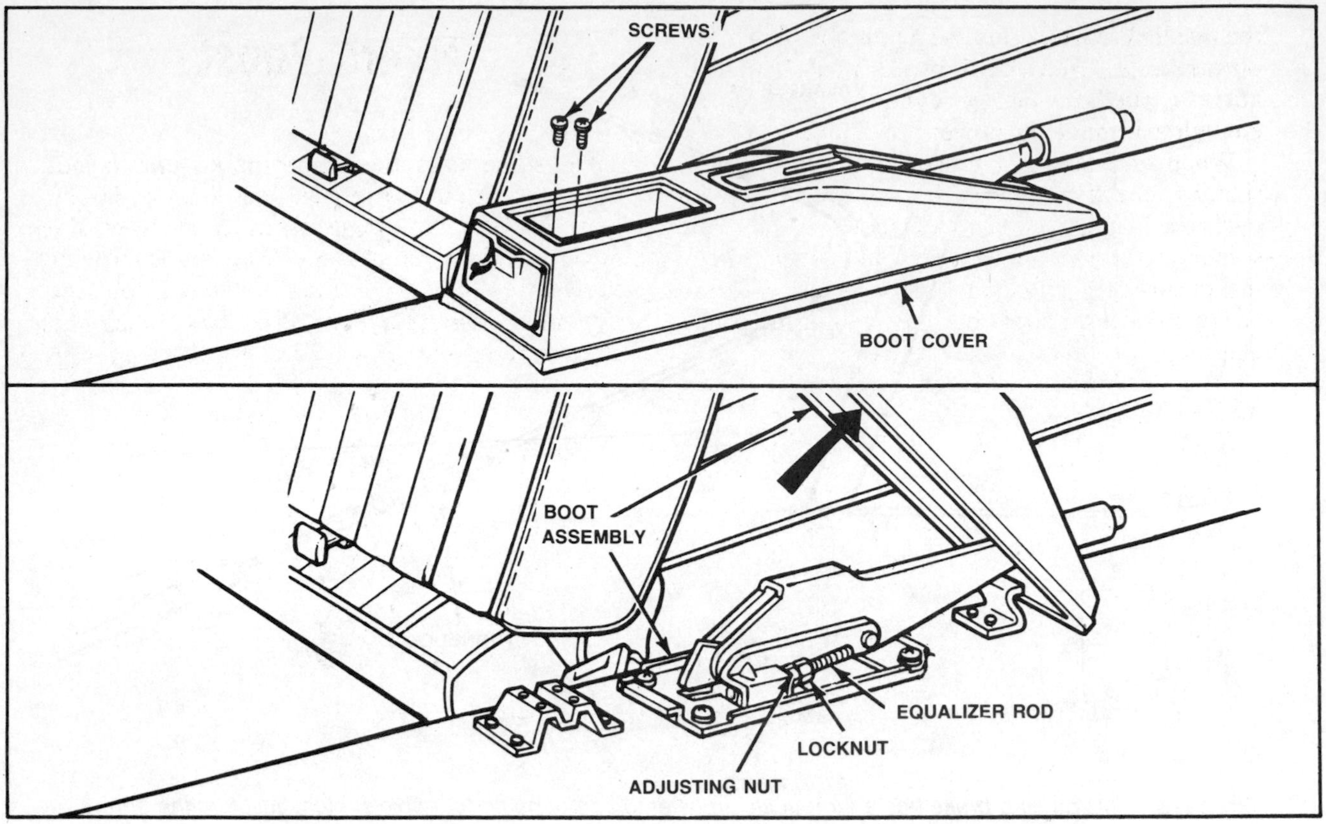

SCREWS
BOOT COVER
BOOT ASSEMBLY
EQUALIZER ROD
LOCKNUT
ADJUSTING NUT

Fig. 11–59. This cable adjuster is under the console. With this type you slacken a locknut, turn the adjusting nut to take up slack, and then tighten the locknut.

Fig. 11–60. This is a typical power brake arrangement. Linkage from brake pedal operates diaphragm assembly. Check valve at top is connected by a hose to the engine's intake manifold, a source of vacuum.

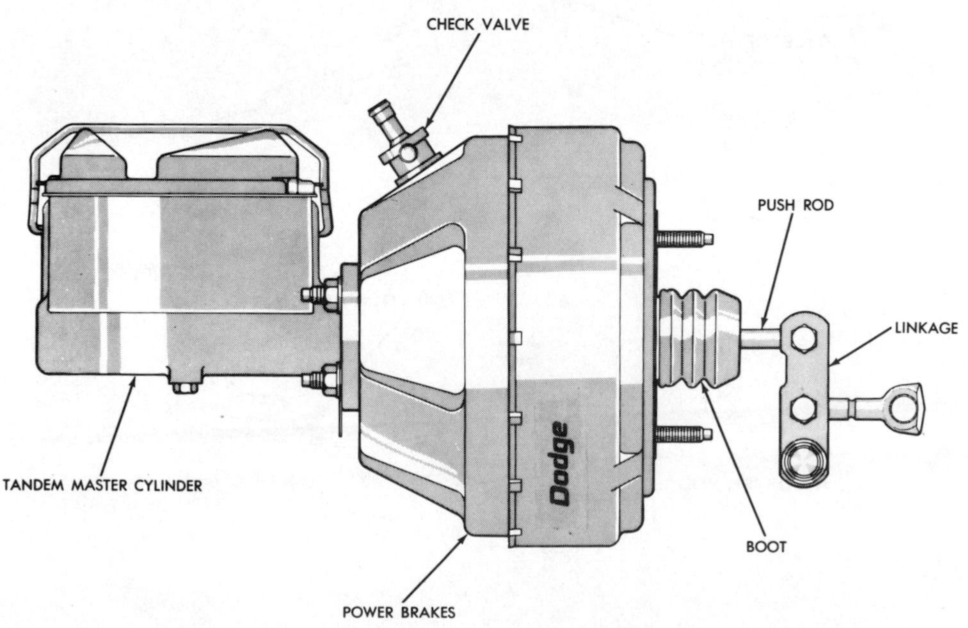

the master cylinder from the power brake and the power brake from the firewall. You should be able to pull the master cylinder forward far enough to remove the power brake unit.

When you obtain a replacement (normally a rebuilt), check to determine if the power brake pushrod length has been correctly preset. On some units it is preset at the rebuilder; on others the pushrod is adjustable and a gauge with instructions is (or should be) included in the package.

Hydro-Boost

On some cars the power brake unit is not vacuum-operated, but is an additional hydraulic unit operated by pressurized fluid from the power steering pump. If the drive belt is loose or fails, or the pump reservoir fluid level is low, the system will not operate properly. Inspection and adjustment procedures are covered in Chapter 7.

Fig. 11–61. Disconnecting hose from check valve and feeling for vacuum with engine idling.

Fig. 11–62. Check valve can be pulled out of grommet in power brake unit for replacement, as shown.

12 BASIC SUSPENSION MAINTENANCE

THE SUSPENSION SUPPORTS the body and chassis as the car rolls down the road. Many different types of automobile suspension systems are used, but they all have certain things in common.

At the front, the suspension is combined with the steering mechanism. Each wheel is mounted on a spindle that is part of the steering knuckle. The knuckle has upper and lower mounting points.

In an upper-lower control arm suspension (see Fig. 12–1), there are two A-shaped arms attached to the chassis by a hinge bolt at the wide end, and to the steering knuckle by a little swivel called a ball joint at the tip end. In between the arms is a coil spring and a shock absorber.

When the wheel hits a road irregularity, the spring-loaded arms rise, the spring flexes to absorb the road shock, and the shock absorber dampens the spring movement to smooth out the ride. The shock absorber, as you can now see, really doesn't absorb shocks—the spring does that. The shock just stops the spring from oscillating, to keep the wheels on the road.

A common variation on this design is the placement of the coil spring between the upper arm and the body (see Fig. 12–2).

Still another arrangement is the use of a torsion bar instead of a coil spring. The torsion bar is a spring in the form of a rod, and it twists to absorb road shock. It is commonly attached to the lower control arm at one end and to the chassis at the other—either to the rear of the control arm, or transversely, toward the center of the car at the front (see Fig. 12–3).

Suspension Basics

Strut Suspension

The use of two control arms is an effective design, but the upper arm takes up space that could otherwise be used to enlarge the engine compartment—valuable real estate in today's smaller cars. The strut suspension (see Fig. 12–4) therefore is very widely used on many cars today.

The strut is a shock absorber to which the upper part of the steering knuckle is clamped (so one ball joint is eliminated at each side).

The steering knuckle swivel is permitted in this case by the ball joint at the lower control arm and the strut upper mounting, which typically includes a bearing assembly.

In most cases, the spring is a coil and it is mounted directly on the strut. However, the spring also may be a coil mounted separately (see Fig. 12–5) or even a torsion bar.

Steering

When you turn the steering wheel you are turning a shaft, at the end of which is a gear. This gear meshes with another in the steering gear box, and through an arrangement of shafts and links it turns the steering knuckle at each front wheel one way or the other (see Figs. 12–6, 12–7).

Suspension Geometry

If you could measure the distance between the front wheels using precision equipment, you would find that the wheels are slightly closer at the forward than the rear end—they toe-in slightly. You also would find that the wheels do

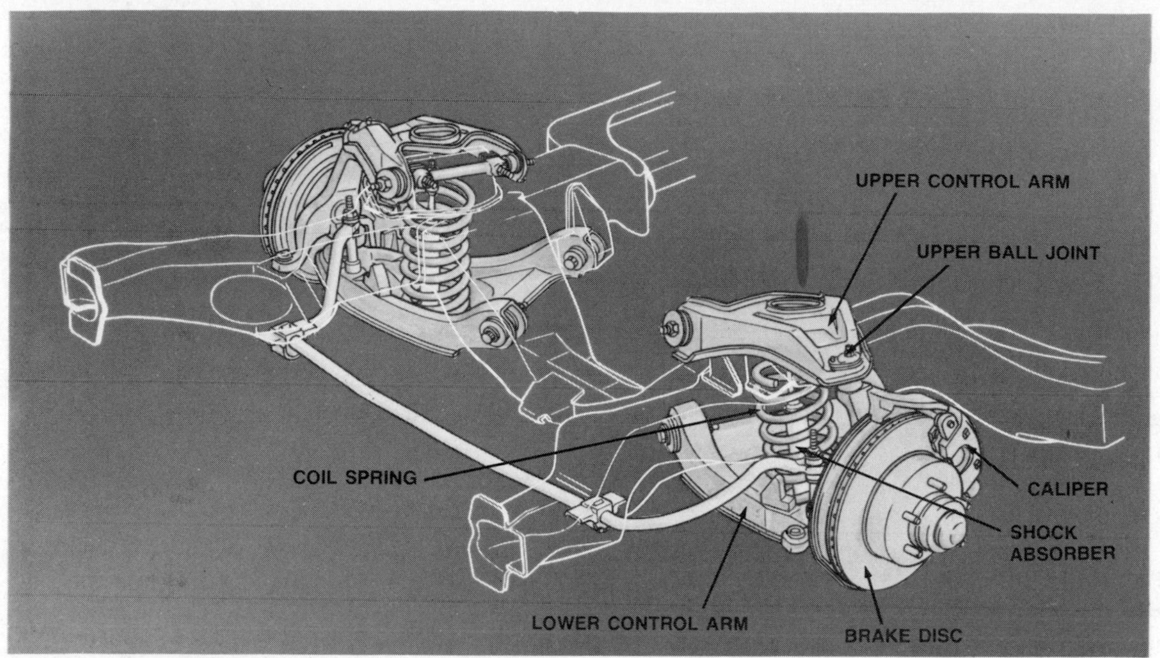

Fig. 12–1. Typical upper-lower control arm suspension, with coil spring between them and shock absorber inside the coils.

Fig. 12–2. In this upper-lower control arm suspension, coil spring is between upper control arm and the body. Shock is also inside the coils.

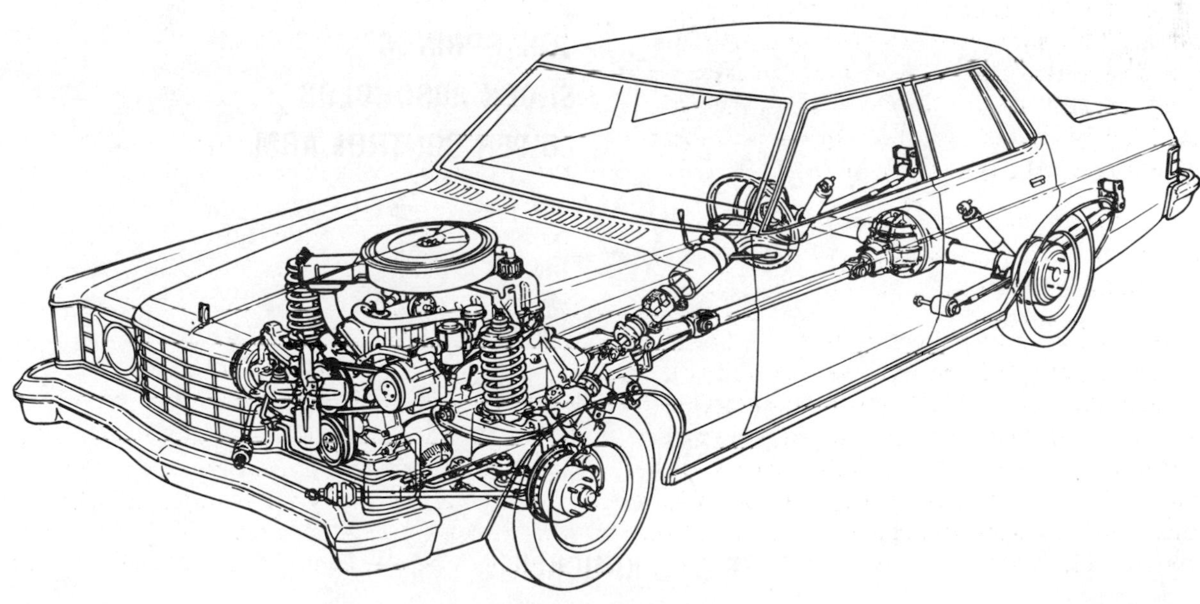

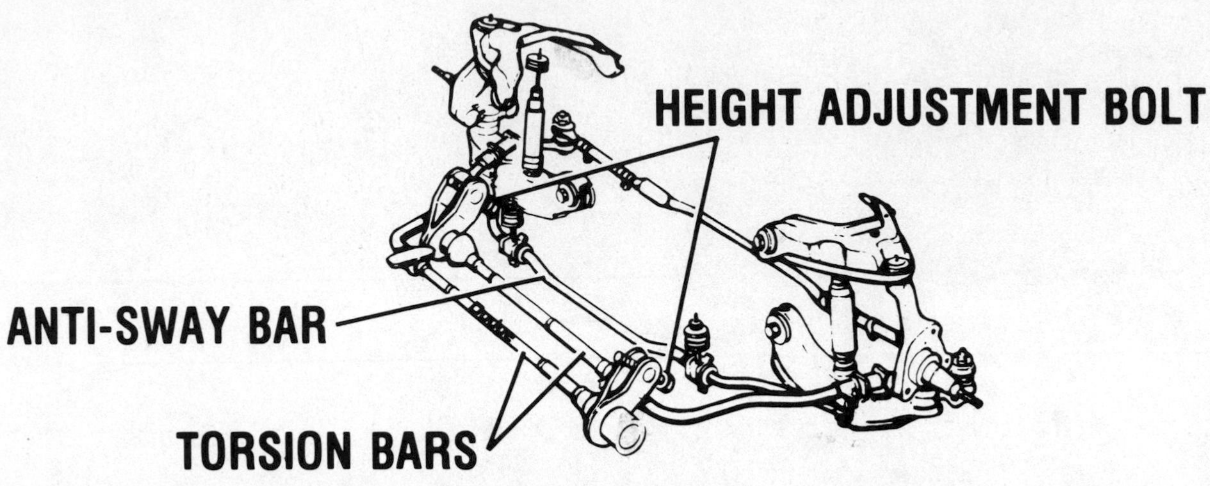

Fig. 12–3. This is a torsion bar suspension in which the springs are bars that twist to absorb road shock. In this type, some spring sag can be taken up by turning height adjustment bolts.

Fig.12–4. This is the most common type of strut suspension, in which the coil spring and shock absorber are combined into a single assembly, eliminating the upper control arm and upper ball joint.

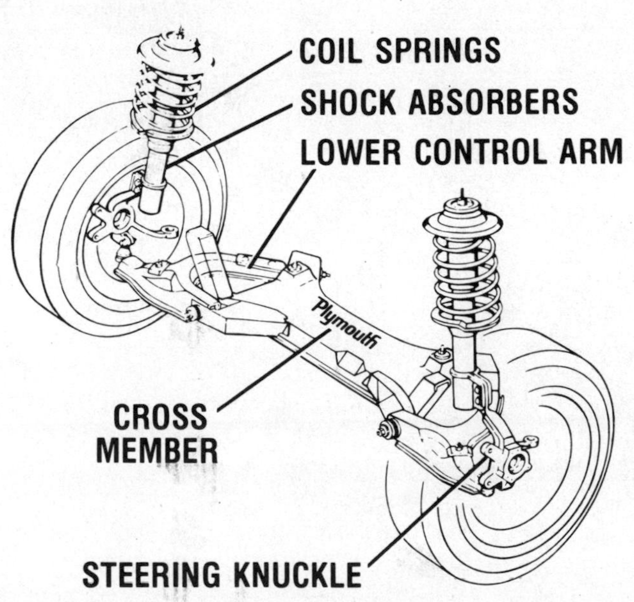

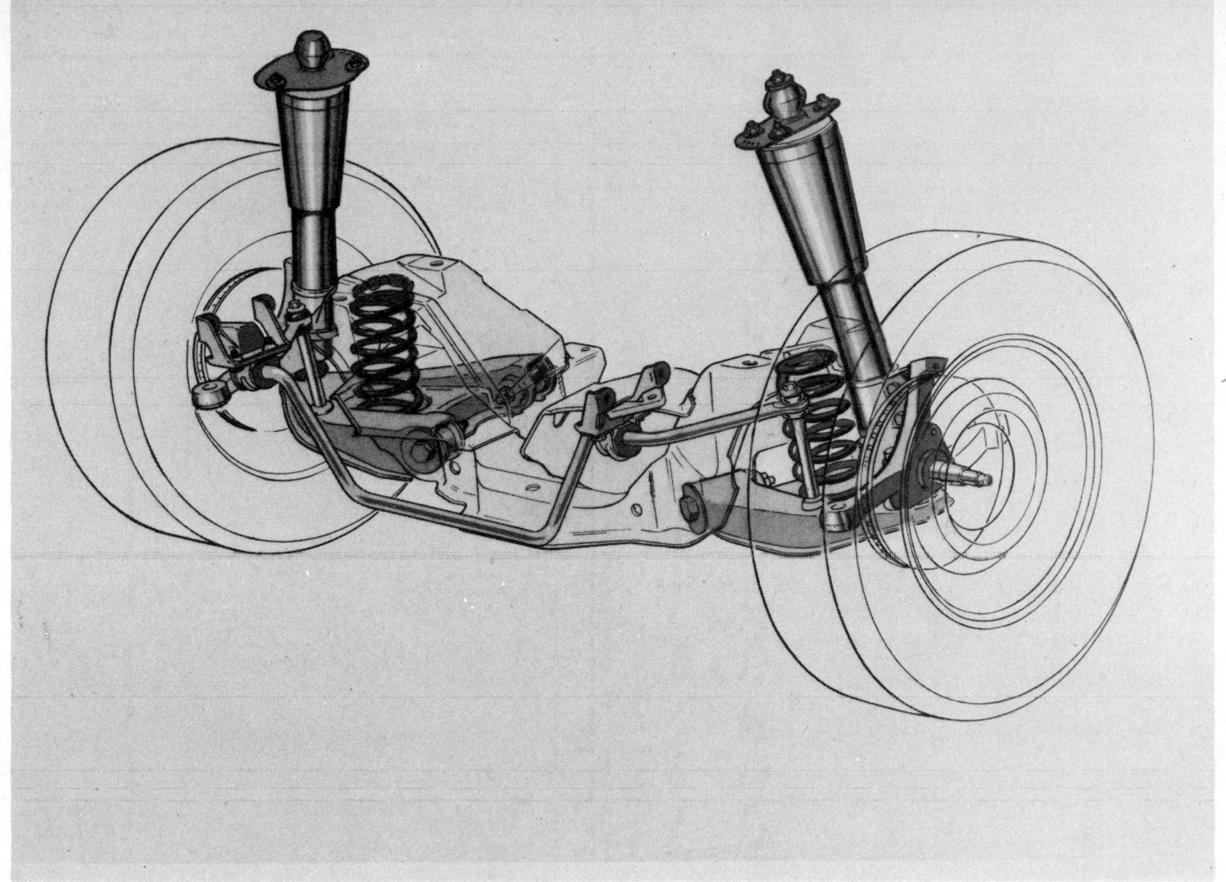

Fig. 12–5. This is a strut design in which the coil spring is a separate part from the shock absorber, as shown.

not hang straight up and down, but tip in slightly at either the top or bottom, depending on the suspension design. Further, you would find that the steering knuckle is not perfectly vertical—that the top mounting (at the upper control arm or strut) is not in a vertical line with the bottom ball joint.

All these differences are called suspension geometry or wheel alignment, and if the slight angles are wrong the car handling is affected. The fact that the front wheels are closer to each other (by perhaps ⅛ of an inch) at the forward end is called toe-in. The slight tip-in of the wheels at top or bottom is called camber, and the steering knuckle inclination is called caster. On many cars, all three are adjustable, but special measuring equipment is required, so the job should be left to a professional. Note: On many strut suspension cars, the caster and camber are fixed, and only toe-in can be adjusted. On newer strut suspensions, only caster is fixed, and both camber and toe–in can be reset. VW, Audi, Dodge Omni, Plymouth Horizon, Chevy Citation, and other GM front–wheel–drive compacts are in this category. GM front drive subcompacts (Chevy Cavalier, etc.) do not have a provision for camber adjustment. However, if one ever proves to be necessary, the strut is made so its lower attachment bolt hole can be precisely enlarged to permit a moderate adjustment.

There are do-it-yourself kits for toe adjustments, but they are not easy to use, and as a result accuracy may be difficult to obtain.

Rear Suspension

There are three basic types of rear suspension: non-independent, semi-independent, and fully-independent. If you look at the front wheels, you will notice that each wheel can rise and drop independently of the other. This is called independent suspension, and it helps ride and handling.

In a conventional front-engine, rear-wheel-drive car, there is less advantage to having the rear wheels rise and fall independently, and because independent suspension costs more it is

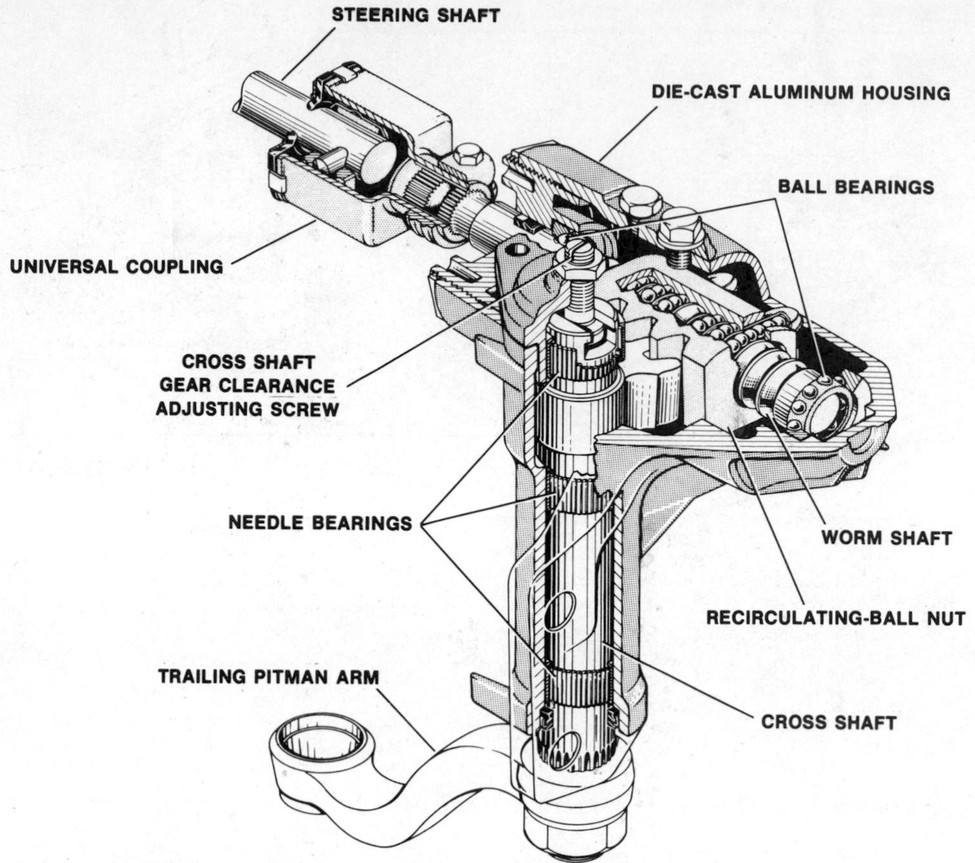

Fig. 12–6. When you turn the steering wheel, you operate a steering shaft that turns a gear inside a steering gearbox. The turning effort leaves the steering gearbox in the form of movement of the trailing pitman arm.

not used. There is a rear axle assembly, with shafts to each wheel, and the whole thing is held to the chassis and body by coil or leaf springs (see Fig. 12–8) and shock absorbers.

If the engine is in the rear, or if the car has front wheel drive, the actions of the rear suspension are more critical, and either semi- or fully-independent suspension is used.

The semi-independent type has a shock and coil spring (perhaps combined as in the strut suspension), and the two rear wheels are connected by a bar or beam that is really a torsion bar (see Fig. 12–9). Therefore, although the wheels are connected they are virtually independent of each other. Each end of that torsion bar or beam can twist to let one wheel respond to road irregularities. Because the two wheels are connected by the torsion bar, the term semi-independent is used.

In the fully-independent suspension, there is a separate control arm (just one, at the bottom for each wheel), plus a spring and shock absorber for each wheel, as in Fig. 12–10.

Rear Drive vs. Front Wheel Drive

Many cars are of the front-wheel-drive type, including an increasing number of American models. Big boosts to the FWD concept came in 1978 when Chrysler introduced the Omni and Horizon, and in 1979 when General Motors converted the 1980 Chevy Citation and its other compacts to the system, joining the luxury Toronado, Eldorado, and Riviera. In 1981 came the Chrysler ''K'' cars (Reliant and Aries) and the

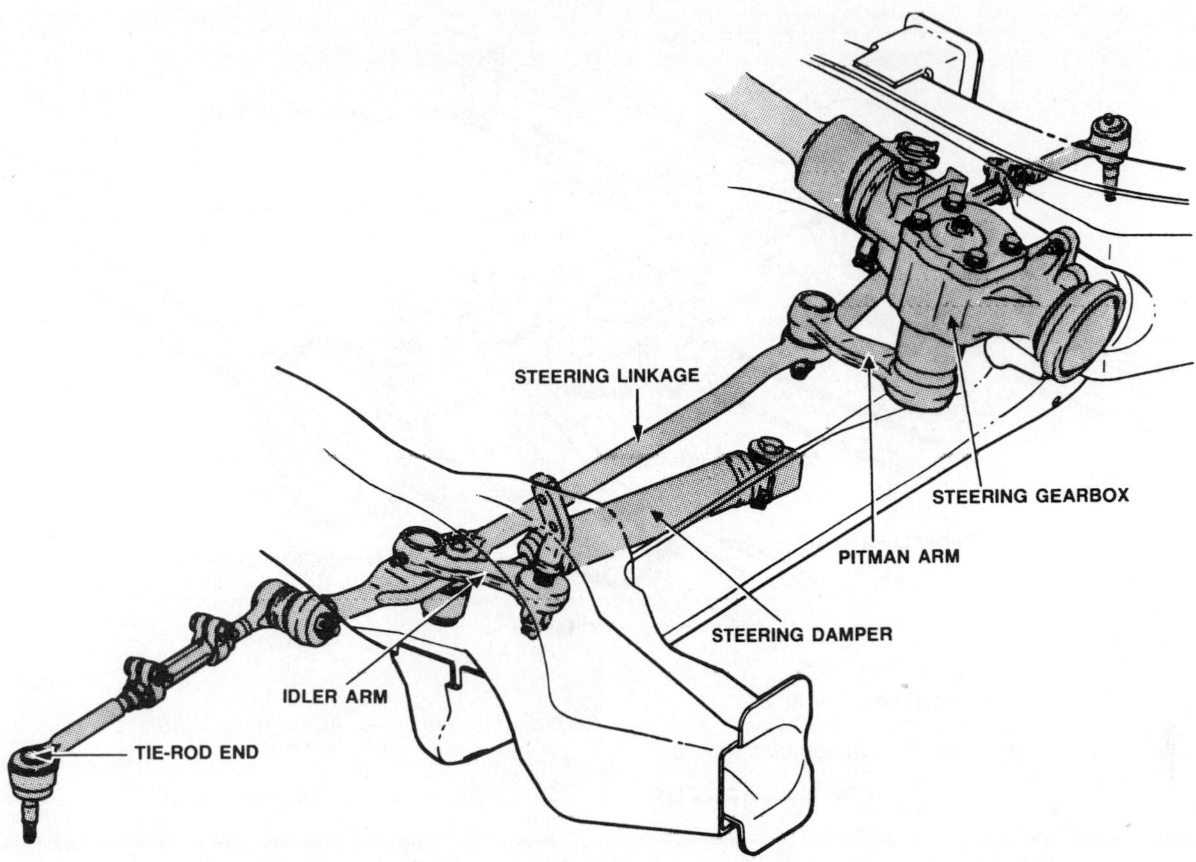

Fig. 12–7. A look at the steering linkage. When the pitman arm moves left or right, in response to turning of the steering wheel, it operates the steering linkage. At each end of the linkage is a joint called a tie-rod end. It engages a part called the steering knuckle, on which the road wheel is mounted. The idler arm is a type of swivel brace for the steering linkage. The steering damper is a type of shock absorber used on a few cars.

Ford Escort and Lynx. Next on the scene are the GM "J" cars (Chevy Cavalier, etc.). For details on underhood service, see *Chapter 14*.

In a conventional rear-drive car (see Fig. 12–11) the engine is in front, followed by the transmission. A long shaft, called the propeller shaft, is next in line (it's under the hump in the middle of the passenger compartment), and at the rear it connects to the rear axle, which houses a little gearbox called the differential. The differential turns the power around 90 degrees (from front-to-rear to rear-to-sides). From the rear axle it is transferred into solid shafts (called axle shafts), on which the rear wheels are mounted. The rear wheels receive the power and force the car to move; the front wheels are just along for the ride.

On a front-wheel-drive car, the rear wheels are the ones along for the ride. Power from the engine goes into a combination transmission-differential that transfers it through shafts to the front wheels. There's no propeller shaft and just a minor exhaust pipe hump in the passenger compartment. Although some FWD cars have the engine and transmission-differential aligned front-to-rear, on most the engine is mounted transversely—across the engine compartment. The transmission is attached to the rear of the engine, but is shaped so that part of it tucks under or to one side of the engine to permit it to fit. The transmission transfers the power off to one side into the differential, and from there the power goes through individual shafts to each wheel (see Fig. 12–12).

Unlike the axle shafts from the rear axle to the rear wheels, the shafts on a FWD are equipped with universal joints and a telescoping arrangement. These are required for two reasons:

(1) The front wheels are an independent sus-

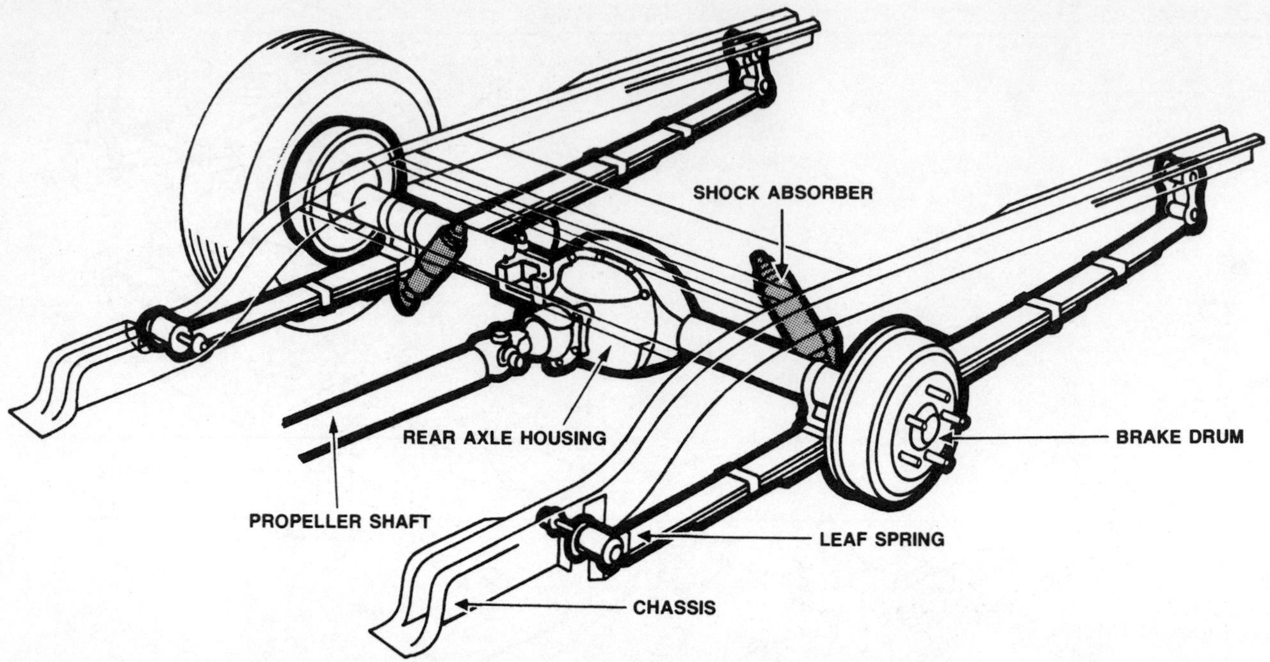

Fig. 12–8. This is a typical rear suspension of a car with front-engine-rear-drive. Although leaf springs are used in this design, coil springs also are a common choice.

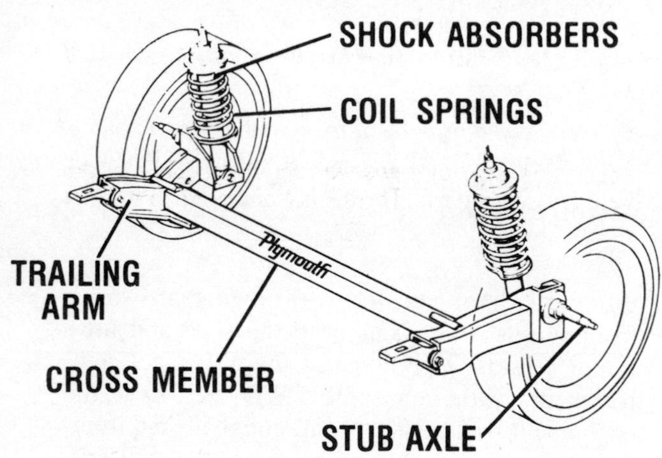

Fig. 12–9. This is a popular type of semi-independent rear suspension, as used on a front-wheel-drive car. The crossmember actually is a form of torsion bar. Therefore, if one wheel hits a road irregularity it can transfer the shock to the crossmember, which will twist to absorb it, while not causing the other wheel to lift. The crossmember mounts to steel members called trailing arms that also hold the stub axle on which the wheel is installed.

pension, and if the wheel hits a road irregularly and rises, the shaft must be able to flex—and it does so at the universal joints. The telescoping joint allows changes in the operating length of the shaft—necessary when it rises and falls because of changes in road surfaces. The axle shafts in these respects are similar to the propeller shaft in a conventional front-engine, rear-drive car.

(2) The front wheels receive power and turn to steer the car. The axle shafts, to which the wheels are attached, must permit them to pivot left or right, and a universal joint on the outboard end of the shaft serves that purpose.

Although many FWD cars have no-service wheel bearings at front and rear, some have conventional wheel bearings. Like the conventional front-engine, rear-drive car, they're on the wheels that are along for the ride. In the conventional setup, they're the front wheel bearings, but in FWD, they're the rears.

Most routine service on a FWD car is similar to that of a front-engine, rear-drive arrangement. The access to parts may be slightly different, however. For example, the ignition timing marks will still be on the front or rear of the engine, but when it's mounted sideways the physical loca-

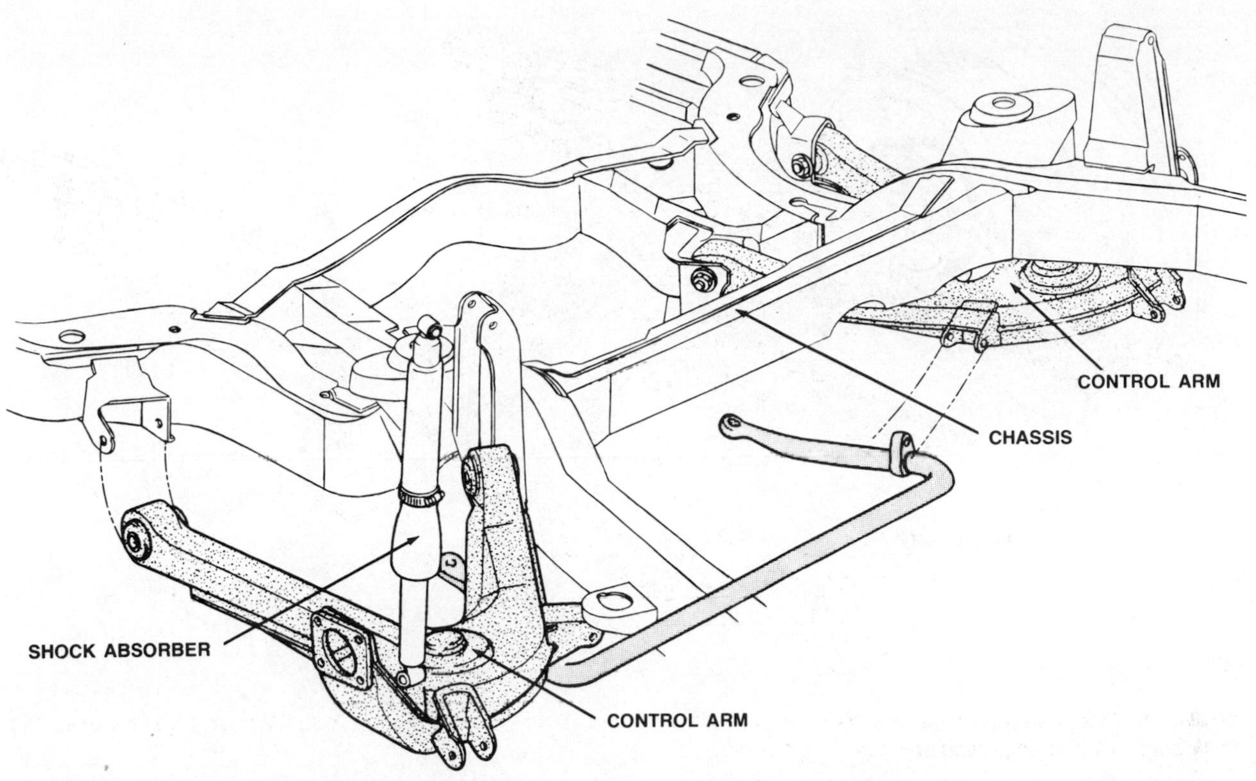

Fig. 12–10. This is a fully-independent rear suspension. Each control arm is separately hinged to the chassis, so if one wheel hits a bump or pothole the other wheel does not also respond.

Fig. 12–11. Layout of a typical rear-drive car. Engine is in front, followed by transmission. At the back of the transmission is a propeller shaft that carries the power to the rear axle-differential. There it is transferred to the rear wheels that drive the car. The front wheels are just along for the ride.

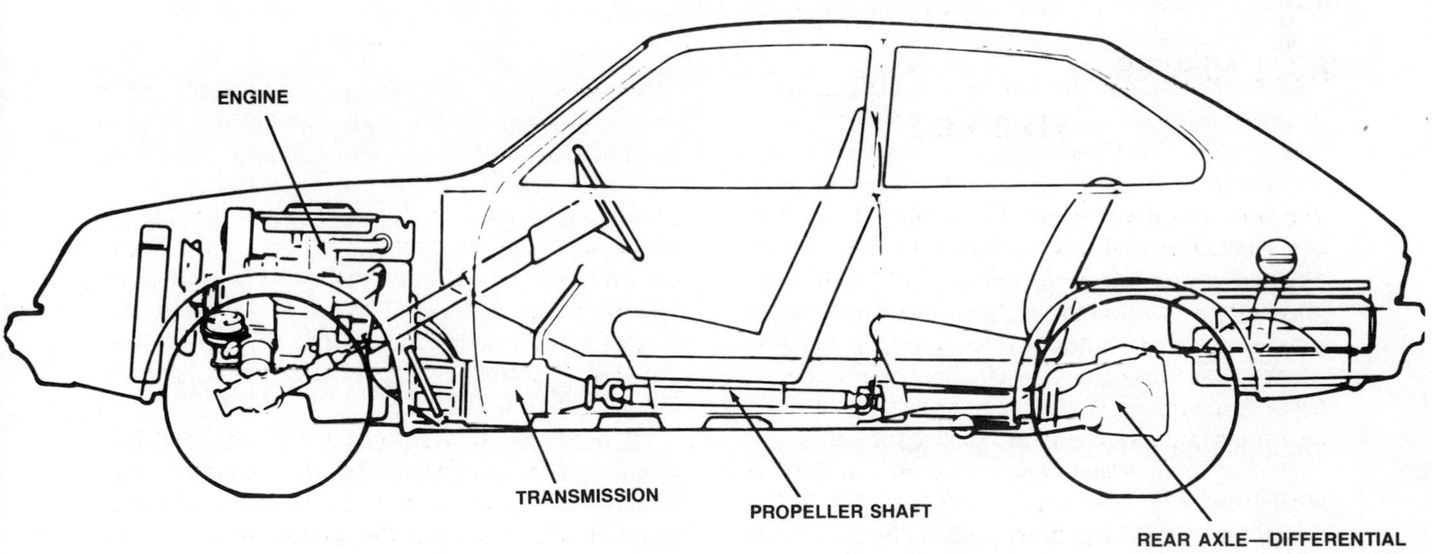

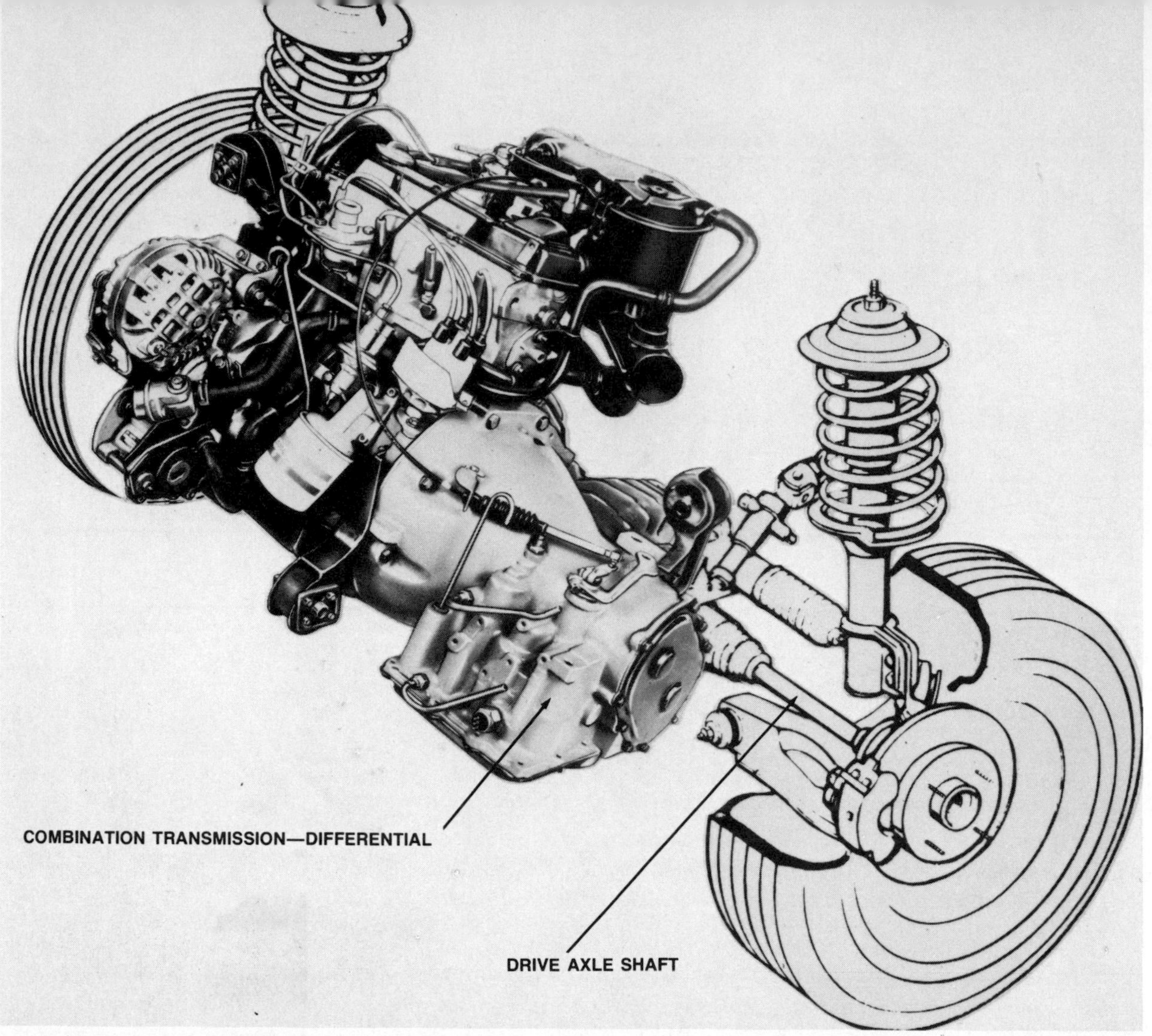

Fig. 12–12. Here is a typical front-wheel-drive car with a transversely-mounted engine, followed by the transmission-differential combination, commonly called a trans-axle. Power is transferred from the transmission out the rear side into the differential, and then across (by drive axle shafts) to each front wheel.

tion in the engine compartment is different. The transverse mounting also makes it impossible to use a mechanical fan because the radiator still is usually at the front of the compartment, whereas the belt-driven pulleys are at the side. Therefore, an electric fan is commonly installed. Some FWD engine compartments are laid out so that the radiator and AC condenser must be side-by-side across the front. This mandates a separate electric fan for each. See Chapter 14.

Suspension Service—Rear Wheel and Front Wheel Drive

The suspension and steering systems do not require a lot of routine service but they do need some, namely replacement of shock absorbers, lubrication and adjustment of wheel bearings (on most cars), and grease application to the steering (and perhaps suspension ball joints) of most cars. Tires also need regular care. The rest of this chapter is devoted to the performance of these services.

Chassis Lubrication

All cars have some chassis and/or steering joints that should be lubricated with grease. The joint will typically have a nipple (also called a grease fitting) for a grease gun or a plug that can be removed and a nipple threaded in (see Figs. 12–13, 12–14).

Fig. 12–13. Unless a steering or suspension joint is pre-packed for life with grease, it has a nipple-type fitting. Shown are the fittings on the upper and lower ball joints. Note the flexible seals that hold in the grease.

Fig. 12–14. This ball joint has a plug instead of a fitting. A box wrench is shown ready to fit on the hex-head of the plug, to remove it for replacement with a grease fitting.

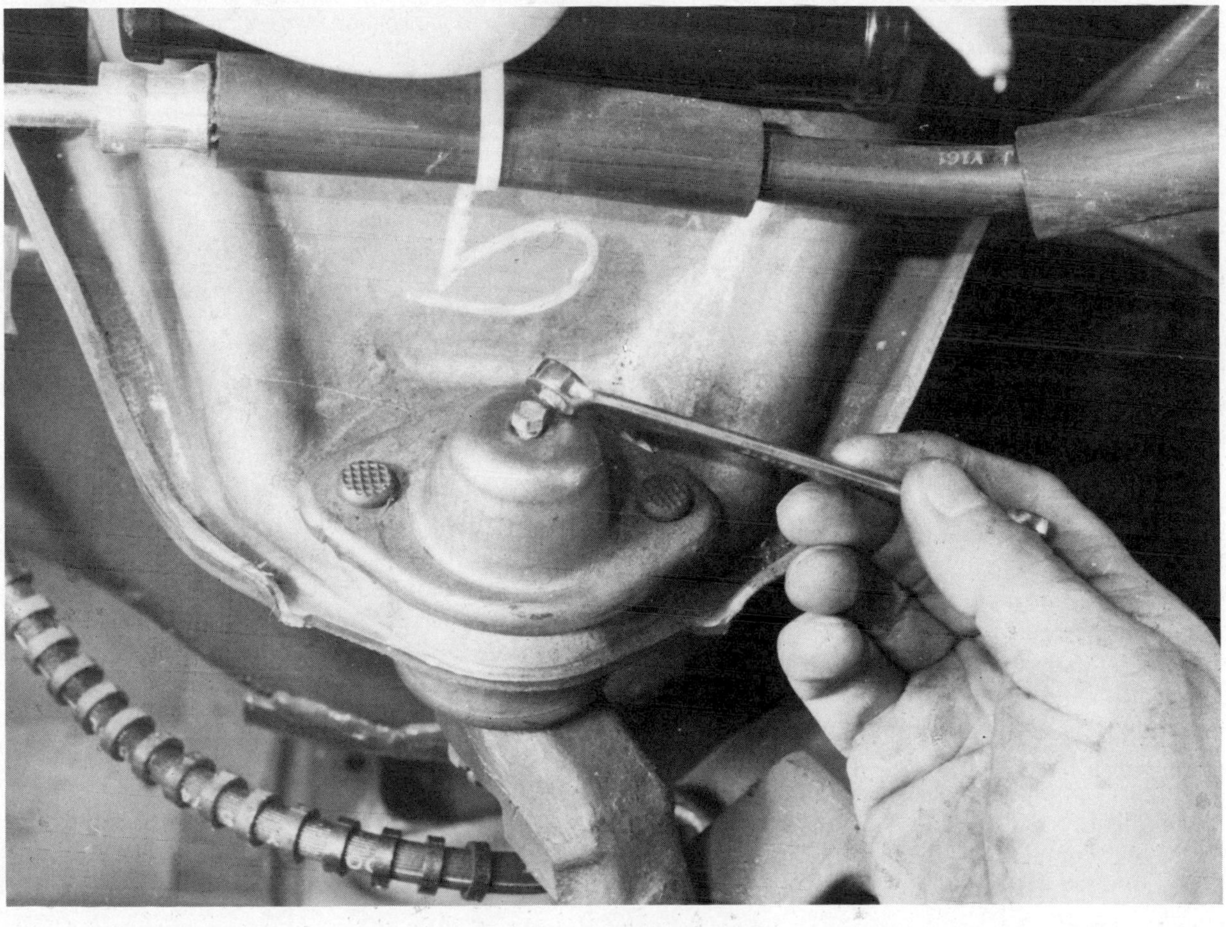

Fig. 12–15. Here's the typical front underbody with the location of grease fittings indicated. Note that there are four fittings on the steering linkage, one on the pitman arm joint with the steering linkage, two on the idler arm and brace arm for the steering linkage. There is one on each of the four spherical (ball) joints.

Many cars today have just a few, in some cases only two, whereas others have perhaps a dozen or more. The typical lubrication points are each of the suspension ball joints and two to seven lubrication points on the steering linkage (see Fig. 12–15). To lubricate them, you will need a grease gun, a tool that accepts a cartridge of grease and has a pump handle for dispensing it. The business end of the gun is a tube with a nozzle that fits onto the grease fitting nipple. The tube itself can be rigid or flexible, and each type has its advantages. Flexible permits you to twist the tubing to get to a less accessible fitting. Rigid means you can apply pressure with the gun to hold the nozzle in place on the fitting while you pump. You also can have the best of both worlds with a flexible tube with a rigid sleeve—just slide the sleeve forward when you want rigidity, backward if you want to be able to flex the tube. Another option is a lock-on adapter for flexible tube—just thread it on to the nozzle end, push it onto the nipple, and turn a knurled lock to secure it (see Fig. 12–16).

Finding the Fittings

Grease fittings are customarily covered by road film and mud and are hard to locate. Find them by feeling with your fingers, pulling off pieces of mud as you do. Look at the top end of an upper ball joint, and bottom end of a lower ball joint, and at each joint of the steering linkage (see Fig. 12–15 for sample locations).

Original equipment universal joints on the propeller are packed with grease at the factory and require no regular lubrication, with the exception of some older full-size General Motors cars (and some Jeep four-wheel-drives) with a double-type rear universal joint. If a GM car has the double-joint, jack up the rear wheels, turn the propeller shaft, and look for a single grease fitting flush with the inside surface. This flush fitting must be lubed with a special pencil-point nozzle on the grease gun (see Fig. 12–17). The nozzle is available at most well-equipped auto supply stores.

The typical replacement universal joint is

Fig. 12–16. Lock-on adapter on end of grease gun flexible hose is pushed onto grease fitting nipple, then knurled knob is turned as shown to lock the hose nozzle to grease fitting.

equipped with a conventional grease fitting (see *Chapter 13,* Fig. 13–23).

Grease

Always choose a brand of chassis grease with a name you know. Grease is so inexpensive that it doesn't pay to take chances. Most of the leading refiners of engine oil also market chassis grease, so you should have no trouble finding a good brand. After finding a name, make sure the type of grease you select is fortified with "moly" (molybdenum disulphide), a mineral lubricant that adheres well to metal and provides lubrication even after much of the grease has been contaminated or extruded from the joint. Moly grease greatly extends the life of the joint and permits longer chassis lubrication intervals. You should grease the fittings every 6000 miles or every six months, but never exceed one year, even if the manufacturer permits longer intervals.

Doing the Job

Wipe each fitting with a clean cloth. If you don't, you may force dirt into the joint. Inspect the rubber seal at the joint. If it is cut through, it must be replaced. You may be able to get a seal kit and install it yourself (instructions included with the kit). In some cases, the seal is not available as a replacement part and you must have the joint changed, a job for a professional.

If the joint has a grease plug in an inaccessible location, remove it and install a right-angle fitting (see Fig. 12–18), available in most auto parts stores. Aim the right-angle fitting down or to the side so you can get the grease gun nozzle on it.

Before you pump, you must know what you are looking for. In some cases the grease will extrude from bleed holes at the base of the joint. In others, the seal will balloon. With this second case, the visual indication will be slight, so squeeze the seal before you start and in between pumping of the gun handle.

Stop pumping as soon as you see any grease extruding from the base of the joint, or if you can see or feel the seal ballooning or feel that the seal has filled with grease.

Ford ball joints are an exception. Operate the lever so as to inject only a two-inch ribbon of grease. Try the gun on the bench so you can see

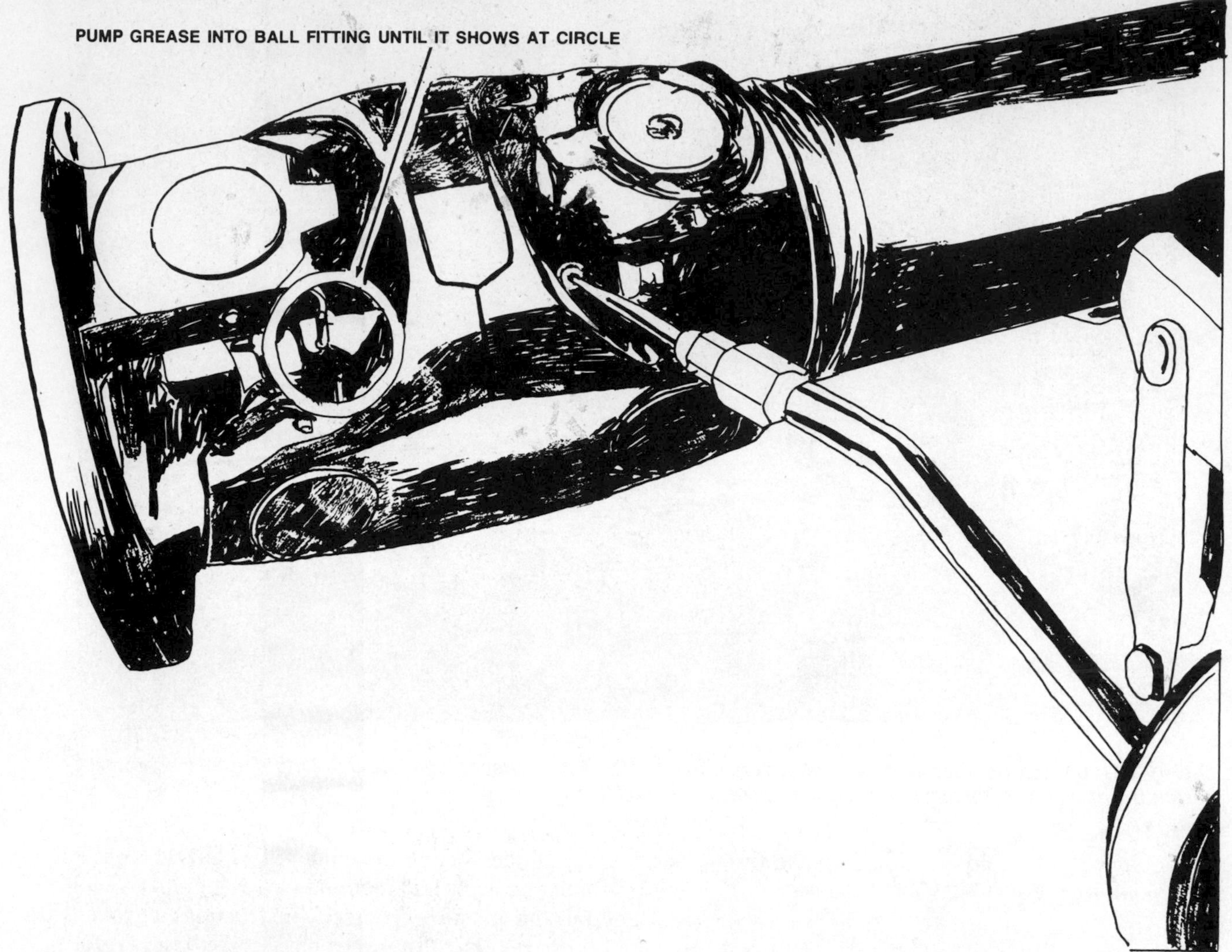

Fig. 12–17. The double-cardan universal joint on older General Motors cars has one fitting. You must obtain a needle-nose grease nozzle for your gun as the fitting is flush with the surface into which it's threaded and won't accept a conventional grease gun nozzle.

how much lever movement is necessary.

Note: If grease extrudes from between the nozzle and the nipple, the fitting is not taking grease. Stop immediately or you'll have a big mess. Remove the old grease fitting and install a new one.

Universal Joints

Although the original equipment universal joints of cars are packed and sealed for life, don't forget the fittings on replacement joints, and the double-type rear universal joint of some General Motors full-size cars. Just a couple of pumps on the handle should take care of the universal joint.

Dab It On

In addition to the grease fittings, chassis lubricant should be applied to joints of door and hood hinge, transmission and clutch underbody linkage, and parking brake levers. Just dab it on with your fingers and try to work it into the joints (see Figs. 12–19, 12–20). If there is dirt or mud on the joint, wipe it off first, or spray it clean with solvent.

Tire Buying and Tire Care

The subject of tires could be an encyclopedia in itself. This chapter, therefore, is intended only as a primer on the important considerations in buying a tire and maintenance tips you should know.

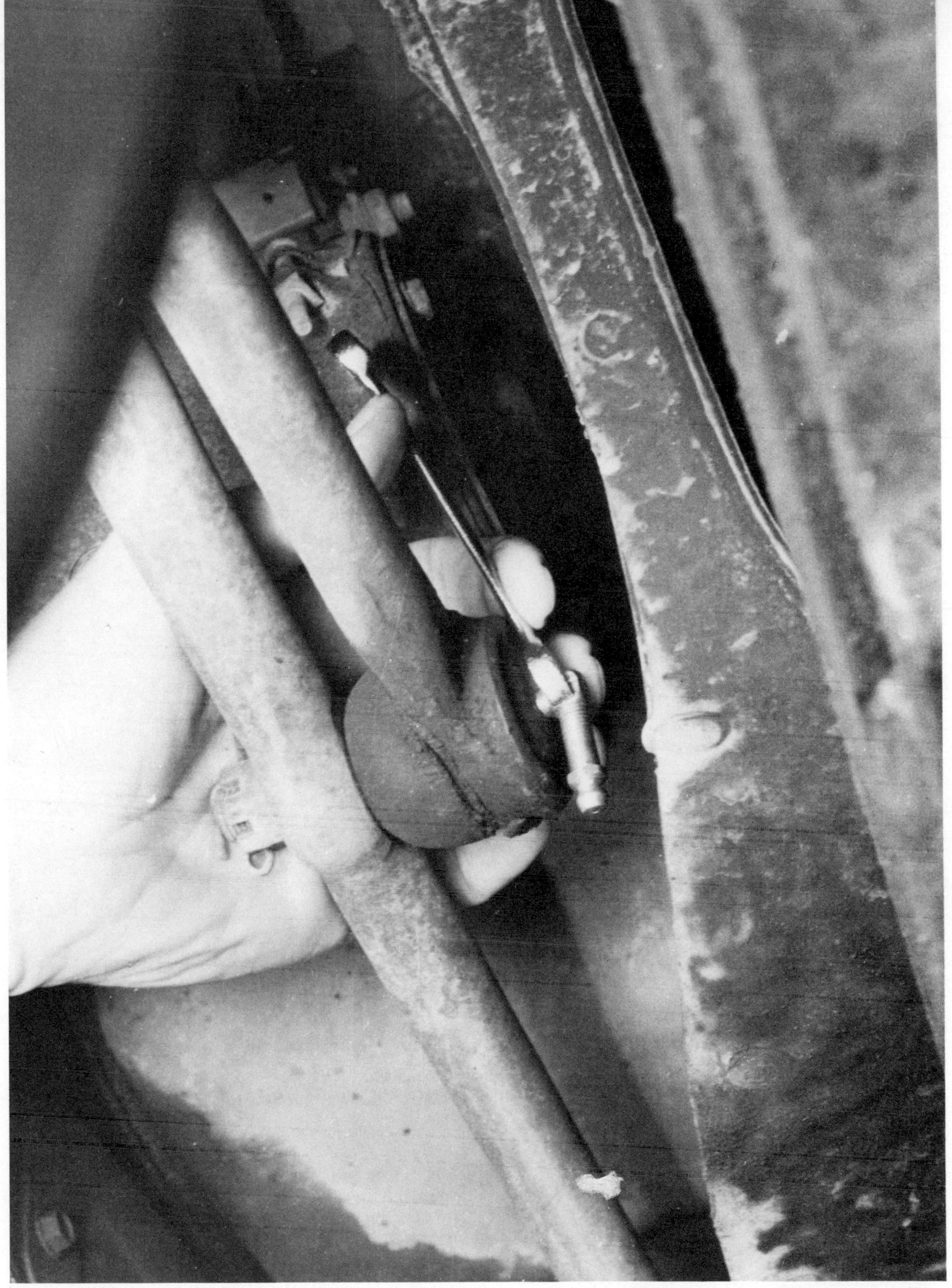

Fig. 12–18. If you can't reach a conventional grease fitting with your gun because of clearance problems, you can remove it and install a right-angle grease fitting as shown. Aim the right-angle fitting down or to an open side so the grease gun nozzle will go on easily.

Fig. 12–19. The indicated hood hinge pivots should be lubricated.

Types of Tires

Tires are made in three basic types of construction: bias-ply, belted bias-ply, and radial. The construction refers to what's under the tread: layers of synthetic fiber cord or steel wire in various forms (see Fig. 12–21).

The bias-ply is the smoothest riding tire, but the simple crossing of the cords results in squirming of the thread and short life because the tread gets hot and literally rubs out against the road. Longer tread life, cooler operation, and less rolling resistance is provided by the belted bias-ply, in which support belts are inserted between the tread and the layers of cord.

The radial is the kingpin. It runs the coolest and has the lowest rolling resistance, and because of the angles of the layers of cord and the belts the sidewalls flex to provide good tread contact with the road at all times. Radials have a tendency to provide a harsh ride, but suspension modifications by the car makers have reduced that. Steel belted radials—belts with steel wire running through them—have a reputation for the greatest durability and resistance to road hazards. The low rolling resistance of the radial reduces fuel consumption, and in some cases radials provide five percent better gas mileage than the simple bias-ply.

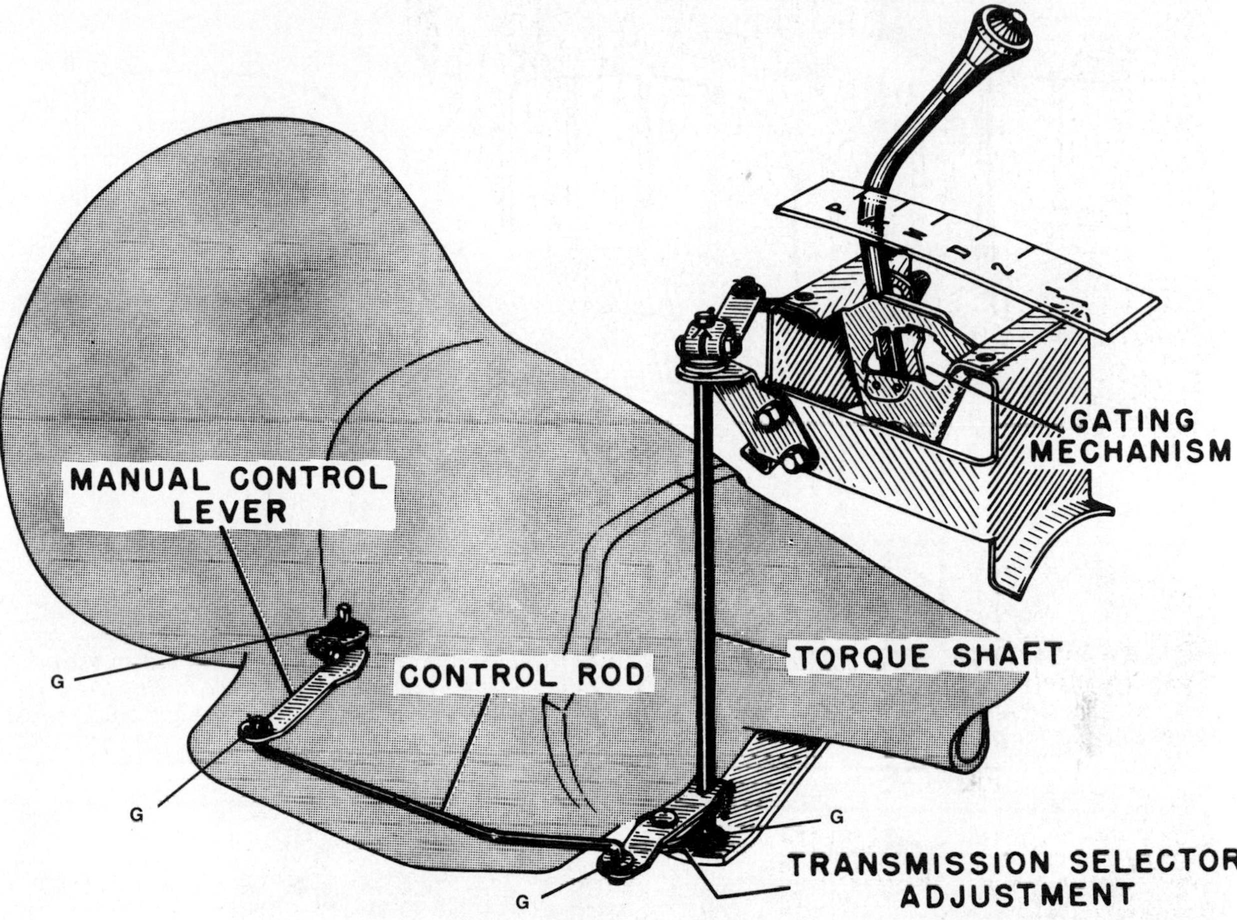

Fig. 12–20. Transmission shift linkage joints, indicated by "G", should be lubricated.

Tire Width

The wider the tire the more tread there is on the road and so the better the traction. Tires have gotten wider over the years, a fact reflected by a decrease in what is called the aspect ratio (relationship of width to height of the tire section). A 70 tire has a section that is equal to 70 percent of the width. A 78 tire (narrower) has a section that is 78 percent of the width. Both tires may have the same section height, but the greater width of the 70 reduces the ratio.

There is a limit to how wide a tire you can install, compared with original equipment. The limits are imposed by the clearances in the fender well and the size of the wheel. You can install a different wheel, but you can't change the clearances. In any case, tread design and construction are equally important factors in traction.

Tread Designs

Many people, particularly owners of recreational vehicles, pick tires with very aggressive-looking treads even if the vehicles are always driven on relatively good roads. The aggressive tread—big, deep lugs in the design—looks good, but they don't provide good highway ride or gas mileage, so unless you really need something special, stick with a conventional ribbed tire (see Fig. 12–22), or year-round type (see Fig. 12–23).

Load Range

Overloading is the most common cause of tire failure. You can overload either of two ways: by underinflating the tires, or by asking them to carry more weight than they are able.

All tires for passenger car use are in Load Range B, which means the tire can carry its

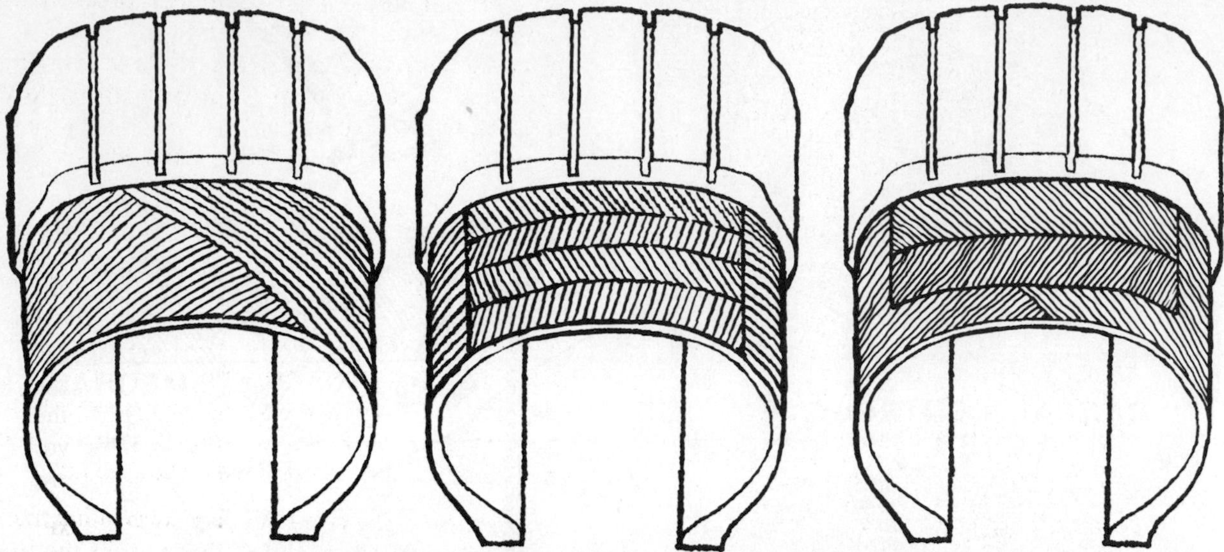

Fig. 12–21. Shown are the three types of tire construction. At left is the bias-ply, in which the tire cords cross each other at an angle. In the center is the radial, in which the tire cords run 90 degrees to the tread and are covered by cord belts that run at very slight angles to the tread. At the right is the belted bias-ply, in which the cords run at angles similar to the plain bias-ply and are covered by belts that run at slight angles.

Fig. 12–22. This is a basic ribbed tire. The ribs are the grooves running around the circumference of the tread.

Fig. 12–23. This is a typical all-year-round tire. The dog-biscuit-shaped lugs help traction in mud and snow.

maximum load (including the weight of the car) at 32 p.s.i. The information is on the tire sidewall and easy to read. If the total load-carrying capability of the four tires or either front or rear tires exceeds the rating of the tires, you must get either larger tires or ones in a higher load range (C,D,E,etc.). The higher load range means that the tire can be inflated to higher pressure to increase its load-carrying capability.

With a passenger car, ride considerations generally limit you to the largest possible tire in load range B. On a recreational vehicle, you can go to truck type tires, which are in the higher load ranges.

Tire Quality Markings

Starting in early 1979, sidewalls of some new tires have important information on them: Ratings of treadwear, traction, and temperature resistance. These markings are being phased in very gradually, so you may not see them on the tires you are looking to buy. If you do see markings, however, here's what they mean:

- *Traction:* The letter A means maximum traction, B intermediate, and C means that the tire did not pass either the A or B tests.
- *High Performance:* This is a temperature resistance grade that involves overloading, underinflation, and high speeds. A tire that passes the stiffest test gets an *A,* the next stiffest a *B,* and the easiest test a *C*.
- *Treadwear:* A *100* means the tire passed a 30,000–mile test, *120* a 36,000–mile test (20 percent longer), an *80* a 24,000–mile test (20 percent lower). The tests are standard ones, but do not constitute a guarantee of these tread mileages to the motorist.

Tire Care

The best things you can do for your tires are to check the pressures regularly, using a gauge, and never overload them. Visual appearances are deceiving, particularly with radial tires, and the gauge on service station air meters is notoriously inaccurate.

Pressures should always be checked when the tire is cold (not driven more than a couple of miles at low speed in six hours). Never bleed air from a warmed-up tire. Note: The typical tire provides the best ride at lower pressures, but you should realize that at those lower pressures the load-carrying capability is reduced.

If you have separate snow tires, buy a spare pair of wheels for them. The spare wheels, available from a wrecking yard, eliminate the twice-a-year service on both the snows and the

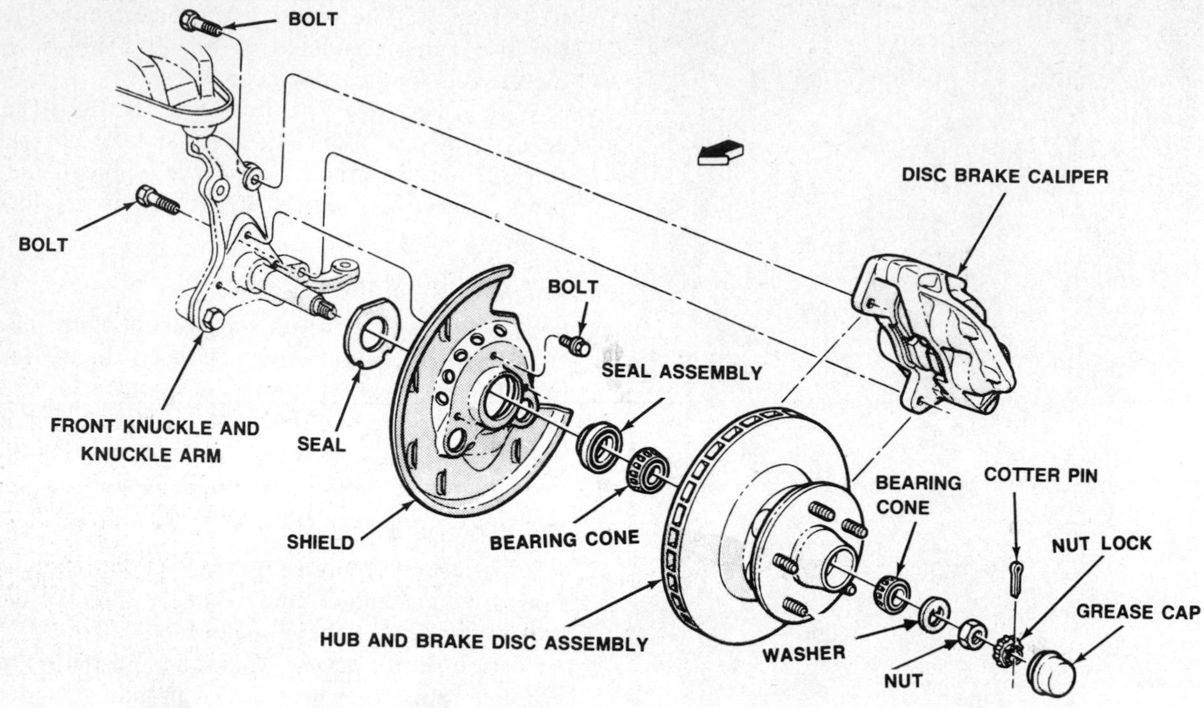

Fig. 12–24. This exploded view of a front wheel assembly shows how the hub and brake disc assembly fit on the spindle section of the front knuckle. Inside the hub and disc are two bearings, one on the inboard side of the disc and the other on the outboard side. The bearings are held to the spindle by the nut and lock. Because the bearings are tapered, retaining them also holds the disc-hub in place, although it can spin freely on the bearings.

regular tires they temporarily replace. This reduces the possibility of damage to the tire bead (the section that bears against the rim of the wheel and seals in the air) during service. Spare wheels also mean you can change the tires yourself.

When storing tires during the off-season, reduce pressure to 15 p.s.i. and keep in a cool, dry place.

Wear Patterns

When your tire treads wear down to 1/16 inch, bars will appear across the width of the tread. These are wear indicators, designed to tell you to replace the tires immediately. With just 1/16 inch of tread, the tire is effectively running bald.

Other wear patterns may indicate a suspension problem. If there is uneven wear on the outside or inside treads only, or scuff marks across the treads, that's wheel misalignment. Wear in the center is usually caused by overinflation; wear on both sides is usually caused by underinflation. Irregular spot wear around the tread circumference may be caused by wobbly wheels (the wheel could be bent or its bearings could be loose), worn shocks, or defective brakes.

Front Wheel Bearings

The front wheels, or the rear wheels in a front-wheel-drive car, do not receive power from the engine. They simply support one end of the car and roll as the car is propelled by power transmitted to the other wheels. It might seem that aside from tire wear they should be of little concern. However, the bearings on which they roll (see Fig. 12–24) are extremely important. If they aren't doing their job, the car will handle very poorly, wander, and even waste gasoline. If the bearings fail completely, the car is stuck.

These wheel bearings, therefore, should be periodically inspected, lubricated, and adjusted on most cars. Surprisingly, it is only the driven wheel bearings you may have to maintain. Those on the drive wheels are lubricated by axle grease (in most rear-drive cars), or are packed for life,

Fig. 12–25. Cutaway of sealed wheel bearing assemblies. The front type is at left, the rear type is at right. Because these are from a front-wheel-drive car, the rears are for driven, not drive wheels. This sealed bearing needs no lubrication or adjustment. If it becomes noisy, you replace the entire assembly.

as in all front-wheel-drive and some rear-drive cars.

Serviceable vs. Non-Serviceable

On all American cars prior to 1979, the driven wheel bearings are serviceable. In 1979, non-serviceable (sealed for life) bearings made their debut on Buick Riviera, Cadillac Eldorado and Oldsmobile Toronado (see Fig. 12–25). The trend toward non-serviceable wheel bearings continues, on GM "X" (Chevy Citation, etc.) and "J" (Chevy Cavalier, etc.) cars, and some imports.

If, however, you have one of the overwhelming majority of cars on which service is required, you should do the job every 20,000 miles or two years.

To determine if your wheel bearings are the serviceable type, just remove the hub cap or wheel cover on a driven wheel (rear on a FWD car, front on a conventional rear-drive car). If you see the bearing hub assembly in the center of the wheel, the bearings are non-serviceable.

Types of Serviceable Bearings

Although some cars have ball-type bearings, most driven-wheel bearings are made with tapered rollers assembled on a circular holder called a cage. The rollers bear against smoothly finished sleeves called races.

The wheel bearings actually don't support the wheel, but a part of the braking system to which the wheel is bolted. If the car has front drum brakes, the drum is supported by the bearings and the wheel is bolted to the drum; if it has disc brakes, the wheel is bolted to the disc.

Each drum or disc has a cylindrical cavity in the center into which the bearings are installed. There is one bearing assembly on the inside half of the cavity, one on the outside half. The drum or disc, with its bearings, fits on a tapered shaft—called a spindle—which is part of the car's front suspension. The outer end of the spindle is threaded to accept a nut that holds the drum or disc in position on the spindle. When you adjust the wheel bearings you actually are setting the position of the nut on the threaded spindle. If the nut is too tight, the bearings are forced tightly against the spindle and, even with adequate lubrication, the friction of the tight fit will cause them to heat up and fail. If the nut is too loose, the vibration of the wheels as they travel over road irregularities will shock the bearings, almost as if you were hitting them with hammers, and again they will fail.

Therefore, not only must you set the position of the nut properly, you must make sure the nut doesn't move. The manufacturer provides one of the following four means to hold the nut in place:

(1) The top of the nut has cross slots and the spindle has one or two holes. A cotter pin, shaped somewhat like a lady's bobby pin, is inserted so it rests in the cross slots of the nut and goes through a hole in the spindle.

(2) A hexagonal sleeve, called a nut lock, fits onto the nut. Like the nut in the previous example, it has cross slots and can be locked with a cotter pin through a hole in the spindle (see Fig. 12–24).

(3) A circular sleeve—another type of nut lock—fits onto the spindle and bears against the nut. The sleeve has slots for a cotter pin, which is installed in the same basic way as the first example.

(4) Adjusting nut has a slit-through section, through which an Allen-head screw goes. To lock the nut, tighten the screw. To permit adjustment, first slacken the screw.

Servicing the Wheel Bearings

• *Testing.* Although some professional mechanics check wheel bearings only with a dial indicator, most still test the adjustment by feel, as follows: Hold the tire at the top and bottom and try to rock the wheel in and out. If you feel any more than just the slightest hint of movement, have a helper do the rocking as you look at the wheel's suspension. If the free play is in the suspension, reduce the rocking effort so the suspension joints do not show movement. If there still is more than barely perceptible free play when the wheel is rocked, an adjustment is apparently required.

Fig. 12–26. Removing the dust cap with "water pump" pliers. Although this job is illustrated on a disc brake car with the wheel already removed, on a drum brake car all you would have to do is remove the hub cap or wheel cover for access to the dust cap.

Fig. 12–27. To remove the cotter pin, straighten the double end as shown, then pull it out from the opposite (eye) end.

Even if you believe free play is barely perceptible, the bearings should be removed for inspection and relubrication as a routine maintenance procedure at least every two years or 20,000 miles. Or if you are doing a brake job on the car, you should service the wheel bearings at that time.

• *Disassembly*. To gain access to the wheel bearings for inspection and lubrication, you must jack up the car and support it on safety stands (see *Chapter 5*).

• *Drum brakes*. Note: Most of the illustrations are of a disc brake car, as this job on a disc system starts with the removal of the wheel, which makes some of the steps photographically clearer, therefore easier to understand.

Begin the job on a drum brake car by removing the hub cap, then pry off the dust cap in the center. If you do not wish to obtain a special tool, you also can pry it off with two thin screwdrivers (inserted 180 degrees apart) or with large "water pump" pliers around the cap (see Fig. 12–26). If you use pliers, be careful not to crush the cap.

With the dust cap off you'll see the nut lock or the wheel bearing nut. Remove the cotter pin (see Fig. 12–27) or the locking screw, the nut lock if used (see Fig. 12–28), and then the wheel bearing nut itself, using a socket wrench and ratchet.

With the nut off, grasp the tire at each side and jostle it toward you. As you jostle, a washer and the outer wheel bearing will pop outward. Remove them (see Fig. 12–29). Continue jostling and the wheel and drum assembly will come off (see Fig. 12–30).

If it sticks, turn the wheel slowly and listen for a scraping sound. If you hear a scraping noise, it indicates that the brake shoes are dragging against the brake drum, preventing the drum from coming off. You will have to turn an adjuster to back the brake shoes out of contact with the drum. The two ways this is commonly done are:

Look at the inner side of the brake backing plate, the flat plate bolted to the suspension at each wheel (see Fig. 11–34). Somewhere in the center, usually near the bottom, is a slot sealed by a rubber or plastic plug. Remove this plug and insert two thin screwdrivers. Use one to push the brake's self-adjusting lever, the other to turn a circular toothed adjuster called the star wheel (see Fig. 11–35). Turning the star wheel one way

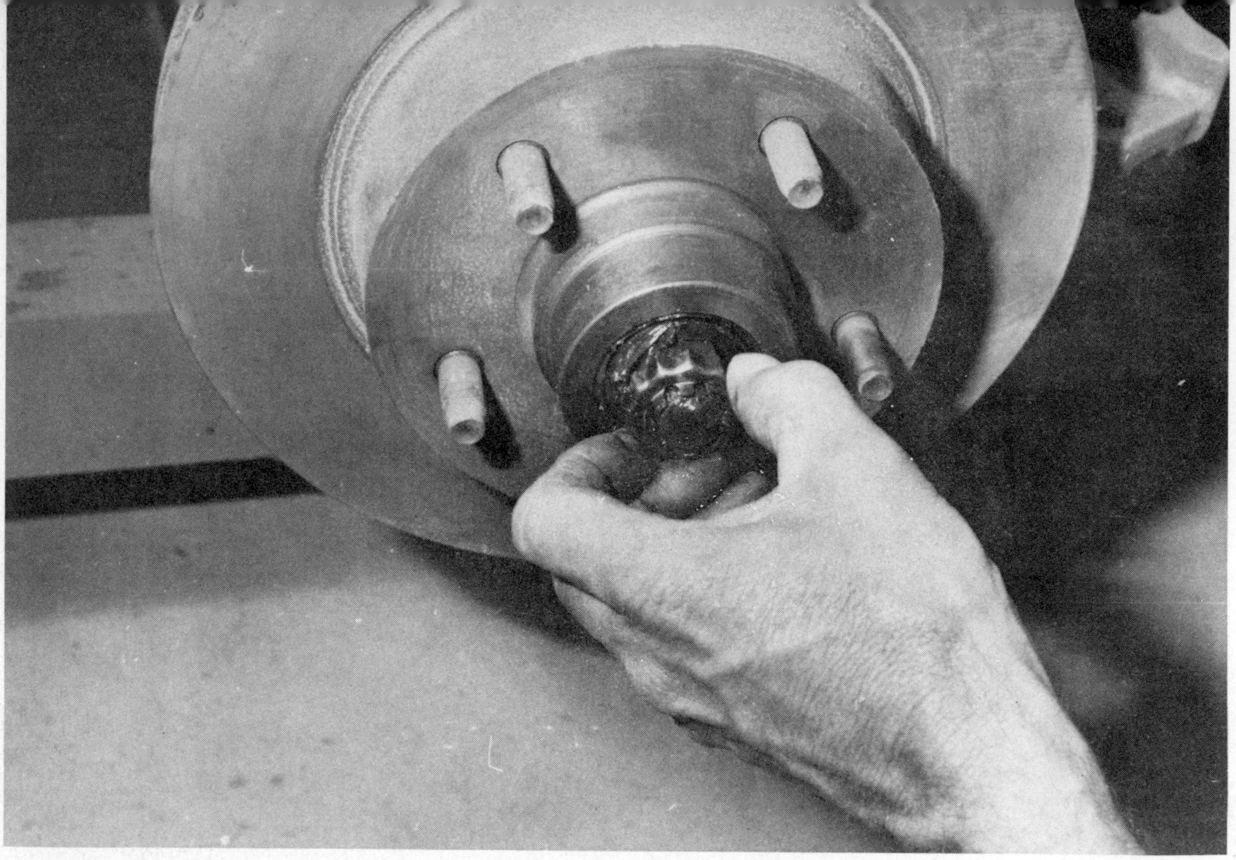

Fig. 12–28. With the cotter pin out, the nut lock can be removed as shown.

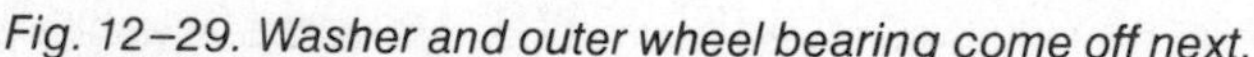

Fig. 12–29. Washer and outer wheel bearing come off next.

Fig. 12–30. Whether it's a wheel and tire assembly, or just a disc, it can now be removed from the spindle. With a disc assembly, however, the caliper must be removed and hung out of the way first, as shown in Fig. 12–32.

Fig. 12–31. Prying out a grease seal with a screwdriver. This job is illustrated with a disc, but it's the same on a drum.

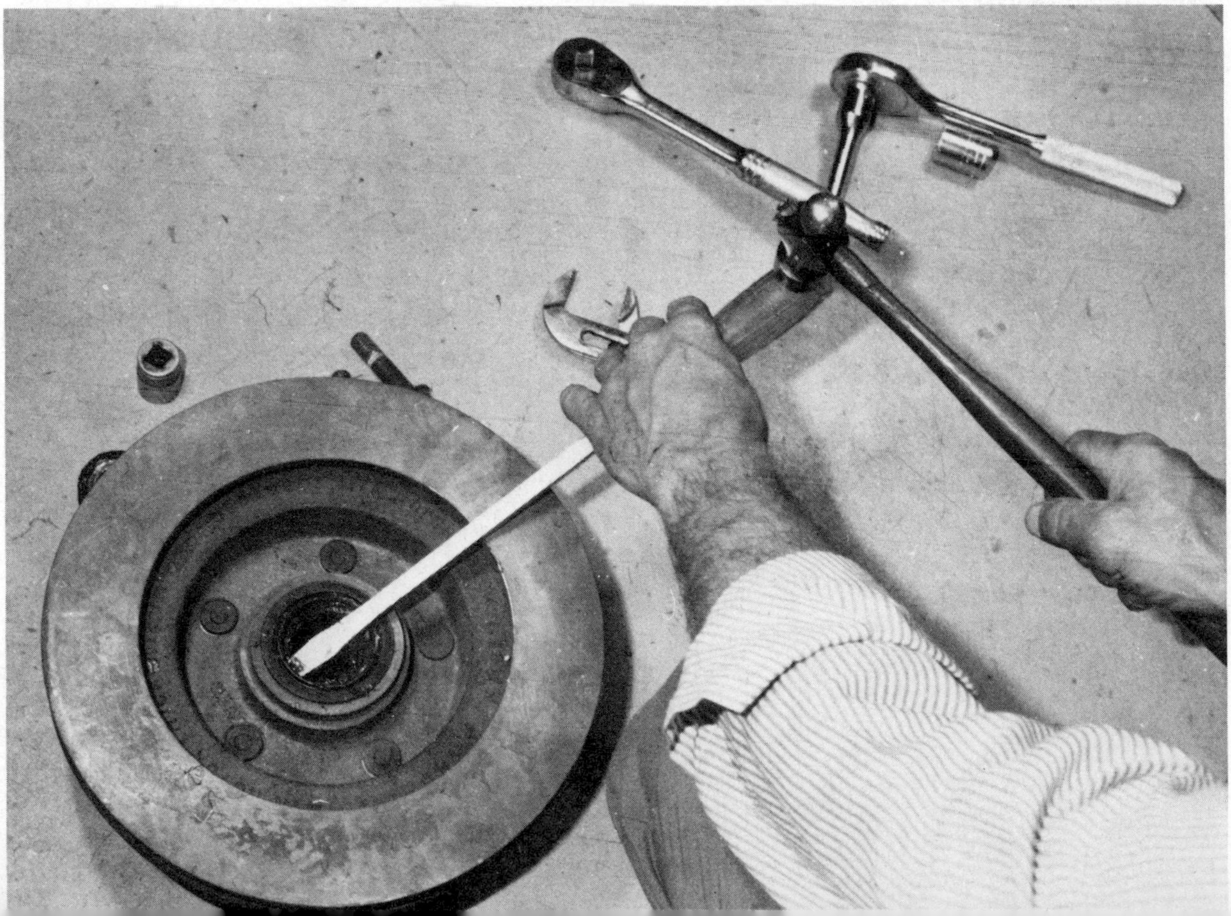

will expand the shoes tightly against the drum (wrong way); the other will contract them inward, away from the drum (the right way). Once the shoes are contracted, the drum will spin freely and you can go back to the jostling procedure to remove the wheel and drum assembly.

On some cars there is no plug in the backing plate, but there is a slot area lanced in the backing plate or drum. Unbolt the wheel and you'll see the one in the drum, if used. With a punch and hammer, whack the lanced area and it will break loose. Insert two thin screwdrivers and proceed as in the previous example.

With the drum off you have complete access. Turn over the drum to inspect the inside, and you'll see a seal on the rim of the cylindrical cavity. Pry out the seal with a screwdriver (see Fig. 12–31), then reach into the cavity and remove the inner wheel bearing.

• *Disc brakes.* Disc brakes pose a special problem, in that the disc must be removed. The caliper, a hydraulic clamp that squeezes the brake shoes against the sides of the disc, is mounted over it, so you must remove the caliper in order to take off the disc.

Begin by removing the wheel-tire assembly. Next, inspect the mounting of the caliper. In most cases, the caliper is held by two rods or by two guide brackets bolted to a mounting cage. If the method of unbolting the caliper is not obvious to you, refer to Chapter 11, which covers the braking system.

Once you have unbolted the caliper, lift it away and, using coat hanger wire, tie it to the front suspension so there is no strain on the brake hose (see Fig. 12–32). Do not let the caliper dangle by the hose, or the hose will be damaged (even if the damage is not something you can see).

Remove the dust cap, then the nut lock, cotter pin, wheel bearing nut, etc. Jostle the disc toward you, as you did with the wheel-drum assembly, and the washer and outer wheel bearing will pop out. Remove them. Continue jostling until the disc is off. Turn the disc over and you will see the same type of seal in the center as with the drum brake. Pry it out and lift out the inner wheel bearing.

• *Cleaning and inspection.* Clean out the bearing cavity in the drum or disc with automotive

Fig. 12–32. Disc caliper has been unbolted from suspension, or from frame on some cars, then hung out of the way with wire, as shown. Disc can now be pulled off the spindle.

solvent and wipe clean. Reach into each end of the cavity and feel the cylindrical bearing race. If the surfaces against which the bearing roller rides are mirror smooth, the races are okay. If they are rough or scratched, they must be replaced. Take a long flat-end brass punch, insert it from the opposite end and, tapping on the inner edge of the race, tap it out. You must tap gently and move the punch around the edge circumference of the race so you don't cock the race in the cavity.

If the race does not budge, take the disc or drum to the machine shop of an auto parts store. The shop can remove a stuck race in seconds by using a press.

Caution: On a very few cars the race is held in place by a C-shaped spring clip. You should inspect the cavity adjacent to the race to determine if such a clip is used. If you find one, carefully pry it out with a thin screwdriver.

• *Inspection*. Clean the roller or ball bearings in automotive solvent, then allow them to air dry. Turn the bearing cage as you hold the inner bearing race (it comes out with the roller or ball bearing cage and normally holds to it). If you feel any roughness, replace the bearing. Also inspect for the following:

• *Cracks in the cage or inner race*.

• *Any wear pattern on the race or rollers that is not completely smooth*.

• *Metal smear marks on the edges of the rollers*. This indicates overheating, possibly from worn out or inadequate lubricant.

• *Little steps worn into the edges of the rollers*. This can be caused by dirt.

• *Wear on the outside diameter of the cage or in the pockets between the rollers or balls*. This is caused by dirt or lubricant failure.

• *Indentations on the rollers or balls, or flaking of surface metal from the balls or rollers*. This is caused by loose bearings, rough roads, and heavy loads.

• *Pockmarks on the bearing surfaces of the races*. Often this is caused by poor lubrication.

• *Heat discoloration*. A faint yellow or dark blue tint on the race and/or balls or rollers is a heat failure caused by heavy loads or poor lubrication. Note: If the color is light brown to black, the problem is not severe, just some staining from moisture and the lubricant. Polish with a clean cloth, and if the bearings are otherwise in good condition they can be re-used. A simple test for heat failure is to draw a file over the roller or ball (while you restrain the ball or roller with your fingers). If the file glides over, the bearing is okay; if it grabs on the roller or ball, the bearing assembly should be replaced.

• *Reassembly*. If any of the bearings have failed, obtain replacements. The inner and outer bearings are not of the same size, so watch when you install. Also obtain a new grease seal and a tube of wheel bearing grease (not chassis grease).

Clean out the bearing cavity in the disc or drum. If you had to remove the old bearing races, tap the new ones into place with a brass drift and hammer, working carefully all around to insure you don't cock them. If you had to take them to a machine shop for removal with a press, have the shop install the new ones.

With the races in place, smear a thin coat of grease over the inside wall of the drum or disc cavity and the races. Apply a generous coat of grease to the cage and rollers, carefully working it onto each roller (see Fig. 12–33).

Install the inner wheel bearing, then tap the new grease seal into place. You can obtain a seal driver (see Fig. 12–34) or use a socket wrench of the same size as the seal, or even a hockey puck. Place the seal driver, socket, or puck onto the seal and tap it with a hammer.

Reinstall the drum or disc, install the outer bearing, the washer, and the adjusting nut. On a disc system, refit the caliper.

• *Adjustment*. Begin the adjustment by tightening the nut to the specified torque (or seven lbs.-ft. if you don't have the specifications), using a torque wrench while spinning the wheel or disc by hand. This will seat the bearings in place on the spindle. Caution: Excessive torque could damage the bearings, so don't overdo it.

Loosen the nut until it is just free, then install the nut lock and cotter pin, or the screw. If the cross slot in the nut or nut lock does not align with a hole in the spindle, you may loosen the nut up to an additional one-half of one flat on the hexagonal nut (that's one-twelfth of a turn, very little). Note: Ford products' wheel bearings are adjusted somewhat differently (see Fig. 12–35).

If you have a cotter pin arrangement, spread the stems of the pin, trim them if necessary with cutting pliers, and bend them back against the nut or lock so they don't touch the wheel cavity.

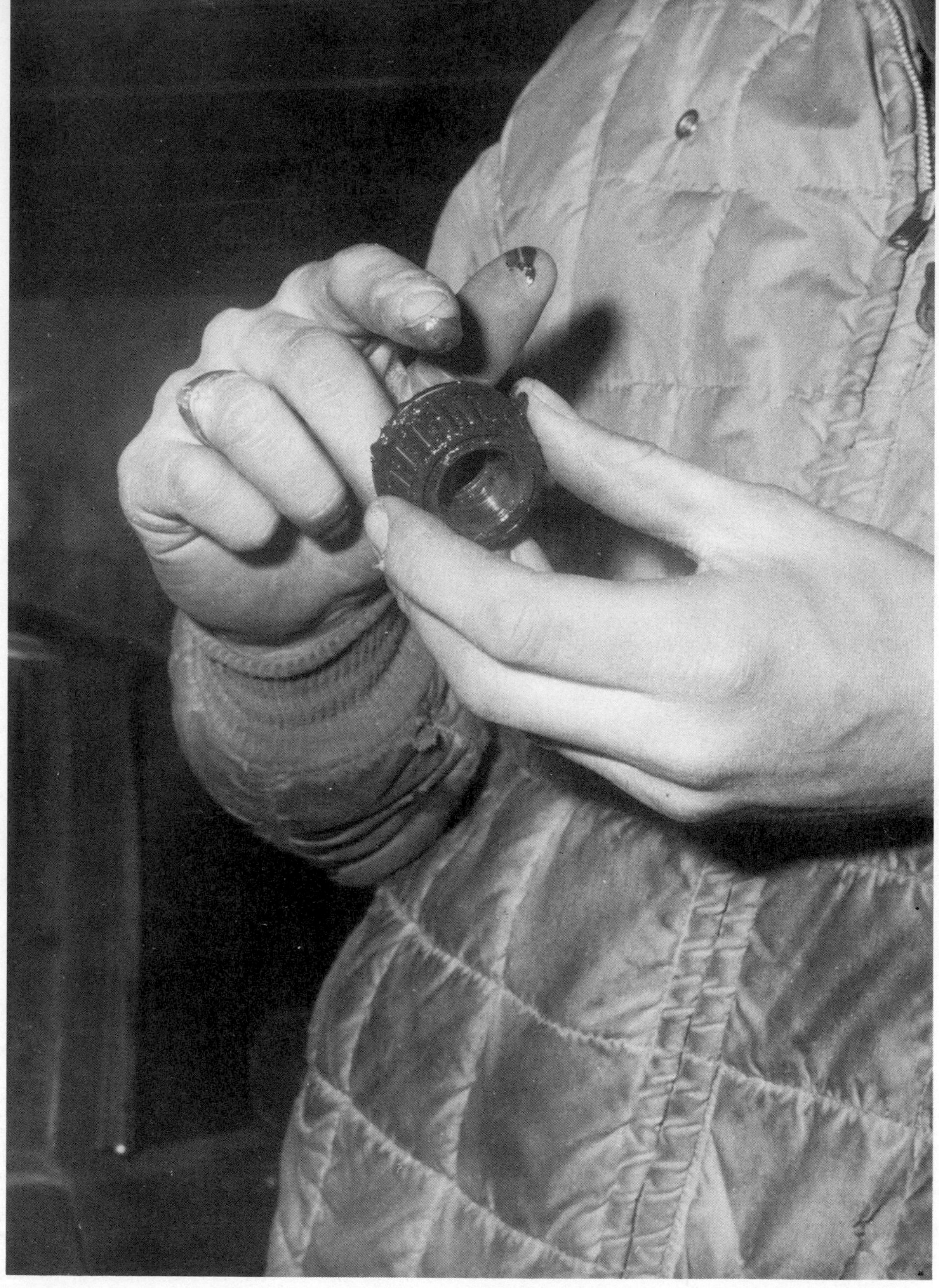

Fig. 12–33. Applying wheel bearing grease to rollers. Use a generous amount. Caution: Do not use chassis grease.

Fig. 12–34. Using a seal driver. Grease seal is in position. Place driver shown squarely on top of it and hammer on driver top to seat the seal. Inner wheel bearing must be in place first, of course.

Fig. 12–35. Illustration shows adjustment procedure for Ford wheel bearings.

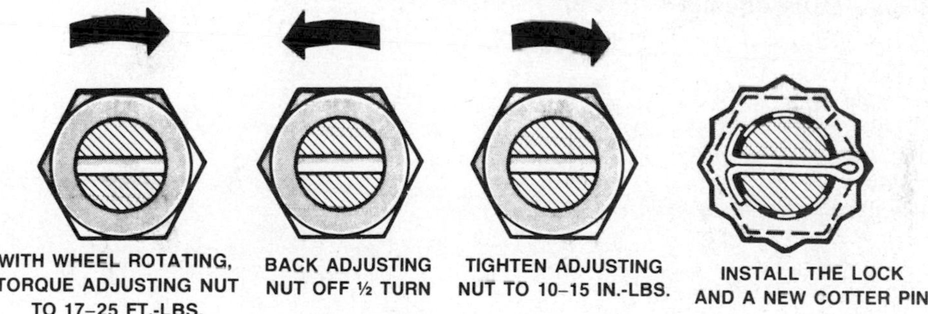

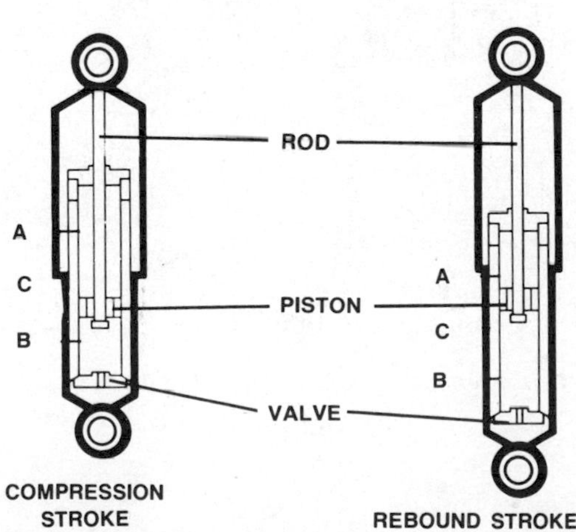

Fig. 12–36. How a shock absorber works. "A" is an oil-filled chamber above the piston, "B" is an oil-filled chamber below the piston, and "C" is an outer oil-filled chamber. When an end of the car goes down, the shock compresses. The piston must force its way through the resistant oil. Some oil passes through calibrated holes and valves in the piston to rise into the upper chamber. Other oil is pushed through the valve at the base of the shock into chamber "C." When the shock extends, as when a wheel bounces up, oil is drawn through the valve from the outer chamber into "B" and some oil passes through the piston holes from "A" to "B." In all cases, the piston movement is slowed by the oil absorbing energy from the piston and rod. In so doing, the piston becomes hot. It dissipates this heat into the atmosphere. The damping action of the shock absorber stops the spring from oscillating.

Check the bearings for free play by rocking the wheel in and out as you did for the pre-disassembly test. If they are too loose, readjust the nut. Smear a coating of grease over the inside surface of the dust cap, then tap the cap into place with a rubber mallet.

Shock Absorbers

The shock absorber, actually a spring damper as explained earlier in this chapter, stops the movement of the spring by dissipating it in the form of heat.

As the spring tries to oscillate, so must the shock absorber, for both are operating in the same location. The shock absorber, however, is a cylinder filled with oil and contains a piston. The piston has to force its way through a cylinder of oil (see Fig. 12–36). This brings it to a quick stop as the energy is converted to heat, and the hot oil quickly dissipates the heat into the atmosphere.

The movement of the piston in the oil-filled cylinder is carefully regulated by valves and holes in the chamber and the piston, making the shock a pretty complicated device. Some shocks are even further complicated by these special features:

- *They are adjustable.* Some shock absorbers have an arrangement in which one of the valves can be pre-set for a normal, firm, or soft ride (see Fig. 12–37).
- *They have air chambers.* Some rear shocks have, in addition to the oil chamber, an air chamber built into the upper half. When air is added, as on a tire, the load-carrying capacity of the spring is increased. On some cars this is done automatically by means of an electric air compressor under the hood and a height sensor in the rear. When the height sensor determines that the car is too low in the rear (because of a heavy load), it triggers the air compressor. When the load is removed and the rear end rises, the height sensor triggers an electric valve in the compressor to deflate the air shocks (see Fig. 12–38).
- *They have assist springs.* The springs on a car may sag after a few years, particularly when heavy loads are carried. Some replacement shocks have a coil spring around them, much like a strut, to help the car's regular spring, whether coil, torsion bar, or leaf type (see Fig. 12–39).

Even the conventional shock absorber is available in several types: standard, heavy-duty, and extra-heavy-duty. The differences are internal, chiefly in the diameter of the piston and the rod to which it is attached, the size of the oil chambers and the quantity of oil, and the type of valves.

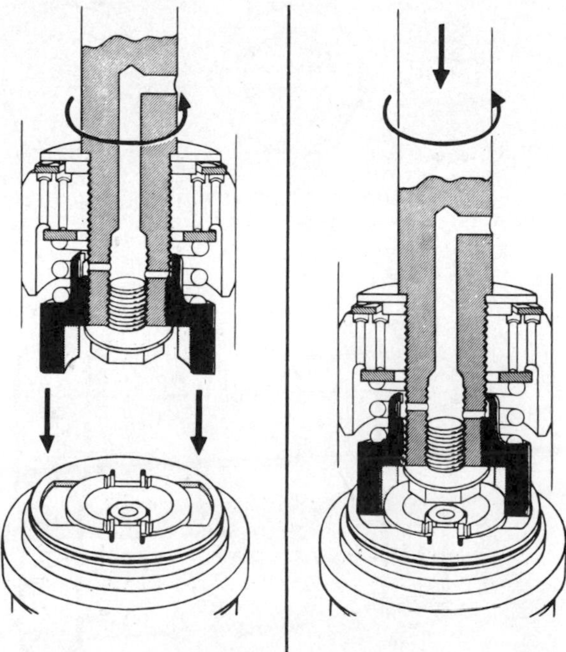

Fig. 12–37. How an adjustable shock absorber works. The adjustable valve in the base of a shock engages specially-shaped base of piston when the shock is completely collapsed. At right, note that when piston rod is turned, the adjustment on the valve is altered.

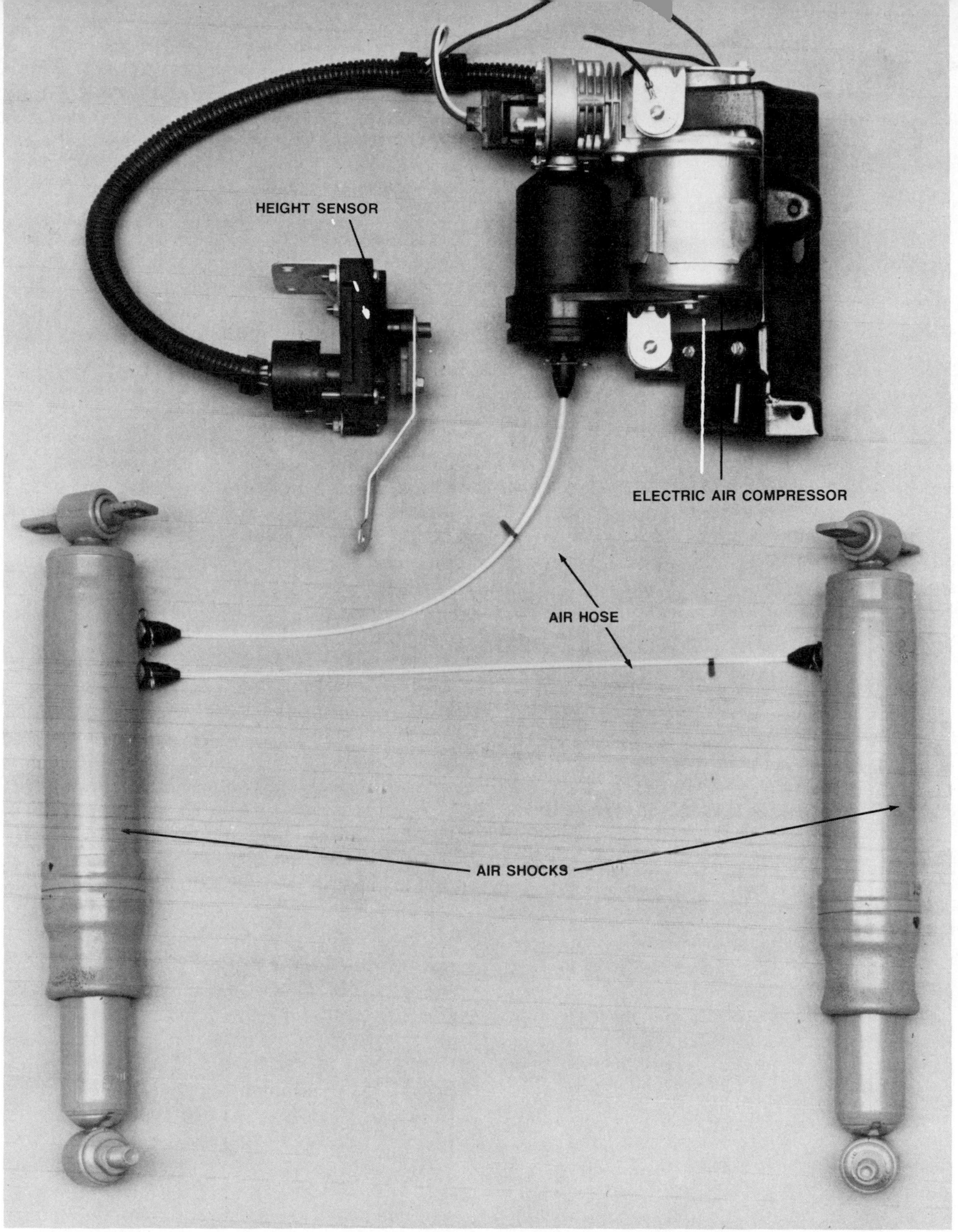

Fig. 12–38. This is an automatic rear-leveling system that uses shocks with built-in air bags, a height sensor, and an automatic compressor. When sensor, bolted to car body, senses that car rear is hanging low, It triggers the compressor which pumps up the shocks to raise the car. When the ride height sensor determines that car rear is too high, it opens a valve in the compressor assembly and shocks automatically deflate.

Fig. 12–39. This shock absorber has a coil spring included, making it suitable for cars with slight spring sag.

Selecting a Shock

For most conditions, a conventional shock is all you need. The standard type duplicates original equipment, and the heavier-duty designs are the most popular for replacement use. The average driver with a suspension in otherwise good condition should find the heavy-duty design provides a good compromise for ride, extra handling control, and extra life over the original equipment type. The extra-heavy-duty, unless it is adjustable, may be too firm a ride, although some of this type are heavy-duty only in construction, not in the ride they provide. Check manufacturer's literature.

If the springs are sagging when the car is normally loaded, install a shock with an assist spring. You can see sag if you look at the car from the profile, and straight on at the front and rear. If the body is not horizontal in all these views, the springs are sagging. The torsion bar type spring can be adjusted to correct height, but coil and leaf springs cannot, and so a shock with an assist spring will be helpful.

If you carry heavy loads in the rear, install shocks with air chambers. They come with extension tubes so you can install the inflation valves in a convenient place, such as the trunk compartment. Or you can install a compressor kit, with dashboard controls, to inflate or deflate the shocks' air chambers from the driver's seat.

Testing the Shocks

The best way to check is with a road test. If the car has poor control on hard cornering and shakes badly on washboard roads, the shocks probably are bad. Check on smooth pavement at 10–15 mph by tapping the brakes repeatedly. If this sets up a rocking motion, in which the front dips while the rear rises, the shocks are defective.

Also, visually inspect the shocks (see Fig. 12–40). Replace them if you see deep dents in the shell, oil leakage (more than minor seepage), or a pitted piston rod (visible when wheels hang down, so when you jack up the car, do it by the chassis).

Most original equipment shocks have had it at 20,000 miles, and the heavier-duty replacements will last up to 30,000. Shock life is not really predictable, however, because of the great variation in the condition of roads on which people drive.

Replacing a Shock

Shocks should be replaced in pairs—both fronts or both rears—for balanced results. Most shocks are outboard of the coil springs. To get the old ones off, begin by looking at the shock's upper and lower mountings to determine if there is enough clearance to work with the wheel in place. If not, loosen the wheel lug nuts.

Next, jack up the car (*Chapter 5*) and support it on safety stands. Remove the wheels if necessary and unbolt the shock at the top and bottom (see Figs. 12–41, 12–42). Some shock studs are specially shaped at the top, so they can be held with a wrench while you loosen the retaining nut with another wrench. If not, you can buy an inexpensive special tool to hold a stud. If the upper mounting is recessed, and the retaining nut is rust-frozen in place, first try spraying with pene-

Fig. 12–40. Visually inspect your shocks for oil seepage, dents, and a pitted piston rod.

Fig. 12–41. Loosening an upper shock mounting bolt in the engine compartment. On other cars the shock mount may be a stud-and-nut type and may only be accessible from under the car.

Fig. 12–42. Loosening a shock's lower mount.

Fig. 12–43. Tightening a shock's lower mount. Rubber bushing under the nut washer should not protrude measurably from the outer circumference of washer, or nut is too tight.

trating solvent.

If that doesn't work, put a deep socket on the stud and with the ratchet wrench try rocking side to side to snap the stud (you're replacing the shock, so breaking the stud doesn't matter). If the stud won't snap, the best remaining choice is a nut cutter designed for shock work. It fits into the recess around the stud, and when turned with a wrench from the top will neatly cut the nut, permitting you to lower the upper portion of the shock and remove it.

If a shock is inside the coil spring, it is lowered out. Look at the lower mounting and you will find some form of plate that can be unbolted to provide clearance for shock removal.

As you remove a shock absorber, notice the use of rubber bushings at the top and bottom. When you install the new shock, you must install the new bushings and any washers in exactly the same manner. When you tighten the shock nuts, you want the shock to be secure, but you also don't want to overtighten. If the bushing under a washer is squeezed so that it protrudes from the exterior circumference of the washer, you've probably overtightened (see Fig. 12–43).

Strut Replacement

The front strut used on so many cars, particularly with front-wheel-drive, is serviced somewhat differently than the conventional shock absorber.

There are many possible replacement procedures, depending on how the strut is attached to the steering knuckle. On the typical American-built transverse-engine, front-wheel-drive car and VW Super Beetles, the strut is attached to the knuckle by two bolts. On all but the GM "J" cars (Chevy Cavalier, etc.) introduced as 1982 models, there is a camber adjustment, and when you replace the strut you must have camber reset by a professional. With the "J" cars, although there is provision for a camber adjustment if necessary, the new strut simply bolts to the knuckle.

To replace this type of strut, you must have a suitable spring compressor. Although inexpensive ones can be purchased, you will do better to rent a high-quality one from a rental center.

The basic procedure is as follows:

(1) Loosen the front wheel lugnuts, then jack up the car and place safety stands under chassis jacking points (see your owner's manual if their location is not obvious).

(2) Remove the wheel.

(3) Remove the screws or nuts that hold the strut upper mount in the engine compartment. Loosen but do not remove the piston-rod nut (the nut in the center).

(4) Remove the nuts and the two bolts that hold the strut to the steering knuckle (Fig. 12–45). Note: If one of the two bolts has an eccentric (cam) built in (see Fig. 12–48), and there is the possibility you will be reinstalling the strut (as explained later, in Step No. 10), make alignment marks on the cam surface and the strut flange adjacent to it. If you reinstall the strut and realign the marks, it will not be necessary to have the camber reset.

(5) Separate the strut from the steering knuckle, then pull the strut down and out of the car. Note: In pulling the strut down on some cars, you could bring it into contact with the axle-shaft universal-joint rubber cover, damaging the cover. To prevent this, wrap a piece of thick, flexible plastic around the rubber cover and tape it in place. In some cases you may have to disconnect a brake line for clearance. If so, bleed the brakes after reconnecting the line, as explained in Chapter 11.

(6) Place the strut in a large vise, and with a spring compressor, compress the spring as in Fig. 12–44. Compress until the spring tension is released.

(7) Remove the piston-rod nut. Most piston rods have Allen slots in the top, so you can hold them while you remove the nut with a box wrench.

(8) With piston-rod nut off, you can lift the upper mount up and off, then take off the coil spring. See Fig. 12–46.

(9) Turn the upper mount bearing by hand, and if you feel roughness, replace the bearing assembly. (See Fig. 12–46.)

(10) Some struts have replaceable shock inserts (Fig. 12–47), others do not. However, you may be able to get a replacement strut with an insert that can be changed, and if you do, the next time you have to service the strut you won't need the complete part. And because you will be refitting the strut, you can make alignment marks as explained in Step No. 4 and not have to get camber reset. Note: On some cars, such as many Hondas and the GM "J" cars (Chevy Cavalier, etc.), the strut cap is welded in position and the

welds must be ground off before the cap can be removed for replacement of the shock insert.

(11) When reassembling, be sure the coil spring fits into its mounting pads at top and bottom. Complete installation basically as reversal of the removal procedure.

Ford cars with struts and rear-drive (Fairmont—Zephyr—Mustang—Capri) are the easiest to service. The strut spring is separate and the strut unbolts from the steering knuckle. Camber is fixed, so no readjustment is needed.

After the lower mounting is free, disconnect the upper, noticing how the parts are assembled so you can install the new ones correctly. Most struts have a bearing assembly in the upper mounting, so when you take out the strut spin the bearing by hand, feeling for any roughness and listening for any bearing noise. If there is either, replace the bearing (see Fig. 12–46).

Fig. 12–44. Using coil spring compressors on a strut-equipped car. Job is shown with strut removed, for clarity, but you should install compressors and tighten them down while wheels are on the ground. Although air wrench is shown being used, ordinary hand tools also will do the job.

Fig. 12–45. This VW strut is held to the steering knuckle by two clamp bracket bolts and nuts. Just remove them and separate strut from steering knuckle. Mark bracket nut and bolt locations on strut and knuckle for precise re-installation and resetting camber may not be necessary if only strut shock insert is changed.

Fig. 12–46. Typical strut has ball bearing in upper mounting.

BALL BEARING

Fig. 12–47. If you want to replace just the shock absorber insert (possible on most struts), remove the cap and then lift it out. It takes a special tool to remove the cap. Buy your replacement insert from an auto parts jobber who has the tool and will install the insert for you.

Fig. 12–48. Cambolt is one of two holding strut to knuckle. Make alignment marks as shown if strut may be reused.

13 DRIVE TRAIN FUNDAMENTALS

Manual Clutch

THE MANUAL CLUTCH is a mechanical device that can disconnect the engine from the transmission to permit stopping the car without turning off the engine and to allow gear changes. Here's how the typical clutch works (see Fig. 13–1):

The key part of the clutch is the friction disc, a disc with friction material (like brake lining) on both sides. It has a splined center hub that fits onto the front of the transmission shaft, which is splined to mesh with those on the hub. The splined arrangement means that the disc will always turn with the transmission shaft but can move fore or aft on the transmission shaft.

A large plate called the pressure plate, with spring fingers, fits over the friction disc and is bolted to the engine flywheel. When the clutch is engaged, the fingers press down on the friction disc and hold it firmly to the flywheel. This provides a friction contact between the engine (the flywheel is bolted to the rear of the engine crankshaft) and the transmission shaft. Thus power from the engine can flow to the transmission.

When you want to separate the engine from transmission, to stop the car or change gears, you press down on the pedal, which operates linkage (or a cable and linkage) that pushes a throwout bearing assembly against the pressure plate fingers, disengaging them from contact with the friction disc. The spinning flywheel then pushes the friction disc rearward (the disc slides along the splines) and the connection between the engine and the transmission is broken.

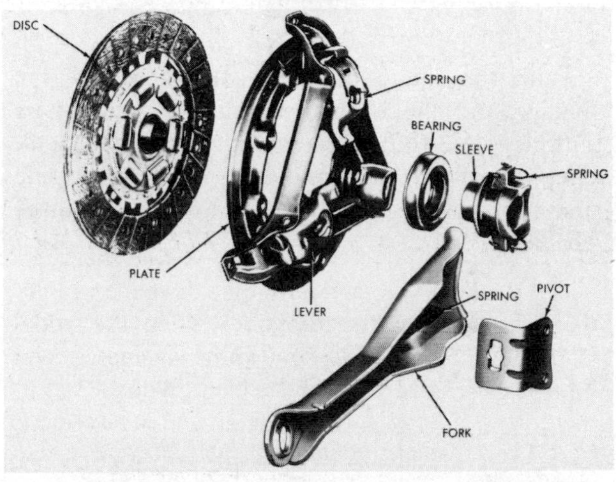

Fig. 13–1. A typical clutch. When you step on the clutch pedal, linkage pivots the fork-pivot, pushing on the sleeve and bearing that, in turn, depresses spring-loaded fingers on the clutch plate. This releases pressure on the disc, a plate coated with friction material on both sides. When the pressure on the friction disc is released, the transmission can be shifted into neutral, separating engine from the transmission and the remainder of the car's drive train.

If the pressure plate spring fingers lose a bit of their grip, or if the friction disc wears down (or becomes contaminated with grease), the friction connection between engine and transmission deteriorates. You step on the gas and the engine speeds up, but the car seems to lag behind. This is called clutch slippage and the only cure is to replace the clutch, a job for professional mechanic.

There is another common cause of a slipping clutch, however, and this one you can do something about if you catch the problem in time.

As the friction disc wears, the spring fingers pivot and come in contact with the throwout bearing assembly that is pressed to disengage them. It is important to maintain some clearance between the throwout bearing assembly and the fingers, in order to allow the other ends of the fingers to keep full spring pressure on the friction disc as the disc wears. This clearance is translated into free play at the clutch pedal. If you slowly depress the normal clutch pedal, you'll feel a point at which you're really pushing down on something. Up to that point is the free play, and it should be ¾ to 1⅜ inches on most cars. If there is no free play, the clutch may not fully engage. If, however, there is too much free play, another problem can occur:

With too much play, the full travel of the linkage and the bearing assembly may not be sufficient to fully depress the spring fingers. Only a partial disconnection between engine and transmission results, and this causes three problems:

(1) Because the friction disc is never really disengaged from the flywheel when the pedal is depressed, the flywheel spins against it and causes excessive wear of the friction disc.

(2) When you try to shift gears, the transmission shaft will not be stopped, as it is when the clutch is fully disengaged. Parts in the transmission clash and wear out prematurely.

(3) The throwout bearing is in constant contact with the fingers and wears out prematurely.

On many new transverse-engine front-drive cars, the clutch-free play is self-adjusting to prevent these possible problems. Cars with self-adjusting design include GM "X" (Chevy Citation, etc.) and "J" (Chevy Cavalier, etc.) cars, plus Ford Escort and Lynx and Chrysler "K" (Reliant and Aries). Note: GM changed its front-drive clutch cable and ratchet system in 1981, and so all front drives require the driver to readjust the clutch every 5,000 miles. Just slip your foot under the pedal, lift it up to a stop, then depress the clutch to complete the adjustment. On other models, there also is a clutch play adjustment you can make to correct or prevent these problems.

You can correct any of these problems with a clutch linkage adjustment. The exact proce-

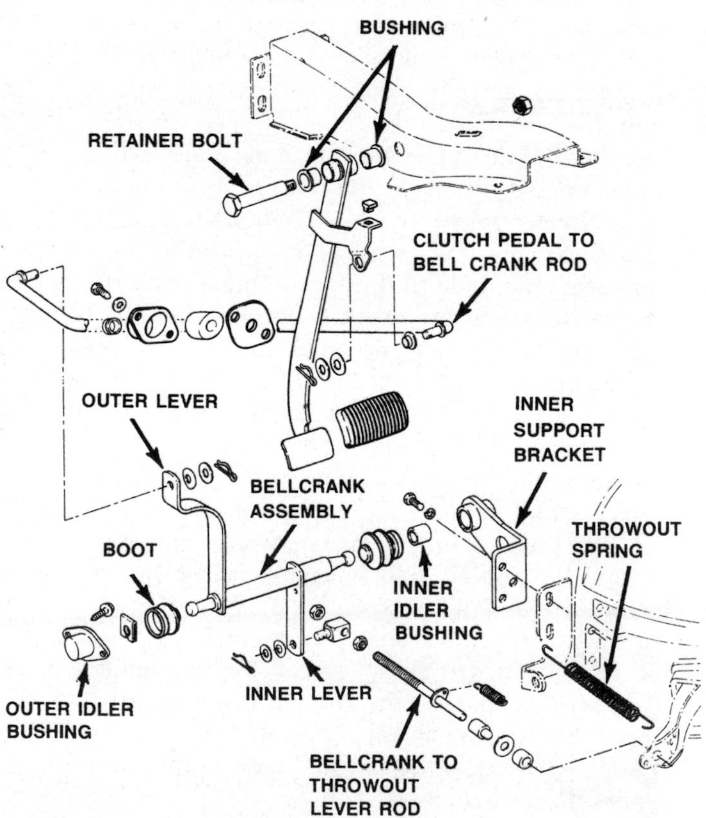

Fig. 13–2. Clutch adjustment for American Motors' cars: loosen lockscrew on outer end of rod that goes from bellcrank to throwout lever, then turn adjusting nut.

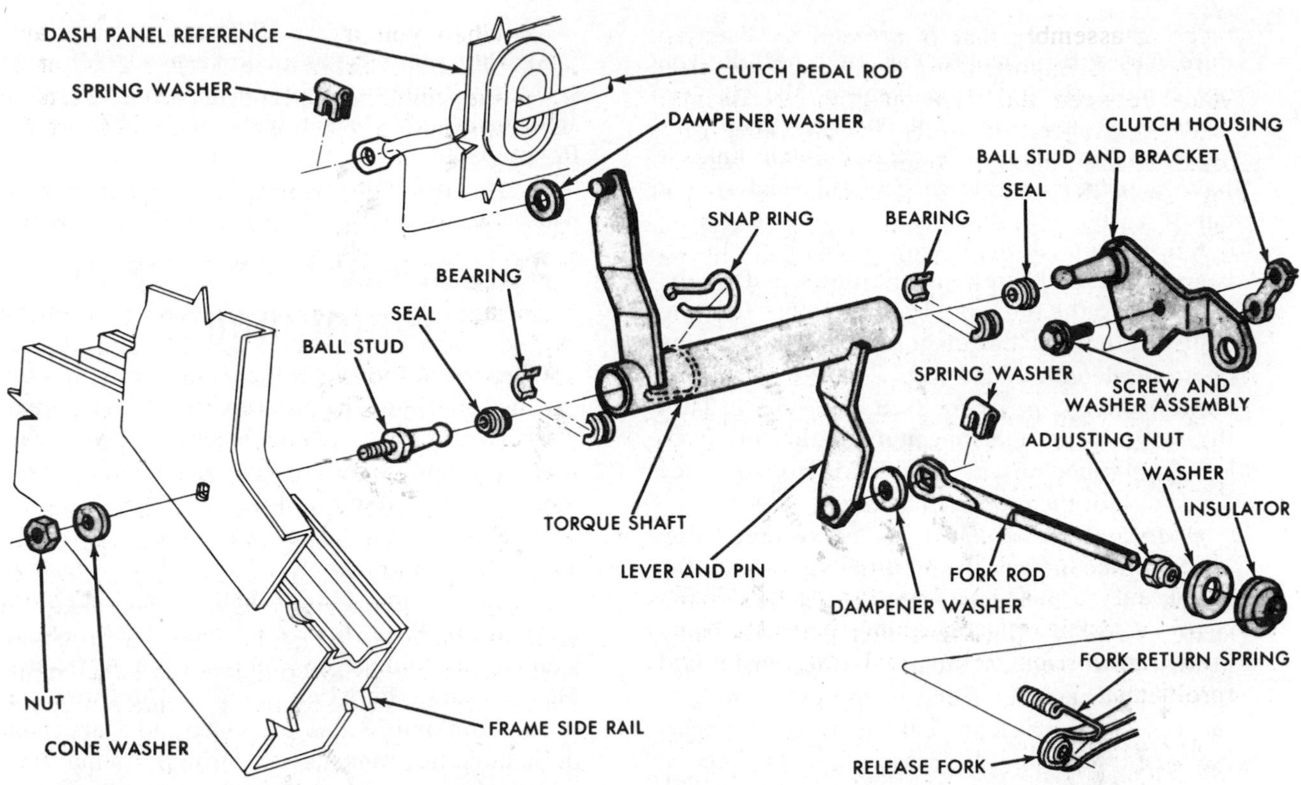

Fig. 13–3. This is a typical Chrysler Corp. clutch linkage. Turn adjusting nut on fork rod to change free play.

Fig. 13–4. On Chrysler's front-wheel-drive Omni and Horizon, lift up on threaded bracket as shown to unseat the bracket, then turn it on the threaded cable end to change pedal free play.

dure varies from car to car, but the following are typical. Jack up the front of the car (see *Chapter 5*) and support it on safety stands. If you have trouble finding the clutch linkage, have a helper operate the clutch pedal so you can trace it.

• *American Motors cars* (see Fig. 13–2). Loosen the lockscrew on the outer end of the bellcrank to the throwout lever rod, then turn the adjusting nut (in the inner side) to change the operating length of the rod.

• *Chrysler Corp. cars* (see Fig. 13–3). Turn the self-locking adjusting nut on the fork rod. On front-wheel-drive Omni and Horizon, see Fig. 13–4 for the proper adjustment.

• *Most General Motors cars* (See Fig. 13–5). Disconnect the return spring at the clutch fork, then rotate the clutch lever and shaft assembly until the clutch pedal is firmly against the rubber stop on the dashboard brace. Push the outer end of the clutch fork to the rear until you can feel the throwout bearing just touching the pressure plate fingers. Loosen the locknut on the rod and shift the rod up to the gage hole. Turn the forward end of the rod until there is no free play in the system. Move the rod back to its original hole, then tighten the locknut, being careful not to change the length of the rod. Install the rod retainer and the clutch fork return spring. Pedal play should be within specifications. Note: Chevy Citation and other GM front-drive compacts have a self-adjusting clutch setup.

• *Ford Motor Co. cars*. Refer to Figs. 13–6 to 13–8.

As with other types of linkage adjustments, such as parking brakes, you may not make the adjustment the correct way on the first shot. Be prepared for a second or even third try. Once you become familiar with it, however, it will be a quick job.

Automatic Transmission Maintenance

When an automatic transmission fails, it's an expensive proposition to repair. Although that work is not for weekend mechanics, you can give your automatic routine maintenance and other care to extend its life.

Most important, check the fluid level regularly. After the car has been driven a few miles, apply the brakes and move the shift lever from Park to Low, then back again, stopping in Neutral on Chrysler and American Motors cars, Park on General Motors and Ford automobiles. Let the engine idle and pull the automatic transmission dipstick. Wipe dry, reinstall (be sure to get it down all the way), then remove and read the dipstick level. Even if the dipstick shows only a half pint low, add that half pint, using a long hose and funnel. Cap the oil can and save for future needs.

Use "Dexron" fluid on all cars except some Ford products, for which Type F fluid is specified. Check your owner's manual to be sure. If the Ford owner's manual lists a Type "CJ" fluid, you can use Dexron in that transmission too. These names appear on the appropriate cans.

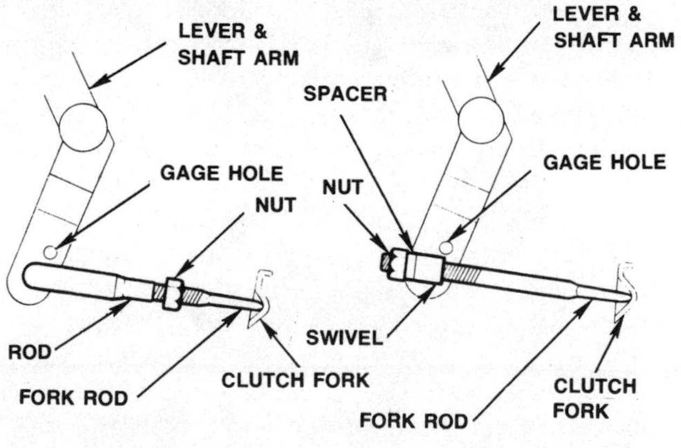

Fig. 13–5. This is typical General Motors clutch free play arrangement. To adjust play, disconnect the return spring (not shown) at the clutch fork. Rotate the clutch lever and shaft assembly until the clutch pedal is firmly against the rubber bumper on the dashboard brace. Push the outer end of the clutch fork rearward until the throwout bearing just touches the pressure plate—something you can feel, not see. Install the pushrod or swivel in the gage hole and adjust the pushrod length until there is no free play in the system. Remove the rod or swivel from the gage hole and refit to the lower (operating hole). Install the washer and tighten the locknut, being careful not to change the length of the rod. Refit spring.

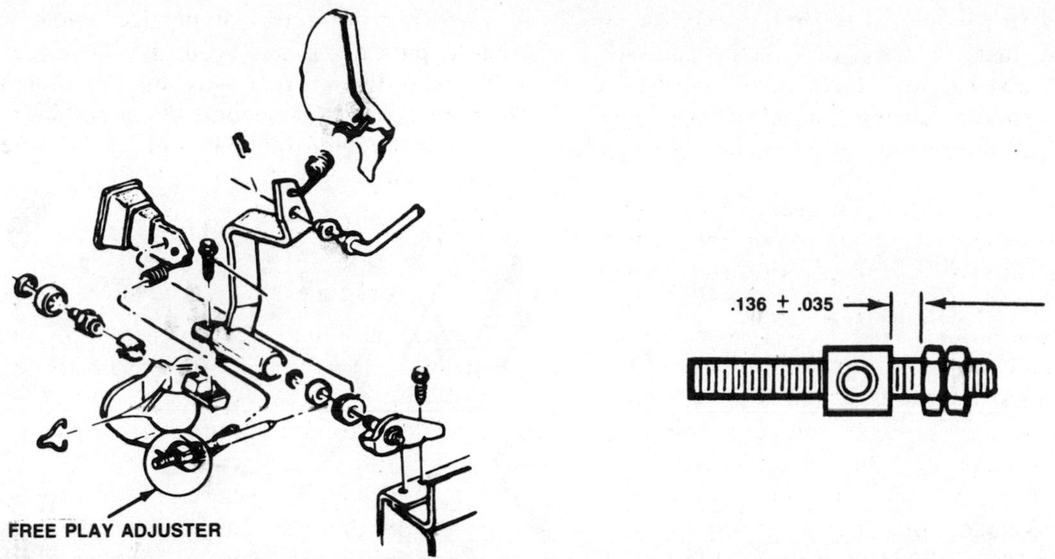

Fig. 13–6. This is a common type of Ford free play adjuster on the linkage. With this type, begin adjustment by disconnecting clutch release lever spring, then move release fork rearward by hand until you feel the throwout bearing just touching the pressure plate. Loosen and back off both nuts at the swivel and place the specified feeler gauge (0.136 inch) between inner nut and swivel. Turn inner nut until feeler gauge is just touching nut, then run outer nut up to inner nut and lock them together. Reconnect the spring.

Save your transmission from a fast burnout—don't try to rock yourself loose if you're stuck in snow or mud. You may eventually get out, but the transmission oil will often have overheated so badly and adversely affected transmission parts that the victory will be short-lived. The cost of a tow truck will be far less than the loss of useful transmission life.

Changing Oil and Filter

Most transmissions have a replaceable filter in the oil pan, and if not that, a cleanable screen. It's good maintenance practice to remove the oil pan, change the filter or clean the screen in solvent, and discard the oil that drains out with pan removal. Only a fraction of the total oil content of the automatic transmission comes out (perhaps 2½ to 3½ quarts) when you drop the pan, but this amount of fresh oil is enough to revitalize the remainder.

Jack up the car (see *Chapter 5*) and support it on safety stands, preferably at all four wheels, but if you only have two stands, jack as high as possible at the front.

Place a large drain pan just under the transmission oil pan (support it on a box if necessary to bring it as close as possible to the oil pan).

Remove all the screws holding the transmission pan (see Fig. 13–9) and carefully lower it so that any spillage goes into the drain pan. Empty the remainder into the drain pan.

Service the oil pan and its mating surface as with any gasketed engine joint (see *Chapter 5*). High-quality rubber-cork or rubber-fiber gaskets are generally supplied with the better brands of automatic transmission filters, so there is rarely a problem of gasket shrinkage or danger of tearing during installation.

Remove the screen or filter, which is held by screws or bolts. If the old filter has a neck with a rubber O-ring, the replacement will have the O-ring too.

Install the cleaned screen or new filter (see Fig. 13–10), refit the pan (see Fig. 13–11), then top up the transmission with fresh oil approximately equal to what was drained. As with other gasketed joints, you should tighten the oil pan screws in a criss-cross fashion or in a spiral manner, starting in the center and working alternately toward the ends.

Run the engine with the brakes firmly applied and move the shift lever through all positions while a helper looks underneath to

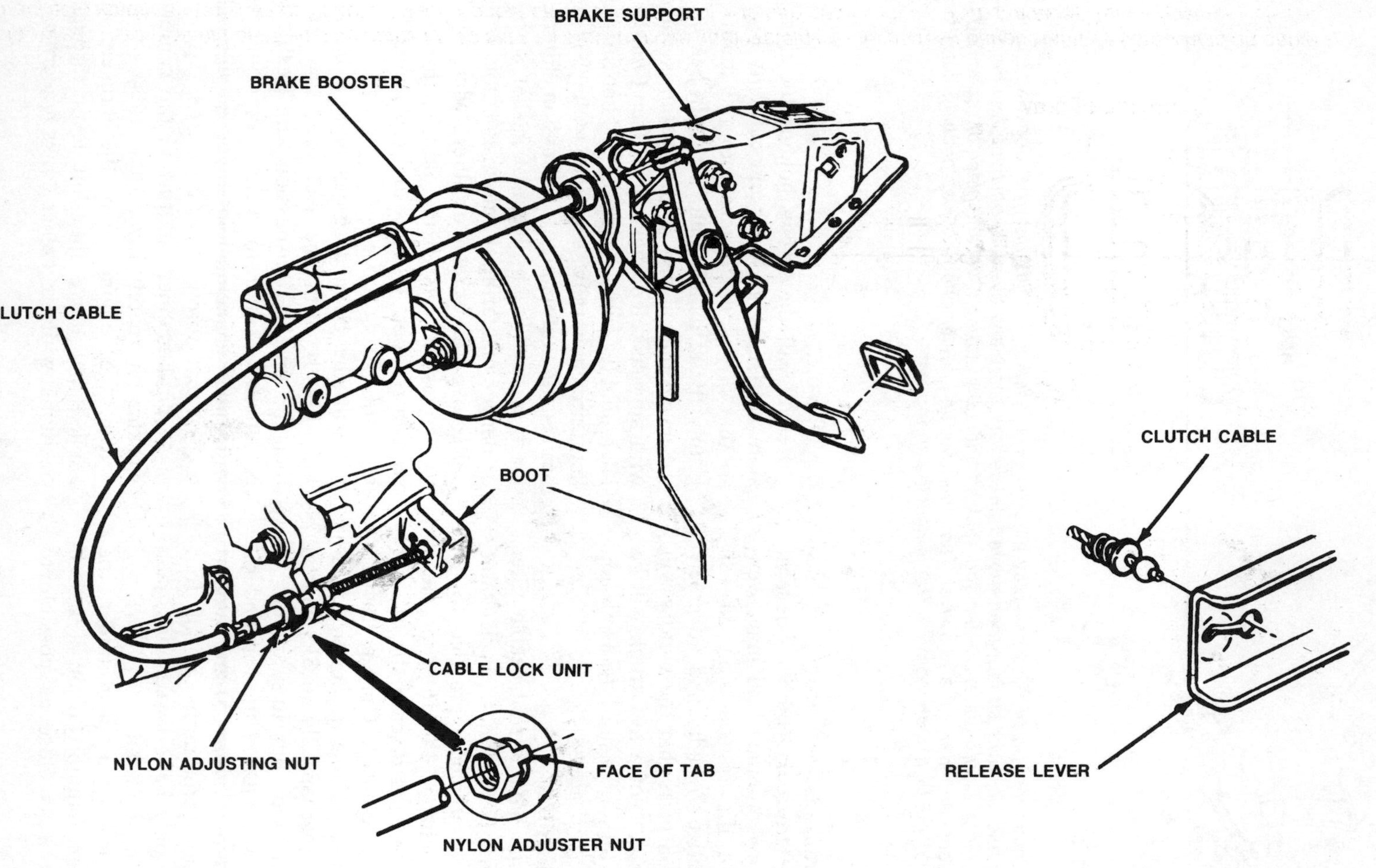

Fig. 13–7. This is another type of Ford clutch linkage arrangement. To adjust free play, loosen locknut for cable and pull cable toward front of car until tabs on nylon adjuster nut are clear of the flywheel housing boss. Then turn adjusting nut about a quarter-inch. Release the cable, then pull it forward again, until there is no free play at the clutch release lever. Turn the adjusting nut until it touches the face of the index tabs and the flywheel housing, then index tabs so they drop into the nearest housing groove. Finally, tighten locknut.

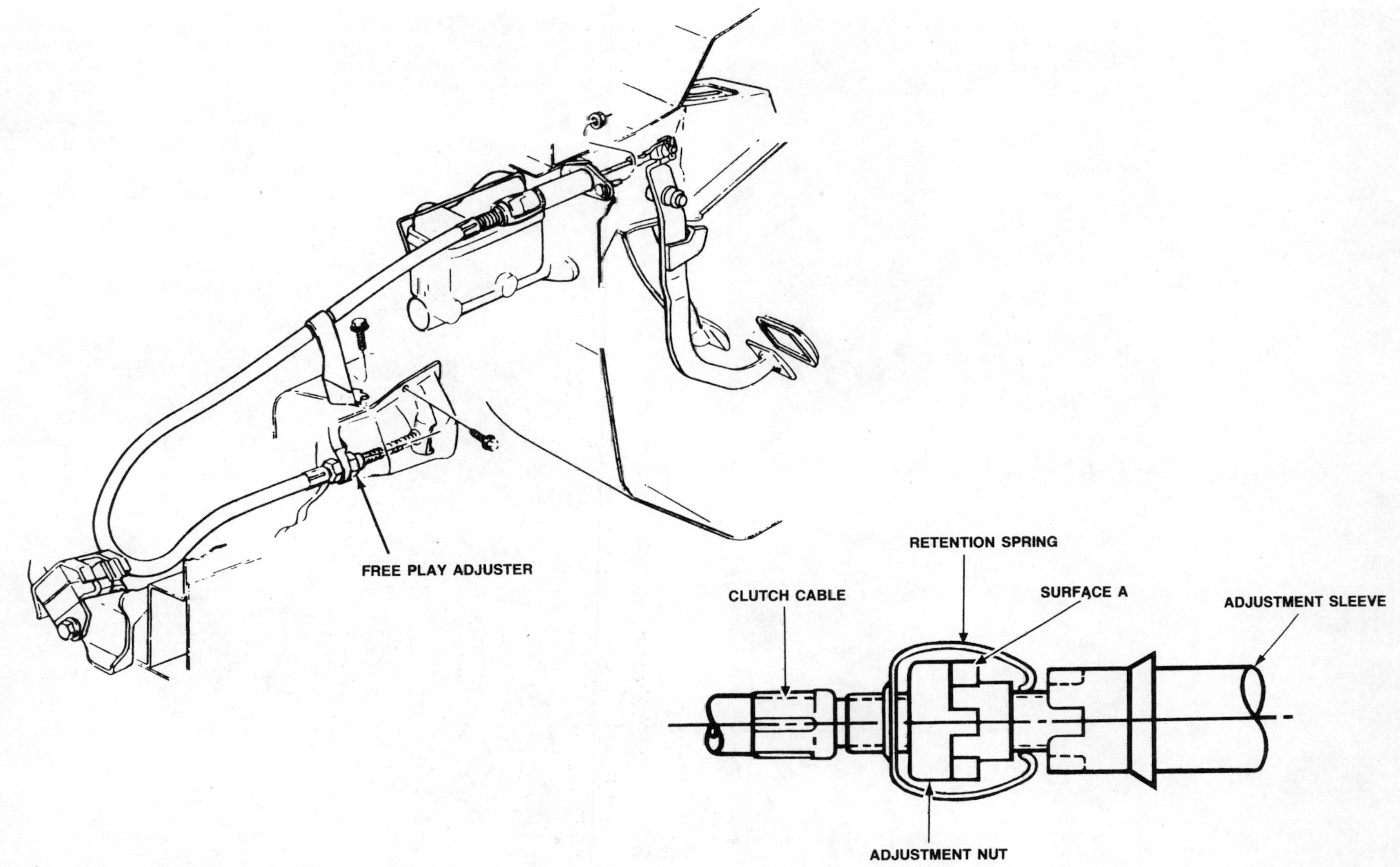

Fig. 13–8. Third type of Ford free play adjuster. To set free play, remove cable retaining clip at dash panel. Remove screw holding cable bracket to fender apron, then pull cable toward front of car until it is possible to turn adjusting nut. Turn nut away from adjustment sleeve about ¼ inch. Release cable, then pull again until free movement of the clutch release lever is eliminated. Turn adjusting nut toward the adjustment sleeve until it just touches, then index into the next notch. Refit cable clip and attaching bracket.

Fig. 13–9. Removing transmission oil pan screws.

Fig. 13–10. Install new filter on transmission valve body.

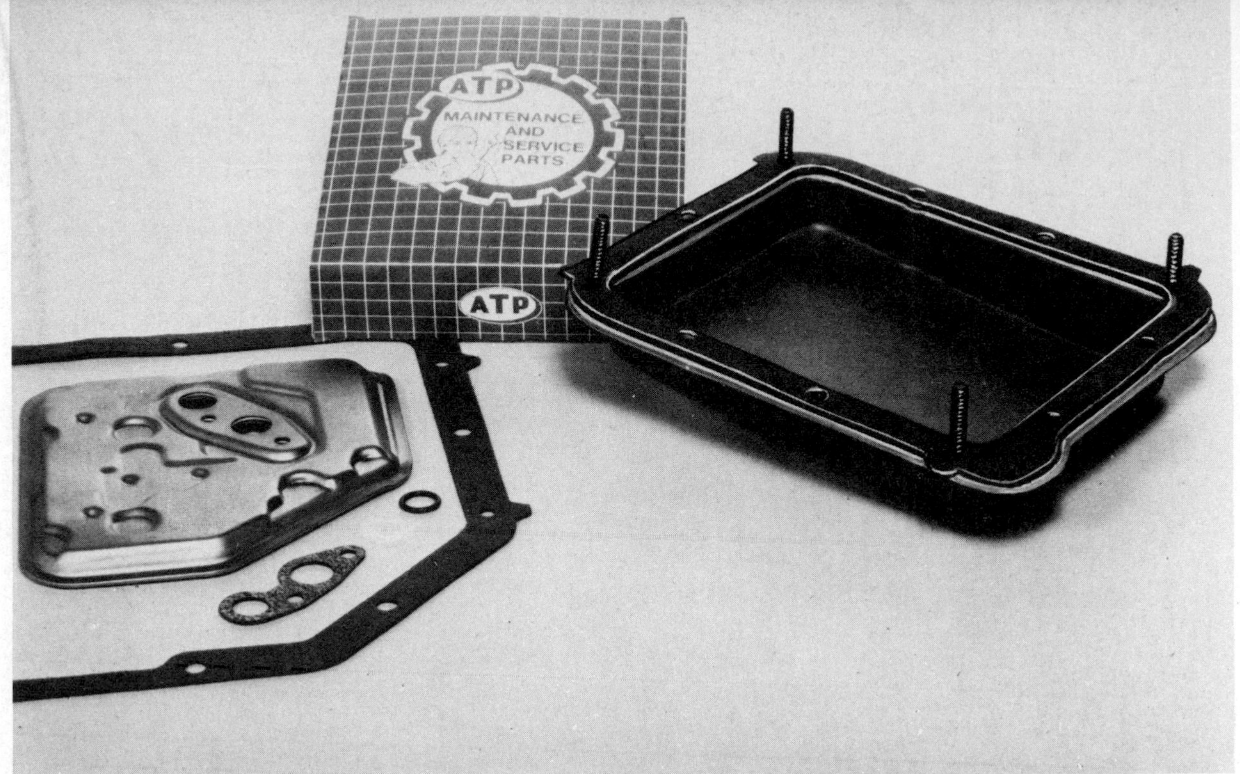

Fig. 13–11. Refitting transmission oil pan with new gasket can be a problem because gasket may slip. The type shown has undersize bolt holes at the corners, so you can thread the bolts as shown and they'll hold the gasket in place.

see if there are any leaks. If there are, stop the engine and recheck the tightness of the oil pan screws.

Make a precise fluid level check as soon as possible and add oil if necessary.

Vacuum Modulator

Some transmissions use a vacuum diaphragm device threaded into the transmission case to help regulate shifting. Called a vacuum modulator, it is connected by a hose to the carburetor base or intake manifold, and because it's the only part threaded into the transmission with a vacuum hose (or two, on some cars), it's easy to spot. If the car suffers from rough or erratic shifting, disconnect the vacuum hose(s) and probe the neck(s) with a pipe cleaner. If the pipe cleaner comes out oily, replace the modulator. It's an inexpensive part that in most cases can be changed with a simple wrench or large pliers. On some Ford products, a special wrench (see Fig. 13–12) is helpful.

Universal Joints

If you hear a clunk in the drive train when you take off or come to a stop, loose universal joints in the propeller shaft may be the problem, if you have a conventional front-engine, rear-drive car. Check by jacking up the car (see *Chapter 5*), at all four wheels if possible, support on safety stands, and check the universal joints.

Hold the propeller shaft on each side of a universal joint and try to turn one part of the shaft clockwise, the other counterclockwise; then reverse the procedure (see Fig. 13–13). If you feel any free play at all, the universal joint is defective and should be replaced. Rebuilding kits are inexpensive and readily available.

Most front-engine, rear-drive cars have simple cross-and-roller universal joints (see Figs. 13–14, 13–15) that you can replace yourself. If a car has what appears to be a double-universal joint at the rear (see Fig. 13–16), even a professional may not be able to service it. Only a few shops are equipped to repair that type of joint, and even they may not be able to guarantee the results. The normal cure is to change the entire propeller shaft.

Removing the Propeller Shaft

To replace the universal joint, you must remove the propeller shaft and service it on a

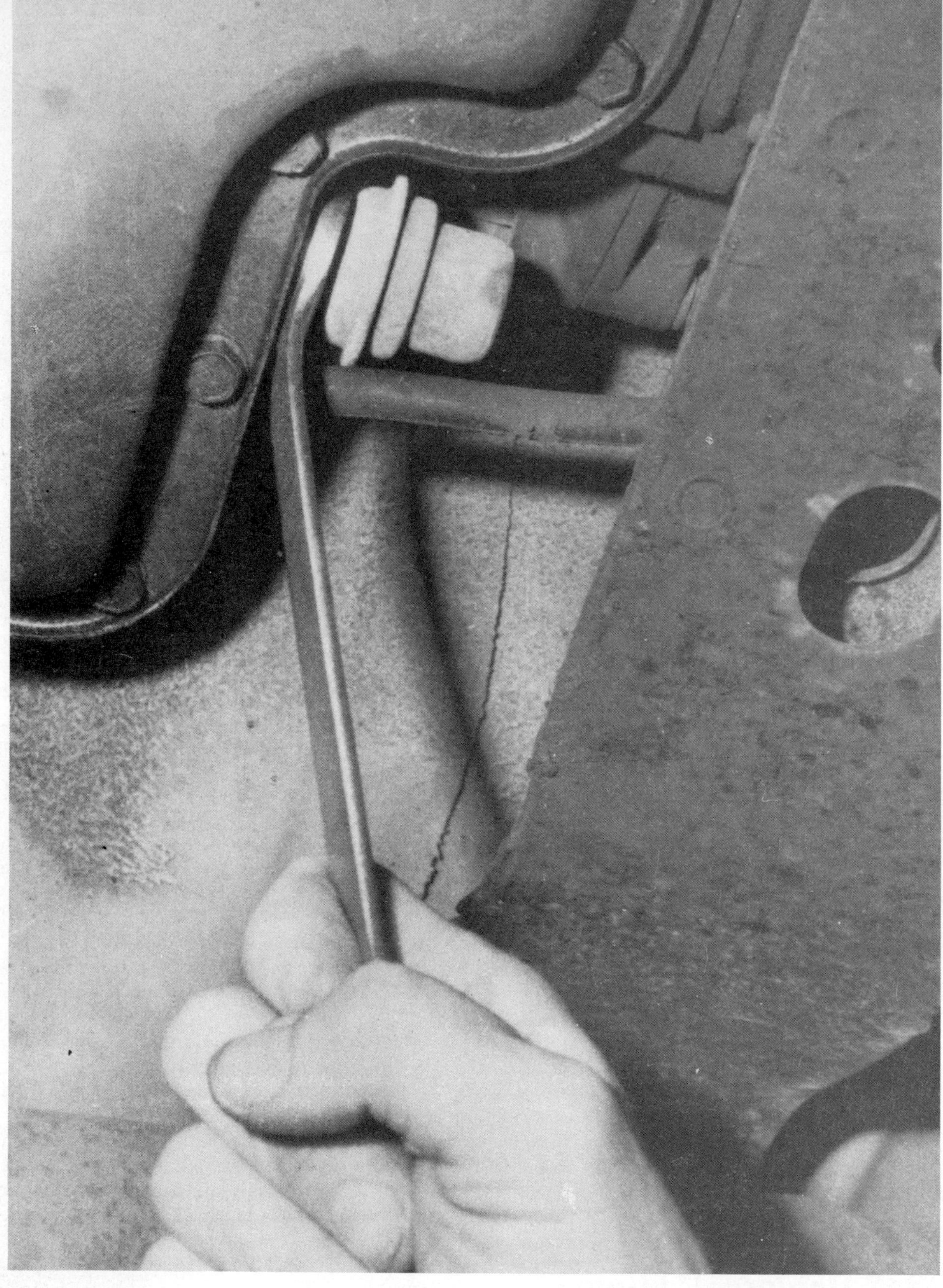

Fig. 13–12. Removing a Ford modulator with a special wrench. This wrench also is useful for tightening the new modulator in place.

Fig. 13–13. Checking a universal joint. If you feel any free play while attempting to turn one section clockwise and the other counterclockwise, the universal joint is defective and should be replaced.

Fig. 13–14. This is a simple front cross-and-roller universal joint. Bushings are bearing cups that contain the rollers—also called needles—that support the cross. Sliding yoke—also called slip yoke—fits on transmission shaft.

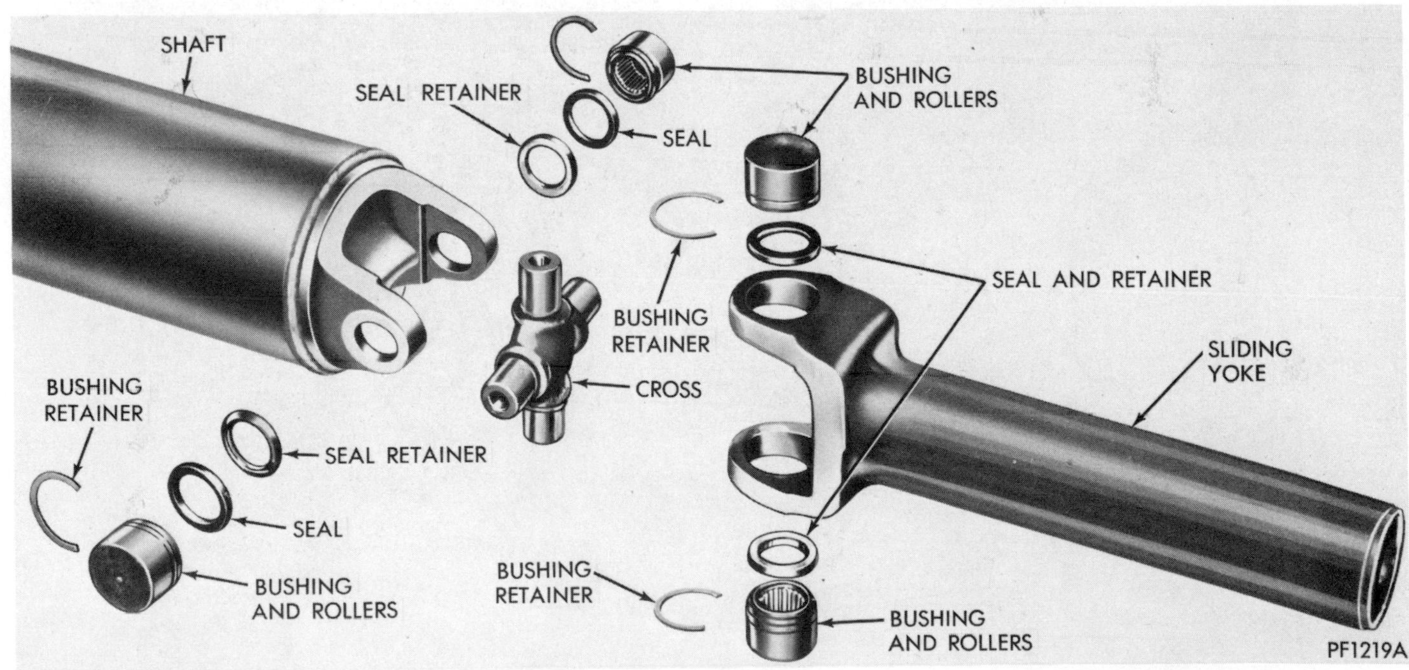

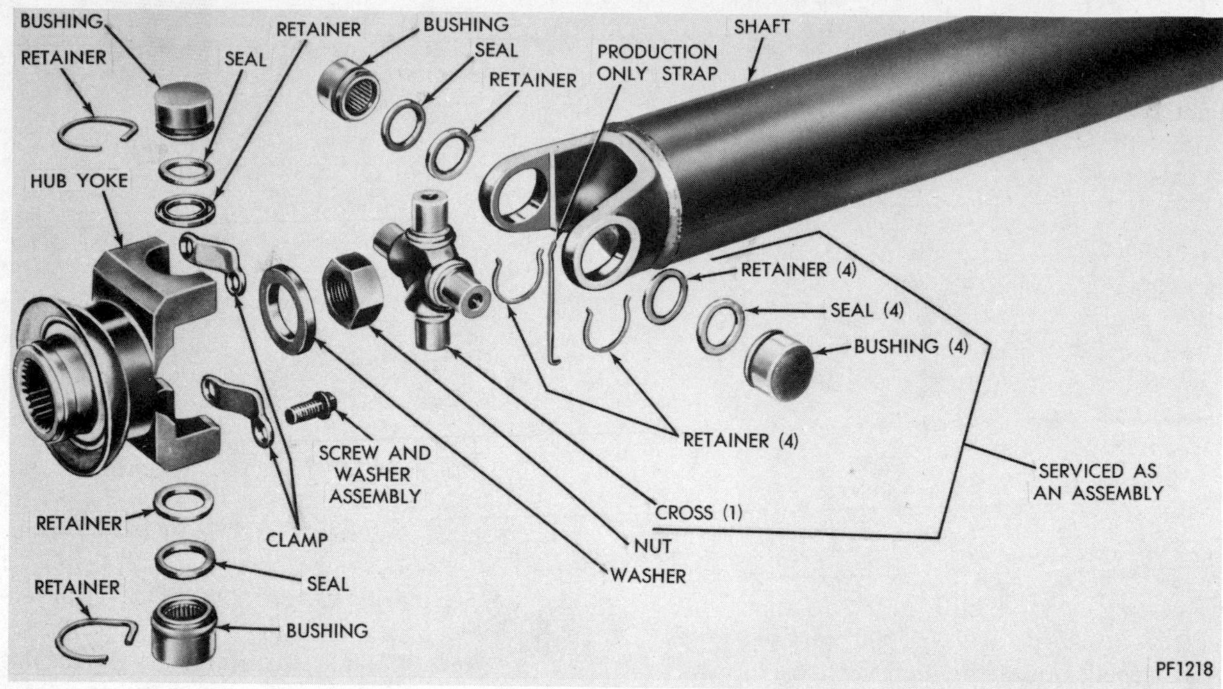

Fig. 13–15. This is a rear universal joint. The hub yoke actually is bolted to the rear axle and when you remove the propeller shaft, you do not remove the nut and washer, but do remove the two clamps (also called straps) and screws. The parts supplied in the repair kit are identified as serviced as an assembly, and the number refers to the quantity supplied.

Fig. 13–16. If you see this type of universal joint at the rear, it is not one you can service. If defective, replace the entire propeller shaft.

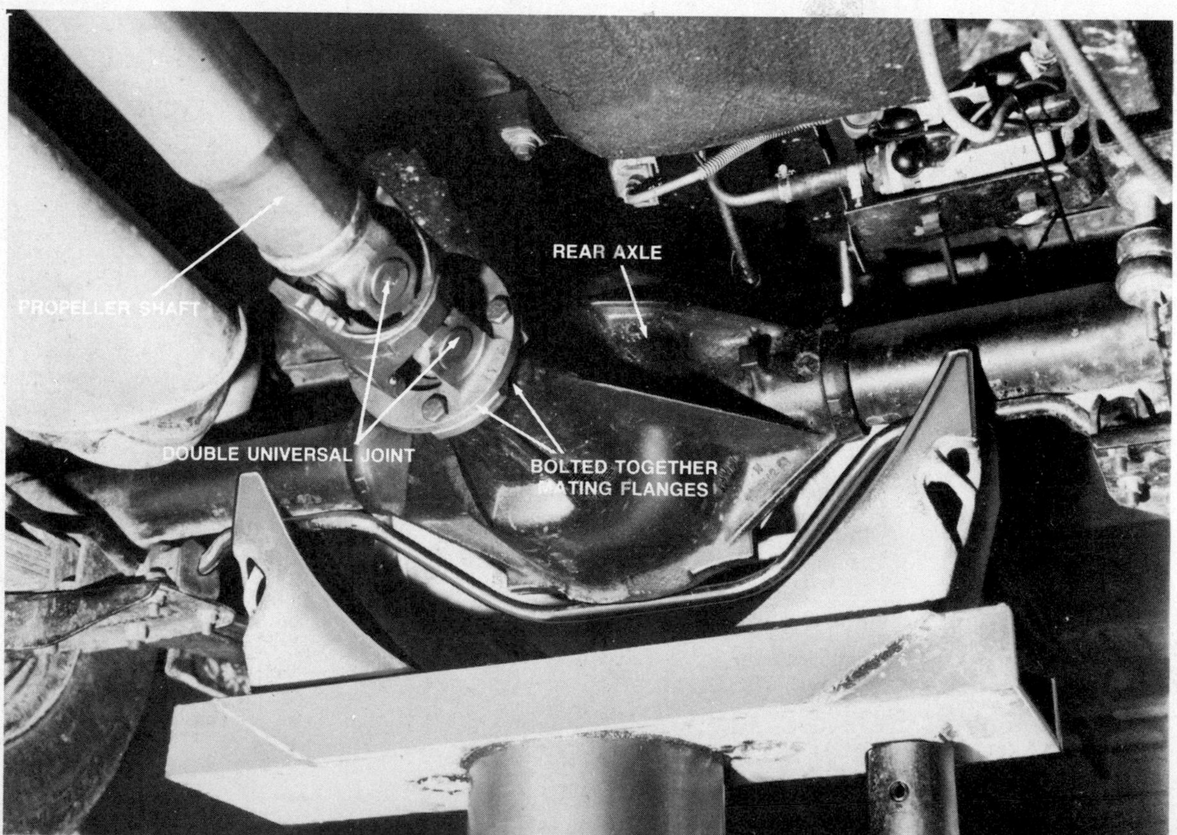

workbench with a vise.

To remove the propeller shaft, you disconnect it at the rear axle. It is commonly held in one of three ways:

(1) by U-bolts and nuts that also secure the rear universal joint (see Fig. 13–17);

(2) by straps and screws that also hold the rear universal joint (see Fig. 13–16);

(3) by a flange on the rear of the propeller shaft (just to the rear of the universal joint) that mates with a similar flange on the rear axle. The two flanges are held together by nuts and bolts (see Fig. 13–18).

Begin by making alignment marks (with dabs of paint, nail polish, etc.) at the rear axle connection, either on the propeller shaft and rear axle flange, or on both mating flanges. These alignment marks will permit you to reattach the propeller shaft to the rear axle in exactly the same position as it was, maintaining the balance. Failure to take this step could result in vibration afterward.

Next, detach the propeller shaft from the rear axle, by removing the bolts and/or nuts. Have a helper hold the shaft at the front, so it doesn't drop to the ground and suffer damage that also could throw it out of balance.

If the rear universal is to be reused, be careful when you remove U–bolts or straps to make sure the two bearing cups they held do not fall off.

Lower the propeller shaft so it is clear of the rear axle, then pull it rearward, out of its slip joint in the transmission at the front. Place a catch pan under the back of the transmission and stuff a clean rag into the back of the transmission, around the transmission shaft, to minimize oil spillage.

Place the shaft on the bench. In most cases you will notice that the bearing cups in the propeller shaft ears are held by snap rings, but in a few cases you may see nothing. Those on which there may be no retainer readily visible actually have injected, formed-in-place plastic retainers, and the replacement universal joint has the snap rings.

If you're working on the front universal, make alignment marks on front and rear sections here too, so they can be reassembled for original balance.

Fig. 13–17. This type of rear universal joint is held together by U-bolts and nuts.

Fig. 13–18. This rear universal joint has a flange that mates with a flange from the rear axle. Just remove the nuts and bolts holding the flanges together.

Remove the snap rings with a small screwdriver and needle-nose pliers (see Fig. 13–19). If you have the design with the plastic retainers, they will break apart later in the disassembly, so no special action is required.

Support the universal joint in a vise and then whack down on the cross with a hammer and chisel (see Fig. 13–20). The bearing cup on one side will be partly pushed out. Next, lock the bearing cup in the vise and tap away the propeller shaft (see Fig. 13–21). Repeat this procedure to remove the other three bearing cups if you are working on a front universal or a rear of the mating flanges type.

If you're replacing a rear universal of one of the other types, just pull off two bearing cups. Only two must be forced out.

Discard the old bearing cups and cross.

Now inspect the sliding (slip) yoke, the part that splines onto the transmission shaft (see Fig. 13–22). If the exterior has any burrs, remove them with fine crocus cloth. Burrs won't come off? Replace the sliding yoke or you may damage a transmission seal when you attempt to reinstall the yoke.

Look at the universal joint side of the sliding yoke, and if it has a vent hole the yoke splines must be lubricated with universal joint grease. Also, check the vent hole to make sure it is open. If necessary, run a pipe cleaner through it. If there is no vent hole, the splines are lubricated by transmission fluid.

Assembly

Inspect the cross of the new universal joint to see if there is a grease fitting (see Fig. 13–23) and if the grease seals are fitted onto the cross stems (called the journals). If they are not, push them on. Do they fit too tight? Position the seals on the outer edges of the cross journals, put sockets of the same diameter against them, and place the cross inside the jaws of the vise. Close the vise jaws against the sockets, and they will push the seals into place.

Next, inspect the bearing cups, which should be well-greased inside. If not, apply a thick coat over the needle bearings (see Fig. 13–24) and make sure all are up against the inner wall of the cup.

Position the propeller shaft in the vise with a pair of ears parallel to the jaws, one ear

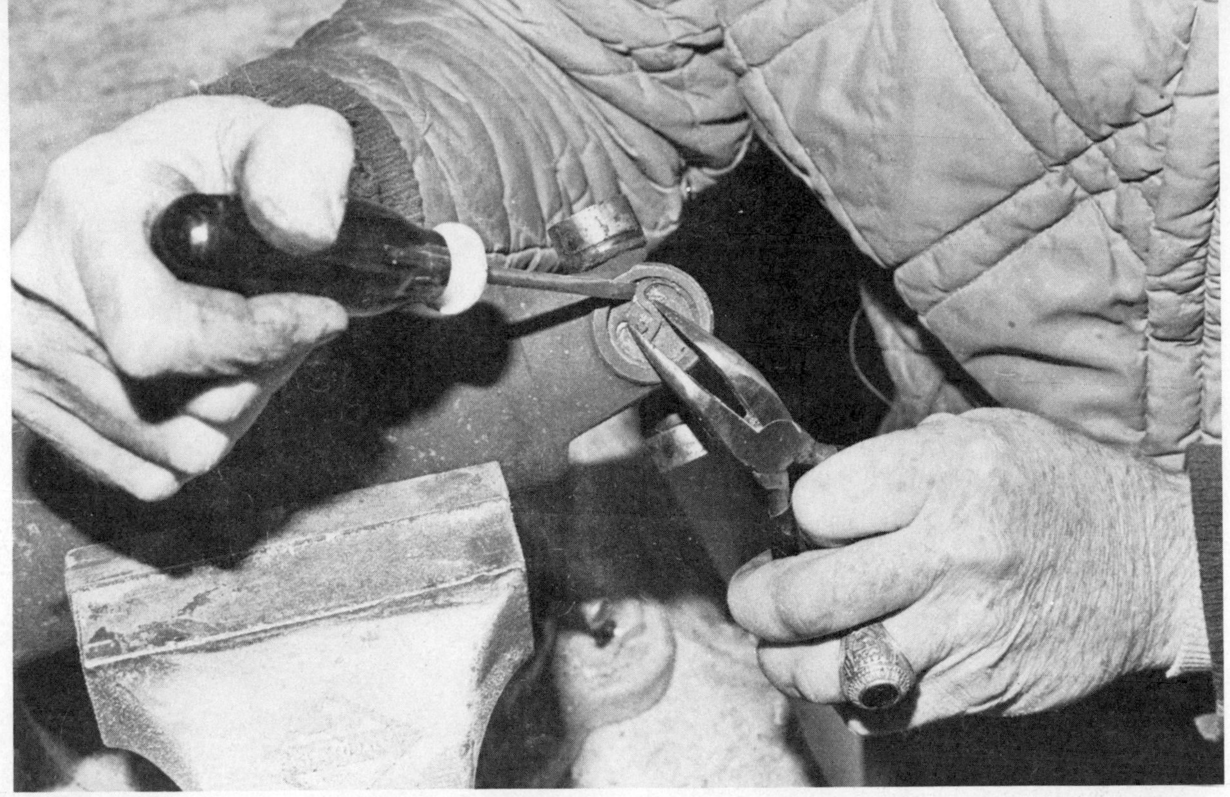

Fig. 13–19. Using screwdriver and needle-nose pliers to remove snap ring from propeller shaft ear. On some cars, the snap rings are on the inboard side of the propeller shaft, and fit in a groove in the cup. Pry out this type with a screwdriver.

against a jaw and the other a couple of inches away. Place a bearing cup between the second ear and the other vise jaw and close the vise jaw and shift the bearing cup so it is aligned with the bearing cup hole in the ear. Continue to close the vise and the jaw will push the bearing cup into the ear. Stop when the cup is about half-way into the ear.

Next, put the cross into place, making sure the grease fitting faces the propeller shaft (so that it can be greased properly when the propeller shaft is installed). Be sure that the cross journal slips smoothly into the bearing cup (see Fig. 13–25).

Now place the opposite side bearing cup up against the propeller shaft ear hole. Close the vise very slowly while you handle the cross, to make sure its journal slips smoothly into the second bearing cup. Continue to close the vise until the cups are flush with the outer ends of the propeller shaft ears.

The bearing cups will not be fully seated, so you can't install the snap rings just yet. Take a socket smaller in diameter than the bearing cup and place it against a cup (or if the vise opens wide enough, place a socket against each cup). Close the jaws and the smaller socket will push the cup in a bit further (see Fig. 13–26). Check repeatedly, so you push the cup in only far enough to expose the groove in the ear for the snap ring. Install the snap ring.

On front universal joints, or rears of the mating flange type, repeat the procedure to install the other two bearing cups. Be careful to maintain the alignment.

Installation

When you are ready to install the propeller shaft, put on the two rear bearing cups that are held by the U-bolts or straps and screws and then tape them lightly into place.

Remove the cloth in the transmission end and have your helper start the sliding yoke onto the transmission shaft (see Fig. 13–27). Raise the rear end of the propeller shaft and push the sliding yoke in as far as it will go. If you've aligned the marks on the rear axle and propeller shaft, then install the straps and screws or U-bolts loosely. Remove the tape and tighten down the rear connections. On a mating flanges type, just align the marks and bolt the flanges together. If the marks do not align, remove the shaft, turn as necessary, and reinstall.

Fig. 13–20. Whack down on cross with hammer and chisel to push out bearing cup at bottom.

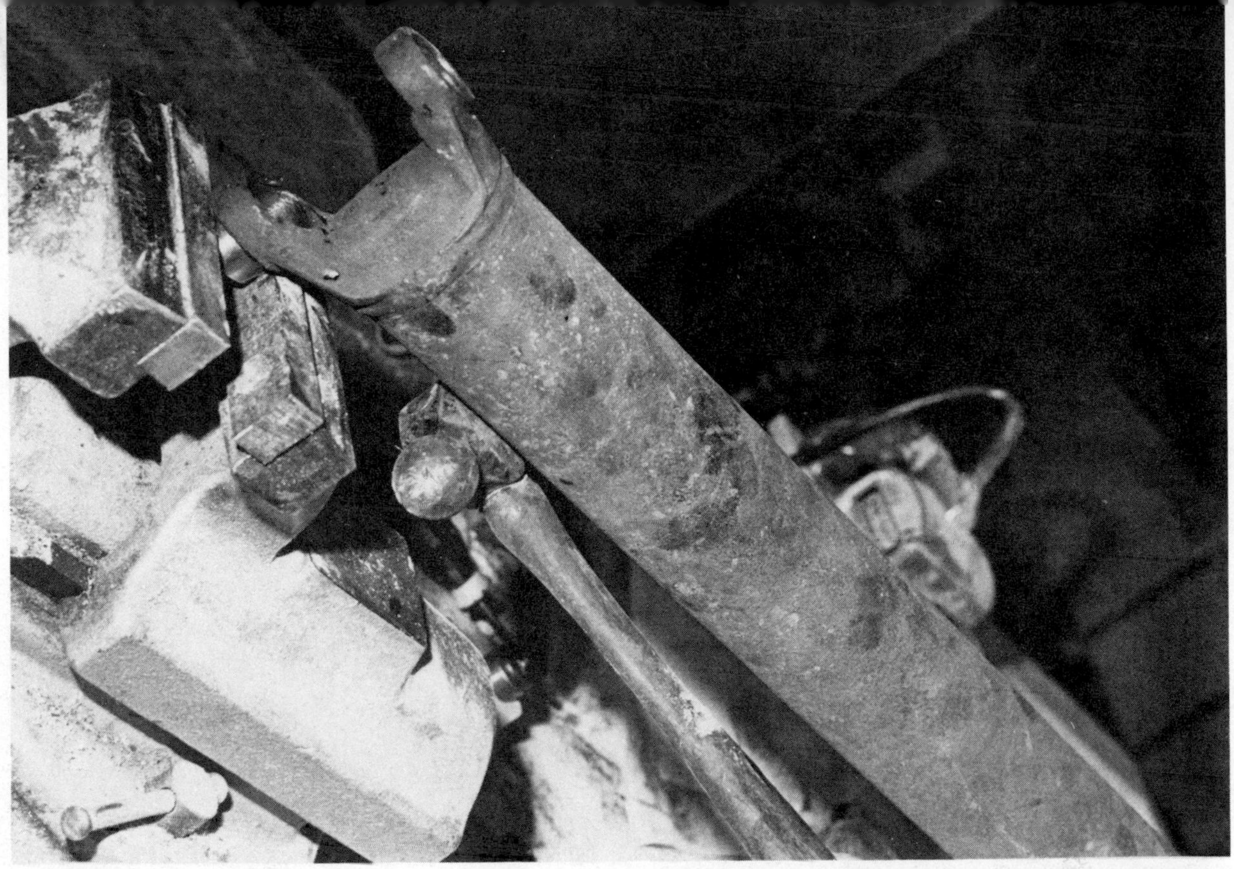

Fig. 13-21. If a bearing cup doesn't come all the way out, just lock it in a vise as shown and tap propeller shaft away with hammer.

Fig. 13-22. The slip yoke end of the propeller shaft. If outer surface is badly scored, or splines inside are damaged, replace it.

Fig. 13–23. This replacement cross has grease fitting. Install cross so you can reach fitting with grease gun when propeller shaft is installed in car.

Fig. 13–24. A look inside a bearing cup. If the needle bearings inside are not properly lubricated, apply a thick coat of universal joint grease. Do not pack the cup with grease.

Fig. 13–25. Pressing bearing cup into propeller shaft ear with vise.

Fig. 13–26. Using socket between bearing cup and vise jaw to seat cup fully, so snap ring can be refitted. If the snap ring goes on the inboard side, into a groove in the bearing cup, you may not have to perform this step. Merely seating the bearing cup flush with the exterior of the propeller shaft ear may expose the snap ring groove in the cup.

Fig. 13–27. Have helper start propeller shaft sliding yoke onto transmission shaft as shown.

14 TRANSVERSE-ENGINE AMERICAN CARS

THE AMERICAN AUTOMOBILE industry is shifting as quickly as possible from front-engine rear-drive cars to transverse-engine front-wheel-drive. The reason is simple: By placing the engine crossways in the compartment, with a combination transmission and differential (called the transaxle) on it at one end, there's a lot more room for passengers and luggage in a small package. This is a big help to fuel economy.

When the car makers go to a transverse-engine arrangement, however, they shorten the engine compartment (actually a complete powertrain compartment with the transaxle). This is one of the ways they get more room for passengers in a given length car. Unfortunately, this means less room under the hood, and with the complete powertrain there, space obviously is at a premium. As a result, the engineers have given very careful attention to underhood layout and have done as much as possible for serviceability.

Here is a review of current American transverse-engine front-drive cars and some things about routine underhood service you should find helpful. Other subjects are covered in the appropriate chapters.

Perhaps the most important thing you should know about all transverse-engine front-drive cars (and some without transverse-engine) is to disconnect the electric fan before you do any underhood work (Fig. 14–1). Although the fan normally is wired so that it will come on only when the ignition is on, on some cars it may come on with the ignition off. Further, many tests require turning on the ignition, and if you forget to disconnect the fan first, you could be injured. So make the fan-disconnect an automatic procedure when you begin work. In most cases, there is a connector very close to or at the fan motor you can unplug to disable the fan. The only exceptions are noted in the discussions on particular cars, which follow.

General Motors "X" Cars (Citation, Etc.)

The "X" cars are offered with two different engines, a V–6 and a four-cylinder in-line. On both, the electric fan connector at the motor is very difficult to unplug. It is easier to make the disconnection at the fan relay, which is to the right of the radiator (see Fig. 14–2).

Four-Cylinder

The four-cylinder engine is an adaptation of an engine that was designed for conventional (fore-aft) installation, and it is more difficult to service.

The oil filter is on the rear side of the engine, and you must reach up between the steering and the engine to get to it. You are working in very tight quarters.

The fuel pump is next to the oil filter, and although you can reach the pump mounting bolts, getting the fuel lines off is extremely difficult and is best done from the top of the engine compartment. First, remove the air pump (beginning with 1981 models) and the alternator to provide some access.

Hydraulic lifters also are reached from the rear side of the engine, after removing a cover plate. However, for access to this cover plate, you must remove the air pump, alternator and intake manifold, making this very difficult work.

The air pump installed as of 1981 is quite accessible, but it makes the alternator just under it inaccessible. You can service the alternator drive belt without difficulty, but to reach the test

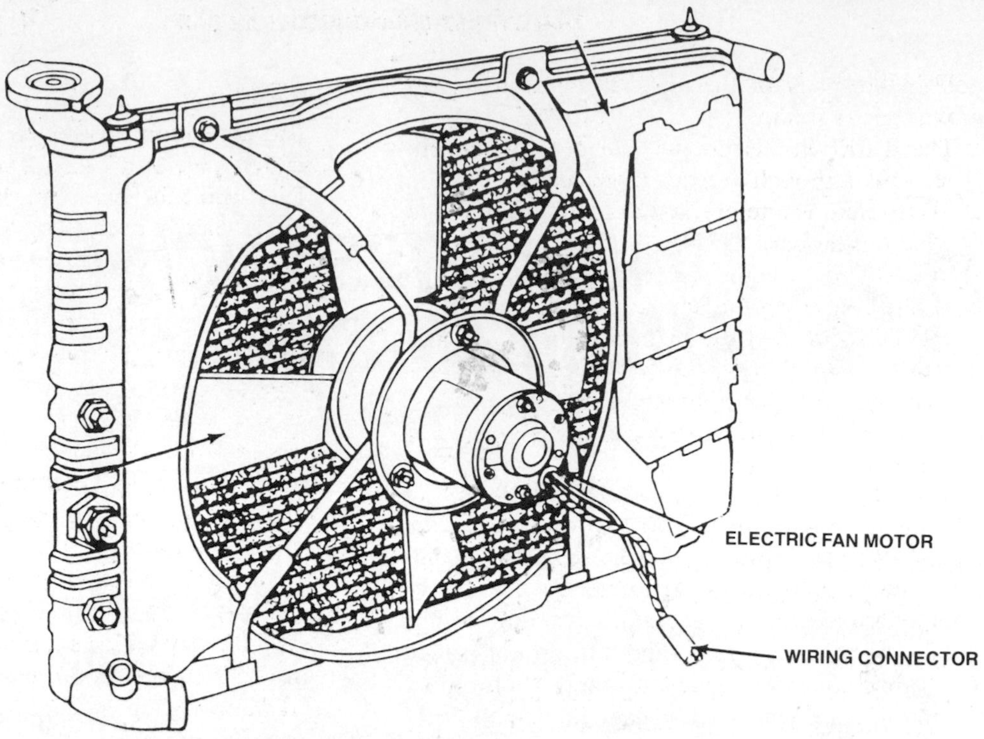

Fig. 14–1. Electric fan motor typically has a connector at or near the motor. If possible, make your safety disconnection there.

Fig. 14–2. If necessary, disconnect the fan by unplugging the relay. This one is on a GM "X" car.

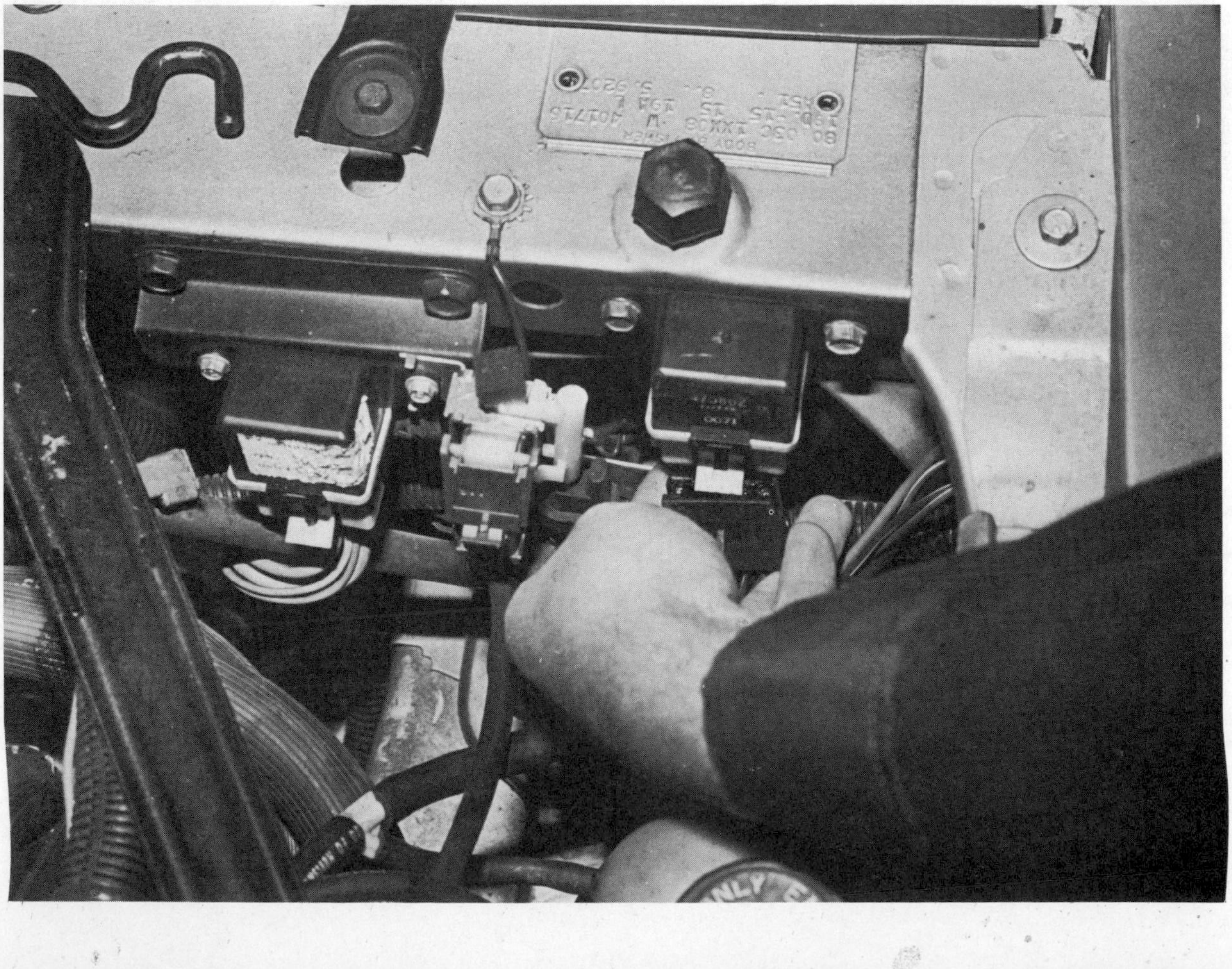

hole in the back of the alternator you must remove the air pump.

The distributor is on the rear side of the engine, and although it was modified starting in 1981, it remains one of the most difficult items to service on the car.

In 1980 models, the coil is on the distributor cap, and removing the cap to check the coil, inspect the rotor, test the pickup coil, etc. is very difficult, particularly if the car has air conditioning (in which case a part called the A/C accumulator is next to it). Even professionals will avoid it as part of a tuneup.

Reaching the distributor timing lockbolt also is a problem. About all you can reach on the distributor without difficulty is the "TACH" terminal in the distributor flange.

There is no "secret way" a professional uses to make the job easier. A C-shaped distributor wrench is available to loosen the lockbolt to change timing, but it isn't easy even with that tool. A short screwdriver is used to disengage and reengage the L-shaped lockrods that hold the cap, but even professionals have struggled to hold the cap in position and reengage the lockrods.

In 1981, the distributor cap was changed and the coil is now a separate part, adjacent to the distributor. It isn't wide open, but you can reach it and its wiring and terminals for tests. Taking the coil out of the cap also eases the job of taking off and refitting the distributor cap, as there is now a bit more room (the job still isn't easy). The timing lockbolt in 1981 models also is a bit easier to reach with a C-shaped wrench.

The tachometer terminal in the distributor cap is harder to reach in 1981 than on 1980 models, but there should be a connector from a wire to that cap, conveniently located near the flywheel end of the engine.

If the oil pan must be dropped for any reason, such as a gasket change, leave the job to a professional unless you are experienced. To do it, you must remove the front engine mount nuts, disconnect the exhaust pipe from the manifold and at its brace on the transaxle mount, disconnect the starter, remove the flywheel housing inspection cover, the upper and lower generator brackets, and raise and support the engine with a hoist. When reinstalling, save the bolts into the timing chain cover for last, as the holes for them do not align until the pan bolts are at least finger tight.

V-6 Engine

The V-6 engine is easier to service than the four-cylinder, but it also has its difficult items.

The 1980 rear bank spark plugs are very difficult, primarily because a Pulsair (see *Chapter 4*) is threaded into the exhaust manifold and sits just above them. To do the job you will need a ratchet, ⅝-inch spark plug socket, universal joint and an assortment of extensions of various sizes (although if you are careful, one six-inch extension will be enough).

In 1981 the Pulsair was replaced by an air pump mounted between the banks at the pulley end of the engine. This opens up the area around the spark plugs a bit, although there is air injection tubing from the air pump system there. The rear plugs, however, can be removed without great difficulty with a short and a long extension.

In order to test compression on the V-6, you must have a gauge with a hose that threads into the spark plug holes.

The water pump also is a problem on this engine. Some of the bolts that hold it also hold the timing cover, and if you just take out the bolts and pull away the pump, you could pull away the timing cover too. In this case, some coolant could get into the engine where it would mix with the oil. Anti-freeze has a very bad effect on piston rings and other parts, so this is to be avoided. To prevent the timing cover from pulling away, you must hold it in place with a brace, which attaches to the engine with a bolt that goes into a hole tapped in the cylinder head for this purpose (see Fig. 14–3).

Although most drive belts are reasonably accessible, the power steering belt poses some difficulty. You must loosen bolts on the pump that must be reached both from the top and underneath. The first time you do the job, you should remove the air pump control valve (bolted to the back of the air pump), the blower motor and a heater hose, so you can see where the wrench has to go.

Like the four-cylinder, the V-6 has an oil pan that is somewhat difficult (on manual transmission cars anyway). To lower it, you must drop the exhaust crossover pipe, remove the clutch housing cover or transaxle/converter/starter shield, then the starter. If it's a manual transmission car, remove the engine mount nuts and jack up the front of the engine about ¾ inch.

Both four- and six-cylinder models have an en-

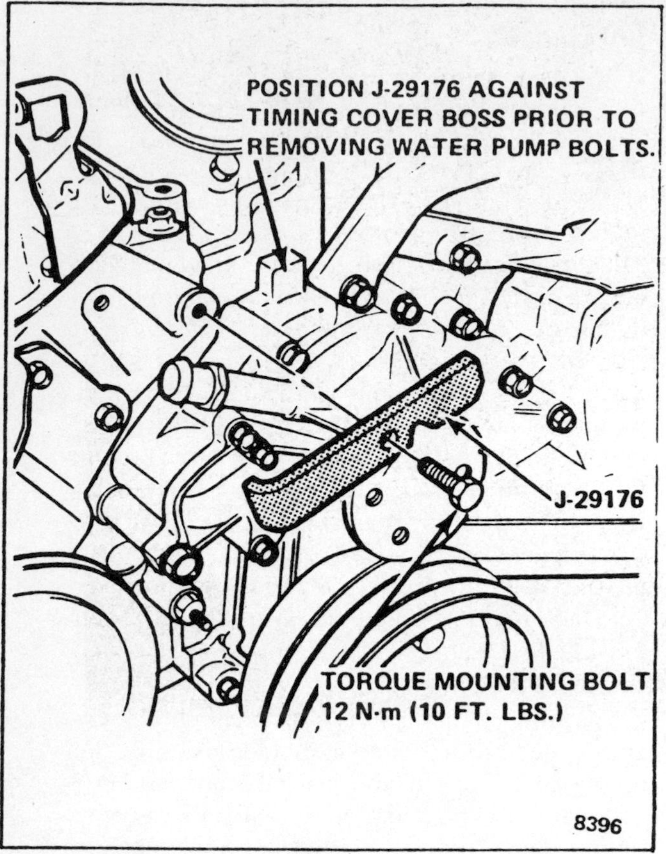

Fig. 14–3. Special bracket to hold timing cover during water pump removal on GM V-6 in X-car can be ordered from many auto parts stores, or you can make one from bar stock.

gine steady brace that bolts to the radiator support (see Fig. 14–4). To remove the radiator, first unbolt it at the front and loosen the bolt at the engine. Then lift it up at the front end and flip it back.

General Motors "J" Cars (Cavalier, Etc.)

With the four-cylinder "J" car it's much easier to work under the hood than on the four-cylinder "X" car, but it too is packed in.

The distributor cap, rotor, etc. are accessible, because the wide part of the distributor body is on top of the engine. However, the distributor shaft is long and the part of the body that necks down also bolts down several inches. As a result, the distributor timing lockbolt is buried. The initial timing adjustment should be correct. If you don't need to change it, the location of the lockbolt doesn't matter. If readjustment is needed, or if you have to take out the distributor, that's another matter. There is a 13-mm C-shaped distributor wrench available (Fig. 14–5), but even with it you have to "feel" the wrench onto the lockbolt. All this wrench can be used for is to loosen the bolt for a timing change. If you have to take out the distributor, first remove the air control valve just to the side of it (held by two bolts) and you have reasonable (not great) access to the lockbolt area, both to remove the bolt and reinstall it.

The ignition coil location is even worse. It's on the underside of the intake manifold, just above the fuel pump (which isn't wide open either). You must jack up the car, support it on safety stands, get underneath, remove the fuel pump, and then you can get to the coil. Fortunately, you should be able to locate an unconnected wire near the distributor, to which you can connect a tachometer, if nothing else.

Unlike the X-car four-cylinder, hydraulic lifters are accessible through the pushrod holes, which are oversize to permit lifter removal. The job can be done from the top of the engine compartment, therefore.

Like the X-car, the fan motor disconnect is difficult at the motor itself, so unplug at the relay, which on the J-car is on the left (driver's) side of the engine compartment, close to the battery.

The J-car water pump is not difficult, but clearances are a bit tight. First take off the alternator, then unbolt the water pump pulley. Unbolt the pump and move it forward, then up and out.

VW Rabbit

The VW Rabbit compartment is one of the least cluttered, and there are no significant accessibility problems.

Most Rabbits are equipped with hydraulic fuel injection (see *Chapter 3*) and the idle speed ad-

Fig. 14–4. Engine steady brace must be disconnected to remove radiator.

justment is not on the linkage. Ignore what appears to be an adjuster on the throttle valve linkage, and instead locate a screw on the back of the intake air distributor, very close to the throttle linkage. Turning this screw clockwise reduces idle speed, counterclockwise increases it, just the opposite of a conventional throttle linkage adjustment. The reason is that the screw controls an air passage around the throttle plate, not the position of the linkage.

The water pump is on the front side of the engine at the bottom, and it also holds the thermostat.

The fuel filter is held to fuel lines by bolts. To loosen them, put a wrench on a hex built into the filter, and use a second wrench on the bolts. The filter is on the left side of the compartment (see Fig. 14–6).

Remove the plastic cover on the pulley end of the engine and inspect the cogged rubber belt, looking for cracks, fraying and excessive looseness. You should be able to twist the belt 90 degrees, not more. Leave replacement to a professional. Belt failure will cause engine failure on the diesel version of this engine.

Ford Escort/Mercury Lynx

The Escort/Lynx is a reasonably serviceable package, but there are some problem areas:

- The cylinder head is aluminum and the bolts that hold it are a special design. Never loosen them for other than a good reason, for once loosened, they must be replaced. If they need minor tightening, however, to bring them to specified torque, you can do this without having to replace them.
- The water pump is recessed in the pulley end of the engine, and is driven by the cogged belt that also spins the overhead camshaft (see Fig. 14–7). Changing a water pump requires loosening and removing of the belt, and once that is done, the belt must be replaced. Ford also recommends replacing the belt at 60,000-mile intervals regardless of appearance.
- The power steering pump belt adjustment requires loosening three bolts, and inserting a ratchet or bar into a square hole in the bracket from underneath (see Fig. 14–8).

All drive belts but the one on the air pump are the grooved type, although the car does not use

Fig. 14–5. C-shaped distributor wrench can reach timing lockbolt on GM "J" cars, but you still have to "feel" the wrench onto the bolt.

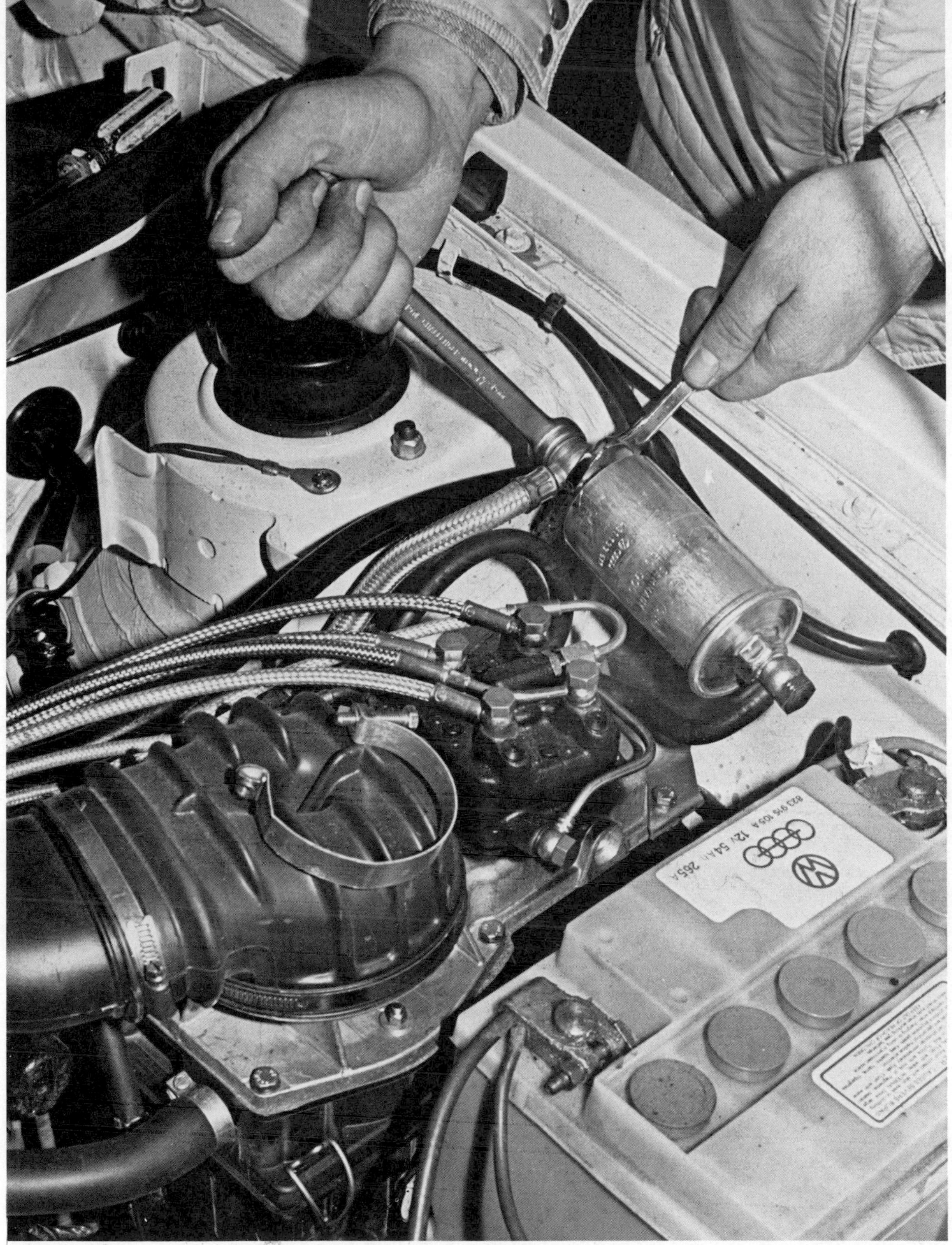

Fig. 14–6. To remove VW Rabbit fuel filter, use wrench on filter hex section to hold it while you loosen bolt that holds the fuel line to the filter. Place dry rag underneath to catch fuel dripping.

Fig. 14–7. Water pump on Escort/Lynx four-cylinder is recessed. If it is necessary to change it, camshaft gear cog belt must be loosened and removed. Whenever belt is loosened it must be replaced, a job for a professional. Therefore, entire water pump job should be left to pro with special tools.

Fig. 14–8. This view of engine lifted from Escort/Lynx shows power steering pump, which is on rear side of engine. Three bolts (to which arrows point) must be loosened (get one from the top, two from underneath). Then insert ratchet ½-inch-drive into square hole (circled) from underneath to apply tension.

the spring housing automatic adjustment of other Ford products. When installing a belt, be sure you have the correct type and that it fits properly into the grooves in the pulleys.

Chrysler Omni and Horizon

Most Omni and Horizon cars have been built with the same engine as the Volkswagen Rabbit, but with a carburetor instead of fuel injection. It also is equipped with more accessories, so underhood service is different in many respects.

The water pump is on the front side and mounted low, as on the Rabbit, but there's an air pump under it that has to come off first to change it. The thermostat housing, however, is conveniently located at the top, front and center of the engine.

The power steering belt adjustment requires working from the top and bottom to get to the two pump bolts. The access to both is good, and there's a half-inch hole in the bracket to simplify setting tension.

The worst problem, however, is with the 1978–80 alternator belt. A special tool (see Fig. 14–9) is absolutely necessary, along with a torque wrench. The tool is available on special order from many auto parts stores that also serve professional mechanics.

Fig. 14–9. Special rectangular tool, used with torque wrench, is necessary to adjust alternator belt tension on pre-1981 Omni and Horizon. Tighten to 70 lb-ft on new bolt, 50 lb-ft on used belt.

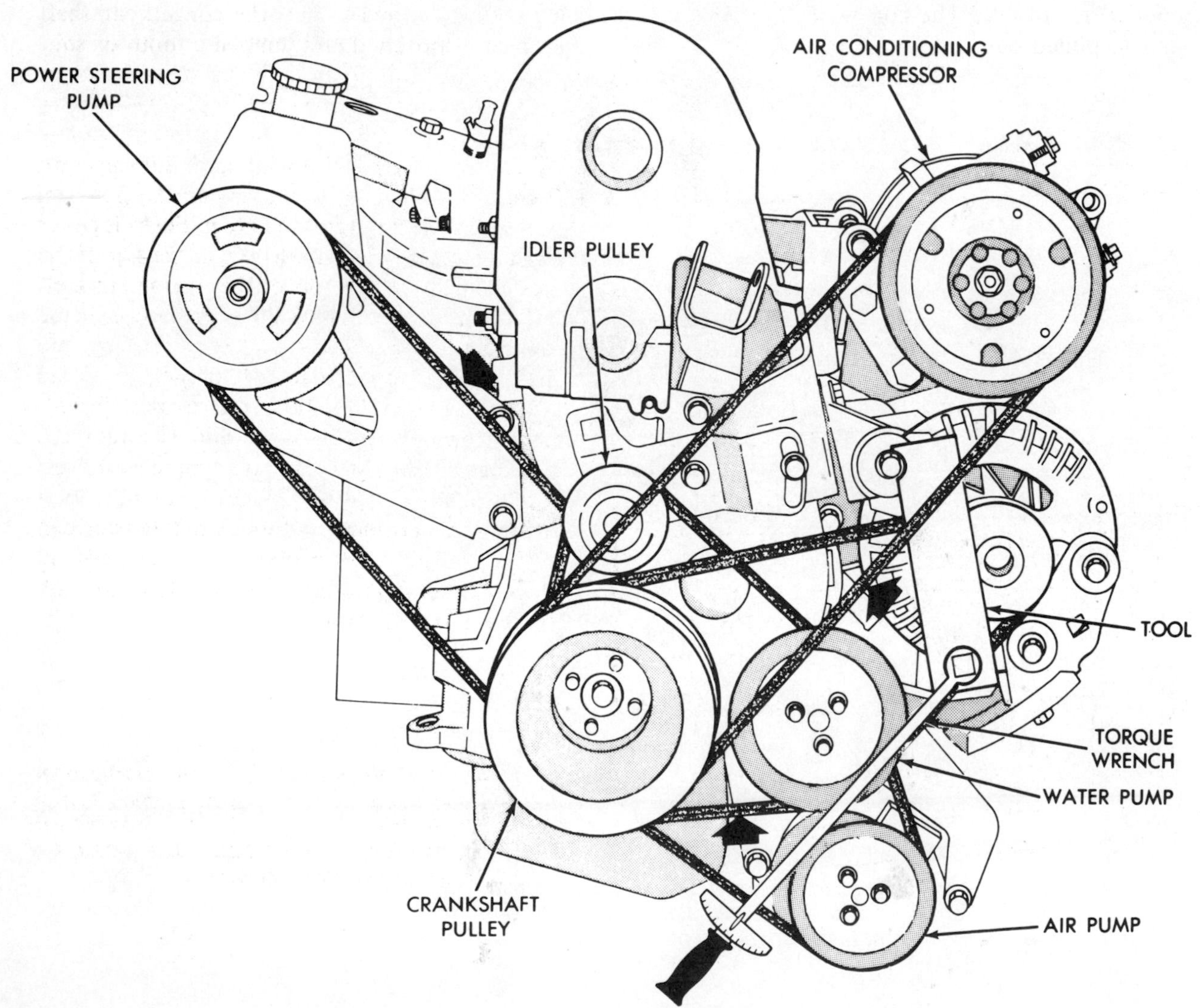

This problem was eliminated on 1981 models with the adoption of a bolt adjuster (see Fig. 14–10). Remove a plastic plug under the front bumper to gain access to the lockbolt (Fig. 14–11), then go through a hole in the top front sheet metal (Fig. 14–12) with a long extension to turn the adjusting bolt.

Different but not difficult are idle speed adjustment and spark plug wires.

The idle speed adjustment is a screw type on a solenoid. To eliminate the need to remove the air cleaner for access, there's an arrow on the air cleaner cover (Fig. 14–13). If you follow the arrowhead, you'll put the screwdriver on the adjuster (Fig. 14–14) without difficulty.

Spark plug wires cannot just be pulled out of the distributor cap. You must remove the cap and disengage their spring clip terminals from inside, using needle-nose pliers to squeeze together the tangs (Fig. 14–15). The coil wire, however, can just be pulled out.

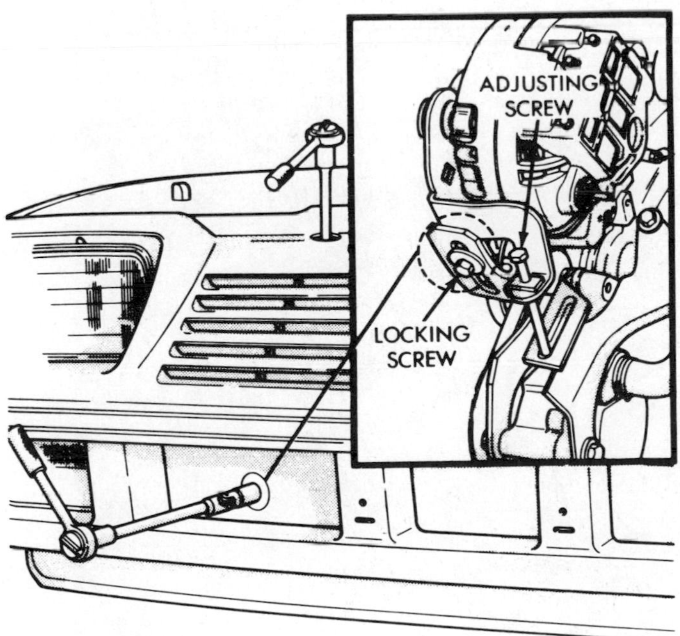

Fig. 14–10. This screw-type belt adjuster, similar to the stud adjuster used on Ford power steering pumps (Fig. 7–31), is found on 1981 Chrysler front-wheel-drive cars, including Omni-Horizon. Just loosen the lockbolt and turn the adjusting screw.

Chrysler "K" Cars—Reliant and Aries

The Reliant and Aries share some engineering features with the Omni and Horizon, such as the spring clip terminals on the plug wires and the screw adjuster for the alternator belt (however, there's good access to the adjuster and its lockbolt, so there are no access holes). And the air cleaner also has an arrow that points to a similar idle speed adjuster underneath it.

The power steering pump belt adjustment also resembles Omni-Horizon in that you reach one bolt from the top, another from underneath, and set tension with a ratchet in a square hole in the bracket.

A feature unique to the "K" car engine (which also may be used on some Omni-Horizon cars) is an easy way to make sure the cogged camshaft belt hasn't loosened and jumped a tooth or so.

Remove a round plug in the plastic belt cover. Turn the engine (with the starter and when you're close, with a wrench on the crankshaft pulley bolt) until the flywheel timing mark lines up with the zero (top dead center) mark on the flywheel housing. Look through the hole in the belt cover and you should see a small hole in the top of the camshaft gear at the top of the engine. In fact, you should be able to look through the hole in the gear and see a rubber tit that projects from the top of the valve cover, just behind the gear. If the hole in the gear and the rubber tit are not aligned, the belt apparently is loose and has jumped, throwing off the engine's valve timing (see Figs. 14–16 and 14–17). More difficult starting can result from this problem. Although belt tension can be reset, a loose belt probably is stretched and worn, and so should be replaced. Have this job done by a professional.

Fig. 14–11. After removing plastic plug on Omni-Horizon, go through hole under front bumper for access to lockbolt with ratchet and long extension. This access hole isn't needed on "K" cars (Reliant and Aries).

Fig. 14–12. Reach screw adjuster from top with ratchet and long extension through hole in sheet metal on Omni and Horizon. There's enough room in engine compartment on "K" cars (Reliant and Aries) so this access hole isn't necessary on those models.

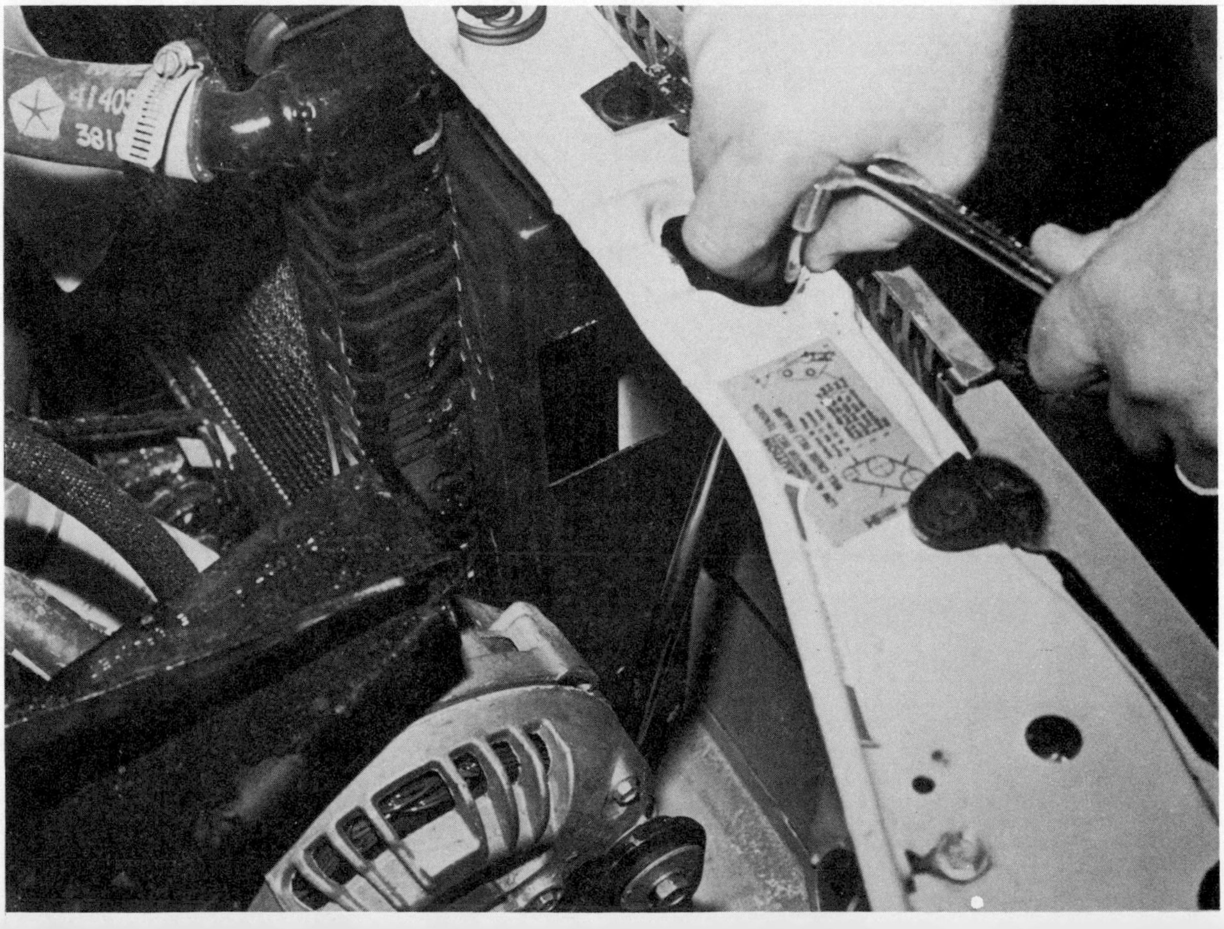

Fig. 14–13. Arrow on top of air cleaner points to idle speed adjuster so you can place screwdriver in right place without taking off air cleaner.

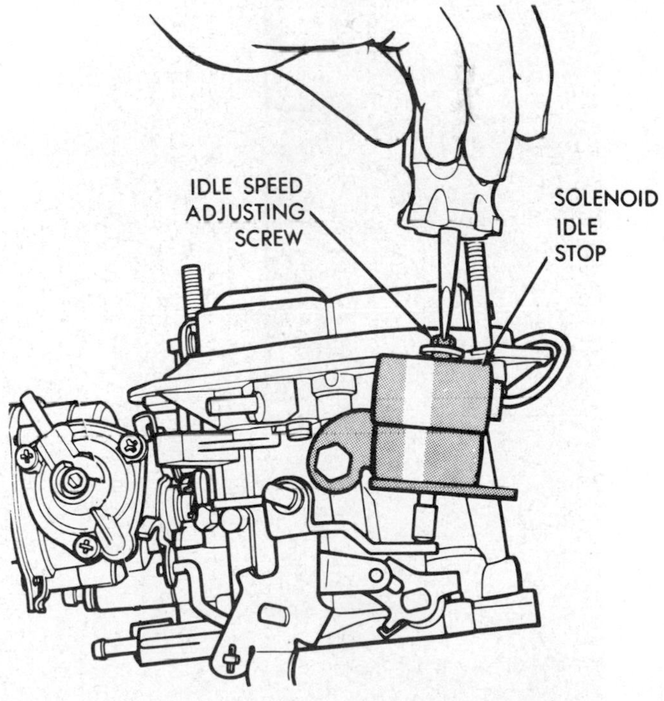

Fig. 14–14. This drawing shows what screwdriver is on—the adjuster of a solenoid.

Fig. 14–15. Distributor cap terminal of spark plug wire has a spring clip. To remove plug wire, take off cap and squeeze together spring clip tangs as shown, using needlenose pliers. To install plug wire, just push its end into cap until you feel spring clip tangs seat.

Fig. 14–16. Pen points to rubber tit projecting from valve cover gasket at top of engine.

Fig. 14–17. Pen points to small hole above one of large holes in camshaft gear. With belt's plastic cover in place, you can remove small plug from cover. Line up timing marks and you should be able to see rubber tit through hole. You can use a flashlight to assist you, or you may push a thin rod through the cover hole and gear hole to the rubber tit.

15 BODY REPAIRS

MINOR DENTS, SCRATCHES TO BARE METAL AND RUSTOUTS are three of the most common body repair jobs you can do with simple tools and easy-to-use patching products available in discount stores and auto supply shops.

Scratches

Surface preparation, as in any body work, is very important. Clean the area around the scratched (or chipped) area with a soap and water solution, then wipe dry and apply a wax and grease remover. Unless these steps are taken, the paint you apply later will not adhere properly.

Using wet-or-dry very fine sandpaper, wet-sand the area (see Fig. 15–1). Sand until you have a featheredge, that is, until there is a continuous smooth surface from bare metal to the paint on adjacent undamaged areas.

Mask off the area to be painted, including a small area beyond the featheredge (see Fig. 15–2). If you are working near chrome, glass parts, wheels, and tires, be sure to mask them off too, using newspaper or wrapping paper and masking tape to protect them from overspray (see Figs. 15–3, 15–4).

Wipe the area to be painted with a tack cloth to remove any dust, then apply a coat of automotive primer from a spray can (see Fig. 15–5). Do not move the can toward and away from the metal, but hold it the same distance (generally eight inches) and move it from side to side or up and down. After the primer has dried thoroughly, wet-sand with wet-or-dry very fine paper, dipping the paper in water repeatedly. Be careful not to sand through the primer coat, but if you do, wipe the area dry and reapply primer.

After sanding, wipe the area with a tack cloth and use a spot putty (sold in squeeze tubes) to fill pinholes and minor scratches (see Fig. 15–6). Wet-sand as before, wipe dry, then apply two additional coats of primer. Permit them to dry, then wet-sand with wet-or-dry super-fine paper. Use rubbing compound to remove overspray. Dry and clean with a tack cloth, then spray on the color coats (see Fig. 15–7). Note: If the paint is available only in liquid form, and you do not have a paint sprayer, you can obtain a sprayer with a replaceable propellant charger and just pour the liquid into the container provided (if not available locally, write to Precision Valve Corp., P.O. Box 309, 700 Nepperhan Ave., Yonkers, NY 10702).

Note: With acrylic lacquers only, you can improve the finish by lightly wet-sanding the color with wet-or-dry ultra-fine paper to remove any overspray and surface irregularities from dust. Finish up by carefully rubbing the repaired area with rubbing compound and a damp cheesecloth (see Fig. 15–8). This procedure should be done only after allowing the color coat to dry overnight.

Minor Dents

Begin by cleaning the surface as for deep scratches (see Fig. 15–9). If the dent is more than ¼ inch deep, it should be knocked out (as best you can) to the normal surface. If you have access from behind, use a hammer. If you can't get behind the dent (such as in a door), drill a $^{1}/_{16}$-inch hole in the center of the dent, or if it's a crease, drill holes along the crease about two or three inches apart (see Fig. 15–10). Thread an oversize sheet metal screw into the hole (because it is oversize it will protude) and lock vise-type pliers onto the head. Pull on the pliers to draw

Fig. 15–1. Using wet-or-dry sandpaper, flood area with wet sponge, to sand scratched area down to featheredge.

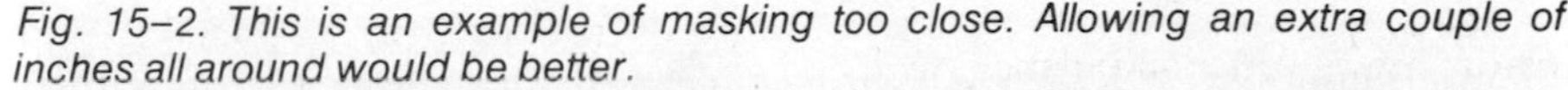

Fig. 15–2. This is an example of masking too close. Allowing an extra couple of inches all around would be better.

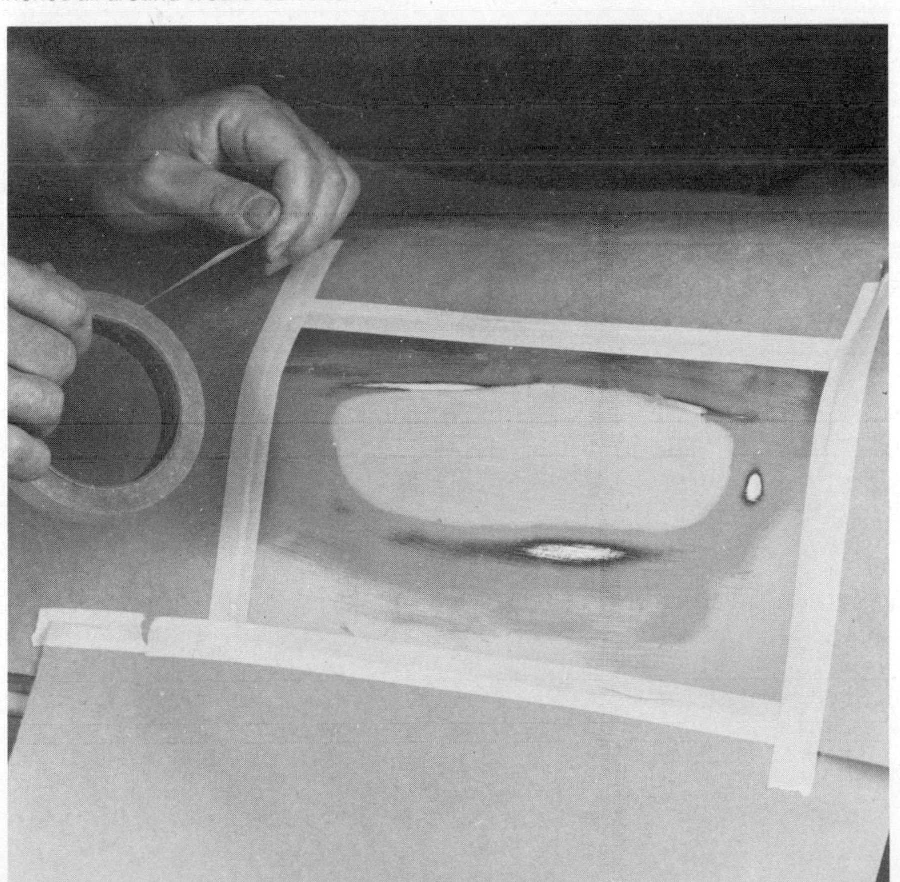

Fig. 15–3. Masking an emblem with tape.

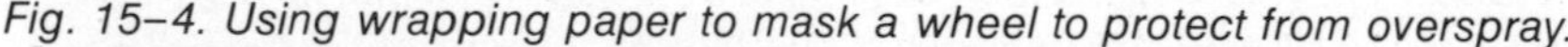

Fig. 15–4. Using wrapping paper to mask a wheel to protect from overspray.

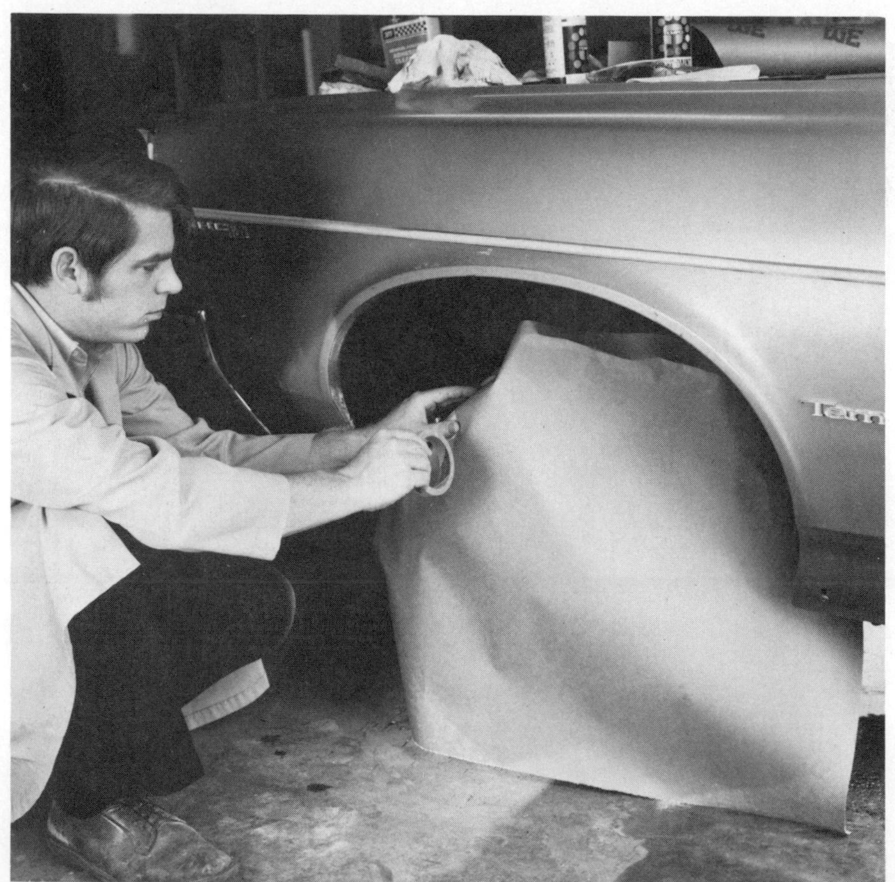

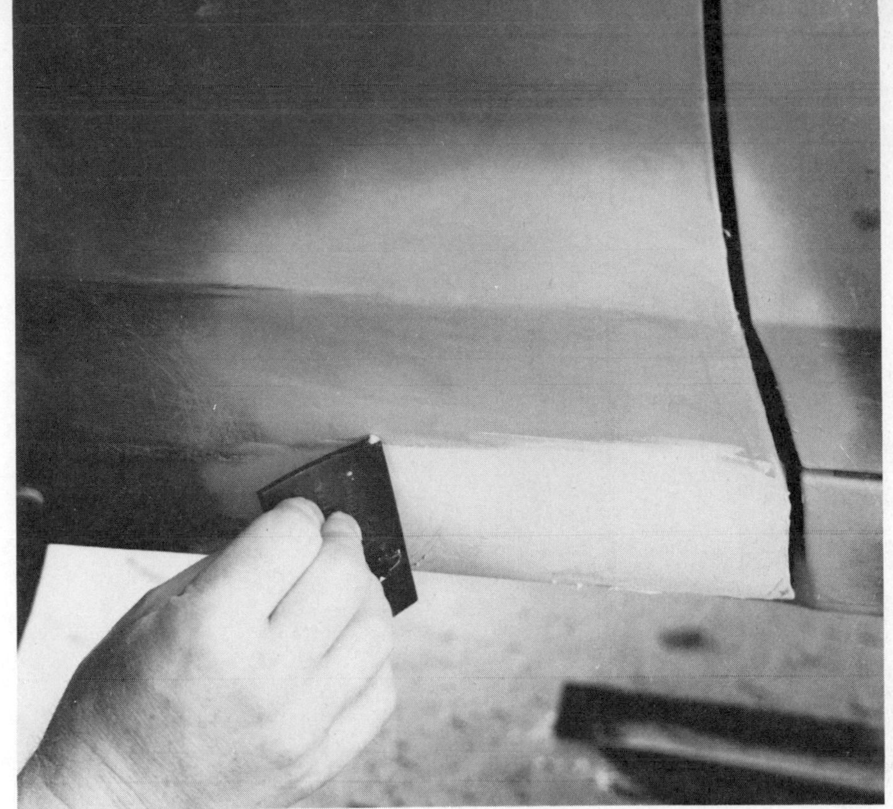

Fig. 15–5. Applying spot putty with plastic squeegee to fill the scratch.

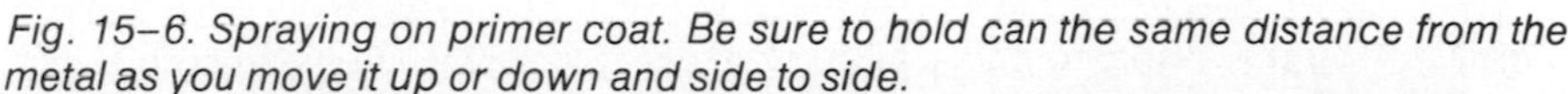

Fig. 15–6. Spraying on primer coat. Be sure to hold can the same distance from the metal as you move it up or down and side to side.

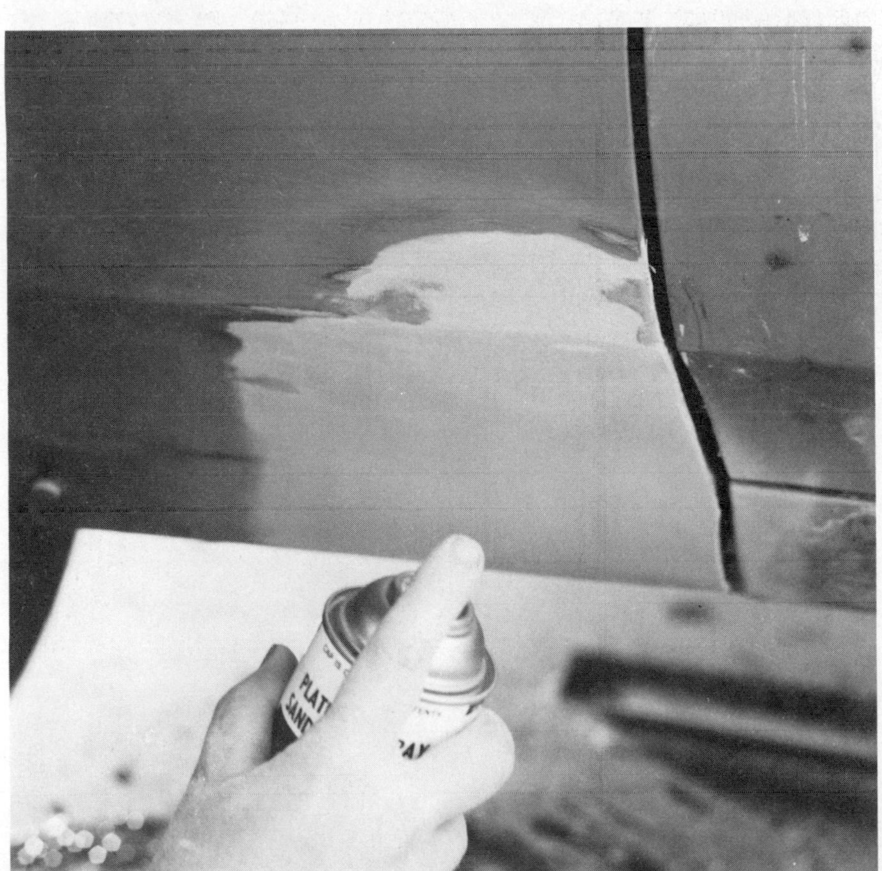

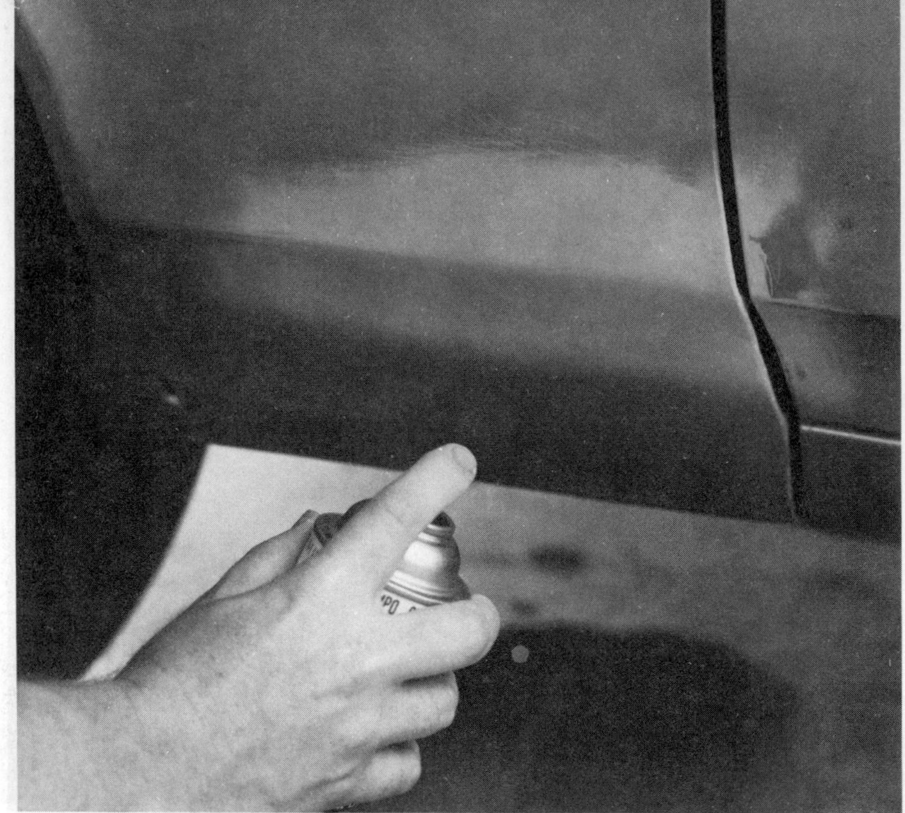

Fig. 15–7. Applying color coat from an aerosol can. Only a spray will give you a professional look. Brushes leave stroke marks.

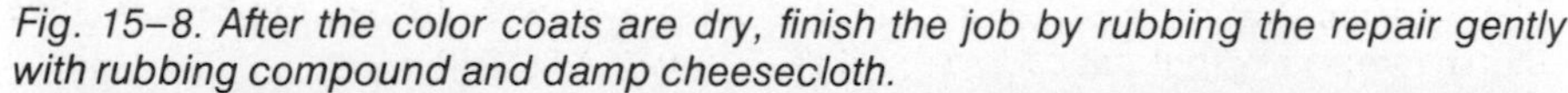

Fig. 15–8. After the color coats are dry, finish the job by rubbing the repair gently with rubbing compound and damp cheesecloth.

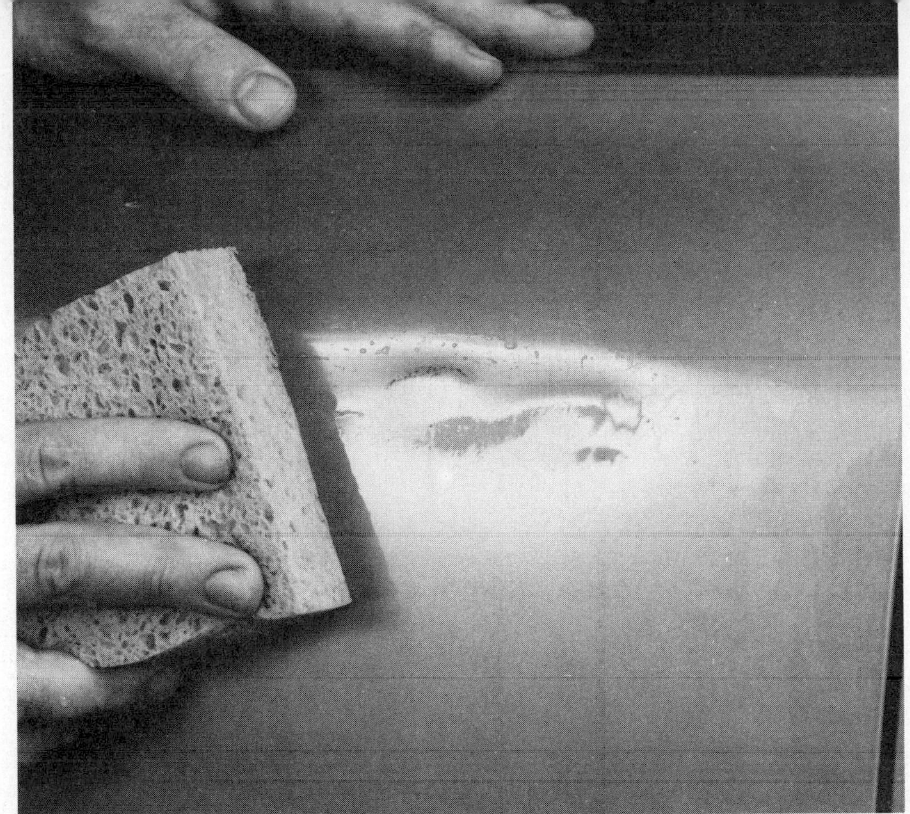

Fig. 15–9. Clean dented surface with sponge and soap-water solution, then apply a wax and grease remover.

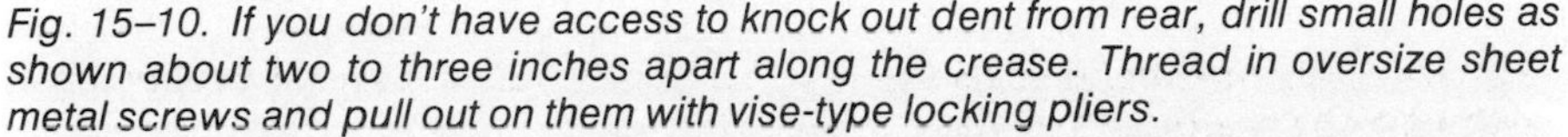

Fig. 15–10. If you don't have access to knock out dent from rear, drill small holes as shown about two to three inches apart along the crease. Thread in oversize sheet metal screws and pull out on them with vise-type locking pliers.

the sheet metal out as close as possible to the normal surface. Follow this procedure along the dent until you have done the best you can.

Next, grind away all the paint in and just around the dent with a coarse sanding disc on a ¼-inch electric drill (see Fig. 15–11), holding the disc at a slight angle to the sheet metal. Wear safety glasses when performing this operation.

Mix a suitable amount of plastic filler, following the manufacturer's instructions for proportions of base material and the hardener (see Fig. 15–12). Mix very thoroughly and apply immediately, for it will start to harden in minutes. Use a plastic squeegee to apply the filler and put pressure on it to force out any air bubbles (see Fig. 15–13). Make the coat as smooth as possible, but don't skimp. After it dries, sand it down to shape, and if there are any low spots mix some more filler and apply. Finally, paint the repaired surface as described for deep scratches.

Rustouts

A small rustout (under 36 square inches) can be repaired in much the same manner as a small dent, using any one of the various types of patches to bridge the opening.

Prepare the surface as for deep scratches (see Fig. 15–14). Sand away rust and the paint immediately surrounding the rustout with a coarse sanding disc on a ¼-inch drill. You should have about two inches of good metal showing around the rustout, and as you sand you may find that more metal weakness from rust has occurred than was obvious before you started. As with any power-sanding operation, wear safety glasses.

Fig. 15–11. Grind away all paint in and just around the dent with a coarse sanding disc on an electric drill. Hold disc at a slight angle to the sheet metal as shown.

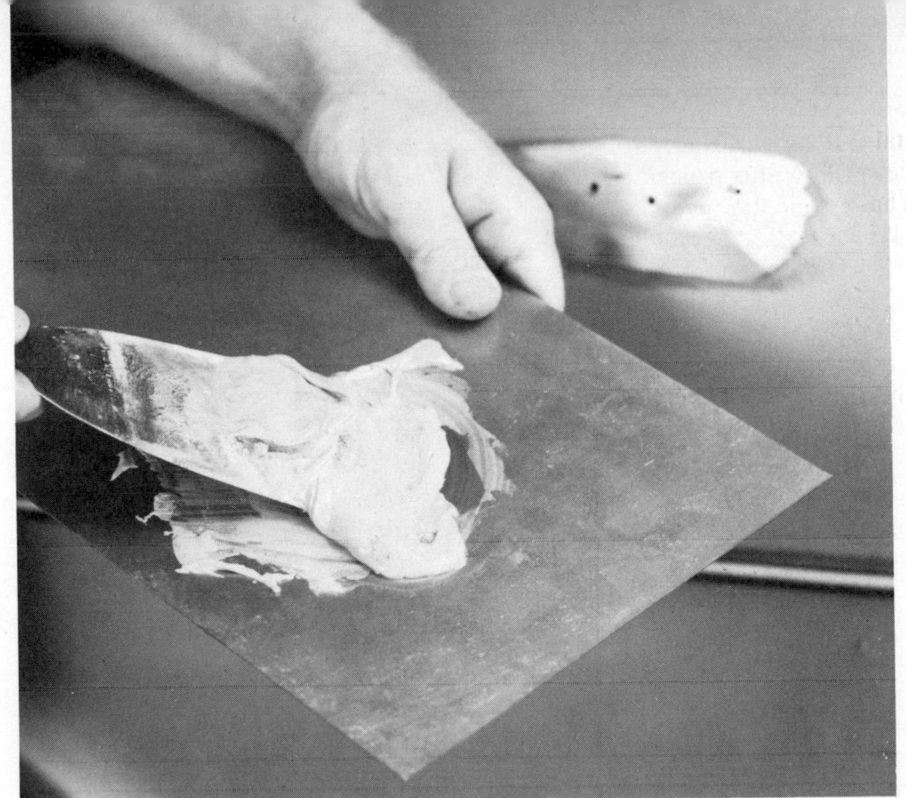

Fig. 15–12. Mix base material and hardener very carefully, following manufacturer's instructions for proportions.

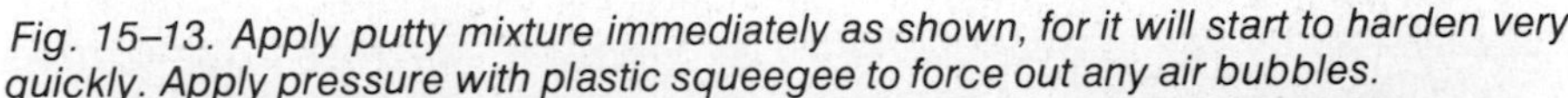

Fig. 15–13. Apply putty mixture immediately as shown, for it will start to harden very quickly. Apply pressure with plastic squeegee to force out any air bubbles.

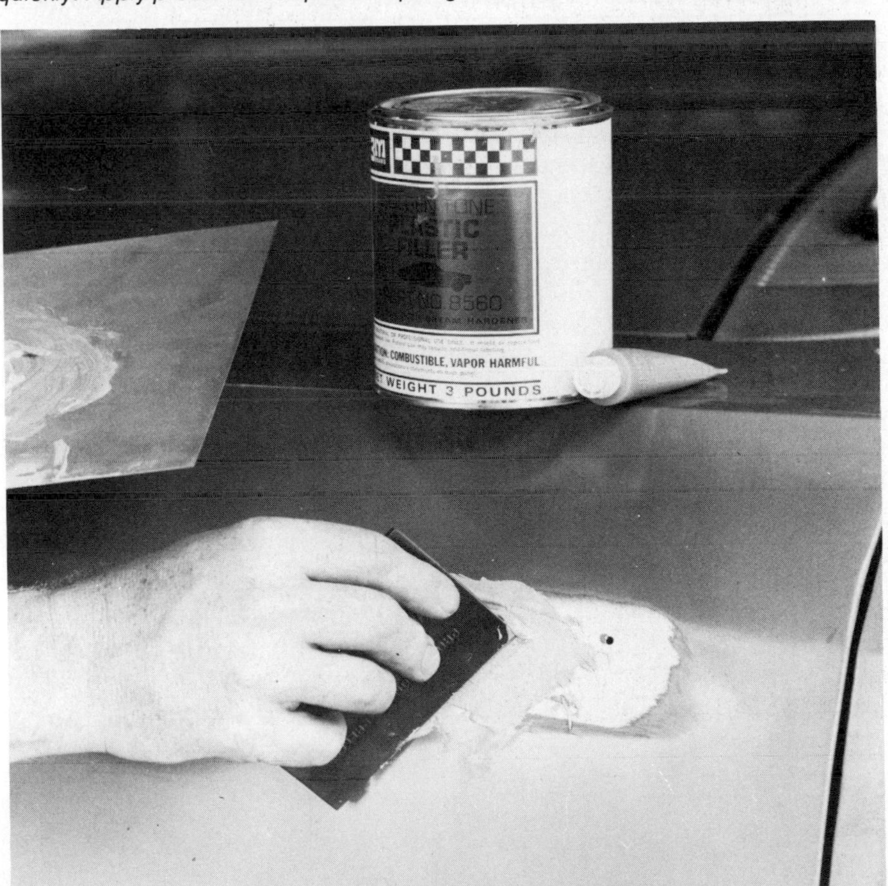

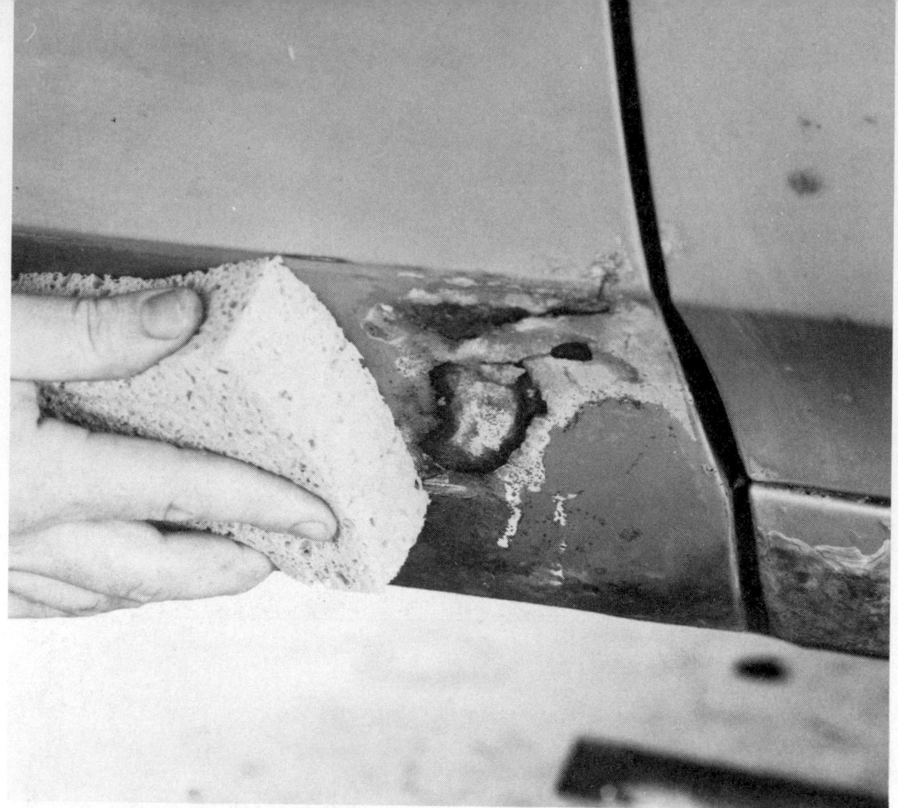

Fig. 15–14. Cleaning rustout area.

Bend in the edges of solid metal to outline the opening, using a small hammer (see Fig. 15–15). This will recess the patch slightly to allow an application of plastic filler.

There are several good choices for bridging the opening: (1) a fiberglass patch; (2) a "Lay-It-On" fiberglass and polyester patch; (3) metallic auto body repair tape or patch. Instructions for application are furnished with each type, but there are special cautions to observe with each.

With the Lay-It-On patch, you cut the patch with scissors to size it to cover the entire sanded area, then moisten your fingers (so the patch won't stick to them), peel the patch from the backing (see Fig. 15–16), and apply (see Fig. 15–17). Cure the patch in direct sunlight (up to two hours at 80° F. in bright sun, as long as eight hours if it's partly sunny). At lower ambient temperatures it takes much longer (up to 24 hours in partial sun, at 50° F.), often an impractical length of time. A better choice is an ultra-violet sunlamp, six inches from the patch, for 12 to 20 minutes (see Fig. 15–18). Heat alone will not work. Be careful not to allow overcuring, or the patch could become brittle.

Once the patch is hard, scuff-sand it with 80 grit paper, from the edges toward the middle, until you have close to a featheredge from the bare metal to the patch (see Fig. 15–19).

With the metallic tape there is no need for curing. However, the tape is only two inches wide, so to bridge most rustouts you must apply slightly overlapped strips (see Fig. 15–20). Cut excess tape (past the bared sheet metal) with a razor blade and use a screwdriver handle to press the tape into good contact with the bared sheet metal. If you can use the metallic patch, you eliminate the need to apply and overlap the strips, and the need to get good adhesion at the overlap points. The single patch should bridge the small rustout completely.

The fiberglass patch is similar to the other types in method of application, and it is heat-cured.

After applying a metallic or fiberglass patch, scuff-sand as with the Lay-It-On type to help the plastic filler adhere.

Complete the job by applying plastic filler as for minor dents and paint as for scratches.

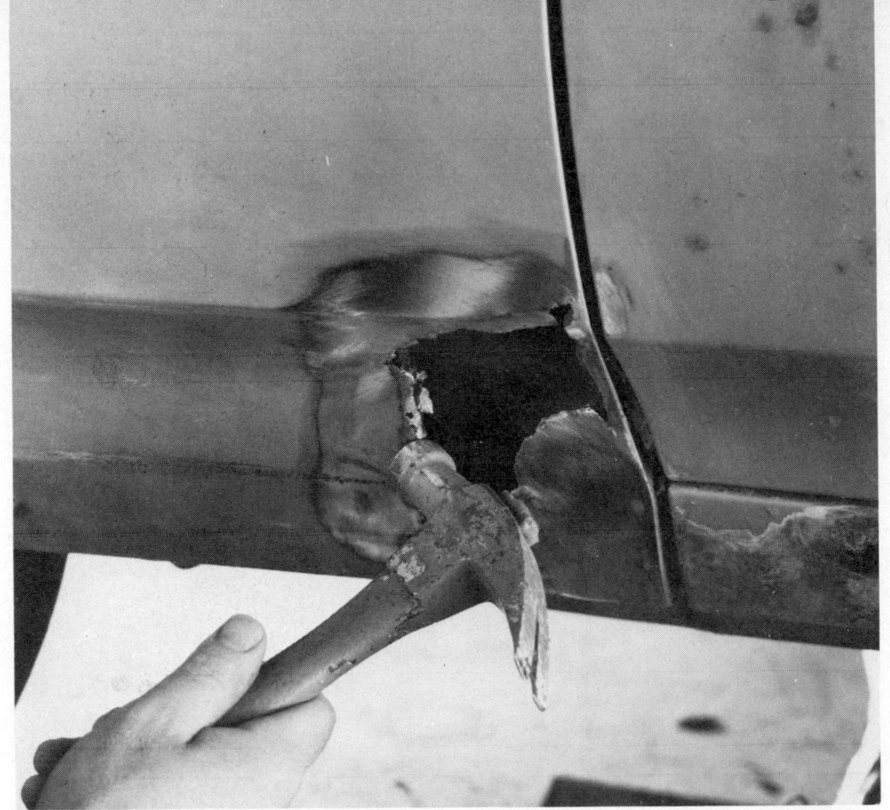

Fig. 15–15. Bending in edges of solid metal to outline the opening. This will recess the patch (to allow application of filler).

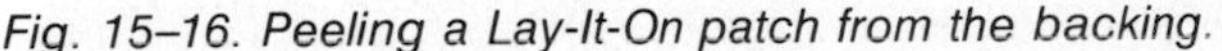

Fig. 15–16. Peeling a Lay-It-On patch from the backing.

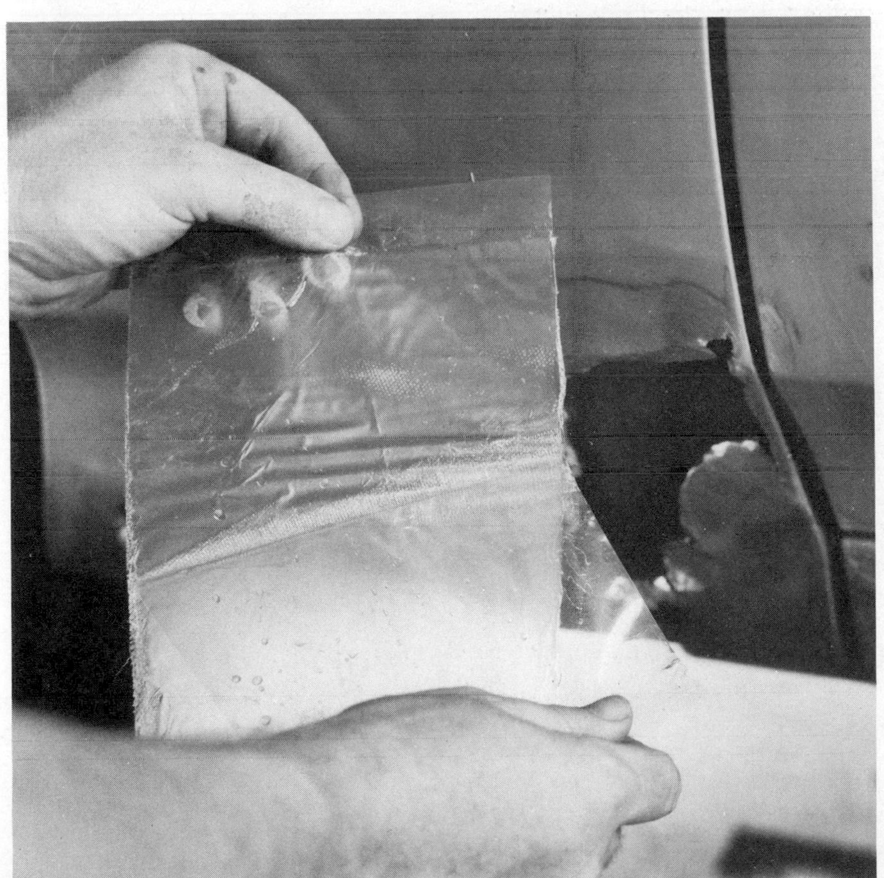

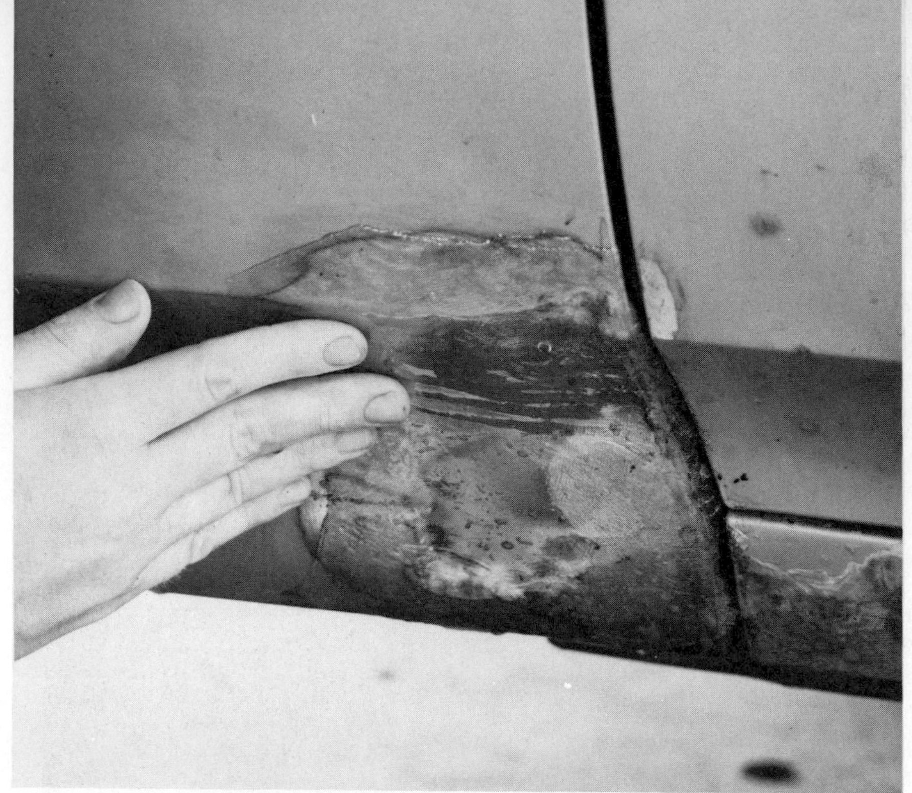

Fig. 15–17. Applying the Lay-It-On patch. Smooth it down with fingers and make sure it adheres all around.

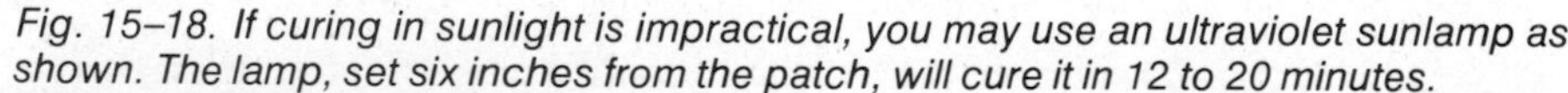

Fig. 15–18. If curing in sunlight is impractical, you may use an ultraviolet sunlamp as shown. The lamp, set six inches from the patch, will cure it in 12 to 20 minutes.

Fig. 15–19. After patch cures (becomes hard), scuff-sand with 80 grit disc as shown until you have close to a featheredge to the bare metal.

Fig. 15–20. If you are using auto body tape, slightly overlap the layers as shown, then use a screwdriver handle to press tape into good contact with the sheet metal, also as shown.

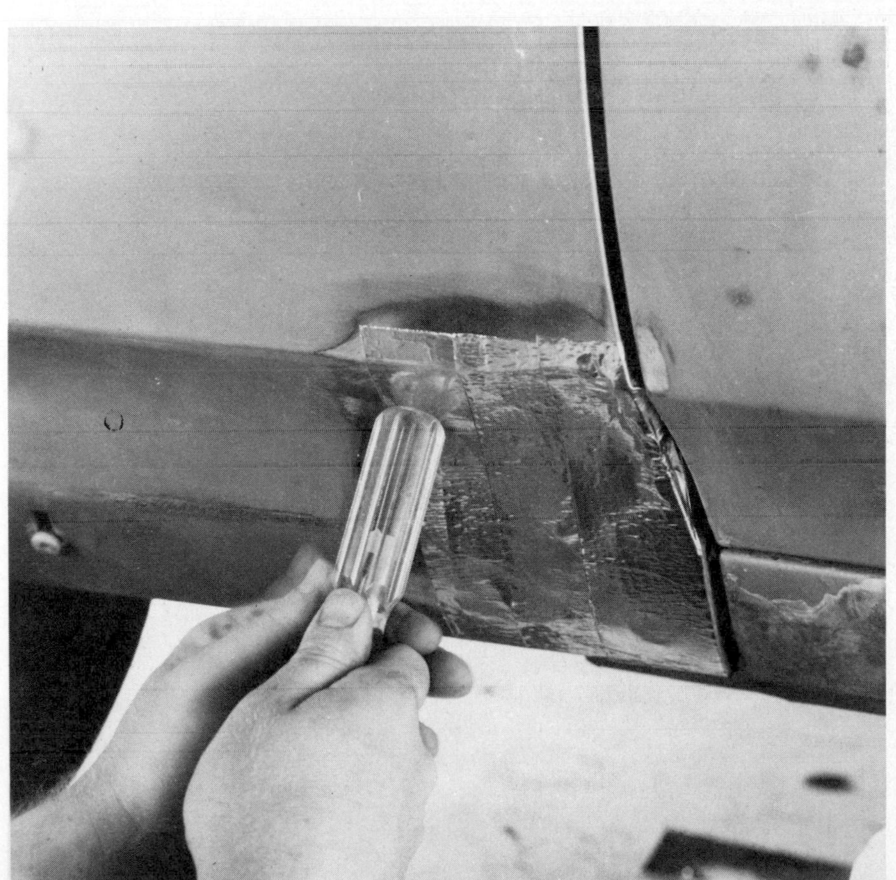

Waxing and Polishing

To get the best results from a wax job, you should start with a clean body. Remove tar with a mild solvent. Touch up any nicks and allow paint to dry for a couple of days. Note: If you have repaired and repainted a panel, or part of it, wait 30 days before waxing the car.

If the body finish is severely weathered (chalky-looking), clean it first with a mild abrasive polish. Do not use rubbing compound, which is very abrasive and will thin out the color coat. If the finish shows only mild weathering, use a combination polish and wax. With a new car, use only a wax, which is commonly advertised for new car use only.

Never use a buffing wheel driven by an electric drill. It is next to useless at sharply-bent edges (of which there are many on the car), and unless handled with great care and skill could result in color coat removal at sharp bends. Your hand still is the best polishing and waxing cloth holder available.

Polishes, waxes, and polish-wax combinations are available in paste, softened paste, and liquid forms. You probably will find the liquid the easiest to use, with the softened paste a close second choice.

16 PARTS BUYING GUIDE

THE KIND OF PART YOU BUY—new, rebuilt or used—is an important factor in the cost of auto repairs. The best choice is not a single one, but depends on the part itself.

A new part is exactly what the adjective says—brand-spanking new. A rebuilt part is one in which all defective contents have either been repaired or replaced by new items. A used part is one that has been salvaged from a car that was wrecked or removed from service by an insurance company for some reason, whether the result of an accident, flooding, fire, or other minor disaster.

It might seem that you could buy almost anything new, but you really can't in many cases without a struggle. The reason: The new part is so expensive that practically no one buys it, so the manufacturers turn out very few in addition to what is needed for the automobile production lines. Rebuilts save you forty to sixty percent off the price of new parts—if you could get new. Some examples: starters, alternators, ignition distributors, and water pumps.

Other parts more commonly sold in rebuilt form than new include: carburetors, brake shoes, clutches, power steering gear, engines, transmissions, windshield wiper motors, power steering pumps, and air conditioning compressors. Rebuilts also are coming on strong for power brake boosters, disc brake calipers, and brake master cylinders. In addition, radiators are commonly rebuilt (a process called recoring) by specialty shops. The saving over a new radiator is small (perhaps less than 15 percent), but recoring reduces the shop's stocking requirements, and the result is about as good as new.

A high-quality rebuilt part can provide very satisfactory service. Only two rebuilders have national distribution—Arrow and Champion; the remainder are regional operations so numerous it would be impossible to list them here. Therefore, when it comes to a rebuilt, you probably will not be able to find a national name. If you buy at a parts store that serves the professional, however, you should get a quality rebuilt, for professional mechanics will not long patronize a parts store that sells unreliable rebuilts.

You should be wary of some rebuilts that you might be tempted to buy. Rebuilt spark plugs and voltage regulators are the two prime examples that just can't compare with a new part. Rebuilt fuel pumps are also available and, although usually reliable, they cost so little less than a new part that their sales are low and, therefore, so is stock in the typical parts store.

Rebuilt batteries are available in some parts of the country, but the workmanship varies so greatly that it represents a risk which is not recommended.

You should be aware that replacement versions of some components do not come complete. You may have to take a pulley off an old water or power steering pump; brackets, solenoid and vacuum diaphragm unit off the old carburetor. This not only is true with new or rebuilt parts, but with a part from a wrecking yard. (Because the same part goes on different cars, the brackets, pulleys, etc. on the part may not match the one on your car, even though the basic part is the same. However, always inspect carefully to be sure the part really is the same.) With a pulley, make sure you can remove it. Some pulleys can only be removed safely with a special tool.

Used Parts

The used part is often a best buy. When it

comes to body parts, you can't beat the wrecking yard. You can buy a fully-assembled door, often in the right color, for much less than the individual parts would cost new—and you save the time of assembly and painting. Interestingly, availability of used body parts is usually much better than for new parts, particularly if the car is a few years old.

After body parts, the wrecking yard is an excellent source for the following:

• *Batteries.* A good used battery from a late-model wreck is a reasonably good risk, at perhaps one-fourth the price of a new one. If you're not planning to keep the car very long, you could save big money.

• *Tires and spare wheels.* These also cost just a fraction of the price for new ones, even less than retreaded tires. Spare wheels for snow tires save the cost of seasonal mounting and demounting, and if you're planning to sell the car soon the much lower cost of used tires makes them a best buy.

• *Electrical parts.* A used electrical part is much cheaper than even a rebuilt, and in some cases the rebuilt may not be available, such as for some imports. Starters, alternators, ignition distributors, wiper motors, horns, power window and seat motors, and some multi-purpose stalk switches (the ones on the steering column) are low-risk items. The wrecker can test the part with jumper wires and a battery, so you know if it works before you buy it.

• *Carburetors and power steering pumps.* These parts cannot be tested before you buy, but they normally are not time-consuming items to install, so if the part doesn't work you haven't invested a lot and can return it for a quick exchange or refund. The prices will be much lower than rebuilts.

• *Brake discs and drums.* You can't get a rebuilt drum or disc, so the choice is either new or used. The used part can be inspected for cracks and other defects, so the risk is really very low. The savings, however, can be seventy percent.

Although major components—engines, transmissions, and differentials—are sold by wrecking yards, there is some greater risk involved. You can see if an engine starts and runs and if a manual transmission shifts into all gears, but that's about it. If you ever attempt a major component replacement, you should, of course, consider the wrecking yard as a source. The wrecker will exchange or refund, as with smaller components, but the labor investment you have is very substantial, so don't go by component price alone.

Wrecking Yard Prices

A wrecking yard doesn't post a price list. The wrecker just quotes what he thinks he can get, based to some vague degree on the price of rebuilts and new parts, whichever is the competition. You can, therefore, do some bargaining and often save some money. If a part is in great demand, however, the price may be very close to a rebuilt, so always know the cost of the alternatives.

Coping With Catalogs

Most of the parts you will buy—plugs, filters, drive belts, lubricants, etc., will be new. You want to get the best price possible, so you'll check with auto parts stores, discount houses, and mail order firms. Generally, the discount house and mail order firm will be cheaper than the auto parts store but, in many cases, you must read the catalogs and application charts to find the part number, then pick the part off the shelf. Parts manufacturers selling to discount houses try to offer easy-to-read catalogs and charts but, in many cases, the listings must be condensed to keep them from being thicker than a phone book. The first thing you can see that you may have to know is what the numbers mean. A 235 six-cylinder is a six-cylinder with a cubic-inch displacement of 235. Or the listing may be given in metric terms, such as 3.8L (for liter). Displacement is the volume of all cylinders added together.

The term 4B carb means four-barrel carburetor. 2B is two-barrel. In some cases 2V or 4V are used, with the V referring to venturi, a section of the carburetor barrel.

In some cases, the displacement number will be on the air cleaner or engine top cover, or on an exterior body emblem. To tell which carburetor you have, remove the air cleaner cover and count the number of holes (one, two, or four). An in-line engine with four or six-cylinders will have a one- or two-barrel carburetor; a V-6 or V-8 will

Fig. 16–1. Top view of a four-barrel carburetor. You will notice that it has four distinct circular sections, or barrels. Some carburetors have only one, others two.

have two or four barrels (see Fig. 16–1).

Let's assume you have a 1973 model with the 325 V-8. In that case, you don't care what carburetor it has, for all take Part No. 1472-D. If you have a 1974 with air conditioning, but only a two-barrel carburetor, the part number also is 1472-D, because No. 1503-R is only for cars with the four-barrel *and* air conditioning.

If you have a 1979 Hummer with the 385 V-8 and power steering, the part Number is 1704-M, not 1828-Z, which is only for those with *both* air conditioning and power steering.

Some catalogs give interchange listings between the manufacturer's make and competitors. If you have a competitive part, you may be tempted to take the number off the part and look at the interchange listing. With many parts, however, particularly spark plugs, this information is often inapplicable. Always use the make-of-car listings only, or you could end up with the wrong part.

Don't be surprised (even with the same brand) if you find that the catalog lists a different part from the one in your engine. Either the old part may have been superceded by the new number, or the last installation could have been in error. It happens all the time.

You should, therefore, know how to read a catalog listing. Here's a sample and what it means:

		Part No.
1971–80	All models with 235 six-cylinder	1471-D
1971–76	All 325 V-8 exc. below	1472-D
1974–75	325 V-8 with air conditioning, 4B carb.	1503-R
1971–77	385 V-8 except below	1468-N
1973–76	Charger, Rammer 325 V-8	1525-L
1977–80	All 325 V-8 exc. below	1516-A
	385 V-8 except below	1704-M
	Charger, Rammer 325 V-8, 385 V-8 exc. below	1851-C
	All 385 V-8 with air conditioning and power steering	1828-Z

INDEX

All *italicized* numbers in this index indicate photos or illustrations that will aid you in understanding the material. On pages where a subject is presented in both text and photo, only one is indexed.

1800
251-7549 [illegible] 2604

CRAMOS CHALO BUS
JADE